The Art of the Helicopter

For Liza

The Art of the Helicopter

John Watkinson

AMSTERDAM BOSTON HEIDELBERG LONDON NEW YORK
OXFORD PARIS SAN DIEGO SAN FRANCISCO SINGAPORE
SYDNEY TOKYO

Elsevier Butterworth-Heinemann
Linacre House, Jordan Hill, Oxford OX2 8DP
200 Wheeler Road, Burlington MA 01803

First published 2004

Copyright © 2004, John Watkinson. All rights reserved

The right of John Watkinson to be identified as the author of this work has been asserted in accordance with the Copyright, Designs and Patents Act 1988

No part of this publication may be reproduced in any material form (including photocopying or storing in any medium by electronic means and whether or not transiently or incidentally to some other use of this publication) without the written permission of the copyright holder except in accordance with the provisions of the Copyright, Designs and Patents Act 1988 or under the terms of a licence issued by the Copyright Licensing Agency Ltd, 90 Tottenham Court Road, London, England W1T 4LP. Applications for the copyright holder's written permission to reproduce any part of this publication should be addressed to the publisher

Permissions may be sought directly from Elsevier's Science and Technology Rights Department in Oxford, UK: phone: (+44) (0) 1865 843830; fax: (+44) (0) 1865 853333; e-mail: permissions@elsevier.co.uk. You may also complete your request on-line via the Elsevier homepage (www.elsevier.com), by selecting 'Customer Support' and then 'Obtaining Permissions'

Nothing in this book overrides any legislation, rules or procedures contained in any operational document issued by HMSO, CAA, equivalent authorities in other countries or by any aircraft or equipment manufacturer

British Library Cataloguing in Publication Data
A catalogue record for this book is available from the British Library

Library of Congress Cataloguing in Publication Data
A catalogue record for this book is available from the Library of Congress

ISBN 0 7506 5715 4

For information on all Elsevier Butterworth-Heinemann publications visit our website at http://books.elsevier.com

Typeset by Newgen Imaging Systems (P) Ltd., Chennai, India
Printed and bound in Great Britain

Contents

Preface

The modern helicopter is a sophisticated device that merges a surprising number of technologies together. This wide range of disciplines is one of the fascinations of the helicopter, but it also makes a complete understanding difficult. The very ability to hover, which sets the helicopter apart, also dooms it forever to vibration, poor performance and economy in forward flight, and thus restricts its numbers.

The unique capabilities, complexity and inherent limitations are part of a helicopter's charm, and have given rise to some wonderful definitions and descriptions, which generally contain more than a grain of truth:

> A helicopter is a mechanical engineer's dream and an aerodynamicist's nightmare.
>
> A helicopter is a collection of ball-races flying in close formation.
>
> A helicopter is a collection of vibrations held together by differential equations.

My own search for an understanding of the helicopter was hampered by the huge gap between books containing beautiful photographs but little information and advanced textbooks full of equations.

This, then, is the book that I couldn't buy when I wanted to learn all this; it could not be a conventional book because I have had such difficulty with them. It combines theory and practice of how helicopters are made, how they fly, how they are powered and how they are controlled. It would be impossible to consider all of that without at least some reference to flying techniques.

Most technical books assume an existing level of knowledge, but the wide readership and wide range of disciplines make that inappropriate here. Instead of making incorrect assumptions about the reader, this book approaches every subject from first principles, and builds up in a clearly explained logical sequence using plain English and clear diagrams, avoiding unnecessary mathematics.

Technical terms and buzzwords are all defined, and acronyms are spelled out. Misnomers, myths and old wives' tales (for these are plentiful) are disposed of wherever they arise. Whilst the contents of this book are expressed in straightforward language there is no oversimplification and all of the content is based on established physics and accepted theory. The student of technology or aerodynamics will find here a concise introduction, leading naturally to the more advanced textbooks on the subject.

The would-be pilot will find clear explanations of the principles to act as a perfect complement to the instruction itself. The experienced pilot will find the detailed descriptions of the characteristics of helicopters an aid to safety. This book covers the theory of the helicopter in more than sufficient depth to enable the reader to pass the helicopter theory examination. There is enough practical information to allow the reader to make sense of the machine's flight manual and to prepare for the helicopter type examination.

The rotary wing flying machine is just part of the larger subject of general aviation. Related subjects such as human factors, meteorology, air law and navigation scarcely differ with the type of flying machine, and are not repeated here save for differences specific to helicopters.

John Watkinson
Burghfield Common, England

Acknowledgements

The author is indebted to the many people and organizations who have found time to help in the preparation of this book by offering information, photographs, constructive criticism or simply encouragement. High on the list must come my publishers, Elsevier. Ray Prouty's columns in *Rotor and Wing* have long been an inspiration that complex subjects can be brought to a wide audience.

Much material was obtained from the International Helicopter Museum in Weston-super-Mare, UK, The Fleet Air Arm Museum in Yeovilton, UK, The Museum of Flight in Seattle, Washington, The Igor I. Sikorsky Historical Archives in Stratford, Connecticut, The Hiller Aviation Museum in San Carlos, California and the American Helicopter Museum and Education Center in Brandywine, Pennsylvania. The kindness and helpfulness of the staff of all of these was exceptional and deeply appreciated.

The universally positive attitude of the manufacturers has been a great delight. AgustaWestland, Bell Helicopter Textron, Boeing Helicopter, Kaman Aerospace, Lockheed Martin and Sikorsky must all be mentioned here.

Reg Austin did a superb job of checking my words for technical accuracy. Mikael Reichel advised on the pedagogical aspects of the text and many of my explanations are the better for that. I would not have obtained all of the necessary photographs and permissions without the assistance of Liza Marshall. I am indebted to Steve Coates for allowing me access to his collection of wartime German helicopter material.

Individuals who have helped include Fred Ballam, Oliver Dearden, Margaret Denley, George Done, David Gibbings, Nick Gribble, Bruce Holben, Russell H. Jones, Bill Kidd, Dan Libertino, Bo Maggs, Tim Price, Renée Renaud, Trevor Scantlebury, Eric Schulzinger, Tom Shenton, Jonathan Simpson, Jay P. Spenser, David Steel, North E. West and Katharine Williams.

1

Introduction to rotorcraft

1.1 Applications of the helicopter

Although helicopters must all follow the same laws of physics, the forms that practical machines take vary tremendously due to the range of tasks to which they can be put. To avoid confusion, this book takes the view that an aircraft is any man-made aerial machine capable of climbing out of ground effect. Thus a helicopter must be an aircraft. Neglecting aerostats (balloons etc.) any aircraft that is not a helicopter will be an aeroplane (USA: airplane).

It will be seen later in this book that the slower an aircraft goes, the more power it needs to maintain height. Hovering is the ultimate case of slow flight, suggesting that helicopters must have a high power to weight ratio. This will require heavy engines and a corresponding fuel capacity. These factors limit load carrying capability and range.

The mechanical complexity of the helicopter and the inevitable vibration demand a lot of maintenance. The airspeed of the true helicopter is forever restricted by fundamental limits. It will generally be more expensive to move a given load by helicopter than by almost any other means, and so if a suitable airstrip exists, a fixed-wing aircraft can do the job at lower cost, and generally at a higher airspeed. In most cases helicopters cannot compete economically with aeroplanes, and so their use is restricted to applications for which aeroplanes, or other forms of transport, are unsuitable.

In remote areas, there will be no airstrips; in wartime, runways are conspicuous targets, and, with the exception of the aircraft carrier, are fixed. The helicopter's ability to hover means that it can land almost anywhere a fairly flat firm surface exists. Some are genuinely amphibious, landing with equal aplomb on water or land. If the ground is unsuitable (or if the waves are too high) many helicopters can transfer goods and passengers whilst in the hover.

This ability makes the helicopter the ideal rescue vehicle. Many lives have been saved because the helicopter can get to places that would otherwise be difficult or impossible to reach. War casualties, the victims of shipwrecks, mountain climbing accidents and natural disasters such as earthquake or flood today have significantly higher chances of survival because a helicopter can get them rapid treatment.

The accounts of helicopter rescues make more thrilling reading than novels because they are true. Helicopters have flown far out of range of the shore by taking fuel from oilrigs and ships, sometimes taking fuel in the hover if a landing was impossible. This would be remarkable in good weather, but emergencies occur in all weather conditions and the helicopter has evolved to handle the worst.

Despite their life-saving ability, most of today's helicopters were originally designed for military use. As military fixed-wing aircraft became faster and faster, they found it harder to attack ground targets or to support ground troops. During the Vietnam War, it was found that the helicopter had a major role. Transport helicopters excelled inserting and withdrawing troops, delivering ammunition and food, evacuating the wounded, recovering the crews of downed planes and even the planes themselves.

Armed helicopters proved to be ideal for attacking ground targets. At first these were general-purpose machines hurriedly fitted with weapons, but later dedicated attack helicopters evolved, complete with armour plating and redundant systems to allow them to withstand ground fire. As their load carrying capability has increased, these machines have virtually rendered the tank obsolete.

Although ideal against ground targets, the helicopter is slow and vulnerable to attack from the air. Fixed-wing planes are necessary to provide the air superiority in which the helicopters operate. As an alternative, helicopters can operate under stealth conditions, avoiding detection by using terrain.

Large military helicopters are very expensive to operate, and armed forces found it worthwhile to have simpler machines specifically for training purposes. A small number of helicopters have been designed specifically for the civil market and these are popular with large companies for executive transport.

For the average private owner, the sheer cost of running helicopters precludes all but the smallest machines with aeroplane-derived piston engines. Virtually all other helicopters are now turbine powered.

1.2 A short technical helicopter history

This is not a history book and this section must necessarily be brief. The reader interested in the US history of the helicopter is recommended to the comprehensive yet highly readable works of Jay Spenser.[1,2] The recent book by Steve Coates, *Helicopters of the Third Reich*,[3] is essential reading to the historian as it shows how far ahead of the rest of the world German helicopter engineers were at that time. For those who read French, two more fascinating volumes are available. *L'Histoire de l'Hélicoptère* by Jean Boulet[4] contains the words of helicopter pioneers themselves. *Les Hélicoptères Florine 1920–1950* by Alphonse DuMoulin[5] recounts the pioneering work of Nicolas Florine.

The history of the helicopter has been very short indeed. In comparison with fixed-wing aircraft, helicopters need more power, have to withstand higher stresses, are harder to understand and control and have more moving parts. It is hardly surprising that the development of the helicopter took place well after that of the fixed-wing aircraft.

Early helicopters lacked enough power to fly. Once helicopters were powerful enough to leave the ground, they were found to be uncontrollable. Once the principles of control were understood, they were found to vibrate and to need a lot of maintenance and so on. Today's helicopters represent the sum of a tremendous number of achievements in overcoming one obstacle after another.

Before World War II helicopters were in an experimental phase. This was the heyday of the gyroplane, invented by the Spaniard Juan de la Cierva and technically refined by Raoul Hafner, an Austrian working in England who would later contribute much to the development of the helicopter.[6]

The first practical helicopter was the Focke-Wulf Fw-61 of 1938 (Figure 1.1), followed in the same year by the Weir W-5 (Figure 1.2) that flew two years before Sikorsky's VS-300 (Figure 1.3). The urgencies of war accelerated all technical development with the emergence of production helicopters, where the work of Anton Flettner (Figure 1.4) and Heinrich Focke was far in advance of anything taking place elsewhere.

Fig. 1.1 The Focke-Wulf Fw-61 was the first helicopter to move beyond the experimental stage and was capable of extended flights. The hull was based on that of an Fw Stieglitz aeroplane. Note the vestigial airscrew that simply cools the engine. (Steve Coates)

Fig. 1.2 The Weir W-5 flew extensively, but development of all helicopters was suspended in the UK during World War II. (AugustaWestland)

The Focke-Achgelis Fa-223 shown in Figure 1.5 became, on 6 September 1945, the first helicopter to cross the English Channel when a captured machine was flown to England by Hans Gerstenhauer. This machine had a payload of 2000 kg.[7]

Helicopter development in the UK was halted by government order during World War II and, being an Austrian, Hafner was locked up, but fortunately not for long.

Fig. 1.3 The Sikorsky VS-300 had an extended development period during most of which cyclic control was not understood and unwieldy auxiliary rotors were used instead. (Igor I. Sikorsky Historical Archives Inc.)

Fig. 1.4 The Flettner Kolibri (Hummingbird) was the first synchropter and was an advanced and capable machine. Kaman developed the concept and produced the successful Husky and K-Max models. (Steve Coates)

Fig. 1.5 The Focke-Achgelis Fa-223 was a large and capable machine that was in production during World War II. Few were produced due to Allied bombing. The machine was far ahead of anything else in the world at the time. (Steve Coates)

In Germany, production was hampered by Allied bombing, whereas US helicopters were unrefined. The result was the same: helicopters made little contribution to the war itself.

After World War II great progress was made in the understanding of helicopter dynamics and stability. This led directly to machines that were less stressful to fly and correspondingly safer. The Bell 47 (Figure 1.6) based on the research of Arthur Young was in 1946 the first helicopter to be certified. The Sycamore of 1952 (Figure 1.7) designed by Hafner, was the first British helicopter to be certified and was noted not just for its performance, but also for its light control forces which needed no power assistance.[8]

Advances in constructional techniques and materials continued to improve the service life of components, especially blades. Possibly the most significant single step

Fig. 1.6 The legendary Bell 47 was based on Arthur Young's research and was the first helicopter to be certified in the USA. Larry Bell never liked the utilitarian appearance, but it outsold all of the more stylish models. The flybar stabilization system was adopted extensively in later Bell machines. (Bell Helicopter Textron)

Fig. 1.7 The attractive Sycamore was Hafner's masterpiece and the first helicopter to be certified in the UK. (AugustaWestland)

Fig. 1.8 The Bell Huey was officially designated the Iroquois, but the Helicopter, Utility designation, HU-1, led to the nickname of Huey, which stuck. In Vietnam the Huey was used in enormous numbers. (Bell Helicopter Textron)

was the introduction of the turbine engine which was much lighter than the piston engine for the same power, yet had fewer moving parts. This allowed greater payload and a reduction in maintenance.

The first turbine-powered helicopter to fly was a modified Kaman K-225 in 1951 and in 1954 the first twin-turbine machine, also a Kaman, flew. The Bell Huey first flew in 1956 (Figure 1.8). It was in the 1960s that the many disciplines in helicopter design were finally mastered allowing the machine to be considered as a system. Some elegant and definitive designs emerged during this period. Such was the validity of their basic concepts that they could accept a steady succession of upgrades that would allow them in some cases to remain in service to the present day. Sikorsky's S-61 and Boeing's Chinook are good examples of longevity.

Since that time there have been few breakthroughs; instead there has been a steady process of refinement. The introduction of composite materials in blades, rotor heads and body parts has reduced weight and extended service life. Refinements in mechanical engineering have produced lighter engines and transmissions having longer life. Manufacturers have used production engineering to reduce the amount of labour needed to build machines.

Instead of a revolution, the employment of electronics and computers in helicopters has seen steady and relentless progress. There is no wear mechanism in electronics and complicated transfer functions can be realized in lightweight parts that use little power. Items such as turbine engine controllers and rotor rpm governors are ideal applications for electronics, along with stability augmentation systems. The flexible control systems that electronics make possible have enabled developments such as the tilt-rotor helicopter.

Helicopters can survive engine failures and the failure of a variety of parts, but there remain some parts such as gearboxes and rotor heads in which failure will be catastrophic. The technique of electronically monitoring critical components has made a great contribution to safety. Spontaneous failures are very rare. Generally there are symptoms such as a change in the characteristic of vibration or noise. These may be too slight to be heard by the crew, but a sensor mounted on the affected part in conjunction with a signal processor which knows what the normal sounds, or signature, from the part are, can give vital warning of a potential problem.

The life of a component may be reduced if it is subjected to higher stress. Modern electronic systems (HUMS: Health and Usage Monitoring Systems) can measure stress and the time for which it has been applied in order to calculate the safe remaining life in major components.

Much fundamental research into helicopters was done using models. De la Cierva's early work was with free flying model gyroplanes. Arthur Young used models extensively (Figure 1.9) and it was his demonstration of a model controlled by a trailing wire that convinced Larry Bell to enter the helicopter business. Irven Culver at Lockheed built what was probably the world's first radio-controlled model helicopter in the late 1950s.

Advances in radio control equipment in the 1960s made the necessary precision available at reasonable cost and this led to the availability of flying model helicopters

Fig. 1.9 Arthur Young flies a wire-controlled model helicopter. Young was a philosopher who argued that designing a helicopter would teach him how to think. He was right. (Bell Helicopter Textron)

for the model building enthusiast. As will be seen in Chapter 9, such models have become highly sophisticated even though, like their full-sized counterparts, they remain expensive to build and operate.

Conceptually somewhere between the model and the full-size helicopter is the UAV (unmanned autonomous vehicle). These machines are designed to perform surveillance tasks, typically carrying cameras and other sensors. As electronic and sensing devices have become smaller, useful equipment can be carried aloft at much lower cost if there is no need to carry a pilot. Unlike the model, which needs actively to be controlled by the pilot on the ground at all times, the UAV carries enough navigational equipment, automatic stabilization systems and processing power to be self-contained. Also unlike the hobbyist's machines, UAVs must be built and operated to professional standards.

1.3 Types of rotorcraft

The definition of a rotorcraft is quite general, embracing any flying machine that produces lift from rotors turning in a plane that is normally close to the horizontal. This definition does not concern itself with the proportion of the machine's weight carried by the rotors, or whether that proportion changes at different stages of flight. Figure 1.10 shows the main classes of rotorcraft that will be defined here.

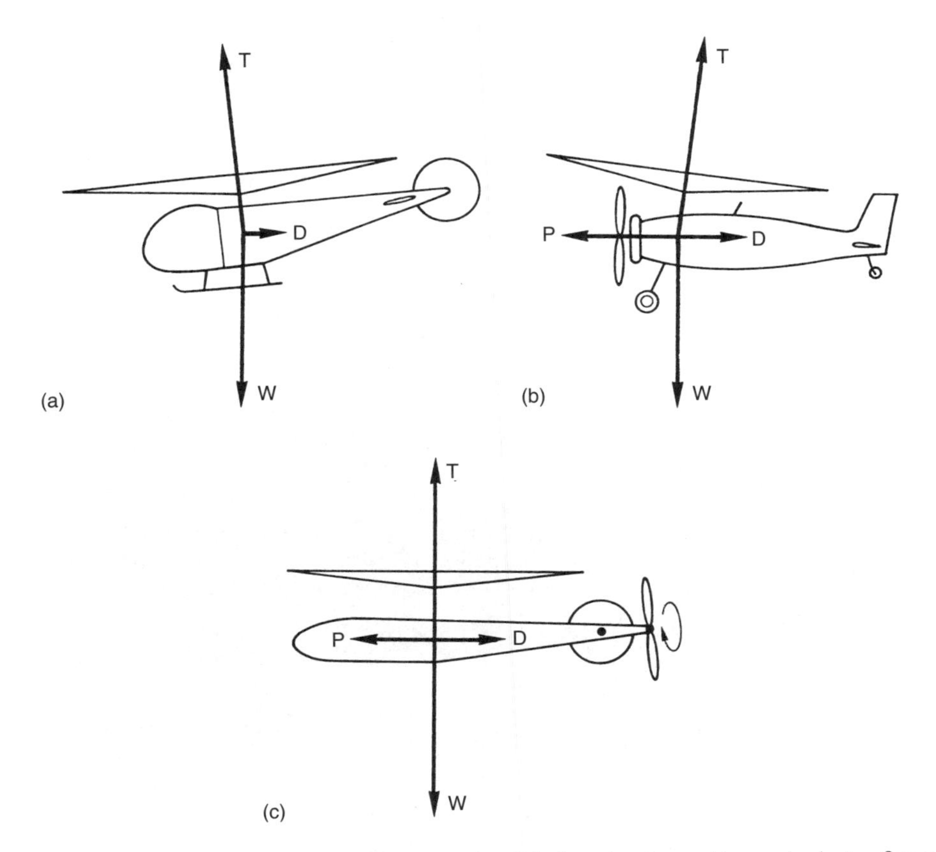

Fig. 1.10 The main classes of rotorcraft. The conventional single main rotor machine predominates. See text for details.

The pure helicopter obtains the great majority of its lift in all modes of flight from a power driven rotor or rotors, and any lift due to airflow around the hull is incidental. The majority of today's rotary wing machines are pure helicopters. The thrust from a rotor is closely aligned with a line drawn perpendicular to the tip path, and the pure helicopter propels itself by tilting the rotor forward to obtain a component of rotor thrust which balances the drag, as shown in Figure 1.10(a).

The gyroplane (also known by de la Cierva's trade name of autogyro) obtains lift from an undriven rotor that must be tilted *away* from the direction of flight to make air flow up through it. The rearward rotor-thrust component, along with the drag, is balanced by the forward thrust of a conventional airscrew as shown in Figure 1.10(b). As the rotor needs to be pulled through the air to maintain height, the autogyro cannot hover in still air, although it can give the illusion of hovering by flying into wind. Simple autogyros must taxi to spin up the rotor, but later machines could spin the rotors with engine power on the ground, and use the stored energy to perform a jump take-off.

Between the pure helicopter and the gyroplane is the gyrodyne, which obtains lift from a power driven rotor. Unlike the pure helicopter, the gyrodyne maintains the rotor disc parallel to the direction of flight, as in Figure 1.10(c) and propels itself with a conventional airscrew. The Fairey Gyrodyne (Figure 1.11) replaced the tail rotor with a side-mounted airscrew to cancel torque reaction when hovering, but also to provide thrust for forward flight. More recently the Lockheed Cheyenne (Figure 1.12) had both anti-torque and pusher rotors at the tail. The gyrodyne offers high speed potential, with the penalties of raised complexity, weight and difficulty of control. Some gyrodynes have wings in addition to the rotor.

Fig. 1.11 The Fairey Gyrodyne had a side-mounted anti-torque rotor that became a tractor propeller in forward flight. (AugustaWestland)

Fig. 1.12 The Lockheed Cheyenne with fixed-wing, anti-torque rotor and pusher propeller was a very fast, highly manoeuvrable machine. (Lockheed Martin)

The compound helicopter is one that hovers like a helicopter, but which may obtain supplementary lift from fixed wings during flight and may incorporate additional means of providing forward thrust.

The convertiplane is a more extreme example of the compound helicopter in that it reconfigures its means of providing lift and propulsion in different flight regimes. The Fairey (later Westland) Rotodyne was a convertiplane. Figure 1.13 shows that it consisted of a twin turboprop aircraft-like structure with a pylon-mounted rotor driven by tip jets. As a helicopter, the tip jet drive provided lift, and yaw control was obtained by differentially changing the pitch of the turboprops. Forward thrust from the airscrews would bring the machine to cruising speed, where much of the lift was developed by the wing, and the tip jets were turned off such that the rotor free-wheeled and the machine became a compound gyroplane.

Reconfiguring can also be done by tilting the whole wing–engine–rotor assembly (tilt wing) as shown in Figure 1.14(a) or by tilting the engine–rotor units on fixed wings (tilt rotor) as in Figure 1.14(b). The Bell-Boeing Osprey is a tilt rotor. As can be seen in Figure 1.14(c), the diameter of convertiplane rotors is usually such that the machine cannot land in the forward flight configuration, but must return to the hover.

The advantages of the convertiplane over the pure helicopter are that using the rotors as airscrews reduces vibration, it is much more efficient and allows a higher airspeed. This reduces fuel consumed and allows greater range. The tilt rotor has its wing in the downwash which reduces hover performance, whereas in the tilt wing the wing is almost always working. However, the tilt wing needs a supplementary mechanism to control the pitch axis, such as a jet or rotor at the tail.

Fig. 1.13 The Rotodyne hovered as a tip jet powered helicopter and cruised as a compound gyroplane. (AugustaWestland)

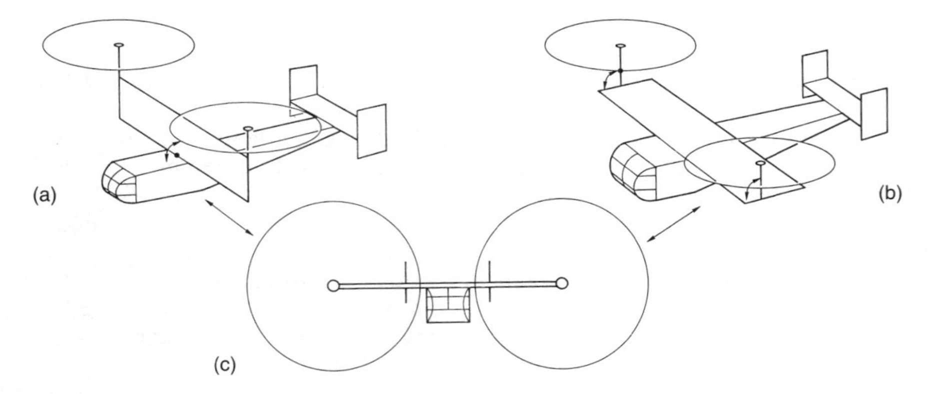

Fig. 1.14 Types of convertiplane. At (a) the tilt wing machine moves the rotors and wing with respect to the hull. At (b) the tilt rotor machine has a fixed wing and the rotors alone tilt. (c) The machine cannot land with the rotors tilted forward.

1.4 Rotor configurations

The various configurations of the pure helicopter will now be considered. The most common configuration is the single main rotor and the anti-torque tail rotor. The remaining configurations, shown in Figure 1.15, handle torque reaction by contra-rotation. Contra-rotating helicopters have no need for the tail rotor, but generally have a tail fin or fins for directional stability at speed.

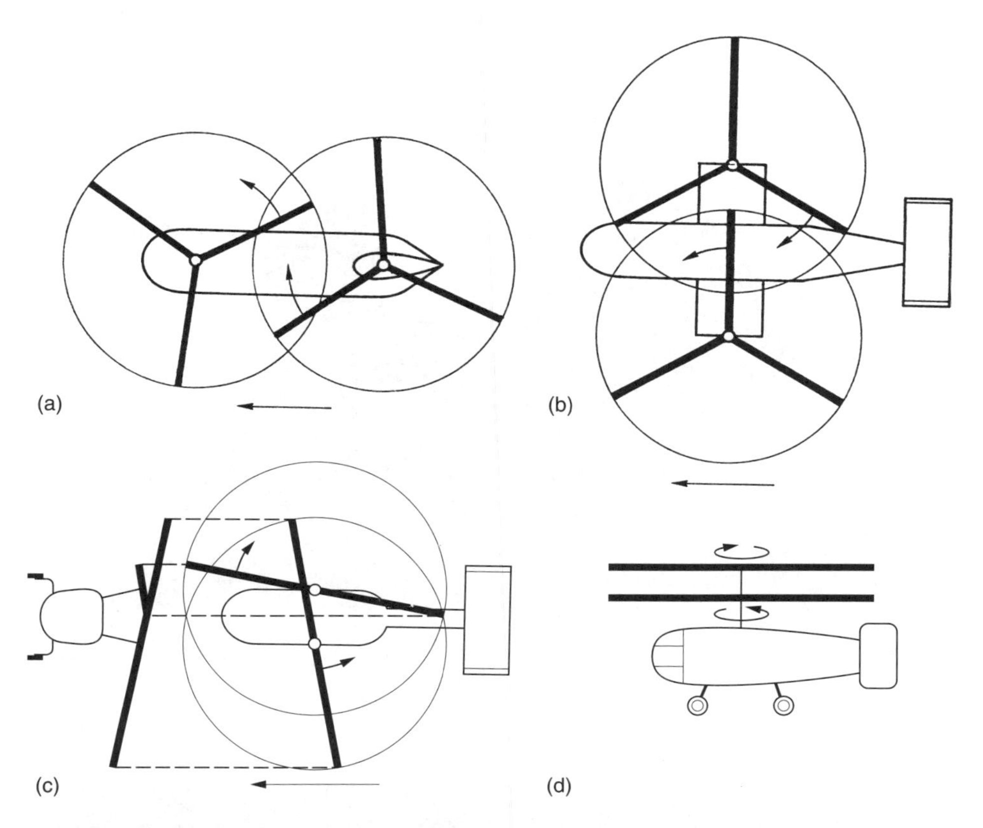

Fig. 1.15 Types of contra-rotating helicopter. (a) The tandem. (b) The side-by-side. (c) The synchropter. (d) The coaxial. See text for details.

In the tandem rotor helicopter, Figure 1.15(a), two rotors turn in opposite directions at the opposite ends of a long hull. The rotors are usually synchronized through a transmission system so that the shafts can be little more than a blade length apart. The Chinook in Figure 1.16 is a good example of the type, the large disc area offering good lifting ability, and the long cabin ample load space.

An alternative to the tandem rotor is the side-by-side twin rotor (Figure 1.15(b)). This has aerodynamic advantages in forward flight because the lifting area has a better aspect ratio, but it is difficult to avoid drag due to the structure needed to locate the rotors each side of the hull and to carry the transmission across the machine. See Figure 1.5. In large machines it is also difficult to make the structure stiff enough to avoid resonance and so the type is rare.

A relative of the side-by-side helicopter which has had more success is the synchropter (Figure 1.15(c)) in which the two rotors mesh so closely that the contra-rotating shafts can be driven by the same gearbox. The close meshing is achieved by tilting the shafts outwards so that the blades of one rotor can pass over the rotor head of the other. The German Kolibri of World War II, designed by Anton Flettner (Figure 1.4), was the first successful machine to use the idea. In the USA, Charles Kaman adopted the synchropter principle, and produced the famous H-43 Huskie which became the definitive crash rescue helicopter of its time (Figure 1.17). The synchropter is the easiest of all

Fig. 1.16 The Chinook is the definitive tandem rotor helicopter and has been produced in large numbers. (Boeing)

Fig. 1.17 The Kaman Huskie is the most successful synchropter design. Note the large fin area needed. (Kaman Aerospace)

Fig. 1.18 The contra-rotating coaxial principle is used extensively by Kamov. (Kamov)

helicopters to fly, as the interactions and second-order effects of the conventional configuration are eliminated, but replaced by some interesting yaw characteristics.

The final approach to contra-rotation is the coaxial helicopter. Stanley Hiller and Arthur Young both built such machines experimentally, but Nikolai Kamov in the USSR put the idea into production (Figure 1.18). The coaxial helicopter places both rotors one above the other on a common shaft, and drives them in opposite directions. Like the synchropter, control interactions are reduced, but yaw control remains an issue. The main advantage of synchropters and coaxial helicopters is that in the absence of a tail rotor, the machine can be much more compact, a crucial factor in naval aviation, where everything has to be squeezed into limited hangar space, although the height needed may increase. The alternative is to fit a folding tail on a conventional machine.

1.5 The essential elements

The helicopter contains a large number of systems and components, but these can generally be broken down into a smaller number of major areas. Figure 1.19 shows a cutaway drawing of a conventional tail rotor type helicopter. The main systems to consider in a helicopter are the hull or airframe, the engine and transmission, the fuel system, the landing gear, the rotors, the controls, electrical and hydraulic power,

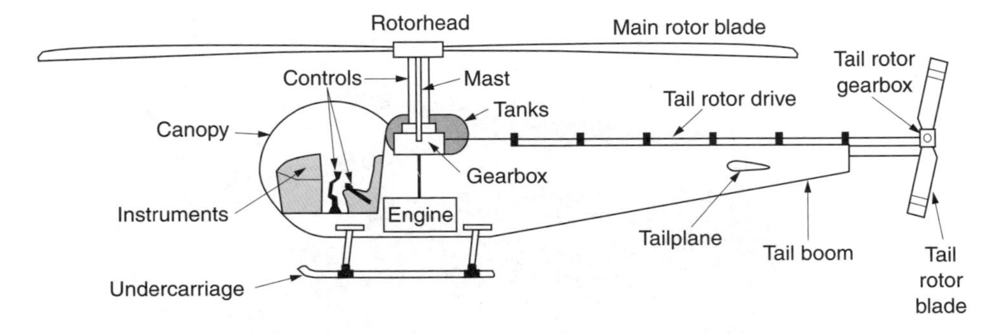

Fig. 1.19 The major components of a conventional helicopter. Note the location of the fuel tanks close to the mast to minimize CM (centre of mass) shifts.

instrumentation and avionics. These subjects will briefly be introduced in the remaining sections of this chapter, and references will be made to more detailed treatments elsewhere in the book.

1.6 The airframe

The fuselage or hull has a number of jobs to perform. One obvious task is to hold all of the components in the correct position and to transfer forces from the rotors, the tail surfaces, the landing gear and any internal or underslung payload. It also protects the occupants and the mechanisms from the elements whilst still allowing good visibility for the pilot. Space has to be found for fuel tanks close to the centre of mass (CM) as single rotor machines are sensitive to trim shifts as fuel burns off. It will also have a more or less aerodynamic shape, although the other requirements often combine to prevent this. In some cases the hull will also be designed to float in case the machine is forced down over water, whereas in others, amphibious operation is planned.

Fuselage construction varies considerably, but the materials and techniques are not much different from those used in any aircraft. Early machines such as the Bell 47 were no more than a steel tube lattice frame with a blown acrylic canopy for the crew. Aerodynamic improvements came when the hull was faired in. The tail cone is often a stressed skin structure, but the centre section has too many doors, windows and access hatches for the skin to carry all the loads, and alloy frames or steel tubes are often used beneath the skin. The main landing gear often shares the same framework as the engine and transmission so that the skin is not unduly stressed on landing.

Figure 1.20 shows the structure of an Enstrom F-28. The landing skids, engine, fuel tanks and transmission are all attached to a welded tube frame known as a pylon. A stressed skin tail cone is attached to the rear of the pylon structure, and the aluminium seat and cabin floor is attached to the front. The cabin is glass fibre with plastics glazing. The cabin lines are faired into the tail cone by unstressed panelling attached to the pylon, so the machine has a sleek outline. Much of the unstressed centre panelling can be removed for servicing.

Composite materials are ideal for airframe construction and are becoming increasingly important since they are less dense than metals and are inherently well damped, which helps to control the inevitable vibration that characterizes helicopters. They can also have an indefinite fatigue life.

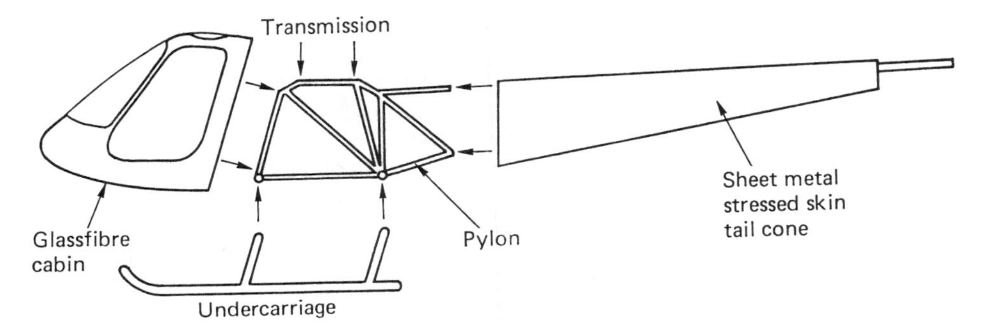

Fig. 1.20 The structure of a light helicopter: Enstrom F-28. See text for details.

1.7 Engine and transmission

The engine or engines and transmission are generally close together. Piston engines are heavy, and they are almost always placed below the rotor head to balance the machine. Turbines are much smaller and lighter and are often built into the roof of the hull to maximize internal space. Both piston engines and turbines turn much faster than the rotor shaft, and the transmission must incorporate reduction gearing.

The gearbox will generate a good deal of heat on a large machine, and require an oil cooler. This will have its own air intake near the rotor shaft; often in the front of the pylon. The gearbox also drives the tail rotor shaft, which runs the length of the tail boom to the tail rotor gearbox. The tail rotor shaft is often mounted outside the tail cone for ease of inspection and maintenance.

The rotors may take some time to come to rest after the engines have been shut down, and this may be inconvenient. The civil user wants to disembark passengers with the rotors stopped. The naval user wants to put the machine on the elevator down to the hangar and the army user wants the rotors stopped quickly on a clandestine mission so that the machine can be camouflaged. This can be achieved by fitting a rotor brake on the transmission.

Helicopters have also been built with jets at the blade tips, and this has the advantages that no gearbox is required and there is no torque reaction. Unfortunately tip jets have short fatigue life, very high fuel consumption and noise level to match and are little used. They will not be considered further here.

As it is a major subject, the whole of Chapter 6 is devoted to helicopter engines and transmissions.

1.8 The fuel system

The fuel system can be simple or complex depending on the type of machine. In early piston engine machines using carburettors, the fuel system was little more than a pair of tanks that fed fuel by gravity through the pilot's cut-off valve and a filter to the engine. The tanks in a gravity fed system are mounted one each side of the mast so that lateral and longitudinal trim is unaffected as fuel is consumed. The tanks are clearly visible on the Bell 47 and the Hughes 300. Larger machines will use the space below the floor to put tanks close to the CM, in which case pumps will be required to feed the engines.

Piston engines burn AVGAS or aviation gasoline, which is basically similar to automobile fuel but made to tighter quality standards. Turbines burn AVTUR that has a similar relationship to kerosene. As a piston engine will stop if AVTUR reaches it, pilots like to know they have taken on the right fuel. AVGAS fillers are marked red and AVTUR fillers are marked black to help prevent a dangerous mistake.

Recently there have been significant advances in the development of Diesel engines, allowing a similar power to weight ratio to a gasoline engine to be obtained. The advantage of the aeroDiesel is that it can burn AVTUR and its improved fuel economy allows better payload or range. A more detailed treatment of fuel systems can be found in Chapter 6.

1.9 The landing gear

The landing gear is subject to considerable variation. Utility and training helicopters are invariably fitted with skids to allow a landing on unprepared ground even with forward speed. Small wheels, known as ground handling wheels, can be fitted so the machine can be moved around. Some skids are broader than usual so they can act as skis for landing on snow or soft ground. Inflatable or rigid floats can be attached to the skids to permit operation over water, but there will be a drag and payload penalty. Larger machines invariably have wheels, as skids would make them too difficult to move. Naval helicopters need wheels to allow them to be moved below decks. In many cases the wheels can be locked so as to be tangential to a circle. The machine can be turned into the wind, but will not roll as the ship heels.

1.10 Oleos and ground resonance

Landing gears often incorporate a telescopic section containing oil and a compressed gas acting as a spring. These are formally known as oleo-pneumatic struts, invariably abbreviated to oleos. When the length of the oleo changes, the oil is forced through a small orifice to damp the movement. The struts that hold up automobile tailgates work on the same principle, but these are sealed units whereas the type of oil and the gas pressure in an oleo may be adjusted to give the correct spring rate and damping.

One obvious purpose of the oleo is to absorb the impact of landing, but a more important role is to control ground resonance. Ideally the rotor blades rotate with perfectly even spacing when run up on the ground, but it is possible for them to be disturbed from that condition. This results in the CM of the rotor moving away from the shaft axis and the rotor tries to whirl the top of the hull in a circular orbit. Under certain conditions this motion becomes uncontrollable unless there is damping to dissipate the energy. The origin of ground resonance will be discussed in Chapter 4 where it will be shown that the rotor head may also need dampers to prevent the problem.

1.11 The rotors

The main rotor takes the place of the wings of a conventional aircraft, and it is not unrealistic to think of a helicopter as being supported by the lift from wings that rotate instead of flying in a straight line. The main difference between rotor blades and wings is that the former are in comparison very thin and flexible, and the forces acting upon them are greater and more rapidly varying. The rotors have more to do than an aeroplane wing, because they are also the control system. Chapter 3 explains how the rotors produce lift and introduces the control of the machine.

1.12 The control system

The control system cannot be treated in isolation, but must be integrated into the design of a machine from the outset. In the pure helicopter, control of the machine is achieved entirely by changing the pitch of the main and tail rotor blades in various ways. This will then determine the amount of engine power needed. The rotors are generally designed to turn at constant speed and the throttle setting will have to be modified whenever the rotor power demand changes so that the speed does not change. Chapter 6 considers engines and power control.

There are two main forces acting on a helicopter, the force due to gravity, which is always downwards, and the rotor thrust vector, which is always at right angles to the tip path plane, otherwise called the rotor disc. Chapter 2 explains how the result of forces acting in various ways can be predicted and Chapter 3 shows how rotors develop thrust. The pilot can control the magnitude of the rotor thrust with the collective pitch lever held in his left hand, and the direction of the rotor thrust with the cyclic stick held in his right hand. The cyclic stick works in two dimensions: if the stick is pushed in any direction, the rotor thrust tilts the same way. These two fundamental controls are illustrated in Figure 1.21.

The blade movements necessary to produce lift and to achieve control will be outlined in Chapter 3, whereas Chapter 4 treats the construction and dynamics of the blades

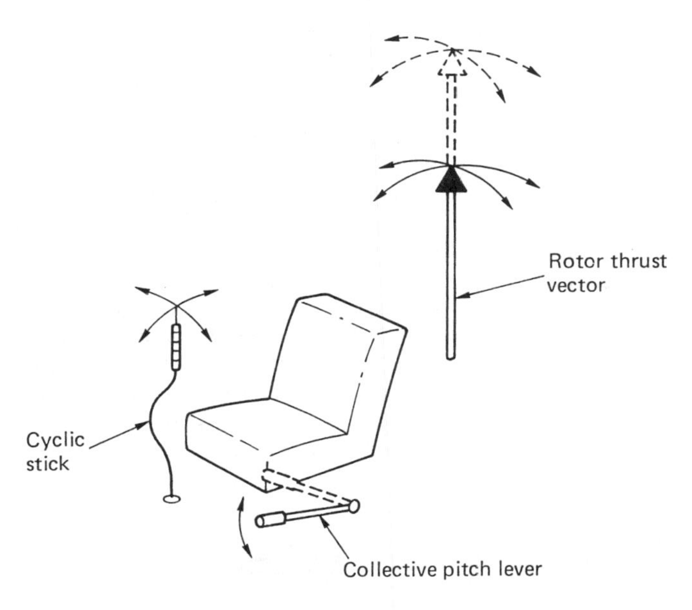

Fig. 1.21 The fundamental rotor controls. The cyclic stick tilts the rotor in the direction it is moved, whereas the collective lever changes the magnitude of the thrust.

and rotor head. Chapter 7 integrates the control system and considers power assistance and auto-stabilization.

The pilot controls the yaw axis by altering the pitch of the tail rotor blades using foot pedals. Chapter 5 considers the tail system of the helicopter.

It should be mentioned in passing that the No. 1 pilot in a helicopter conventionally occupies the right hand seat; the opposite of aeroplane practice. In some early machines, only one collective lever was provided, between the seats. The test pilots would sit in the left seat, as per aeroplane practice. Because of the difficulty of reversing the function of left and right hands, a pilot under training would be placed in the right seat and so the right seat became the conventional location for a helicopter pilot. A further factor is that in early helicopters, the forces fed back to the cyclic stick from the rotor were such that it was not safe to let go of the stick for an instant. However, the collective lever could be released temporarily. Given that most secondary controls such as the radio and instruments are centrally disposed so that pilot and co-pilot can both see them, it made sense to put the pilot in command on the right where his free left hand would be of most use.

1.13 Electrical and hydraulic system

The power systems of a typical helicopter are not that much different from those of any aircraft. Electrical power is needed for instruments, radios, autopilots, lighting, engine starting, navigation and intercom systems, as well as a host of further avionics that might be needed for special purposes.

A light helicopter may have no power assistance, but as machines get larger the control forces may cause fatigue and in very large machines they will be beyond the strength of the pilot. Power-assisted controls then become essential. Power controls are also needed if some kind of automatic stabilization or autopilot is fitted so that the low powered electronic signals can operate the controls.

When powered controls operated by electrical signals have faultless reliability, the mechanical controls from the pilot can also be replaced by electrical controls, leading to the concept of fly-by-wire. The pilot operates controls having no mechanical connection to the rotors, but which instead produce electrical signals. This concept is explored in Chapter 7.

Electric motors are useful for low powered control purposes such as the trim mechanism, but hydraulics allows greater forces to be developed within small actuators, and so they will be used for powered flying controls.

Electrical and hydraulic power is vital to the safety of the machine, and the hydraulic pump and the generator may be driven from the rotor shaft so that power is still available even in the case of engine failure. In small machines the generator is driven from the engine, as battery capacity is enough to keep the electrical system working in the case of an engine failure. Larger machines may have two or more engines and each will have a generator so that electric power is still available in case of engine failure.

The electrical system is discussed in Chapter 6, and an explanation of hydraulic controls will be found in Chapter 7.

1.14 Instruments and avionics

Many of the instruments fitted to helicopters are the same as those used in other aircraft, but in addition to the usual engine-related gauges the helicopter will also need a rotor tachometer. The response of a helicopter to control inputs depends on rotor speed, and this needs to be controlled to close limits. The rotor tachometer and engine-related instruments are detailed in Chapter 6, whereas flight instruments are covered in Chapter 7.

References

1 Spenser, J.P., *Whirlybirds*, University of Washington Press, Seattle, ISBN 0-295-97699-3 (1998)
2 Spenser, J.P., *Vertical Challenge*, University of Washington Press, Seattle, ISBN 0-295-97203-3 (1998)
3 Coates, S., *Helicopters of the Third Reich*, Classic Publications, Hersham, ISBN 1-903223-24-5 (2002)
4 Boulet, J., *L'Histoire de l'Hélicoptère*, Editions France-Empire, Paris, ISBN 2-7048-0040-5 (1982)
5 DuMoulin, A., *Les Hélicoptères Florine 1920–1950*, Fonds National Alfred Renard, Brussels (1999)
6 Everett-Heath, J., *British Military Helicopters*, Arms and Armour Press, London, ISBN 0-85368-805-2 (1986)
7 Nowarra, H.J., *German Helicopters*, Schiffer, West Chester, ISBN 0-88740-289-5 (1990)
8 Dowling, J., *RAF Helicopters: the first twenty years*, HMSO, London, ISBN 0-11-772725-3 (1992)

2

Technical background

This chapter provides an introduction to concepts that will be needed to follow the technical explanations in the rest of the book.

2.1 Introduction to mechanics

Mechanics is the study of how objects interact with the forces applied to them. Designers are concerned with resisting the forces generated in flight so that the machine stays intact. Using the controls of a helicopter the pilot changes the forces applied to it and thereby determines the path it takes. It is important for a pilot to understand some mechanics so that he can predict what control movements will be necessary to make the machine do what is wanted. When this happens, the pilot is said to be in control. Any other situation is not recommended.

2.2 Mass and density

The amount of matter in a body is specified by its mass, measured in pounds (lb) or kilograms (kg). Since the amount of matter in a body cannot readily change, the mass is the same wherever the body is and however it moves. The density of a substance is the mass of unit volume. In the old imperial units, density was expressed as pounds per cubic foot. In the SI system, the term 'relative density' is used, where water has a relative density of one.

2.3 Force and acceleration

Weight is the force a body exerts on its supports. In orbit, objects have no supports: they are weightless. Back on earth the gravitational field pulls them down and they have weight. Weight is equal to mass multiplied by the strength of the gravitational field. This is why astronauts can bounce around on the moon; gravity is much weaker there. Since the strength of gravity doesn't change much from one place on earth to another, earthlings tend to buy things by weight not by mass, and often confuse the two.

All real objects, helicopters included, have their mass distributed over their dimensions. Whilst each part has its own mass, for some purposes this can be replaced by a single mass located at the centre of mass (CM) (Figure 2.1(a)). The older term centre

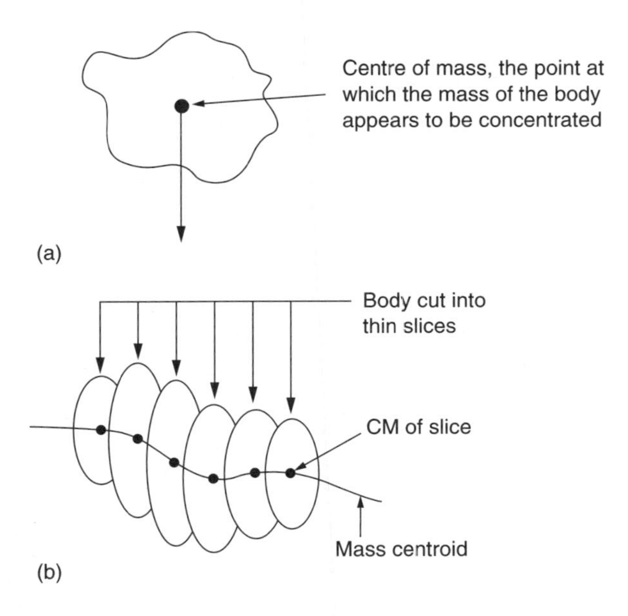

Fig. 2.1 At (a), the distributed mass of a body can be replaced by a point mass acting at the centre of mass (CM). The mass centroid (b) is a line joining the centres of mass of thin slices through the body.

of gravity (CG) will also be found, but this is incorrect as objects still have a centre of mass in the absence of a gravitational field. If a body is imagined to be divided into a series of slices, each slice will have a CM. Figure 2.1(b) shows that the line joining all of these CMs is known as the mass centroid. The location of the mass centroid is important in blade design and in balancing rotating assemblies.

The speed of an object is the rate at which it covers distance, and the direction is immaterial. In contrast, velocity is the rate at which distance is covered in a specific direction. Any quantity that also has direction is called a vector. Acceleration is the rate of change of velocity, so it must be a vector. Acceleration can come about by keeping the direction the same and changing the speed, or by keeping the speed the same and changing the direction. The force necessary to change velocity is equal to the mass multiplied by the acceleration:

$$F = m \times a$$

In SI units, the unit of force is the Newton (N). This is defined as the force that will cause a mass of one kilogram to accelerate at one metre per second per second.

On the surface of the earth, a mass of 1 kg experiences a gravitational attraction of about 9.81 N. Consequently any object released at a height will accelerate downwards at 9.81 metres per second per second. This downward acceleration, commonly called falling, can be prevented by opposing gravity with an upward force of 9.81 N. Figure 2.2(a) shows what happens. The object is supported by upward force opposing its weight. Clearly force is a vector quantity. When the two forces are exactly equal and opposite, the object is in equilibrium: the resultant force and the acceleration are both nil.

Consider an elevator in a high-rise building. When the elevator starts, the motor applies a greater upward force to the occupants than gravity applies downwards.

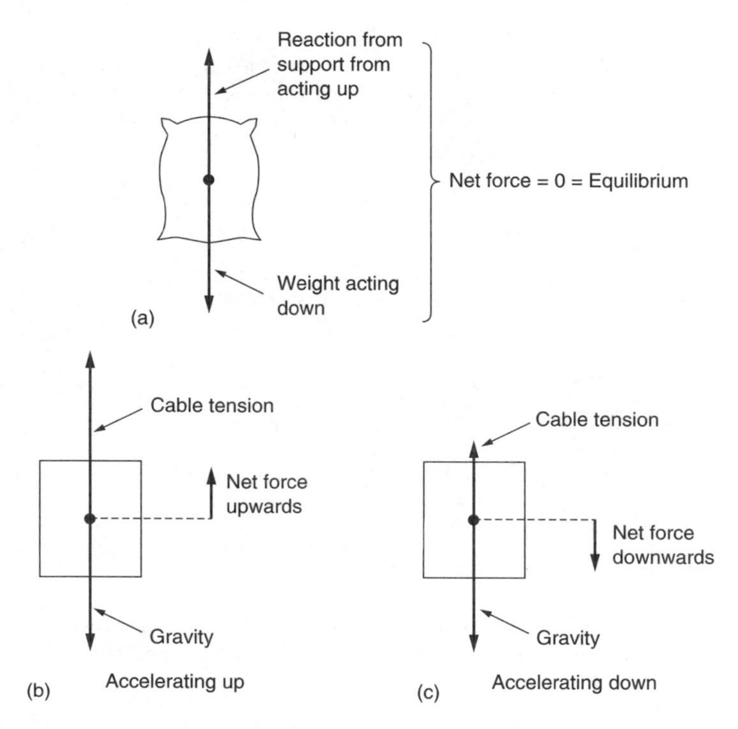

Fig. 2.2 Weight is the force a body exerts on its supports. At (a) the weight is equal to the supporting force and there is no resultant. At (b) upward acceleration requires a greater force than in (a) so the weight appears to increase. At (c), downward acceleration causes weight to reduce. Note that the mass does not change, but the weight does.

Figure 2.2(b) shows that there is no longer a balance of forces, so there is no equilibrium. The difference between the force of gravity and the greater force from the floor of the elevator accelerates us upwards. We feel the extra load on our feet: temporarily our weight has increased. As the elevator approaches the desired level, the motor will reduce power, and the force it applies will be less than that due to gravity. Figure 2.2(c) shows that the resultant force is now downwards, and our ascent is slowed until we stop. Momentarily our weight is reduced: we feel light on our feet. Note that as the elevator slows we are going upwards but accelerating downwards. There is no contradiction here; acceleration is the rate of change of velocity.

Figure 2.3 shows what happens to an object having mass when forces act on it to accelerate and decelerate it. As there is always a reaction to the application of a force, if force is applied to an object in order to accelerate it, the reaction will attempt to accelerate whatever is supplying the force the opposite way. Figure 2.4 shows some examples. The recoil of a gun is the reaction to accelerating the shell. The thrust of a ship's propeller is the reaction to accelerating water backwards. A helicopter stays airborne by accelerating air downwards: the reaction is upwards, and if it is equal to the force due to gravity, the helicopter is in equilibrium. Note that the helicopter is not weightless, its weight is acting on the air around it as the substantial downwash indicates.

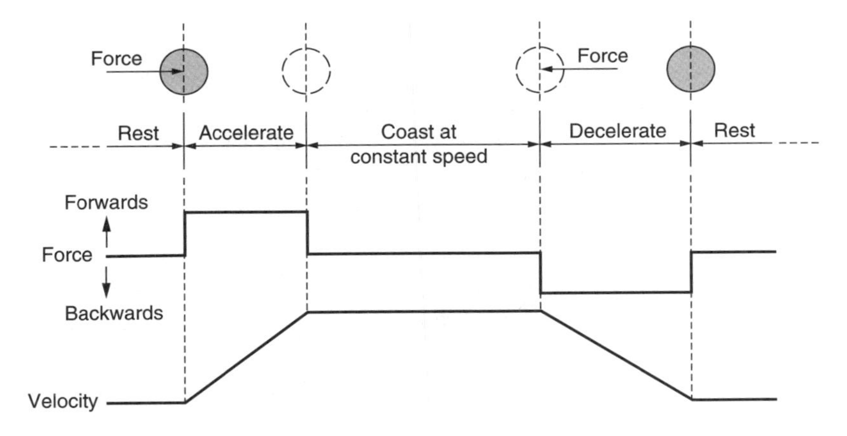

Fig. 2.3 Effect of force on velocity of a body.

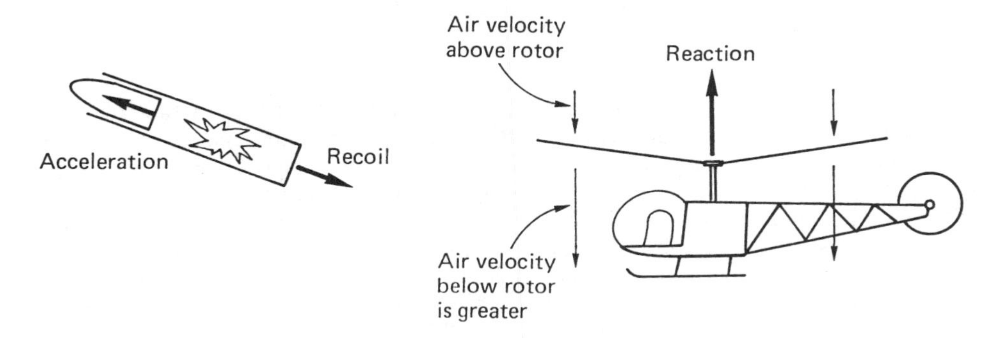

Fig. 2.4 Reactions to acceleration include the recoil of a gun and the thrust of the rotor that sustains a helicopter.

2.4 Strength and rigidity

In aircraft, various forces must be resisted by the structure. The ability of a material to resist forces is measured with respect to unit cross-sectional area, typically a square metre. The force applied per unit of area is known as the stress. Figure 2.5 shows that as the force applied to unit area increases, initially there will be a proportional elongation according to Hooke's Law. The elongation per unit length is known as the strain. The constant of proportionality is known as the stiffness. In the linear region, the material recovers its dimensions when the stress is removed. If the stress is further increased, the strength of the material may be exceeded and an irreversible extension, known as plasticity, may take place, leading finally to breakage. In brittle materials, the plastic region may be negligibly small.

A structure that can resist a given static load may eventually suffer fatigue failure if the load is repeatedly applied and removed. In practice a safety factor must be applied to the maximum stress and the life of the component may be restricted to a certain number of cycles. Fatigue primarily occurs in parts under tensile stress, probably because this condition encourages the spreading of cracks. Highly stressed components

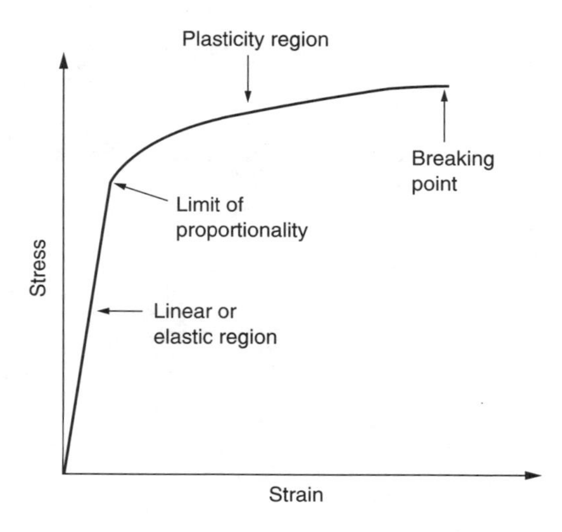

Fig. 2.5 Elongation is initially proportional to stress, followed by differing characteristics according to whether the material is ductile or brittle.

such as rotor heads may need extremely fine finishing and surface treatment to minimize any irregularities from which a crack might propagate. Periodic inspection of highly stressed parts may be needed. Cracks can be detected by X-rays, ultrasonic testing or by penetrating dyes. At high temperatures, materials working within their elastic limit may creep if the load is sustained. This phenomenon limits the life of turbine blades.

In many cases it is not the ability of the component to resist the load that matters, but the amount of deflection the load is allowed to cause. Unwanted deflections are generally unwelcome in aircraft as they may result in flutter or imprecise control. Consequently stiff materials are advantageous. All aircraft have to lift themselves as well as any payload, so it is an advantage if the weight of the structure can be minimized. It is often thought that this will be achieved using low density materials, but this is not always the case. In practice the lowest weight will be achieved by using materials with a high stiffness to relative density ratio. This ratio also controls the speed of sound in the material, which is generally high in aerospace materials as can easily be established by tapping them. The result is a sound of a higher pitch than that found in everyday materials. Such materials also make good loudspeaker cones, where the same criteria apply.

In practice the designer would like to have a range of materials all having high stiffness to relative density ratio, but with a range of relative densities and strengths. Where stresses are concentrated, a high strength, high density material is used. Where stresses are reduced, a lower density material will be superior. For example, if a component were made from an excessively strong material, the thickness required to carry the design load might be so small that the component might suffer handling damage. Honeycomb and foam cored materials are one approach to reducing density. In a helicopter, the density of materials used tends to fall with distance from the rotor axis.

2.5 Resultants and components of forces

Earlier in this chapter, the forces were either up or down in order to introduce one concept at a time. In the real world forces can act in arbitrary directions, except for the

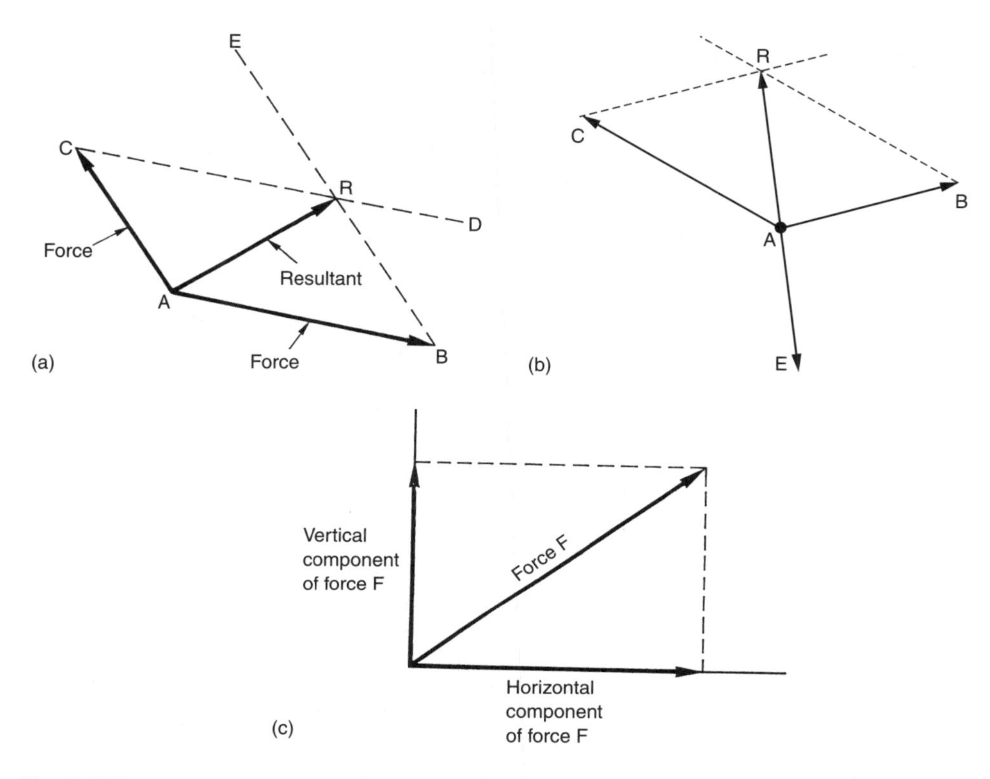

Fig. 2.6 Forces are vectors having direction and magnitude represented by the length of an arrow. At (a) the resultant of two forces AB and AC can be found by drawing CD parallel to AB and BE parallel to AC. This completes the parallelogram CABR and the diagonal AR represents the resultant. At (b) a force AE equal and opposite to the diagonal will cancel the other two forces. At (c) a single force can be resolved into forces in two orthogonal directions.

force due to gravity, which doggedly remains pointed downwards. When several forces are acting, it is possible to find out what happens by determining the direction and magnitude of the resultant force.

Figure 2.6 shows how forces are represented. An arrow is drawn in the direction of each force, and the length of the arrow is proportional to the magnitude. In this case two different forces act on the same point, and in different directions. We can find the resultant force, which is a single force having the same effect as the others together, by completing the parallelogram as shown, and drawing in the diagonal. The body will accelerate in the direction of the resultant. Alternatively, by drawing a force of equal length to the resultant but in the opposite direction, we have found the force needed to maintain equilibrium with the other two.

The opposite of finding a resultant to two forces is splitting a single force into two different ones. This is known as resolving, and the two forces are known as components. It is common to resolve an arbitrary force into horizontal and vertical components. The figure shows that it is easy to do. Horizontal and vertical lines are drawn from both ends of the force arrow. The intersection of the lines shows the magnitude of the horizontal and vertical components of the force.

2.6 Moments and couples

So far only forces that all conveniently act at one point have been considered. In reality forces can also act at a distance from a point. When a force does not pass through a point the result is called a moment with respect to that point. The moment is equal to the force multiplied by the distance at right angles to the force, as shown in Figure 2.7. Applying a moment to an object will cause it to turn and accelerate along simultaneously.

If two equal and opposite forces act a distance apart, the result is a pure turning effort known as a couple or as torque. A helicopter engine produces a couple at the output shaft in order to drive the blades. The reaction to this couple attempts to twist the engine against its mountings. The transmission conveys the engine power to the rotor shaft. The rotor shaft exerts a couple on the rotor head in order to drive the blades round. The torque reaction to this couple attempts to rotate the transmission, and the helicopter hanging from it, the opposite way. One of the jobs of the designer is to find a means to prevent this rotation.

When a body is supported at the CM, it will remain in the initial attitude. Figure 2.8 shows that if the support is not at the CM, the body will rotate. It may swing from side to side, but eventually will come to rest with the CM below the point of support.

It is often assumed that the CM of the helicopter is below the rotor shaft, but this may not be the case. Figure 2.9 shows how a helicopter hovers with the CM displaced. The CM tends to hang directly beneath the rotor head, and the pilot has to displace the cyclic control in order to keep the disc horizontal so the rotor thrust vector is straight up. Clearly if the CM is a long way from the mast, a substantial amount of cyclic control will be used up to compensate, and there may not be enough left to control the machine. In addition the bearings and joints in the rotor head find it harder to transmit power when the shaft and the disc are not aligned.

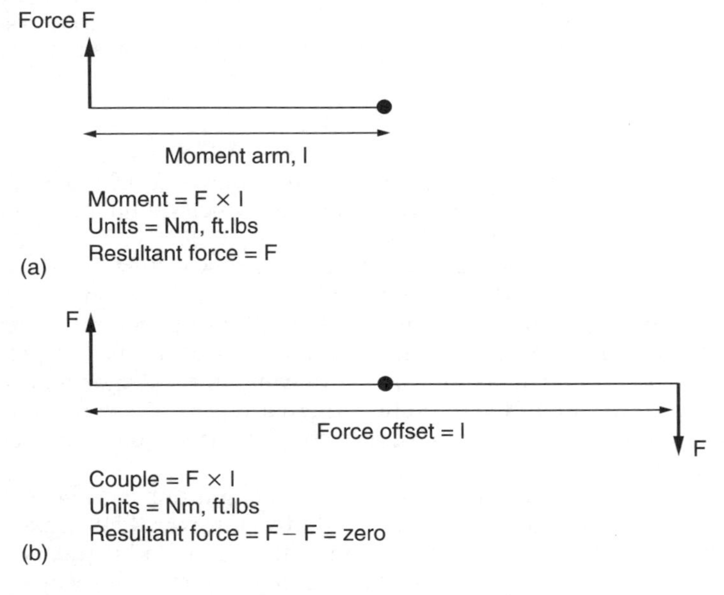

Fig. 2.7 (a) A moment about a point is the product of the force and the distance. (b) A pure couple results from two equal and opposite forces that do not coincide.

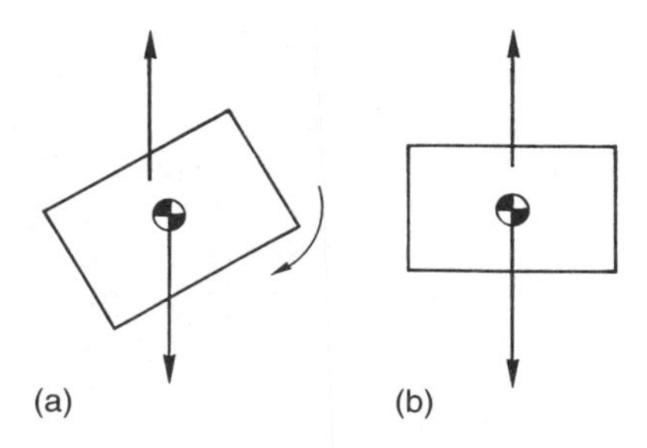

Fig. 2.8 (a) A body supported at a point not above the CM will swing until the CM is below the support as in (b).

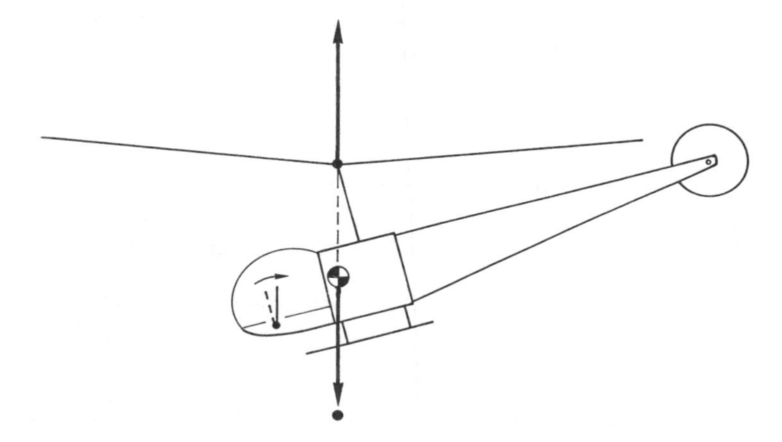

Fig. 2.9 In a hovering helicopter, if the CM is displaced, the mast will have to tilt until the CM is below the thrust line of the rotor.

For all these reasons, the CM position of a helicopter is important. Taking on passengers, fuel and payload can move the CM of the machine. It is the responsibility of the commander to ensure that the CM of the helicopter is within the limits laid down by the authorities for the machine in question.

2.7 Work, energy, power and momentum

In everyday speech these terms are used with a variety of meanings, but in mechanics their meanings are much more closely defined. Work is done when a resultant force moves through a distance. A table supporting objects placed on it does no work on them because although there is a force holding them up, there is no movement. A crane lifting an object does work because the force of gravity is opposed throughout the distance the load is lifted.

Energy is a measure of the ability to do work. Sometimes the interchange between work and energy is reversible. Burning fuel releases energy that allows an engine to do work. This is not reversible. A wound-up spring holds the energy that drives a clock. A helicopter battery holds enough energy to start the engine. These are both reversible. Work is done winding the spring or charging the battery and this is turned into stored energy, most of which can later be released to do work.

Energy can also be stored in the position and motion of objects. If a weight is lifted against gravity, it stores potential energy that can be released when the weight is lowered again. Potential energy is proportional to the height through which the object is lifted. If a weight is accelerated, the resultant force acting on it moves through a distance and does work which is stored as kinetic energy. Figure 2.10 shows a mass being accelerated from rest by a constant force F over a distance s in time t up to a velocity v. The kinetic energy of the mass is equal to the work done on it. This is equal to the product of force F multiplied by distance s.

$$\text{Since } F = m \times a, \text{ then KE} = m \times a \times s.$$

Acceleration is the rate of change of velocity, thus $a = v/t$. Since the velocity increases uniformly then distance s is the time multiplied by the average speed, thus $s = \frac{1}{2}v \times t$.

$$\text{KE} = m \times a \times s = m \times \frac{v}{t} \times \frac{v \cdot t}{2} = \frac{m \cdot v^2}{2}$$

Kinetic energy is proportional to the square of the velocity. Power is the rate at which work is done or the rate at which energy is released, which is the same thing. Thus a high powered engine will burn fuel at a higher rate in order to release more energy in a given time.

The final quantity to be considered is momentum. This is equal to the mass multiplied by the velocity. The use of momentum will better be explained when the topic of lift generation is dealt with in Chapter 3.

A helicopter in flight has both kinetic and potential energy. This energy was stored in the helicopter by doing work against the earth's gravitational field. Thus the helicopter's potential and kinetic energy exists with respect to the earth and Newton's laws determine what the helicopter will do with respect to the earth when forces act upon it. The earth is the inertial frame of reference.

Other than gravity, the forces when airborne can only come from the movement of the machine and its rotors with respect to the air. In still air the aerodynamic frame of reference and the earth's frame of reference are the same and this simplifies matters. However, when considering a machine flying in the presence of a wind, it is important to realize that the wind means that the aerodynamic frame of reference is moving with respect to the inertial frame of reference.

The reader is cautioned against texts that erroneously claim that the helicopter flies only with respect to the air because it doesn't know what the ground is doing and that all that matters is airspeed. If this were true artificial horizons and inertial navigators wouldn't work.

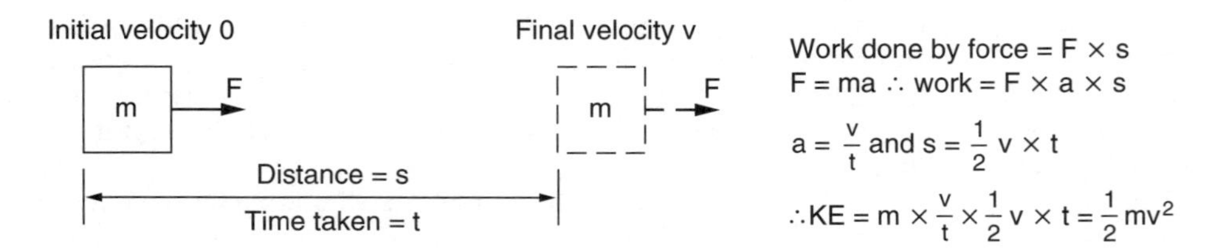

Fig. 2.10 A mass being accelerated will gather kinetic energy (KE) as derived here.

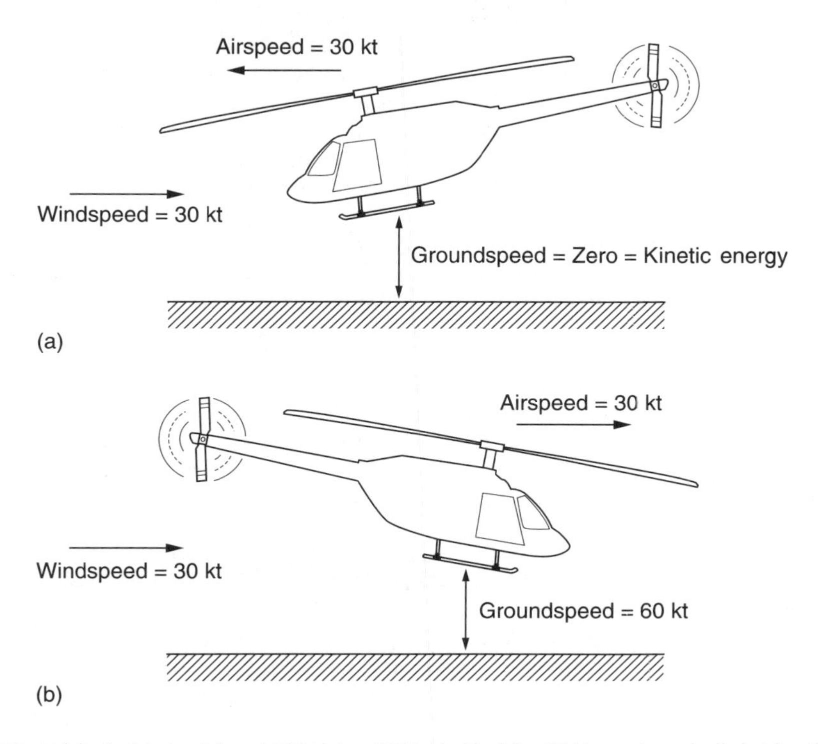

Fig. 2.11 At (a) a helicopter flying at 30 kt into a 30 kt wind is doing 30 kt aerodynamically but inertially it is hovering. At (b), flying downwind the aerodynamic conditions are the same but the machine has KE corresponding to 60 kt.

Figure 2.11(a) shows a helicopter flying at 30 knots directly into a 30 knot wind. Aerodynamically it is doing 30 knots but inertially it is stationary and has no kinetic energy. Inertially it is hovering. Figure 2.11(b) shows a helicopter flying at 30 knots down wind in a 30 knot wind. Aerodynamically the conditions are the same, but inertially the machine is now doing 60 knots and possesses the kinetic energy due to that groundspeed. The difference in kinetic energy becomes obvious when an attempt is made to change heading. If the helicopter in (b) does a 180° turn, it will conserve its kinetic energy and will exit the turn with 60 knots of groundspeed and 90 knots on the ASI. The result will be that the machine will tend to climb as the surplus airspeed is converted to potential energy.

If the helicopter in (a) tries a 180° turn it will be in difficulty and will lose height because it lacks 60 knots worth of kinetic energy. If the machine isn't powerful and doesn't have height to lose it could crash. And the pilot who thought that the helicopter doesn't know what the ground is doing wouldn't know what he did wrong.

2.8 Efficiency

Efficiency is generally defined as the ratio of useful output power to input power in any mechanism. In the case of a gearbox the useful power is the output shaft power. This will always be less than the input power and the difference will be converted to heat.

In the case of an engine, the input power is the heat released by burning the fuel. In practical engines, this exceeds the output shaft power considerably. The percentage of the thermal energy in the fuel that emerges as shaft power is the thermal efficiency of the engine. The waste heat has to be removed by a cooling system to prevent the temperature of the engine rising to the point where components are damaged. Cooling systems waste further power in driving fans and pumps and usually increase the drag of the airframe.

Given the necessary high power to weight ratio in helicopters, the power plant and fuel form a significant part of the all-up weight. It is beneficial to explore means to improve the thermal efficiency of the engine. Not only will this reduce the weight of the fuel to be carried for a given range, but it may also allow the cooling system to be lighter, to consume less power and to cause less intake drag. Thus a small improvement in thermal efficiency may result in a significant increase in performance.

Passenger aircraft may be compared using specific air range (the mass of fuel used per unit of distance), but in a hovering helicopter this figure is meaningless. In helicopters, it may be better to compare the power actually used to hover with the theoretical power needed by an ideal rotor under the same conditions.

2.9 Gases and the atmosphere

The atmosphere is the medium in which helicopters fly but it is also one of the fuels for the engine and the occupants breathe it. It is a highly variable medium that is constantly being forced out of equilibrium by heat from the sun and in which the pressure, temperature, and humidity can vary with height and with time and in which winds blow in complex time- and height-variant patterns. The effect of atmospheric conditions on flight is so significant that no pilot can obtain qualifications without demonstrating a working knowledge of these effects.

The atmosphere is a mixture of gases. About 78% is nitrogen – a relatively unreactive element – whereas about 21% is oxygen, which is highly reactive. The remainder is a mixture of water in the gaseous state and various other traces. The reactive nature of oxygen is both good and bad. The good part is that it provides a source of energy for life and helicopters alike because hydrocarbons can react with oxygen to release energy. The bad part is that many materials will react with oxygen when we would rather they didn't. Chemically, combustion and corrosion are one and the same thing. The difference is based on the human reaction to the chemical reaction.

Gases are the highest energy state of matter, for example the application of energy to ice produces water and the application of more energy produces water vapour. The reason that a gas takes up so much more room than a liquid is that the molecules contain so much energy that they break free from their neighbours and rush around at a high speed, which is a function of absolute temperature. As Figure 2.12(a) shows, the innumerable elastic collisions of these high speed molecules produce pressure on the walls of any gas container. In fact the distance a molecule can go without a collision, the *mean free path*, is quite short at atmospheric pressure. Consequently gas molecules also collide with each other elastically, so that if left undisturbed, in a container at a constant temperature, every molecule would end up with essentially the same energy and the pressure throughout would be constant and uniform.

Pressure is measured by physicists and by engineers in units of force per unit of area using imperial units of pounds per square inch or SI units of Newtons per square metre. At sea level, the atmosphere exerts a pressure of about 15 pounds per square inch and has a density of about 0.075 pounds per cubic foot or in metric units about

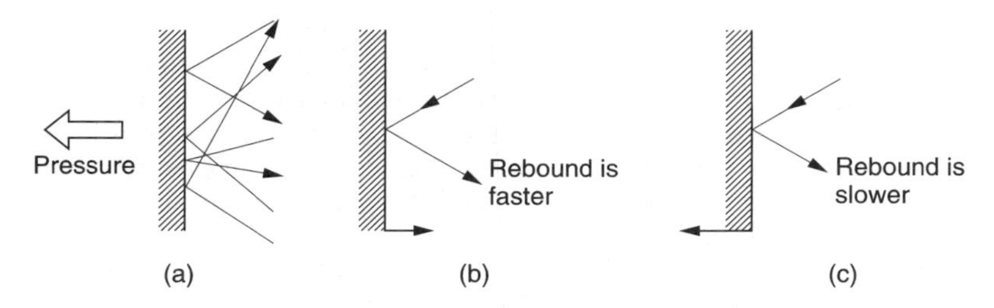

Fig. 2.12 (a) The pressure exerted by a gas is due to countless elastic collisions between gas molecules and the walls of the container. (b) If the wall moves against the gas pressure, the rebound velocity increases. (c) Motion with the gas pressure reduces the particle velocity.

100 000 Newtons per square metre with a density of 1.225 kg per cubic metre. Over the years, many other units of pressure have evolved, some from meteorology. One of these is the bar (after barometry) where one bar is the average atmospheric pressure at the place where the bar was defined. In practical use, the bar is divided into one thousand millibars. The bar is slowly being replaced by a numerically identical unit known as the hectoPascal (hPa). The bar and hPa are commonly used in aviation altimetry. The principle of the mercury barometer is that atmospheric pressure supports a column of mercury exposed to a vacuum at the top. Consequently the length of the column is proportional to pressure and can be expressed in inches or cm of mercury. At sea level a reading of about 26 inches of mercury is obtained. This unit may be found in use in altimeters originating in the United States.

The Gas Law states that the product of pressure and volume is proportional to temperature. Reducing the volume means that external work has to be done to oppose the pressure. This work increases the temperature of the gas. The Diesel engine obtains ignition in this way. Conversely if the volume is increased, work is done by the expansion of the gas and the temperature must fall. This is why carburettors are prone to icing on part throttle because the air expands on entering the manifold. Air conditioners work in the same way.

If the volume is fixed, as temperature rises, the velocity of the molecules increases and so the impact at each collision with the walls of any container is greater and the pressure rises. Alternatively the same pressure can be exerted in a given volume with a smaller mass of gas. Thus in the atmosphere where pressure increases can be released by free movement, the result of an increase in air temperature is that the density goes down. Density is also affected by humidity. Water molecules are heavier than those of atmospheric gases and increase the pressure due to molecular collisions. Thus in the presence of water vapour a given pressure can be sustained with a smaller mass of air and the density goes down.

2.10 Sound

Sound is simply an airborne version of vibration which is why the two topics are inextricably linked. Sound disturbs the equilibrium of a gas, as does the creation of lift. Not surprisingly, aerostats excepted, all flying objects make a noise as a by-product of creating lift. Consequently there is a great deal of overlap between aerodynamics and acoustics.

Figure 2.12(b) shows that a solid object that moves *against* gas pressure increases the velocity of the rebounding molecules, whereas in (c) one moving *with* gas pressure reduces that velocity. The average velocity and the displacement of all the molecules in a layer of air near to a moving body is the same as the velocity and displacement of the body. Movement of a body results in a local increase, or decrease, in pressure. Thus sound is both a pressure and a velocity disturbance. Integration of the velocity disturbance gives the displacement.

Despite the fact that a gas contains endlessly colliding molecules, a small mass or *particle* of gas can have stable characteristics if molecules leave and are replaced by others having identical properties. As a result aerodynamics and acoustics seldom need to consider the molecular structure of air and the constant motion can be neglected. Thus when particle velocity and displacement is considered in aerodynamics or acoustics, this refers to the average values of a large number of molecules. In an undisturbed container of gas the particle velocity and displacement will both be zero everywhere.

When the volume of a fixed mass of gas is reduced, the pressure rises. The gas acts like a spring; it is compliant. However, a gas also has mass. Sound travels through air by an interaction between the mass and the compliance. Imagine pushing a mass via a spring. It would not move immediately because the spring would have to be compressed in order to transmit a force. If a second mass were to be connected to the first by another spring, it would start to move even later. Thus the speed of a disturbance in a mass/spring system depends on the mass and the stiffness. Sound travels through air without a net movement of the air.

After the disturbance had propagated the masses would return to their rest position. The mass/spring analogy is helpful for a basic understanding, but is too simple to account for commonly encountered acoustic phenomena such as spherically expanding waves.

Unlike solids, the elasticity of gas is a complicated process. If a fixed mass of gas is compressed, work has to be done on it. This will generate heat in the gas. If the heat is allowed to escape and the compression does not change the temperature, the process is said to be *isothermal*. However, if the heat cannot escape the temperature will rise and give a disproportionate increase in pressure. This process is said to be *adiabatic* and the Diesel engine depends upon it. In most acoustic cases there is insufficient time for much heat transfer and so air is considered to act adiabatically. Figure 2.13 shows how the speed of sound c in air can be derived by calculating its elasticity under adiabatic conditions.

$$\text{Velocity } V = \sqrt{\frac{\gamma R T}{M}}$$

γ = adiabatic constant (1.4 for air)
R = gas constant (8.31 J K^{-1} $mole^{-1}$)
T = absolute temp (K)
M = molecular weight (kg $mole^{-1}$)

Assume air is 21% O_2, 78% N_2, 1% Ar

$$\text{Molecular weight} = 21\% \times 16 \times 2 + 78\% \times 14 \times 2 + 1\% \times 18 \times 1$$
$$= 2.87 \times 10^{-2} \text{ kg mole}^{-1}$$

$$V = \sqrt{\frac{1.4 \times 8.31\, T}{2.87 \times 10^{-2}}} = 20.1\sqrt{T}$$

at 20°C $T = 293$ K $V = 20.1\sqrt{293} = 344 \text{ ms}^{-1}$

Fig. 2.13 Calculating the speed of sound from the elasticity of air.

If the volume allocated to a given mass of gas is reduced isothermally, the pressure and the density will rise by the same amount so that *c* does not change. If the temperature is raised at constant pressure, the density goes down and so the speed of sound goes up. Gases with lower density than air have a higher speed of sound. Divers who breathe a mixture of oxygen and helium to prevent 'the bends' must accept that the pitch of their voices rises remarkably.

The speed of sound is proportional to the square root of the absolute temperature. At sea level the speed of sound is typically about 1000 feet per second or 344 metres per second. Temperature falls with altitude in the atmosphere and with it the speed of sound. The local speed of sound is defined as Mach 1.

As air acts adiabatically, a propagating sound wave causes cyclic temperature changes. The speed of sound is a function of temperature, yet sound causes a temperature variation. One might expect some effects because of this. Sounds below the threshold of pain have such a small pressure variation compared with atmospheric pressure that the effect is negligible and air can be assumed to be linear. However, on any occasion where the pressures are higher, a situation not unknown in aviation, this is not a valid assumption. In such cases the positive half cycle significantly increases local temperature and the speed of sound, whereas the negative half cycle reduces temperature and velocity. Figure 2.14 shows that this results in significant distortion of a sine wave, ultimately causing a *shock wave* that can travel faster than the speed of sound until the pressure has dissipated with distance. This effect is responsible for the sharp sound of a handclap and for the sonic boom of a supersonic aircraft.

Whilst the hulls of helicopters do not approach the speed of sound, their rotor blade tips certainly do and the effects of this have to be taken into account. The drag on a rotor blade at moderate speeds increases as the square of the relative airspeed but as Mach 1 is approached the drag increases disproportionately; an effect called compressibility. This is intuitively understandable as the air has less time to get out of the way when the disturbances that propagate ahead of the blade are travelling a little faster than the blade itself. In extremely cold weather, or at high altitude, the speed of sound falls and helicopters with high tip speeds will suffer a loss of efficiency.

Sound can be due to a one-off event known as percussion, or a periodic event such as the sinusoidal vibration of a tuning fork. The sound due to percussion is called transient whereas a periodic stimulus produces steady-state sound having a frequency f.

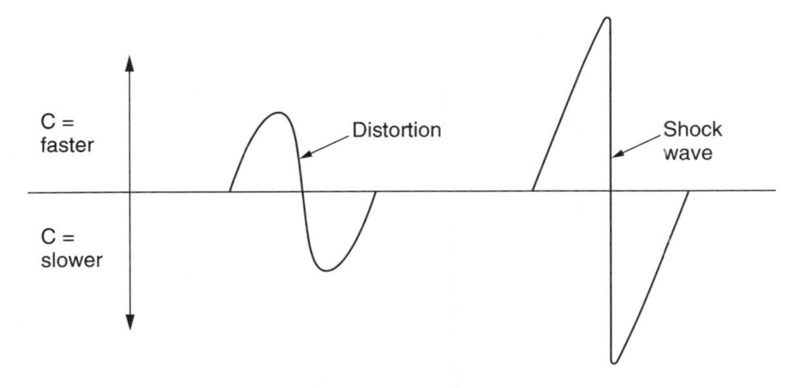

Fig. 2.14 At high level, sound distorts itself by increasing the speed of propagation on positive half cycles. The result is a shock wave.

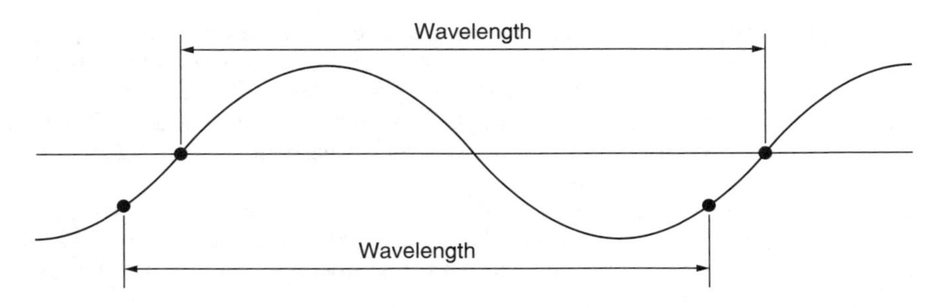

Fig. 2.15 Wavelength is defined as the distance between two points at the same place on adjacent cycles. Wavelength is inversely proportional to frequency.

Because sound travels at a finite speed, the fixed observer at some distance from the source will experience the disturbance at some later time. In the case of a transient, the observer will detect a single replica of the original as it passes at the speed of sound. In the case of the tuning fork, a periodic sound source, the pressure peaks and dips follow one another away from the source at the speed of sound. For a given rate of vibration of the source, a given peak will have propagated a constant distance before the next peak occurs. This distance is called the wavelength lambda. Figure 2.15 shows that wavelength is defined as the distance between any two identical points on the whole cycle. If the source vibrates faster, successive peaks get closer together and the wavelength gets shorter. The wavelength is inversely proportional to the frequency. It is easy to remember that the wavelength of 1000 Hz is a foot (about 30 cm) at sea level.

2.11 The mechanics of oscillation

By definition helicopters contain a lot of rotating parts and for a proper understanding of their characteristics, knowledge of the mechanics of rotation is essential. In physics, the engineering quantity of RPM is not used. Instead the unit of angular velocity is radians per second and the symbol used is ω. Figure 2.16 shows that a radian is a natural unit of angle which is the angle subtended by unit circumference at unit radius. As the circumference is given by 2π times the radius, then there will be 2π radians in one revolution, such that one radian is about 57°.

Figure 2.17 shows a constant speed rotation viewed along the axis so that the motion is circular. Imagine, however, the view from one side in the plane of the rotation. From a distance, only a vertical oscillation will be observed and if the position is plotted against time the resultant waveform will be a sine wave. The sine wave is unique because it contains only a single frequency. All other waveforms contain more than one frequency.

Imagine a second viewer who is at right angles to the first viewer. He will observe the same waveform, but at a different time. The displacement is given by multiplying the radius by the cosine of the phase angle. When plotted on the same graph, the two waveforms are *phase shifted* with respect to one another. In this case the phase shift is 90° and the two waveforms are said to be *in quadrature*. Incidentally the motions on each side of a steam locomotive are in quadrature so that it can always get started (the term used is quartering). Note that the phase angle of a signal is constantly changing with time whereas the phase shift between two signals can be constant. It is important that these two are not confused. In a three-bladed rotor, the motion of each blade

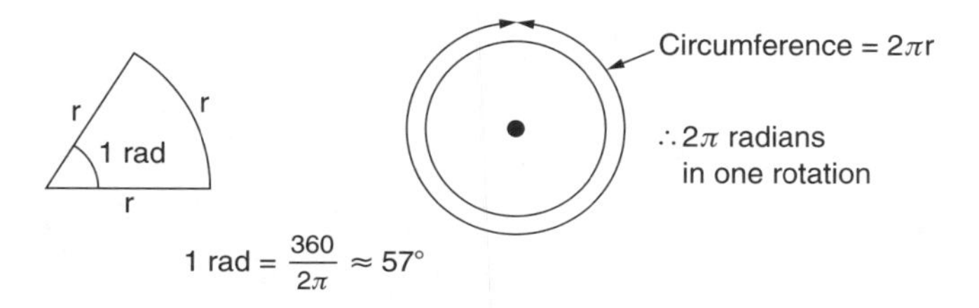

Fig. 2.16 The definition of a radian. See text.

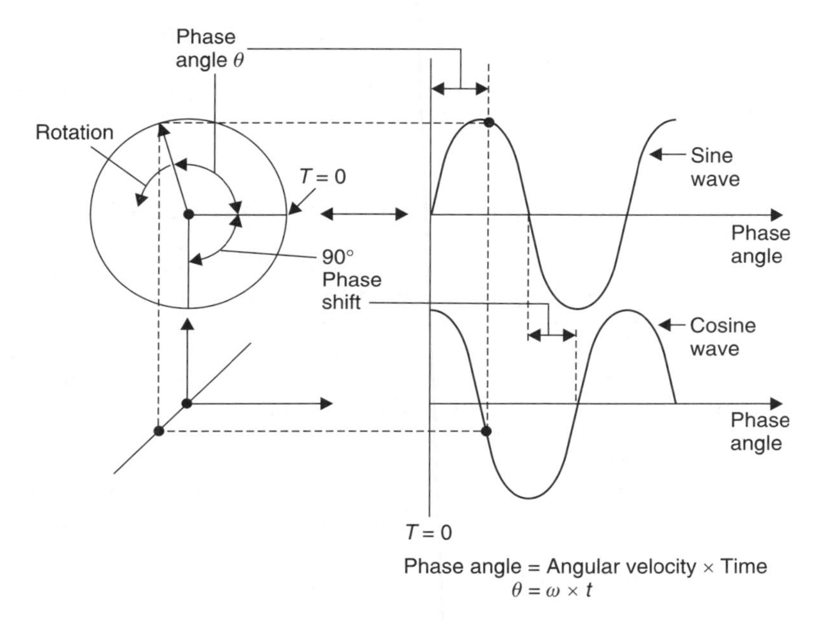

Fig. 2.17 A sine wave is one component of a rotation. When a rotation is viewed from two places at right angles, one will see a sine wave and the other will see a cosine wave. The constant *phase shift* between sine and cosine is 90° and should not be confused with the time variant *phase angle* due to the rotation.

would be phase shifted by 120° from that of the next blade, whereas the phase angle of a blade changes continuously as it rotates. In helicopters the phase angle is taken to be the angle between the blade and a line drawn directly aft from the rotor hub.

Geometrically it is possible to calculate the height or displacement of the sine wave in Figure 2.17 because it is given by the radius multiplied by the sine of the phase angle. The phase angle is obtained by multiplying the angular velocity ω by the time t. Frequency f is measured in rotations per second or Hertz (Hz). Thus the phase angle at a time t is given by ωt or $2\pi ft$.

A rotating object is not in static equilibrium even if the RPM is constant. With the exception of those elements of the object which are on the axis of rotation, all other elements are in a constant state of acceleration because their velocity is constantly changing as they follow a curved path. Figure 2.18(a) shows that if an arbitrary increment of time is taken, the velocity at the beginning is different from that at the end and a small resultant vector is necessary to close the triangle. As the time increment

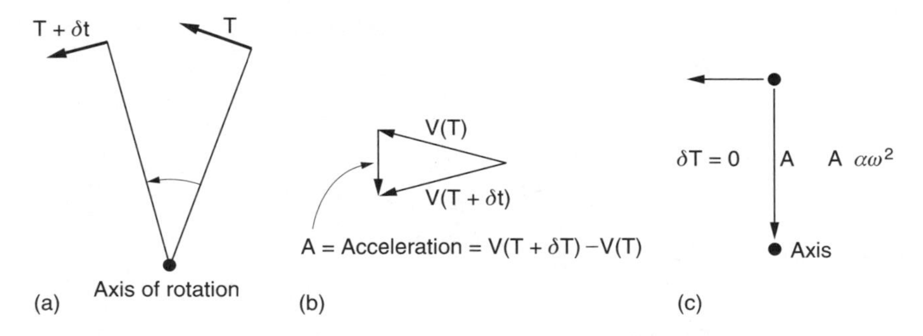

Fig. 2.18 A body following a circular path is at constant acceleration and if this is calculated, it can be used to derive the forces acting. (a) shows that after a short time δT, the velocity has changed direction. (b) The rate of change of velocity is the acceleration. (c) shows that if δT is allowed to fall to zero, the acceleration points to the centre of rotation.

becomes vanishingly small this vector will be seen to point to the axis of rotation as can be seen in (c).

As the RPM increases, the length of the velocity vectors in Figure 2.18 increases in proportion to RPM, as does the angle between them. As the resultant for small angles is the product of the vector length multiplied by the angle between them, it is easy to see that the acceleration or rate of change of velocity is proportional to the square of the RPM. The advantage of the use of ω to measure angular velocity is that if this is done the velocity at any radius r is simply $\omega \times r$ and the acceleration is simply $\omega^2 \times r$. There are no constants or conversion factors to remember which is a major advantage of the MKS metric system.

It was shown above that $F = m \times a$, it can be seen that the inward or centripetal (Latin: centre seeking) force needed to accelerate an object in a circular path is simply:

$$F = m \times \omega^2 \times r \quad \text{as } \omega = v/r \quad \text{then } F = m \times v^2/r$$

When a mass moves along, or *translates*, it has kinetic energy. When an object rotates, it also has kinetic energy, but the amount is less easy to calculate. This is because all elements of a translating mass move at the same velocity, whereas in a rotating mass different elements move at different velocities according to how far from the axis they are. By adding up all of the kinetic energies of blade elements at different velocities, the kinetic energy of the rotor can be found. If an elemental mass δm is rotating at radius r with angular velocity ω, its velocity is $\omega \times r$ and so its kinetic energy must be:

$$\frac{\delta m \times \omega^2 \times r^2}{2}$$

An arbitrary body can be treated as a collection of such masses at various radii and the integral of the kinetic energies of all of these gives the total kinetic energy. If this is divided by the square of the angular velocity the result is the moment of inertia; the rotational equivalent of mass. Any rotating body could have the same moment of inertia if all of its mass instead were concentrated at one radius from the axis of rotation. This is known as the radius of gyration.

The rotor blade requires an inward or centripetal force to accelerate it into a circular path. Each element of the blade is at a different radius and so calculating the overall force would be complicated were it not for the fact that the blade mass appears for this purpose to be concentrated at the radius of gyration. The figures for root tension are quite impressive. In a typical model helicopter a tension of 500 N is normal. In a full size helicopter 100 000 N is not unusual.

When a body of arbitrary shape is rotated, if the centre of mass is not on the axis of rotation, forces must come from the bearings to accelerate the CM into a circular path. This is the origin of vibration due to imbalance. In space no such forces can be applied and the only way the rotation can occur is for the CM and the axis to align. Thus rotating devices try to achieve such alignment and vibration results when bearings prevent it.

Figure 2.19(a) shows that a simple balancing process consists of supporting the assembly with the axis horizontal. If the CM does not coincide, gravity will turn the assembly until the CM is beneath. Balance weights may be added until the assembly will stay where it is left. In this condition the assembly is statically balanced, with no net moments, but it may still vibrate when rotated. Figure 2.19(b) shows that a statically balanced assembly can result with masses at different places along the axis of rotation. This will be imbalanced when rotating.

The statically balanced body in Figure 2.19(b) will try to rotate along its mass centroid. To eliminate vibration, the body must be dynamically balanced. This means

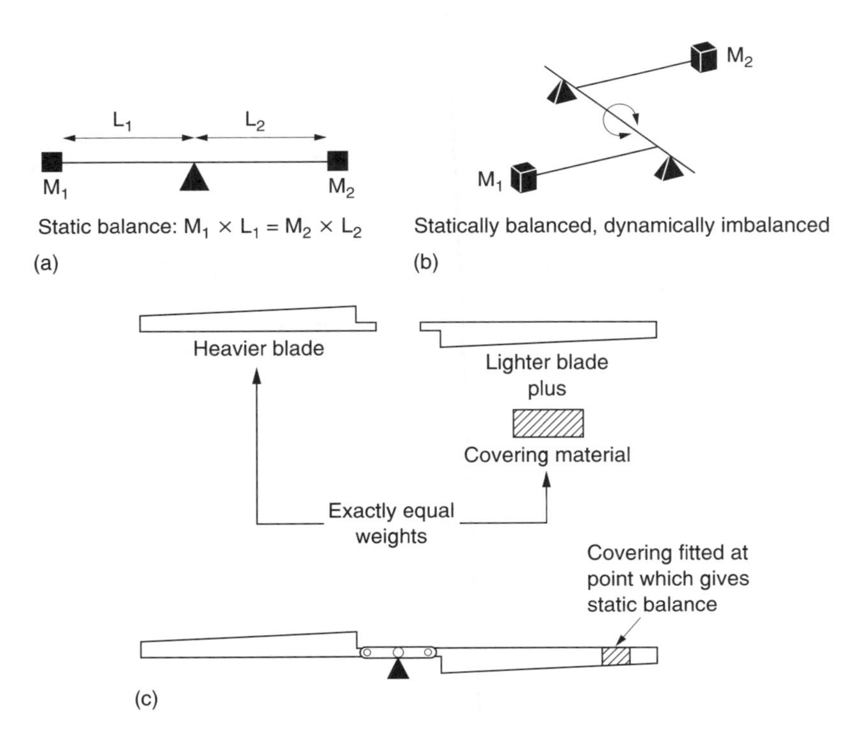

Fig. 2.19 Static balance (a) can be achieved when no overall moment results about the axis of rotation. (b) A statically balanced system that will vibrate when rotated as it is not dynamically balanced and will tend to turn about its mass centroid. (c) In model helicopters, dynamic balance is achieved by adding a small piece of covering material. The mass needed is equal to the difference in mass of the two blades, and the position is such that the blades balance in the same place.

that weights are added at various places along the axis to bring the CM of the slices onto the axis. When balancing car wheels, note that weights may be fitted both to the inner and outer rims to achieve this.

In the case of a helicopter rotor, the blades will be in the same plane and so the effects due to Figure 2.19(b) will be small. However, a statically balanced rotor could still be achieved if one blade were heavier than the other, if its CM were nearer the shaft. Thus to dynamically balance a rotor, it is necessary that all of the blades should have exactly the same mass, and the same radius of gyration. This means that the distribution of mass along and across each blade should be identical and the CM should be at the same point in three axes. In full size rotor blades, weight pockets are provided at various places to allow these conditions to be met.

In model helicopters dynamic balancing is just as important, but a simplified approach can be used. Figure 2.19(c) shows that the blades must be carefully weighed, and a piece of covering material is cut to have exactly the same mass as the difference in masses of the blades. The blades are assembled to the balanced rotor head, and the covering material is applied to the lighter blade at a point where static balance is achieved. It will then be found that the two blades have their CM in the same place and so the rotor will be dynamically balanced. The same result will be obtained if the covering is moved until both blades balance in the same place. In full-size helicopters the leading edge of the blade may be protected with a replaceable plastic film. If part of this comes off or the film is not fitted identically to each blade, vibration may result.

2.12 The mechanics of rotation

Figure 2.20(a) shows a steady rotation which could be a mass tethered by a string. Tension in the string causes an inward force that accelerates the mass into a circular path. The tension can be described as a rotating vector. When viewed from the side, the displacement about the vertical axis appears sinusoidal. The vertical component of the tension is also sinusoidal, and out of phase with the displacement. As the two parameters have the same waveform, they must be proportional. In other words the restoring force acting on the mass is proportional to the displacement.

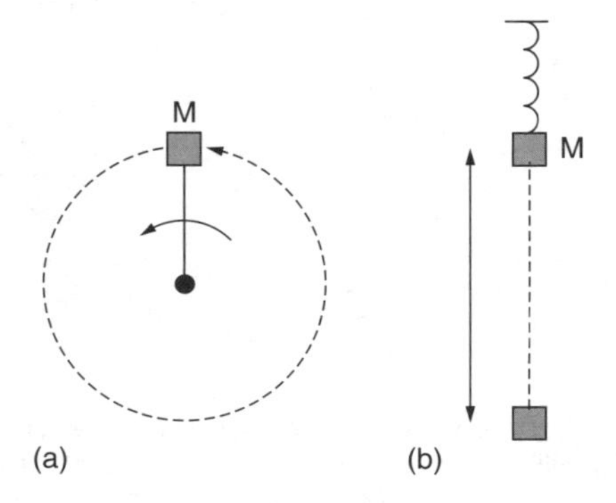

Fig. 2.20 (a) A tethered mass is moving in a circle. If the component of this motion in one axis only is resolved, the motion will be found to be the same as a mass supported on a spring, (b).

Figure 2.20(b) shows that the same characteristic is obtained if the mass is supported on a spring. The mass supported by a spring is found widely in engineering because structures have mass and can never be infinitely rigid. An ideal spring produces a restoring force proportional to the displacement. The constant of proportionality is called the *stiffness* and is the reciprocal of *compliance*. When such a system is displaced there is sustained resonance. Not surprisingly the displacement is sinusoidal and is called *simple harmonic motion* or SHM and has all of the characteristics of one dimension of a rotation as shown in Figure 2.17.

The only difference between the mass on a string and the mass on a spring is that when more energy is put into the system, the mass on a string goes faster because the displacement cannot increase but more tension can be produced. The mass on the spring oscillates at the same frequency but the amplitude has to increase so that the restoring force can be greater.

The velocity of a moving component is often more important than the displacement. The vertical component of velocity is obtained by differentiating the displacement. As the displacement is a sine wave, the velocity will be a cosine wave whose amplitude is proportional to frequency. In other words the displacement and velocity are in quadrature with the velocity lagging. This is consistent with the velocity reaching a minimum as the displacement reaches a maximum and vice versa. Figure 2.21

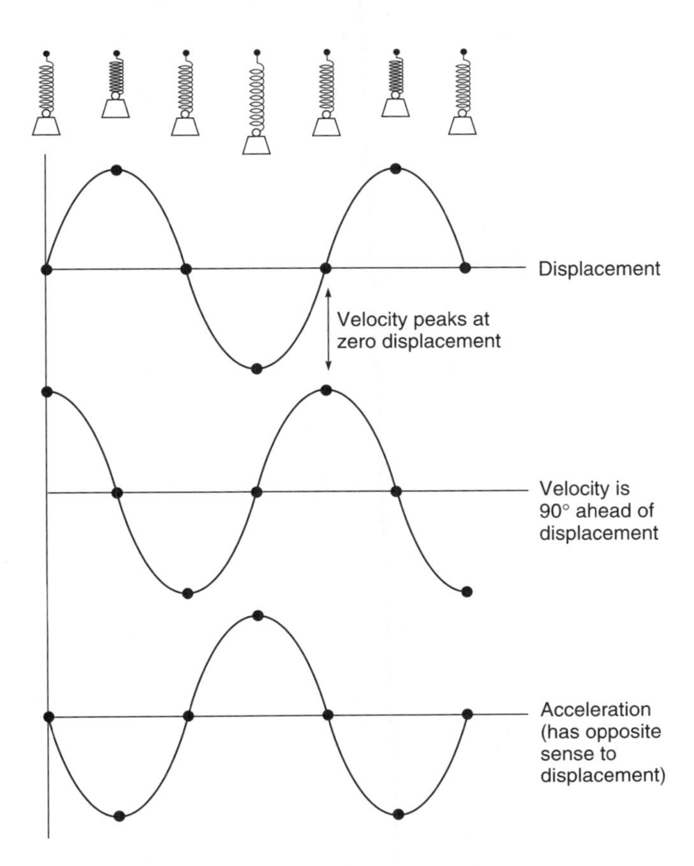

Fig. 2.21 The displacement, velocity and acceleration of a body executing simple harmonic motion (SHM).

shows the displacement, velocity and acceleration waveforms of a body executing SHM (simple harmonic motion). Note that the acceleration and the displacement are always anti-phase.

Eventually the resonance of a mass on a spring dies away. The faster energy is taken out of the system, the greater the rate of decay. Any mechanism that removes energy from a resonant system is called *damping*.

The motion of a rigid body can be completely determined by the mass, the stiffness and the damping factor. It is important to consider what happens when resonant systems are excited at different frequencies. Figure 2.22 shows the velocity and displacement of a mass-stiffness-damping system excited by a sinusoidal force of constant amplitude acting on the mass at various frequencies. Below resonance, the frequency of excitation is low and little force is needed to accelerate the mass. The force needed to deflect the spring is greater and so the system is said to be *stiffness controlled.* The amplitude is independent of frequency, described as *constant amplitude* operation, and so the velocity rises proportionally to frequency below resonance.

Above resonance the inertia of the mass is greater than the stiffness of the spring and the response of the system is described as *mass controlled.* With a constant force there is constant acceleration yet as frequency rises there is less time for the acceleration to act. Thus velocity is inversely proportional to frequency. As the displacement is the integral of the velocity the displacement curve is tilted by an amount proportional to frequency so that below resonance the displacement is constant and in-phase with the

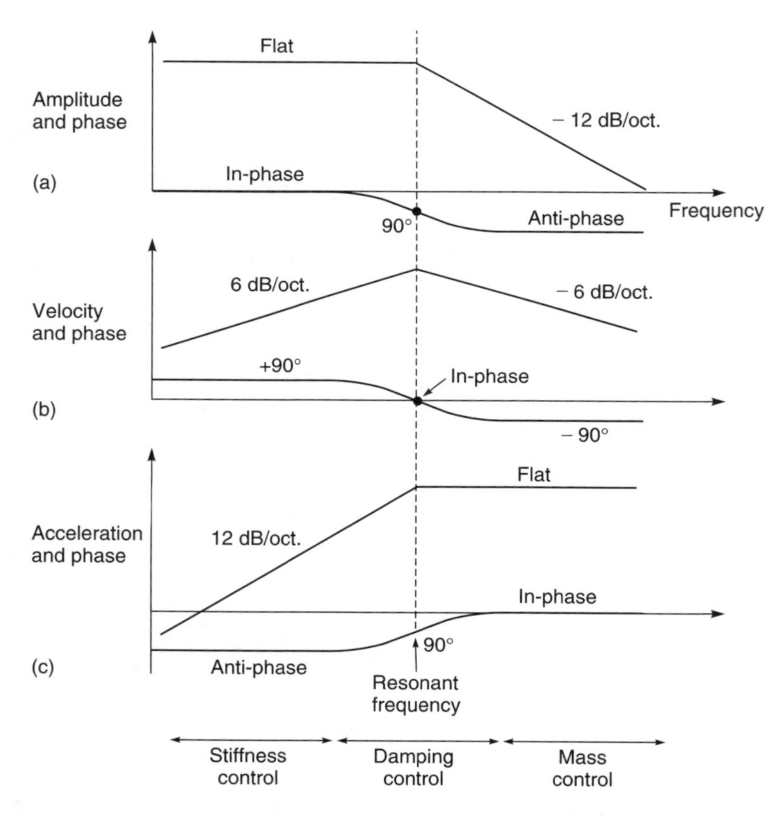

Fig. 2.22 The behaviour of a mass-stiffness-damping system: (a) amplitude, (b) velocity, (c) acceleration.

force whereas above it is inversely proportional to the square of the frequency and is anti-phase to the force.

In the vicinity of resonance the amplitude is a function of the damping and is said to be *resistance controlled*. With no damping the Q-factor is high and the amplitude at resonance tends to infinity, resulting in a sharp peak in the response. Increasing the damping lowers and broadens the peak so that with high damping the velocity is nearly independent of frequency. Figure 2.23 shows the effect of different damping factors on the step response, i.e. the response to a sudden shock. The underdamped system enters a decaying oscillation. The overdamped system takes a considerable time to return to rest. The critically damped system returns to rest in the shortest time possible subject to not overshooting.

Below resonance the displacement of the spring is proportional to the force. Here force and displacement are in phase. Above resonance the acceleration of the mass is proportional to the force. Here force and acceleration are in phase. It will be seen from Figure 2.22 that the velocity leads the displacement but lags the acceleration. Consequently below resonance the velocity leads the applied force whereas above resonance it lags. Around resonance there is a phase reversal so that at the precise resonant frequency there is no phase shift at all. Figure 2.24 shows that the rate of phase change in the vicinity of resonance is a function of the damping.

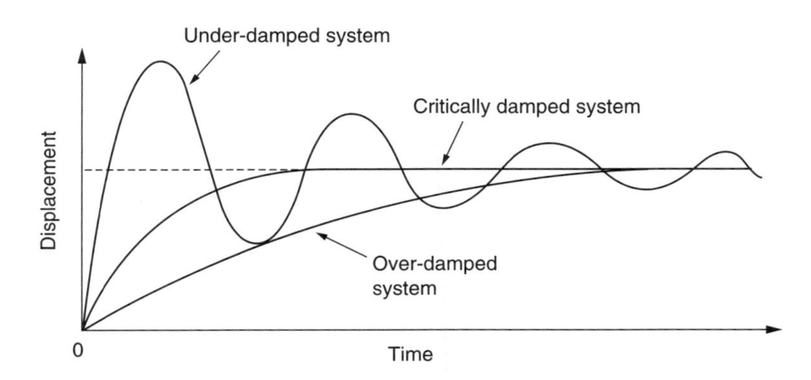

Fig. 2.23 The effect of different damping factors on the response to a step input.

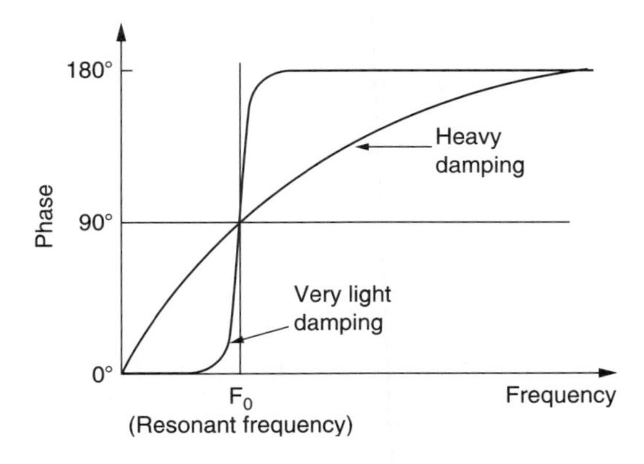

Fig. 2.24 Damping factor affects the rate of phase change near resonance.

It will be clear that the behaviour just noted has a direct parallel in the behaviour of an electronic damped tuned circuit consisting of an inductor, a capacitor and a resistor and the mathematics of both are one and the same. This is more than just a convenience because it means that an unwanted mechanical resonance or phase change in a control system can be suppressed by incorporating at some point a suitable electronic circuit designed to have the opposite characteristic. Additionally by converting mechanical parameters into electrical parameters the behaviour of a mechanism can be analysed as if it were an electronic circuit.

2.13 Sidebands

It was seen in Figure 2.17 that a sinusoidal function is a rotation resolved in one axis. In order to obtain a purely sinusoidal motion, the motion on the other axis must be eliminated. Conceptually this may be achieved by having a contra-rotating system in which there is one rotation at $+\omega$ and another at $-\omega$. Figure 2.25(a) shows that the sine components of these two rotations will be in the same phase and will add, whereas

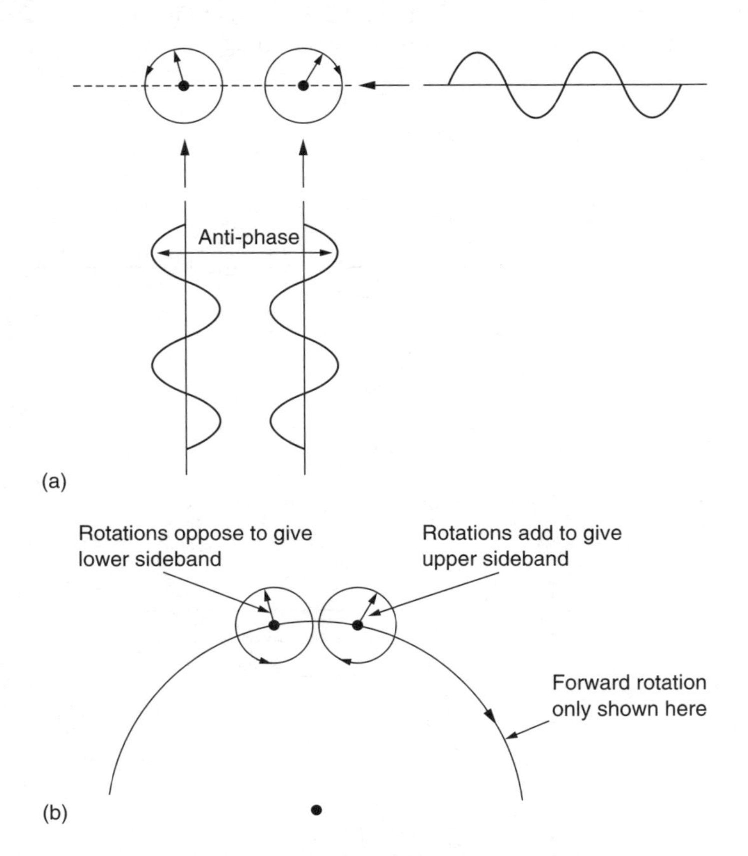

Fig. 2.25 (a) A sinusoidal motion can be considered as the sum of contra-rotations which cancel the cosine components. (b) When two sinusoids are multiplied, the effect is that one rotation is carried around by the other. A difference frequency exists because the forward rotation of one signal carries round the backward rotation of the other which makes it slow down. The sum frequency is obviously rotating faster.

the cosine components will be in anti-phase and will cancel. Thus all real frequencies actually contain equal amounts of positive and negative frequencies. These cannot be distinguished unless a modulation process takes place. Figure 2.25(b) shows that when two signals of frequency $\pm\omega_1$ and $\pm\omega_2$, are multiplied together, the result is that the rotations of each must be added. The result is four frequencies, $\pm(\omega_1 + \omega_2)$ and $\pm(\omega_1 - \omega_2)$, one of which is the sum of the input frequencies and one of which is the difference between them. These are called sidebands. Sidebands are found extensively in avionics, where the deliberate use of the process is called heterodyning. In a communications radio the carrier frequency is multiplied or 'modulated' by the audio speech signal, called the baseband signal, and the result is a pair of sidebands above and below the carrier.

The rotation of the helicopter rotor has a certain frequency. Any vibration due to periodic motion affecting the movement of the blades within the rotor plane may have a characteristic frequency with respect to the rotor. When referred to the hull of the helicopter, the frequency of the vibration may have been heterodyned by the rotor frequency and the frequencies experienced in the hull may then be the frequencies of the sidebands. This will be considered in Chapters 3 and 4.

In a voice radio system, the carrier frequency is much higher than the frequencies in the speech, whereas in other systems this may not be the case. In digital systems, continuous signals are represented by periodic measurements, or samples, and sidebands are found above and below the frequency of the sampling clock F_s. Figure 2.26(a) shows the spectrum of a sampling clock which contains harmonics because it is a pulse train not a sinusoid. Figure 2.26(b) shows that the sidebands of the sampling frequency can be rejected using a low-pass filter which allows through only the original baseband signal. The sampled representation of the signal is returned to a continuous waveform. This is exactly what happens in a Compact Disc player.

As the baseband frequency rises, the lower sideband frequency must fall. However, this can only continue until the base bandwidth is half the sampling rate. This is known as the Nyquist frequency and it represents the highest baseband frequency at which the original signal can be recovered by a low-pass filter.

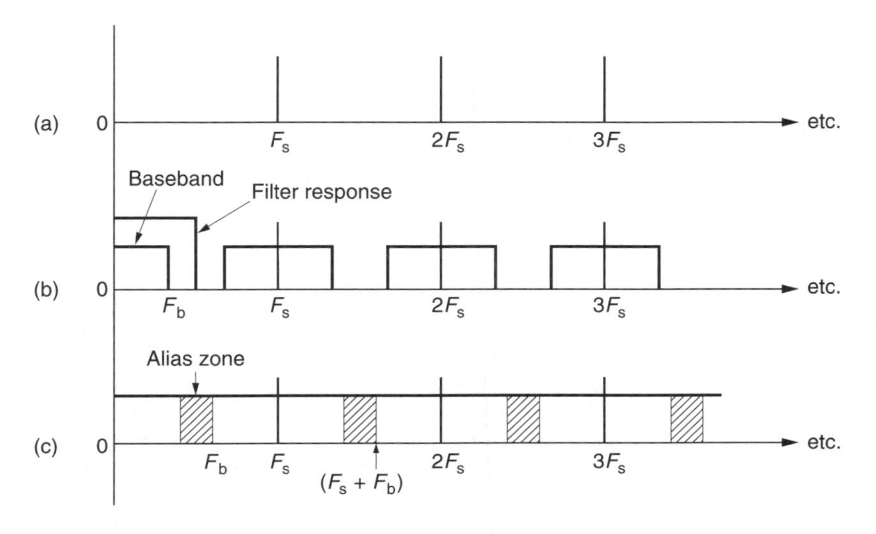

Fig. 2.26 (a) Spectrum of sampling pulses. (b) Spectrum of samples. (c) Aliasing due to sideband overlap.

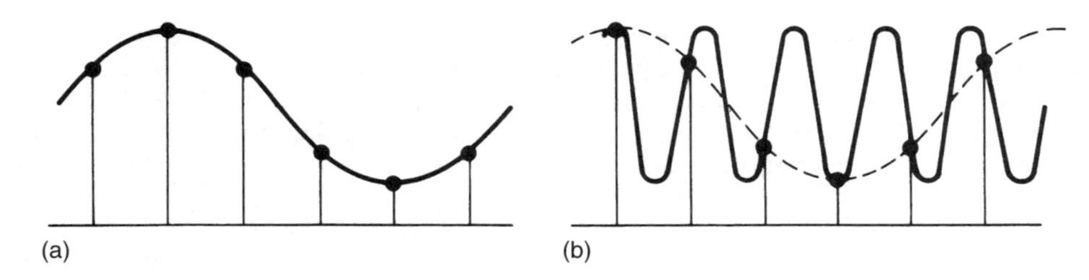

Fig. 2.27 In (a) the sampling is adequate to reconstruct the original signal. In (b) the sampling rate is inadequate, and reconstruction produces the wrong waveform (dashed). Aliasing has taken place.

Figure 2.26(c) shows what happens if the base bandwidth increases beyond the Nyquist frequency. The lower sideband now overlaps the baseband signal. As a result, the output frequency appears to fall as the input frequency rises. This is the phenomenon of aliasing. Figure 2.27 shows aliasing in the time domain. The waveform carried in the envelope of the samples is not the original, but has a lower frequency. There is an interesting result when the two frequencies are identical. The lower sideband frequency becomes zero. This forms the basis of Fourier analysis seen in the next section. It is also exactly what happens when a stroboscope is used, perhaps to check the tracking of a helicopter rotor. The frequency of the flashing light (the sampling rate) is adjusted until it is the same as the rotational frequency of the object to be studied. The lower sideband frequency becomes zero and the object appears stationary. The upper sideband frequency is usually visible as flicker.

Small variations in the frequency of the strobe light above and below the rotational frequency will cause the rotating object to turn slowly forwards or backwards. The forward rotation is due to the lower sideband having a low frequency. However, when the strobe frequency is above the rotational frequency, the lower sideband frequency becomes negative, hence the illusion of reversed rotation. On a spectrum analyser, the negative frequency would fold about zero Hz to become a positive frequency. This folding phenomenon is particularly important to an understanding of ground resonance that will be considered in Chapter 4.

2.14 Fourier analysis

Fourier was a French mathematician who discovered that all periodic or repetitive phenomena, however complex, could be described as the sum of a number of sinusoidal phenomena. As the rotation of a helicopter rotor is periodic, then Fourier analysis is a useful tool to study rotor motion.

It was shown above that the sine wave is the waveform of a single frequency. Musically such a waveform would be called a pure tone or fundamental and the frequency would determine the pitch. If musical instruments only produced pure tones they would be hard to tell apart. They sound different partly because of the presence of harmonics. Harmonics are frequencies given by multiplying the frequency of the fundamental by a series of whole numbers. Figure 2.28 shows the waveforms of such harmonics. The final waveform is the sum of all of the harmonics. The relative amplitude of the different harmonics determines the tonality or timbre of an instrument. What Fourier did was to show how to measure the amplitude of each harmonic in any periodic waveform. The mathematical process that does this is called a transform. Reversing the process

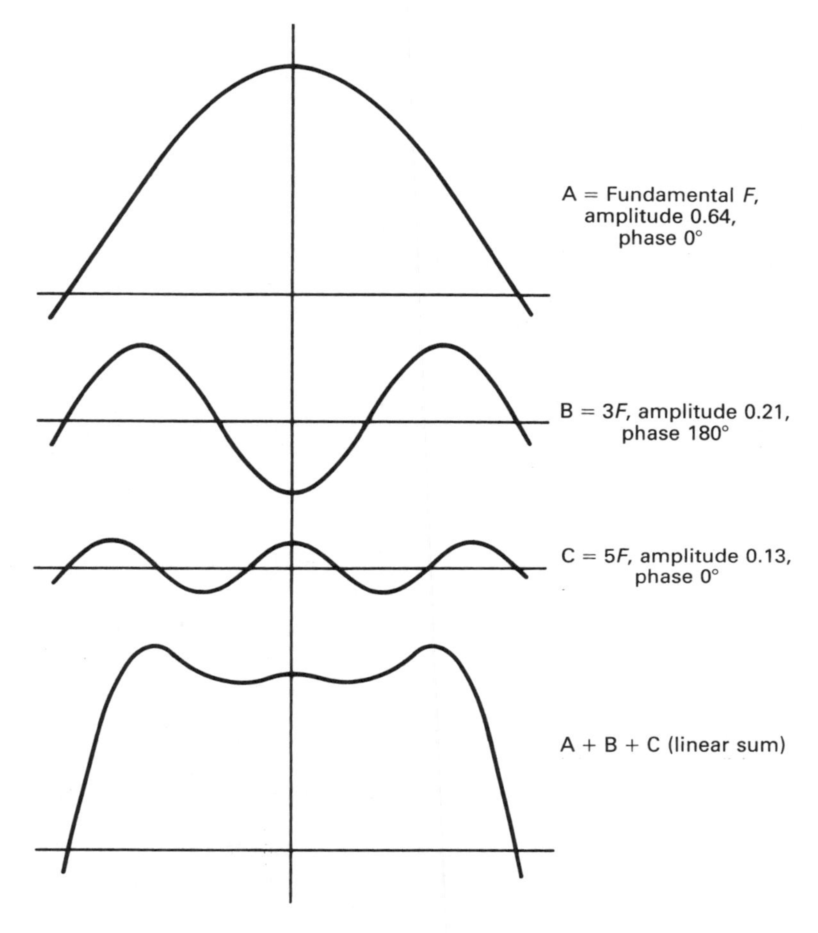

Fig. 2.28 Fourier analysis allows the synthesis of any waveform by the addition of discrete frequencies of appropriate amplitude and phase.

by creating all of the right harmonics and adding them is called an inverse transform. Figure 2.29 shows how a square wave can be synthesized in this way by adding together different harmonics of the correct amplitude and phase.

Fourier analysis is based on multiplying the waveform to be analysed by sine waves called 'basis functions'. In order to analyse an arbitrary waveform to see if it contains a particular frequency, it is multiplied by a basis function at that frequency and the product is averaged. Figure 2.30(a) shows that if the signal being analysed contains a component having the same frequency as that of the basis function the product will have a zero frequency component that will give a finite result after averaging. The value of the result after averaging is called a coefficient. Components at all other frequencies will average to zero. Thus a complete Fourier analysis requires the process to be repeated at all of the frequencies of interest.

Figure 2.30(c) shows that if by chance the input and the basis function have a phase difference of 90° the product will be zero even though the frequencies are identical.

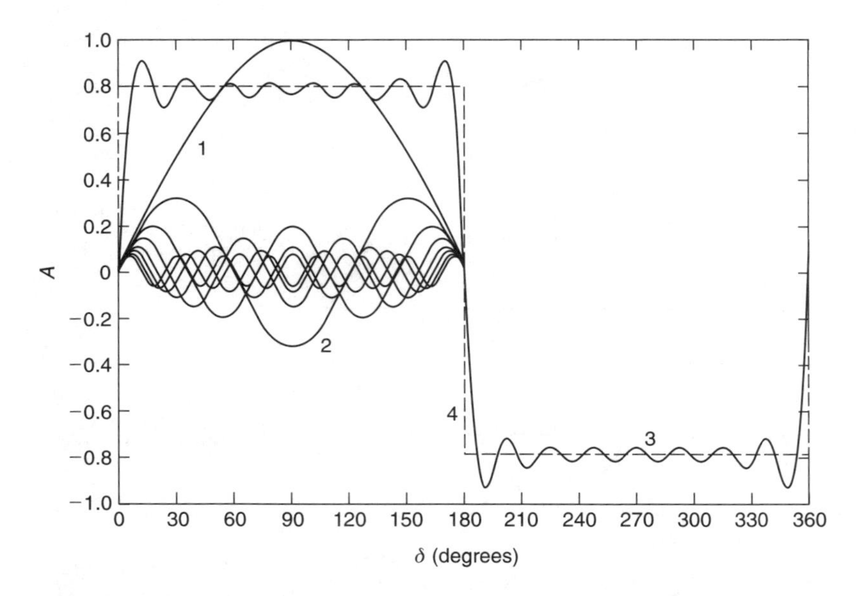

Fig. 2.29 Fourier analysis of a square wave into fundamental and harmonics. A, amplitude; δ, phase of fundamental wave in degrees; 1, first harmonic (fundamental); 2 odd harmonics 3–15; 3, sum of harmonics 1–15; 4, ideal square wave.

To overcome this problem the Fourier analysis searches for each frequency using both a sine wave and a cosine wave basis function. Thus at each frequency two coefficients will be obtained. The ratio of the two coefficients can be used to determine the phase of the frequency component concerned.

The full frequency accuracy of the Fourier transform is only obtained if the averaging process is performed over a very long, ideally infinite, time. In practice this may cause difficulties, not least with the amount of computation required, and the averaging time may need to be reduced. If a short-term average is used, the same result will be obtained whether the frequency is zero or very low, because a low frequency doesn't change very much during the averaging process. Thus the short-term Fourier transform (STFT) allows quicker analysis at the expense of frequency accuracy. For best frequency accuracy, the signal has to be analysed for a long time and so the exact time at which a particular event occurs would be lost. On the other hand if the time when an event occurs has to be known, the analysis must be over a short time only and so the frequency analysis will be poor. In the language of transforms, this is known as the Heisenberg inequality. Werner Heisenberg explained the wave-particle duality of light. When light is analysed over a short time, its frequency cannot be known, but its location can. Light can then be regarded as a particle called a photon. When light is analysed over a long time its frequency can be established but its location is then unknown and so it is regarded as a wave motion.

In helicopters the frequencies of interest will generally be known from the rotor speed and usually only the first few harmonics contribute to the aerodynamic result although higher harmonics may result in vibration. These harmonics are normally sufficiently far apart in frequency that the time span of the analysis is not critical.

One peculiarity arises with Fourier analysis of helicopter rotors. The behaviour of a typical rotor is such that the coefficients of the harmonics are always negative.

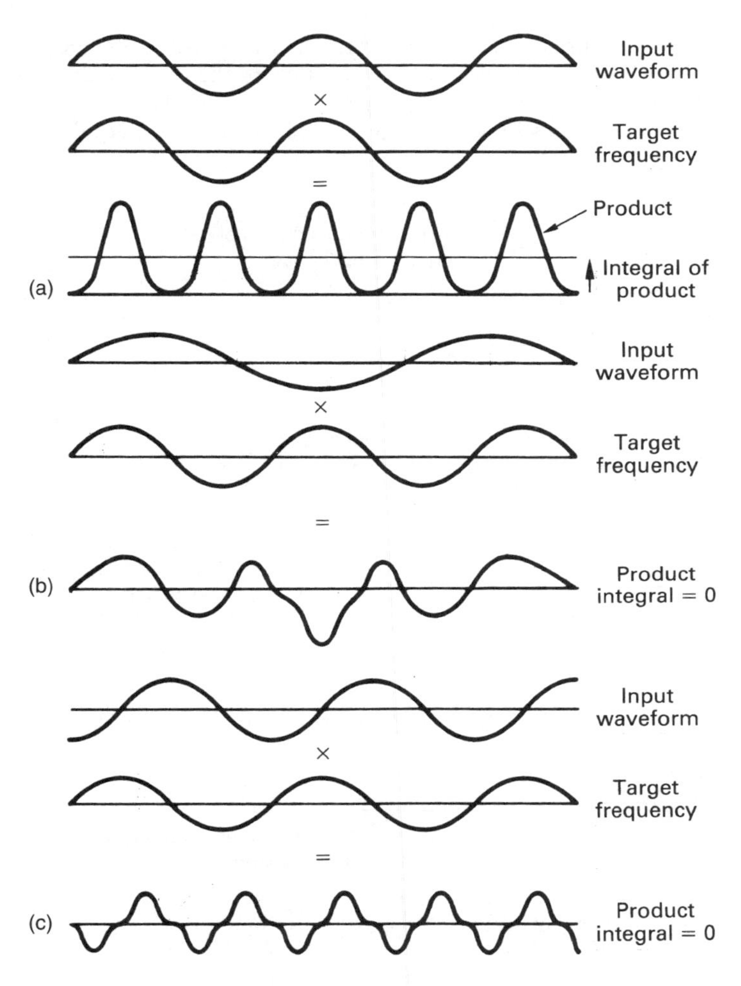

Fig. 2.30 The input waveform is multiplied by the target frequency and the result is averaged or integrated. At (a) the target frequency is present and a large integral results. With another input frequency the integral is zero as at (b). The correct frequency will also result in a zero integral shown at (c) if it is at 90° to the phase of the search frequency. This is overcome by making two searches in quadrature.

For simplicity the traditional practice has been to redefine the transform so that the numbers come out positive. This causes confusion for those who are familiar with transforms but are studying helicopters for the first time.

2.15 Centrifugal and Coriolis forces

Most calculations in mechanics involve a stationary frame of reference. However, as helicopters contain some major rotating parts, designers sometimes find it convenient for analysis to use a rotating frame of reference turning at the same angular velocity as the blade. The blade is then usefully more or less stationary with respect to the frame of reference. Figure 2.31(a) shows that when the frame of reference rotates at the same angular velocity as the component of interest, the rotation disappears and so that component cannot be accelerating due to the rotation. However, the centripetal force due to the rotation is still acting and in order to maintain equilibrium with the

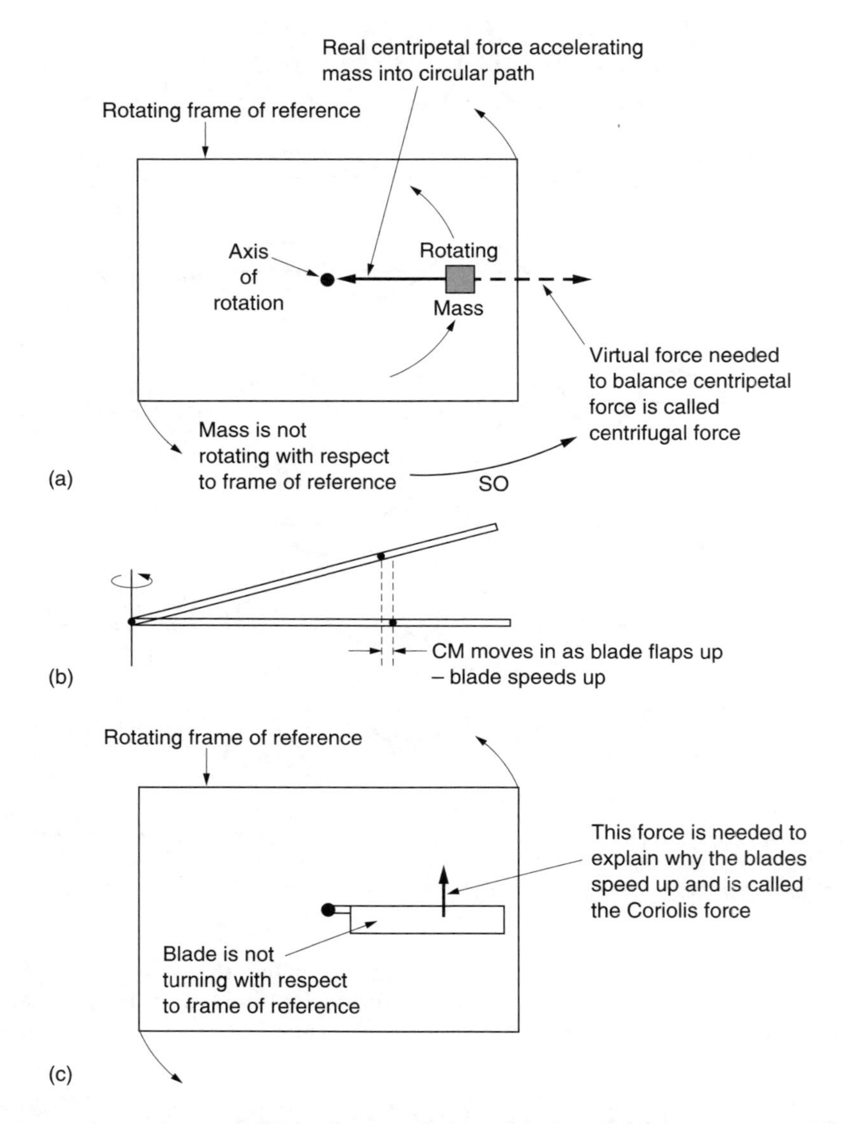

Fig. 2.31 (a) A rotating body is not in equilibrium. Centripetal force accelerates it into a circle. When the frame of reference rotates at the same speed to arrest the rotation, there is no acceleration to balance centripetal force and so a virtual force, called centrifugal force, is imagined to exist. (b) If a blade flaps up its CM moves closer to the shaft and the MoI (moment of inertia) falls. It will conserve momentum by swinging forwards. (c) In a rotating frame of reference a blade has no momentum and a virtual force called a Coriolis force is imagined to cause the forward swing.

centripetal force in the absence of acceleration, a new virtual force of equal magnitude but opposite direction has to be imagined to act. This is called the centrifugal force. In the real world centrifugal force does not exist, it is a mathematical convenience needed to allow an apparent state of equilibrium in a rotating frame of reference.

In a fixed frame of reference, consider a turning rotor. Figure 2.31(b) shows that if one of the blades should move upwards, the centre of mass of the blade will move closer to the axis of rotation, reducing the moment of inertia. Conservation of momentum requires that the reduced moment of inertia be balanced by an increase in velocity. Consequently the blade will accelerate in the direction of rotation as it moves up. Naval gunners discovered a related effect which is that when a shell is fired a distance of several miles, the shell moves significantly with respect to the earth's axis in a rotating frame of reference and so it will also experience conservation of momentum which takes it to the left or right of its intended target. These are examples of the Coriolis effect, named after the mathematician who derived the mechanics of rotating frames of reference.

Figure 2.31(c) shows that when the frame of reference rotates at the same angular velocity as the component of interest, the rotation apparently disappears. However, the conservation of momentum effect is still occurring and the blade appears to accelerate with respect to the frame of reference. In order to account for this acceleration, a new virtual force is imagined to act. This is called the Coriolis force. In the real world the Coriolis force does not exist, like centrifugal force, it is a mathematical convenience needed to allow the use of a rotating frame of reference.

The uninformed will often make reference to both centrifugal and Coriolis forces as though they really existed, usually as a result of reading an advanced textbook without understanding it. Such observations must be treated with suspicion. Centrifugal force and Coriolis force are the technical equivalent of the unicorn and the mermaid. They are in the imagination and the literature but they don't exist.

There is a further consequence of the use of a rotating frame of reference that affects vibration frequencies created in the rotor and transmitted to the hull. The frequencies generated in the rotor may have the rotational frequency of the rotor both added and subtracted. The phenomenon is known as heterodyning and was treated in section 2.11.

2.16 Rotating masses and precession

The rotors of a helicopter at flight RPM contain a large amount of stored energy and this alters the way they respond to forces. In order to control the helicopter, the entire rotor has to be tilted in the direction the pilot wishes to go. This will *not* be achieved by applying a couple in that direction. Figure 2.32(a) shows that if one blade is considered, when a couple is applied to the rotor, as it rotates the couple will alternately try to move the blade up and down. In fact in a rotating frame of reference the blade experiences a sinusoidal forcing function at the rotational frequency of the rotor. The way in which the blade responds to this is non-intuitive.

Figure 2.32(b) is drawn in a rotating frame of reference with respect to which the blade is stationary and it becomes correct to refer to centrifugal force. The figure shows that if it is assumed that the blade is displaced from its normal plane of rotation and released, there will be a component of centrifugal force trying to return the blade to the original plane. For small angles this force is proportional to the displacement, so the condition for SHM is met. The blade will oscillate like a pendulum about the normal plane at some natural frequency. This frequency can be calculated from the mass of the blade and the centrifugal force. It turns out that the frequency is nearly identical to the rotational frequency.

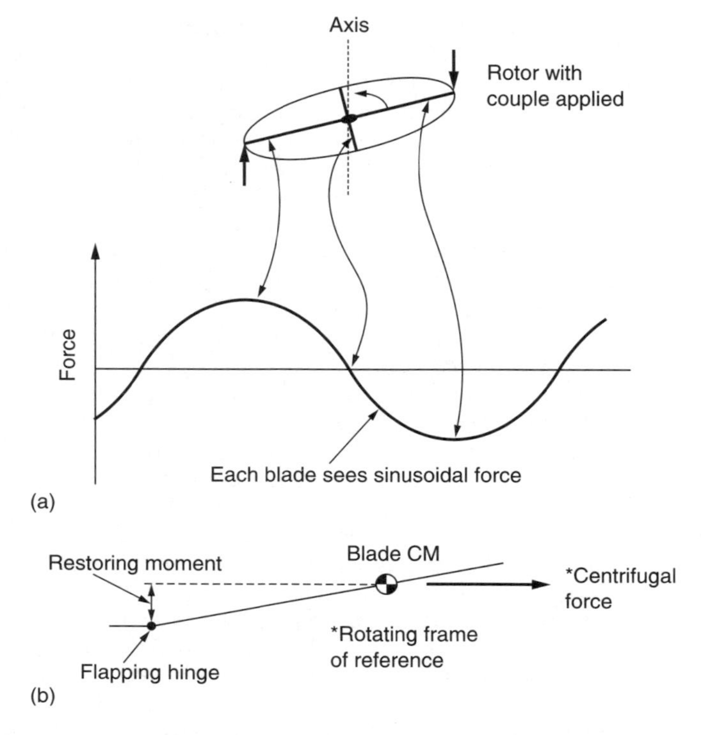

Fig. 2.32 (a) A couple applied in the plane of the rotor is experienced as an alternating force by each blade. (b) In rotating co-ordinates, centrifugal force tends to return a flapping blade to the flat position. The restoring force for small angle is proportional to the angle and so the blade performs SHM.

As a result when an external couple tries to tilt the plane of a rotor, the sinusoidal force on each blade is driving the blade at its natural resonant frequency. Section 2.9 showed that at resonance the driving force is in phase with the velocity. This means (from Figure 2.22) that the displacement lags both the velocity and the force by 90° of rotation. Thus the rotor does not tilt in the direction of the applied couple, but on an axis that is 90° further around the rotor in the direction of rotation. This is a fundamental characteristic of gyroscopes and is called gyroscopic precession (invariably changed to procession by the ill-advised use of spelling checkers). This 90° lag in the response of a rotor to applied forces has a significant bearing both on the control of the helicopter and on how it responds to external forces. The controls have to be connected up at right angles to the intuitive way.

As the blade tension is proportional to the square of RRPM (rotor revolutions per minute), when the rotors are turning slowly, the acceleration experienced by the blades due to rotation is small and the phase lag due to precession will be considerably reduced. The helicopter cannot fly at such low RRPM, but the effect may be noticed during rotor start or shutdown in a strong wind. When the pilot tries to minimize blade sailing by using the cyclic control, it will be found that the response is not as expected. Instead at low speeds the blades will respond with an advance with respect to the cyclic control. On a clockwise-from-the-top helicopter if the blades sail up on the right, the cyclic stick will have to be moved backwards.

2.17 The gyroscope

When bodies move in a straight line, all of the body is going at the same velocity and energy calculations are easy. However, when a body rotates, those parts furthest from the axis are going faster than those parts near the axis. The amount of rotational energy stored is a function of the distribution of mass with respect to the axis of rotation. This is measured by the moment of inertia (MoI). In a flywheel, as much mass as possible is concentrated at the outside of the rotor to give the greatest MoI.

Angular momentum is the product of inertia and the rate of rotation about a given axis. Earlier in this chapter it was seen that a body may be accelerated by changing its speed or its direction. The same is true of a gyroscope. Changing rotational speed is acceleration but so also is changing the direction of the rotational axis. If the rotor is spinning rapidly it has a large momentum, and even a very small rate of movement of the rotational axis represents a large rate of change of momentum and so a large force is necessary. Put simply a gyroscope tends to resist disturbances to its axis in a manner disproportionate to its weight. That makes it attractive for aviation where weight is at a premium. The tendency to resist axis disturbance is called *rigidity*.

Figure 2.33(a) shows a gyroscope suspended by the shaft in a pair of nested cages called gimbals. The axis of rotation is vertical and the gimbals allow turning in two

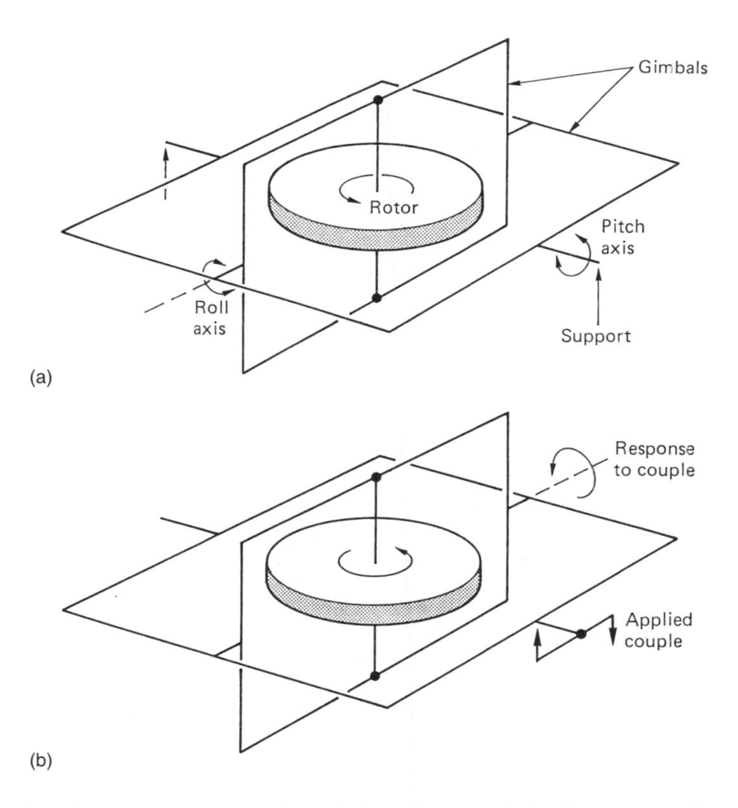

Fig. 2.33 (a) A vertical axis gyro supported by nested rings called gimbals. (b) A couple applied to a gimbal is resisted by rigidity. The gyro axis turns with a 90° phase lag. This is called precession.

orthogonal (at 90° to one another) horizontal axes. If the base of the system is pitched and rolled, the rigidity of the gyroscope will be much greater than the friction in the gimbals and the result is that the gyroscope axis remains vertical so that its angular momentum stays the same. Figure 2.33(b) shows that a couple is being applied to the outer gimbal, which tries to twist it in the direction shown. The couple is transmitted through the bearings of the inner gimbal and to the rotor shaft. If a small element on the periphery of the flywheel is considered, this reacts in the same way as the blade in the previous section. As the flywheel is really a collection of such elements, all of which behave in the same way, it is clear that in response to a couple on the outer gimbal axis, the rotor responds by turning around the inner gimbal axis.

The gyroscope rotor has a 90° phase lag just like a helicopter rotor. The outer gimbal will not turn because the applied couple is perfectly opposed by a reaction due to the rigidity of the gyro. The gyro can only generate the rigidity reaction by changing its angular momentum and to do this it must precess as the rotational speed cannot change. The rate of precession is proportional to the applied couple and inversely proportional to the angular momentum.

2.18 Piezo-electric and laser gyroscopes

The mechanical gyroscope held sway for a long time but has recently been supplanted by two alternatives. A piezo-electric material is one that deflects when an electrical voltage is applied to it and which creates an electrical voltage when it is distorted. The quartz crystal in an electronic wristwatch works on the same principle. In the quartz watch the natural mechanical resonant frequency of the crystal forms the timing reference. Signals from sensing electrodes measure the deflection and are amplified and fed to drive electrodes that cause the crystal to resonate. By dividing down this highly stable frequency a count of seconds can be obtained.

In the piezo-electric gyroscope a vibrating element is made to oscillate as shown in Figure 2.34(a). If the assembly then rotates, the Coriolis effect causes a component

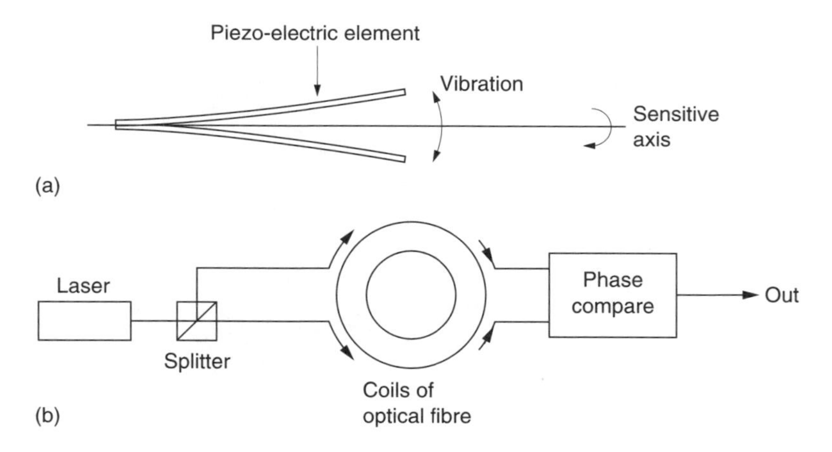

Fig. 2.34 (a) In the piezo-electric gyro there is a vibrating beam. Rotation about an axis along the beam causes the beam to twist as it tries to conserve momentum. This can be measured to sense the rotation. (b) In a laser gyro, light is sent in both directions around an optical path. If the system rotates, there will be a phase change between the two components of light.

of oscillation to occur at right angles to the main oscillation. Further electrodes and suitable electronic circuits can measure the amplitude and sense of the Coriolis effect that is proportional to the angular velocity.

A piezo-electric gyroscope may be only 20 mm long and weighs a few grams. The power consumed is a few milliWatts. They were initially developed for image stabilizers in consumer video camcorders. These gyros have already eclipsed the mechanical gyro in model helicopters on account of the low power and weight, and may find certain applications in full-size machines in due course. The small size allows completely redundant units where two or more sensors can be used in case one fails.

In the laser gyroscope, shown in Figure 2.34(b), light from a laser is split into two components and sent in both directions round an optical path. As the distance in both directions is the same, the two components will emerge in the same phase. However, if the optical path rotates about the axis shown, light travelling in one direction will take a slightly longer time and that travelling in the other direction will take a shorter time. The two components will no longer be in phase and the phase difference is proportional to the angular velocity. This can be sensed electronically. These alternative gyros both have the advantage of freedom from wear.

2.19 Feedback

Control signalling conveys the desired position of a flight control or a subsidiary control from one part of the airframe to another. The next essential is to ensure that when a command is received it is carried out accurately. Feedback is a useful tool to reach this goal. Feedback is a process that compares the current condition of a system with a desired condition and tries to make that difference smaller. This is exactly the characteristic needed to make a remote load follow a control signal, hence the extensive use of feedback in control systems. Feedback may be implemented in mechanical, hydraulic and electrical systems and in the case of the latter may be implemented with analog or digital techniques, although the principles remain the same in each case.

Figure 2.35(a) shows a basic electrical feedback servo system. The *desired position* of the load is defined by setting a control potentiometer that creates a proportional electrical voltage. The *actual position* of the load is measured by a second potentiometer. The actual position voltage is subtracted from the desired position voltage to create a signal called the *position error*. This is a bipolar signal whose polarity indicates the sense of the error and whose magnitude indicates how far the load is from the desired position.

The position error is amplified and, in this example, used to power a DC electric motor that drives the load and the feedback potentiometer. The polarity of the system must be such that the motor runs in a direction to cancel the position error. In other words the system has negative feedback: the information regarding the position of the load flows to the error sensor and the error information flows to the load forming a *closed loop* or *feedback loop*.

If the control potentiometer is set to a new position, a position error will result which will drive the motor until the error becomes zero again. The power to the motor will then be zero. In most cases this will not cause the motor to stop. The motor and the load have inertia and once in motion may continue even if the motor power is removed. As a result the load may overshoot the desired position. This will cause a reverse position error so the motor is driven back. Again the load may overshoot and an indefinite or decaying oscillation known as *hunting* takes place.

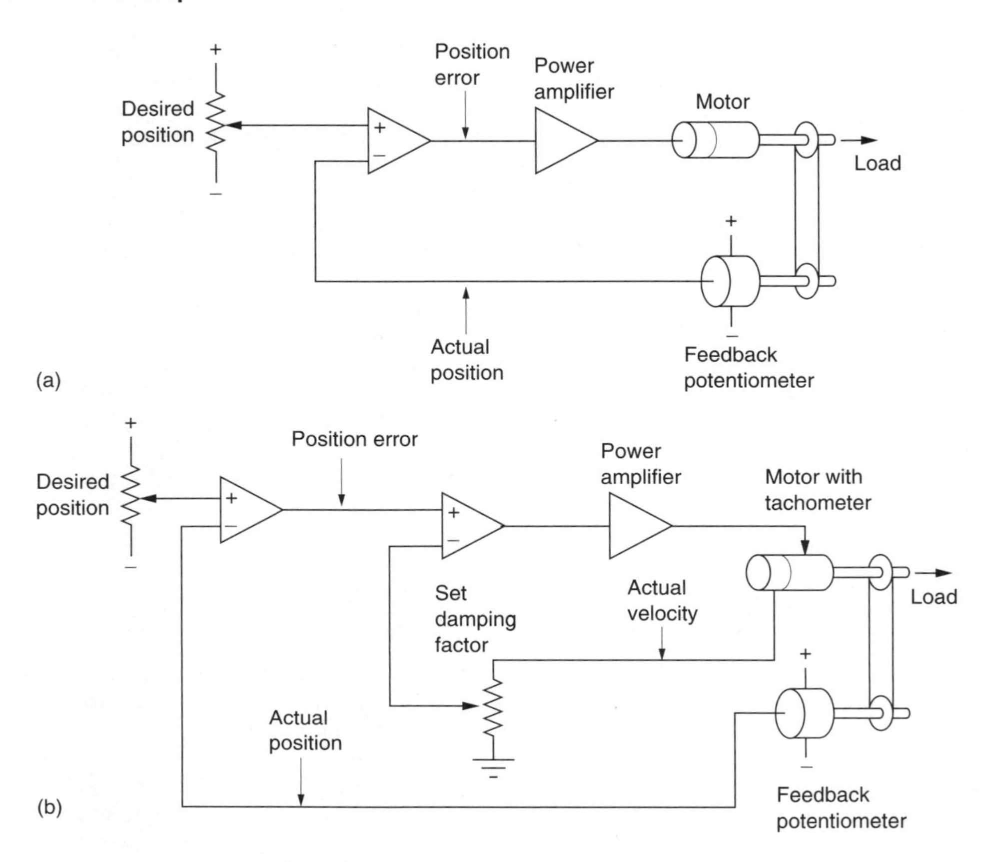

Fig. 2.35 (a) A simple feedback system compares the actual load position with the desired position. To prevent hunting, the load velocity may be measured with a tachometer (b) or by differentiating the actual position signal.

This is not acceptable in any precision system. The problem of hunting is due to considering only the position of the load. The solution is to consider the load's velocity as well. In this way the control system knows how much kinetic energy is stored in the moving load so that it can be brought to rest by applying a suitable reverse drive. Figure 2.35(b) shows that in this example the load velocity can be measured with a tachogenerator connected to the motor shaft. Many servo motors are designed with an integral tachometer.

If the control potentiometer is set to a new position, the position error will drive the motor to cancel the error. As the load gathers speed, the load velocity signal will be subtracted from the position error. As the load approaches the desired position, the position error becomes smaller, and when the position error is less than the velocity, the signal driving the motor will actually reverse polarity so that the load is retarded. This retardation process damps the hunting tendency. Instead of having a separate tachometer the velocity of the load can be obtained by differentiating the actual position signal.

The performance of a servo may be tested by measurement of the response to a step input. The speed of response can never exceed the limits set by the motor power and the inertia of the system that determine the maximum *slew rate*. Figure 2.23 showed the effect of different damping factors on system response. In the case of a servo the amount of velocity feedback determines the damping factor. With little or no velocity feedback the slew rate will be high, but hunting will be excessive and the step response will be a decaying oscillation. With too much velocity feedback the response will have no overshoot, but will be very slow. In between these is the condition known as *critical damping*. This is defined as the amount of damping which gives the fastest response without any overshoot. The critical damping condition is simply a definable condition, and it is not necessarily the optimum response. In most cases the response speed can be increased significantly with the penalty of a very small overshoot.

In an unpowered system, if the pilot experiences resistance to a control input, he will automatically apply more force to obtain the desired control position. In a fully powered system, the pilot (or the autopilot) does not feel this resistance and cannot compensate. The feedback system must be designed to overcome resistance automatically. In a simple servo, the load will stop at a position where the force from the motor is balancing the disturbing force. Clearly in a simple feedback system, the motor can only produce a force if there is a position error. The result is that the external force pushes the load out of position until the position error is large enough to oppose further motion. The ability of a servo to oppose external force in this way is called *stiffness*.

The stiffness of a servo can be improved by increasing the gain given to the position error. Then a given resistance to an external force will be obtained with a smaller position error. Increasing the gain may also speed up the step response, but not beyond the slew rate limit. If a control input exceeds the slew rate limit, for a period of time the position error will be large and the system is not feedback controlled. In this state the system is said to be in an *open loop* condition.

Clearly high gain is desirable in a servo because it increases response speed and stiffness. However, high gain can also cause instability. During servo development, a control input is created which is sinusoidal and the frequency is swept upwards from a very low value. The amplitude and phase response are plotted against the frequency. As the frequency rises, the servo has to work harder to follow the rapid movement and so the position error may increase. This causes the load to lag the input. At some high frequency, the phase lag may reach 180°.

With a 180° phase shift, a negative feedback system has turned into a positive feedback system. In other words the sense of the correcting action is wrong. If the loop gain is above unity at the frequency where the phase response has gone to 180°, the system will oscillate at that frequency. This may be spontaneous upon applying power or result from a small control input. Clearly this would be catastrophic in any control system. In practice feedback systems must contain a filter to reduce the loop gain at high frequencies to prevent oscillation.

Another possibility that allows higher loop gain is to have a signal processor containing an inverse model of the phase characteristics of the servo loop. When the servo loop lags, the processor will introduce a phase lead to balance out the lag. The use of these filters and processors in a servo loop is called *compensation*.

If high gain cannot be used for stability reasons, it is possible to improve the accuracy of a servo in the long term by integrating the error. Figure 2.36 shows a feedback system with integral control. A small error at the input to the integrator will become larger as the integrator operates until the loop acts to cancel it. Integral control is useful for overcoming friction in mechanisms.

Feedback loops can be *nested* which means that one operates inside another. For example, Figure 2.37 shows an automatic navigation system. The flight director error loop provides an error signal to the autopilot in order to keep the machine on track.

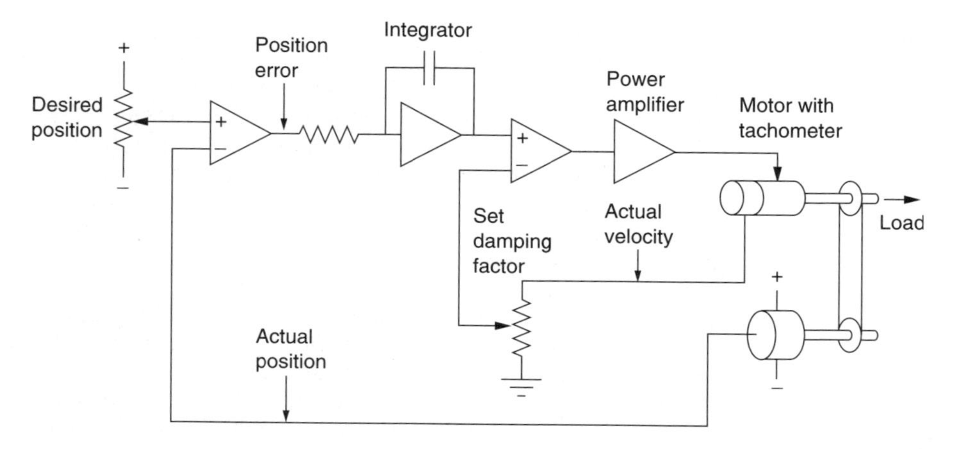

Fig. 2.36 Servo error may be reduced by integration of the error. This is ideal for overcoming the effects of static loads.

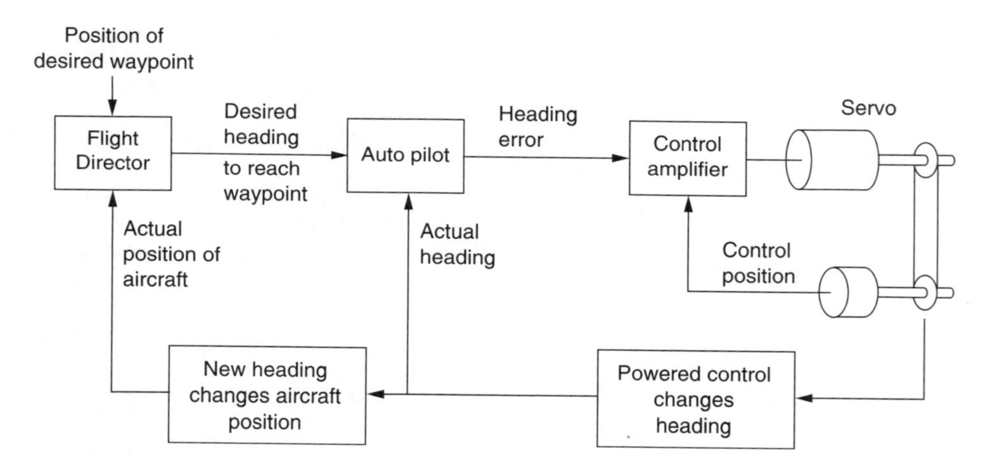

Fig. 2.37 Nested feedback loops. Here a flight director compares the actual aircraft location with the desired location to produce a navigational error. This is supplied to the autopilot that corrects the course by changing the attitude. The autopilot's commands are executed by feedback servos that operate the controls until the new attitude is correct.

The autopilot loop responds by producing an attitude error that keeps the machine in the attitude needed to hold the new track. The attitude error is supplied to the powered control system loop that actually controls the rotors.

The advantage of nested feedback is that each loop is stabilized in its own domain. The powered control loop is compensated to prevent oscillation or instability in the rotor systems, the autopilot loop is compensated to prevent instability in the attitude of the machine and the navigation loop is compensated to prevent the machine following a serpentine track to the waypoint. A further advantage of the nested loop is that different parts can be changed. An inertial navigator could be replaced with a GPS navigator, for example. Also in the case of a failure, only the affected loop needs to be shut down and replaced by pilot action. For example if the autopilot failed, the pilot could still use the outputs of the navigator and the power assistance, but he would have to move the controls to stabilize the attitude of the machine on the course stipulated by the flight director.

Feedback loops have many advantages, but they do behave badly in the event of certain failures. If the position feedback sensor of Figure 2.35 failed or jammed, the movement of the load would not reduce the servo error and the servo would inadvertently keep driving the load in the direction that would normally reduce the error, until the load reached a mechanical limit or stop. This is a characteristic of feedback loops, which is that when they fail they generally go to one control limit. This is known as a *hardover*.

Helicopters have crashed because a stability augmentation system had a hardover failure. Subsequently systems were designed with limited authority so that even if a hardover occurred, the amount of incorrect control travel or the force produced was not so great that the pilot could not overcome it. The end stops of the servo could be brought closer to the neutral position and a slipping clutch could be fitted in the drive train. Today part of the certification process is to prove that the pilot can overpower the servo so that control can be retained if a stability augmentation system has a hardover failure. If this is impossible, the system will have to be made failsafe by incorporating redundancy.

Certain helicopters use a redundant system in which the autopilot or stability augmentation signals are generated in two independent systems and fed to two motorized servos that drive the controls through a differential gear. In the event that one of the servos experiences a hardover failure, the tachometer of the failing system will output an unusually large signal. Detection of the hardover condition will shut down the failing servo by removing power and apply a shaft brake to lock the motor. The remaining servo will be able to retain control through the differential gear and the performance will be identical if the signal which shuts down the failing servo also doubles the travel of the surviving servo.

In a fully powered feedback servo system, the pilot simply moves a transducer and the feedback loop will do whatever it can to carry out the command. It will still try even if enormous resistance is met. This may result in overstressing and a solution is a system of force feedback. The control stick is not centred by a spring, but by a force motor. The resistance felt by the pilot comes from the force motor that produces a force proportional to the force being exerted by the servo. In this way the pilot produces a small force that is used to control the machine, but he also experiences a scaled down replica of the resistance to the control efforts.

This gives the pilot a good sense of how much stress the machine is under. Force feedback systems of this kind are also known as *artificial feel* systems. Artificial feel of this kind can also be used with autopilots. If an altitude error is suddenly experienced due to turbulence, the autopilot could react very quickly and powerfully, but to do so would stress the airframe and cause discomfort to the passengers. It may be better to take longer to cancel the error by reacting more gradually.

Feedback is used in servos designed for radio control of models. These work on the principle of pulse width modulation where the length of a pulse, typically 1 millisecond at the neutral position, is increased or reduced as a means of control signalling. A servo operating on pulse width signalling can use feedback based on comparison of pulses. The feedback potentiometer controls a pulse generator and the servo error is obtained by determining the difference in pulse width between the input and the feedback pulses. The error pulses are used directly to drive the motor amplifier. As the error becomes smaller, so does the width of the pulses, reducing the drive to the motor. Velocity may be sensed by measurement of motor back EMF whilst the amplifier is switched off. Pulsed amplifiers are electrically efficient because they are either on or off and so dissipate little heat. This is important in models where the power source will be a battery.

Although this section has considered a simple feedback system in which the controls are signalled by analog voltages or variation of pulse width, it is possible to build feedback systems using any type of signalling or even a mixture of types. Chapter 7 considers various methods including digital signalling.

3

Introduction to helicopter dynamics

3.1 Creating and controlling lift

The basic principle of wings, rotors and airscrews is that they accelerate a mass of air and that the resultant lift or thrust is the Newtonian reaction to that acceleration. Producing sufficient lift to permit a helicopter to fly is only a matter of having enough power to accelerate the air mass without excessive weight. A practical helicopter must, however, be able to control that lift precisely or it would be dangerous.

Any structure placed in a flow of air for the purpose of generating lift is called an aerofoil or airfoil. Aeroplanes and helicopters can fly when there is a wind blowing and can climb or lose height in the process. In helicopters the rotor blades have airspeed due to their rotation. Thus there can be situations where the airfoil moves through the air, or where the air moves past the airfoil. All that matters from the point of view of generating lift is the relative velocity and the direction from which it appears to be approaching the airfoil. Figure 3.1(a) shows that this is known as a relative airflow (RAF). RAF is a vector quantity as it has speed and direction. Figure 3.1 further shows that a flat plate will produce lift if it is slightly inclined to the RAF at an *angle of attack*. When the airfoil changes the direction of air flowing by, this represents a change of velocity and so is classed as acceleration. The reaction to that acceleration will point in the opposite direction.

Figure 3.1(b) shows how the direction of the reaction is found. For the time being, the air is assumed to be inviscid (having no viscosity). The relative airspeed does not change, only the direction, so the velocity vector V is changed to V'. The acceleration to change V to V' must have been in the direction of the vector V_a so the reaction must be in the opposite direction. It can be seen that this is at right angles to the average airflow direction. The blade reaction is the only actual force present. In fixed-wing aircraft studies, the reaction is traditionally resolved into two components: that which is measured at right angles to the RAF and called the *lift*, and that which is measured along the direction of the RAF and called the *induced drag*. In cruise lift will be vertical and drag will be horizontal.

Figure 3.1(c) shows that the lift can be increased by increasing the angle of attack since this has the effect of increasing the change of velocity. In the aeroplane, the angle of attack is controlled by alteration of the attitude of the entire machine with the elevators. In a helicopter the blades are mounted on bearings in the rotor head that

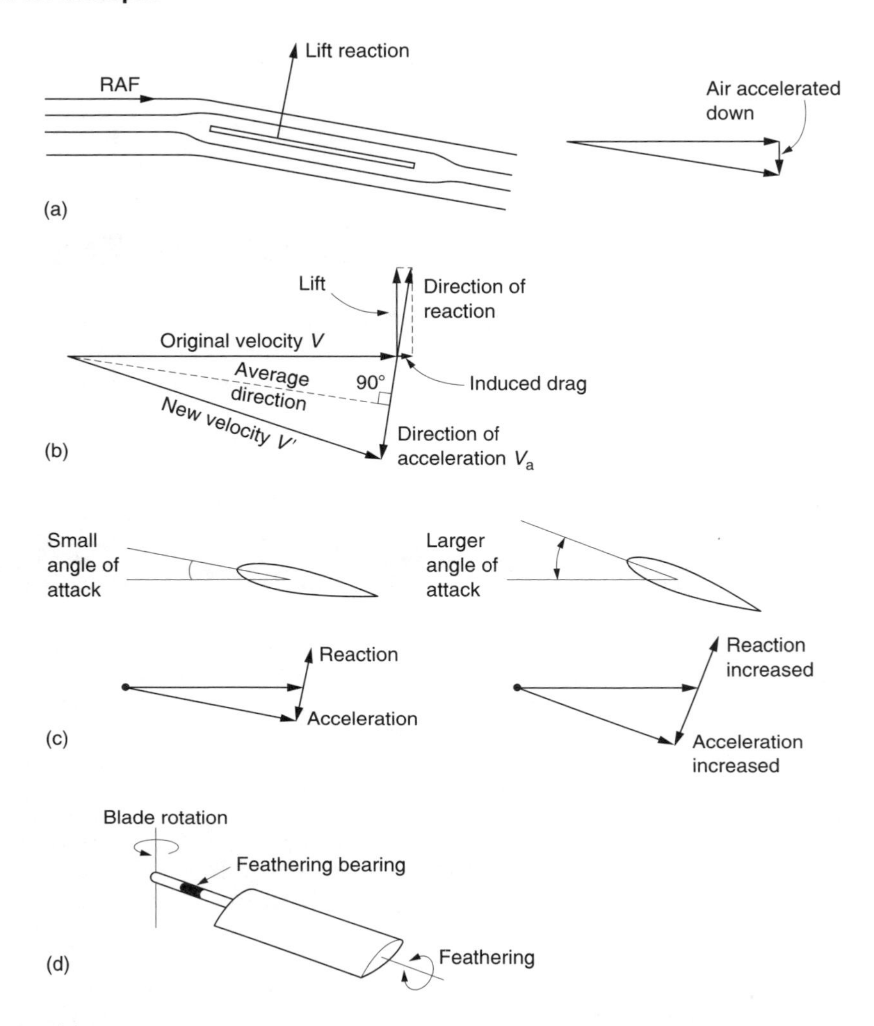

Fig. 3.1 (a) Lift is the reaction to the acceleration of air into a new direction. (b) The direction of acceleration of the air is derived as shown. The reaction must be in the opposite direction. (c) An increase in the angle of attack increases the lift. (d) In the helicopter, lift is controlled through twisting the blade about a span-wise axis; a mechanism called feathering.

allows them to turn about a radial axis as can be seen in Figure 3.1(d). This mechanism is known as *feathering*.

When the relative velocity alone is doubled with respect to the case in (b), the vector V_a representing the change of velocity is doubled. As the air is moving twice as fast, the wing accelerates the air in half the previous time. Achieving double the velocity in half the time means that the reaction is increased by a factor of four. In other words, the reaction is proportional to the *square* of the velocity. In practice four times as much lift would be quite unnecessary to balance the weight of the machine and instead, as the relative velocity doubled, the angle of attack would be reduced to about one-quarter of the previous value. This is the mechanism by which aircraft obtain the same lift over a range of airspeeds.

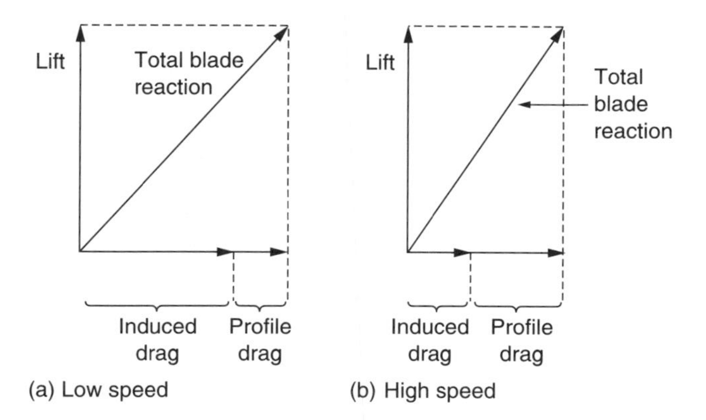

Fig. 3.2 Profile drag must be included in the behaviour of any real airfoil. The induced drag and profile drag add to tilt the total blade reaction backwards. Conditions at low airspeed are shown in (a). At a higher speed, (b), the profile drag is greater, but the induced drag is less. This is only possible because it is not really a drag.

In practice the air also has viscosity that resists movement as a function of the surface area of any body moved through it. The loss due to this effect is called *profile* drag. The profile drag tilts the blade reaction further to the rear as in Figure 3.2(a). There is a boundary layer between a moving object and the stationary air. At the leading edge, the speed difference across the boundary layer is maximal and the drag is greatest. Further back, the viscosity has been able to get the boundary layer moving and the drag is reduced. Thus the drag experienced by a wing or rotor blade is a function not just of the surface area but also of the chord. Very small objects possess disproportionately high profile drag, which is why the wind can blow sand but not rocks. Reynolds numbers are used to describe the relationship between dimensions and profile drag. It is important in model testing that the airspeed used is scaled correctly. Scale speed is not used. Instead the speed is chosen to produce the same Reynolds number in the model as in the full-sized device and then the drag of the model will be representative.

If an airfoil is not producing lift, the induced drag will be zero and only the profile drag will be observed. In a full-size helicopter in the hover roughly 30% of the rotor shaft power is lost to profile drag, the rest goes in accelerating the air downwards. In model helicopters and full-sized tail rotors profile power is proportionately greater as the small-chord blades operate with adverse Reynolds numbers.

Figure 3.2(b) shows the production of the same lift at a higher relative velocity. The profile drag has increased but the induced drag has reduced. This is not what one would expect from a true source of drag. The explanation is that the airfoil is an actuator or transducer, which accepts mechanical power as the product of the mechanical impedance and the relative velocity in order to drive the air against aerodynamic impedance. Accelerating air downwards constantly provides lift. The actuator reflects the aerodynamic impedance back as mechanical impedance. The variation in angle of attack is analogous to changing the turns ratio of a transformer or the ratio of a gearbox. Clearly if the relative velocity goes up, to keep the power the same the impedance must fall.

Once the concept of an airfoil as an actuator is understood, it is a small step to appreciate that most types of actuator are reversible. Some helicopters use an electric motor to start the engine when that motor becomes a generator. In a turbine engine, the rotating blades in the compressor are putting energy into the air, whereas the rotating blades in the power turbine are taking energy out of the gas flow. Given this reversibility, we can immediately see how in a glider or in an autorotating helicopter the power flow through the actuator reverses. In descending through the air the wing or rotor delivers power extracted from loss of potential energy and this power is used to overcome profile drag. The glider can maintain airspeed and the helicopter can maintain rotor speed with no engine power.

The act of turning the airflow also results in rotational energy being imparted to the passing air. Conceptually, the airflow can be divided into two flows. One is the steady flow past the blade and the other is the rotational component that in the absence of the steady flow would rotate around the blade. This is known as *circulation* and it is proportional to the lift. In vortex theory, the circulation of an airfoil is calculated and the lift follows from that.

A further consequence of the change of direction is that the horizontal component of the relative velocity is reduced. It follows from the rearward inclination of the blade reaction that there must be a forward component of the acceleration imparted to the air. Air does not go straight down from a hovering rotor; it also revolves in the same direction the rotor turns, but much more slowly: a phenomenon known as swirl. Swirl is considered in section 3.14.

It is more efficient gradually to accelerate the air down rather than have it suddenly find an inclined plate. Air passing the top edge of a flat plate cannot change direction quickly enough and the flow separates and becomes turbulent. These are the reasons for curving or cambering an airfoil. Furthermore, practical wings must contain structural members such as spars, and the thickness will be increased to accommodate them. The thickness is contained within a streamlined shape to reduce drag.

Figure 3.3(a) shows a streamlined cambered section. The mean camber line is half way between the upper and lower surfaces. The chord line joins the ends of the camber line. The angle of attack is the angle between the chord line and the relative airflow. The camber of the airfoil can be optimized for the speed range of interest. For high lift at low speeds, the camber will be heavy in order to make V_a as large as possible.

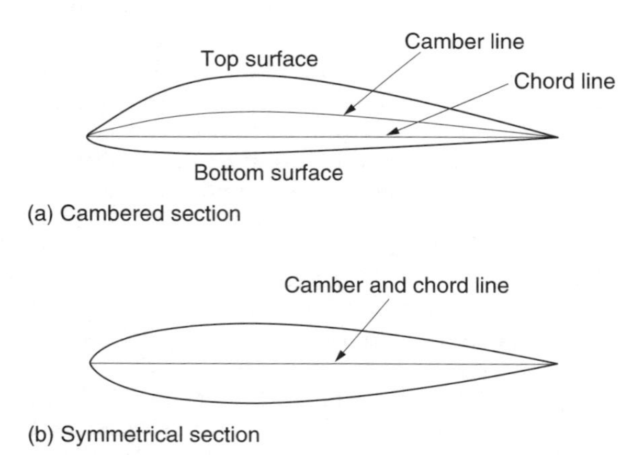

Fig. 3.3 Both cambered and flat airfoils will produce lift. In practice they will be thickened to reduce separation and to allow structural components to be incorporated. Cambered airfoils work best at low speeds. In the cambered section (a) the camber line will curve away from the straight chord line. In the symmetrical section (b) the chord line and camber line coincide. A symmetrical airfoil is equally capable of producing lift upwards or downwards. With zero angle of attack there is no induced drag, only profile drag.

Clearly such a wing will be creating a lot of drag at the small angles of attack needed at high speeds. For high speed operation, the camber will need to be very small, but this kind of wing will be inefficient if used at large angles of attack at low speed, hence the use of flaps on fixed-wing aircraft. With a cambered section, air is still accelerated downwards leaving the section even when the angle of attack is zero, and so a slight negative angle of attack is necessary to obtain the zero lift condition.

Figure 3.3(b) shows a symmetrical section in which the camber line and the chord line are one and the same. A symmetrical section is a streamlined flat plate. The curvature prevents separation over the leading edge. At zero angle of attack, the airflow is also symmetrical, and no net air reaction results. There is thus no induced drag, only profile drag. If the angle of attack is made positive, air is accelerated down, and the reaction is upwards. If the angle of attack is made negative, air is accelerated up and the reaction is downwards.

The reader is cautioned against explanations of the origin of lift based on Bernouilli's theorem. Bernouilli made it quite clear that his theorem relates to conservation of energy in flowing air such that the sum of the static and dynamic pressures remains constant. Bernouilli's theorem only applies if no energy is put into the air. However, a wing or a rotor blade is an actuator that is exchanging energy with the air. This is clear from the presence of induced drag. When energy is put into the airflow, Bernouilli's theorem simply doesn't apply and the explanations based on it are flawed and should be disregarded.

3.2 The centre of pressure

The lift developed by the blades is distributed over the chord, but not uniformly. The centre of pressure is where a single force would act producing the same effect as the distributed lift. In real airfoil sections, the centre of pressure is ahead of the mid-chord point. If the blade is made of a material with uniform density, the centre of pressure will be ahead of the centre of mass and, as Figure 3.4(a) shows, a couple results. This couple would tend to twist the blade and increase the angle of attack, making the lift greater and increasing the twist further. In extreme cases the blade will flutter; a violent condition that will usually destroy any structure suffering from it.

In aircraft, wings can usually be made rigid enough to prevent flutter, but this cannot be done with the long thin blades of the helicopter. The solution universally adopted is to construct the blade so that the mass centroid (see Chapter 2) of the blade is ahead of

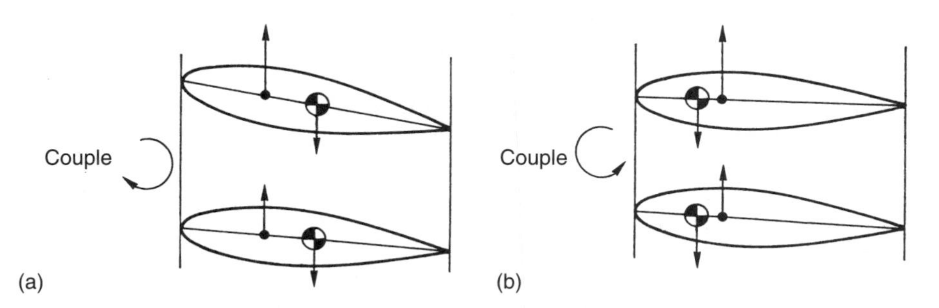

Fig. 3.4 (a) If the centre of lift is ahead of the CM, lift tends to increase the angle of attack. (b) If the centre of lift is behind the CM, lift tends to reduce the angle of attack: a stable condition.

the local centre of pressure. As Figure 3.4(b) shows, this results in a stable blade, since an increase in angle of attack producing more lift tends to generate a couple *reducing* the angle of attack.

The airfoil section selected for a helicopter blade will be a compromise to satisfy a number of conflicting requirements. One of these is a minimal migration of the centre of pressure over the normal range of operating conditions so that excessive feathering couples are not fed back into the controls.

In a cambered airfoil, the centre of pressure moves fore and aft with changes in angle of attack. The downward twist at high speed was enough to twist the blades against the pilot's efforts on early machines and caused some crashes. For some time, helicopters used little or no camber in the blade section, but subsequently cambered sections having reflex trailing edges were developed which reduced the centre of pressure movement. This along with the development of structures with greater torsional stiffness allowed cambered blades to return to use, although almost invariably in conjunction with powered controls.

3.3 The coefficient of lift

Lift is equal to the rate of change of momentum induced downwards into the surrounding air. For unit airfoil area, it is proportional to the air density, the square of the relative air velocity and, as was seen in section 3.1, it is a function of the angle of attack. When an airfoil section is tested, the reaction is resolved into horizontal and vertical directions with respect to RAF so that two factors of proportionality, called the coefficient of lift C_L and the coefficient of drag C_D are measured. Figure 3.5(a) shows how these coefficients change as a function of the angle of attack. For small angles C_L is proportional as it was seen in Figure 3.1(c) that the acceleration of the airflow is proportional to the angle through which it is deflected. As the angle of attack increases the reaction rotates back and the lift component of the reaction will be smaller than the reaction. The graph curves away from proportionality. At a large angle of attack, separation takes place, and the C_L drops sharply, accompanied by a sharp increase in C_D. When this occurs the airfoil is said to be *stalled*.

An airfoil does not stall immediately the angle of attack is raised, and if the rate of angular change is sufficiently high the lift momentarily available might be double the amount available in steady-state conditions. This is variously called the Warren effect or *dynamic overshoot* and is shown in Figure 3.5(b). In the case of a helicopter the angle of attack is changing at rotor speed and dynamic overshoot has a considerable effect. In practice it means that helicopter rotors may not lose significant amounts of lift through stall.

In this respect the helicopter has the advantage over the fixed-wing aircraft. The latter has to be flown at all times with the probability of a stall in mind. Even when the pilot has sufficient airspeed, a sudden gust due to wind shear or a microburst can reduce that airspeed and cause a stall. It was this concern that led Juan de la Cierva to develop the gyroplane. He correctly argued that a rotary wing aircraft that could not stall would be safer.

The coefficient of lift cannot be controlled directly; in order to obtain a certain C_L the airfoil must be set to the appropriate angle of attack. The drag is also a function of angle of attack. It is more useful to know the ratio of lift to drag, since the peak value of this, $L/D_{max.}$, indicates the most efficient mode in which the airfoil can be used.

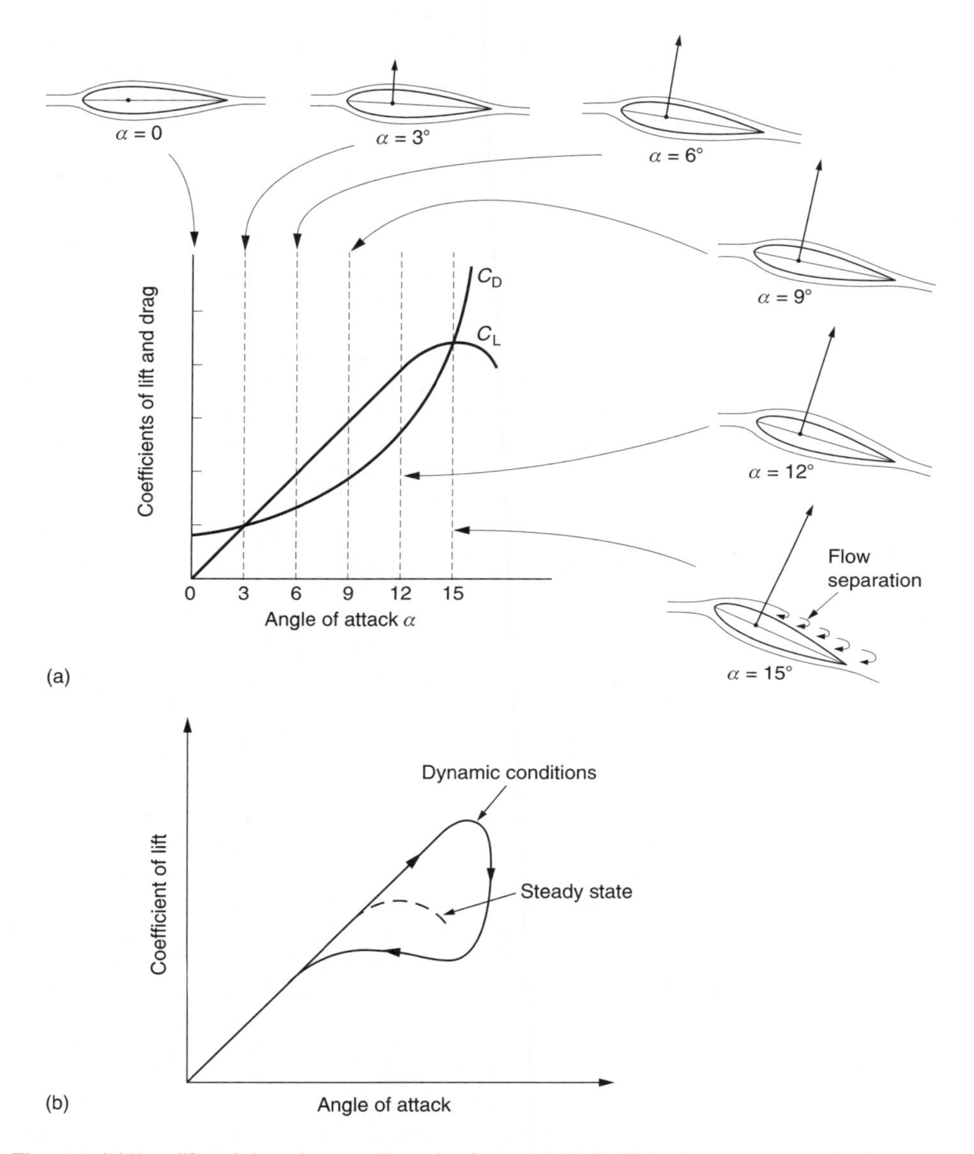

Fig. 3.5 (a) How lift and drag change with angle of attack. Initially lift is almost proportional whereas drag changes with the trend shown. As there is profile drag at all angles of attack, the curve does not start from zero. The most efficient use of the airfoil will be where the L/D ratio is greatest. At steady large angles of attack the airfoil stalls. (b) Stall does not occur instantaneously and for a short time after a large angle of attack is applied, lift increases. With their rapidly feathering blades, helicopters can exploit this phenomenon.

For a given engine power, the greatest lift in the hover would be obtained by operating the blades at the maximum lift over drag ratio. However, a practical helicopter cannot operate in this condition because it would be unsafe in forward flight. Many manoeuvres, such as a banked turn, require the rotor to produce a thrust which exceeds the static weight of the machine. The ratio of thrust to static weight is called the *load factor*.

A rotor hovering at $L/D_{max.}$ has a load factor of unity and would stall in any real manoeuvre. It is thus a characteristic of helicopters that the rotor must operate below $L/D_{max.}$ of the blade section in the hover in order to allow a reasonable load factor for forward flight.

Turning at a given rate (number of degrees per second) requires a lateral acceleration proportional to the square of the airspeed. As aeroplanes tend to fly faster than helicopters, they will need high load factors to obtain reasonable rates of turn. Aeroplanes automatically have high load factors at speed because available lift increases as the square of the airspeed. In helicopters lift is dominated by rotor speed, not airspeed and so load factors in helicopters tend to be small although helicopters can still easily out-turn most aeroplanes.

As the relative airflow in helicopters is dominated by the rotor speed, the gust response of helicopters is much reduced in comparison to that of aeroplanes. This is one reason why helicopters can operate in bad weather. A further small luxury is that the flexibility of the rotor gives in gusts a decoupling effect similar to that given by the suspension of a car.

3.4 Collective control

In a hovering helicopter, the only source of lift is the rotor. Obtaining sufficient lift is only a matter of providing a suitable combination of power and efficiency, and is much easier than controlling lift. As the lift is proportional to the square of the speed, it is in principle possible to control lift by changing the rotor speed. Some early machines did just that. Unfortunately, the inertia of the rotor means that speed changes cannot rapidly be accomplished. For practical reasons a constant rotor speed is much to be preferred.

The proportionality between the coefficient of lift and the angle of attack is the solution. As was shown in Figure 3.1, the rotor blades are mounted on feathering bearings. Figure 1.21 showed that the pilot holds in his left hand a collective pitch lever pivoted near the back of his seat. The lever gets its name because by lifting it, the pitch angle of all of the rotor blades is increased by the same amount, and the rotor lift increases immediately. This feature of the helicopter contributes enormously to its safety. When a fixed-wing aircraft flies slowly, sudden loss of lift can only be countered by raising the airspeed and this takes time. In the helicopter the airspeed is always present due to the rotating blades and as this speed is many times higher than any normal windspeed, the airspeed seen by the blades is always adequate.

In addition to instantaneous response to changing lift conditions, the helicopter rotor stores in kinetic energy the equivalent of full engine power applied for several seconds. The pilot can transiently increase the available power by using so much collective pitch that the rotor slows down and converts its kinetic energy into lift.

3.5 In the hover

Figure 3.6(a) shows a helicopter with flight RRPM, but with the collective lever lowered. The thrust vector is small, so gravity keeps the machine firmly on the ground. If the collective lever is raised, eventually the rotor thrust vector will exceed the weight, and the machine will rise. By adjusting the collective lever, the machine can be made to hover with the thrust exactly balancing the weight as in (b). If the thrust is directly upwards, there is no resultant and the machine stays still. In the still-air hover, the inflow through the rotor is axial, and if, for simplicity, it is assumed that the centre of gravity of the machine is at the mast, the forces on the blades will not vary as they turn.

The collective control can be used to make the rotor thrust greater than or less than the weight of the machine. The force imbalance causes the machine to accelerate up or down. However, this acceleration does not continue indefinitely. Figure 3.7(a) shows that as the machine rises, the angle of relative airflow experienced by the blades is

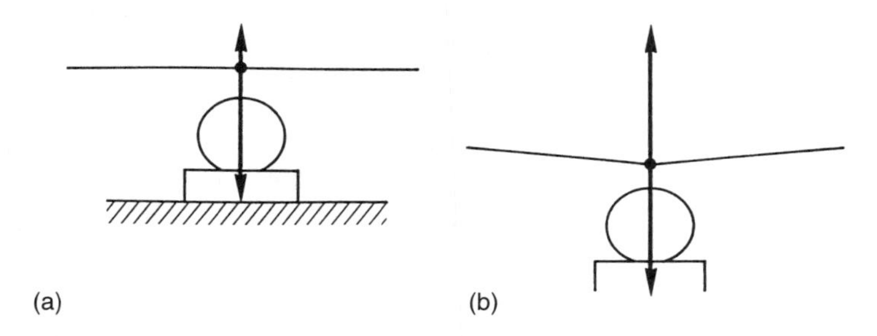

Fig. 3.6 (a) With the collective lever lowered, the rotor thrust is less than the weight and the machine remains on the ground. (b) With an appropriate setting of the collective lever, the rotor thrust can be made exactly equal to the weight.

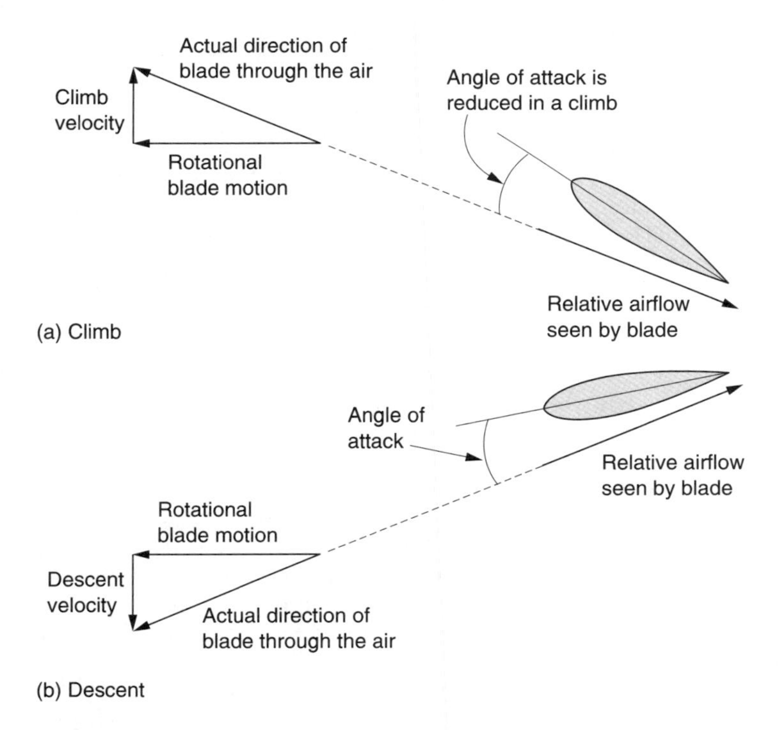

Fig. 3.7 (a) With increased collective, the machine rises, but in doing so changes the angle at which the air approaches the blades and reduces the angle of attack. (b) With reduced collective, the opposite happens. Thus the collective lever primarily controls vertical velocity.

reduced by the vertical velocity. This tends to reduce the angle of attack and hence the rotor thrust. As a result the vertical velocity reaches a value where the rotor thrust once more balances the weight and the velocity then remains constant.

Figure 3.7(b) shows that if the collective control is lowered, the machine accelerates downwards, but the change in relative airflow is such that the rotor thrust is increased. The vertical velocity again stabilizes at a value where the thrust balances the weight. Thus the effect of the collective control in the hover is primarily to control the vertical velocity.

In most powered flight conditions, but particularly when hovering, the rotor is working in the descending airflow it has itself caused. This self-inflicted problem is called *inflow*. When inflow is present, there is a vertical component of airflow vectorially to be added to the local blade velocity in order to find the direction of the RAF.

The pilot controls the pitch angle using the collective lever and this, along with the velocity of inflow through the rotor, determines the angle of attack. As a result there is no one correct position for the collective pitch lever and the pilot will need constantly to adapt the collective pitch setting for changes in speed, fuel burned and altitude. As there is no correct position for the collective lever, there is no spring centring. Instead there may be a friction mechanism that can be adjusted to hold the lever at the last position without unduly adding to the force needed to move it. This allows the pilot to release the lever for short periods in order to operate secondary controls.

3.6 Forces on the blades

In a helicopter having a hingeless rotor, the displacement between the area where lift is generated and the hub where the load is applied causes upward bending stresses that increase dramatically towards the rotor head. The underside of the blade is placed in tension, and the upper surface is placed in compression. The rotation of the blades produces a tension along their length that also increases towards the rotor head. The stress caused by this tension is of the same order as the stresses due to bending, and the stress on the underside of the blade is roughly doubled, although the compression on the upper surface is relieved. The drag of the blades causes a rearward bending stress that is constant in the hover.

There is a tendency for the blade rotation to force the blade to twist to a pitch angle of zero. Figure 3.8(a) shows that when the blade is set to zero pitch, the leading and trailing edges of the blade are further from the feathering axis than when pitch is applied. In order to set the blades to a working pitch, some of the mass of the blade must be brought closer to the shaft axis, and work will need to be done against the rotational forces that will be reflected as a reaction seen by the control system.

The effect can be reduced by the use of so-called Chinese weights that are placed on a rod at right angles to the chord line as shown in Figure 3.8(b). These will tend to move away from the rotor shaft axis as a result of rotation. The term comes from the tales told by sailors to the gullible alleging the orthogonality of certain parts of the anatomy of the oriental female at a time long before the development of either the helicopter or political correctness.

A less colourful alternative is to incorporate a spring into the collective linkage that is adjusted to supply a steady force to hold the blades at a typical positive pitch, relieving the pilot of the need to apply a constant force to the collective lever. When the blades are not turning this spring may be powerful enough to force the blades to maximum pitch and so the collective lever is fitted with a lock that the pilot can apply prior to shutting

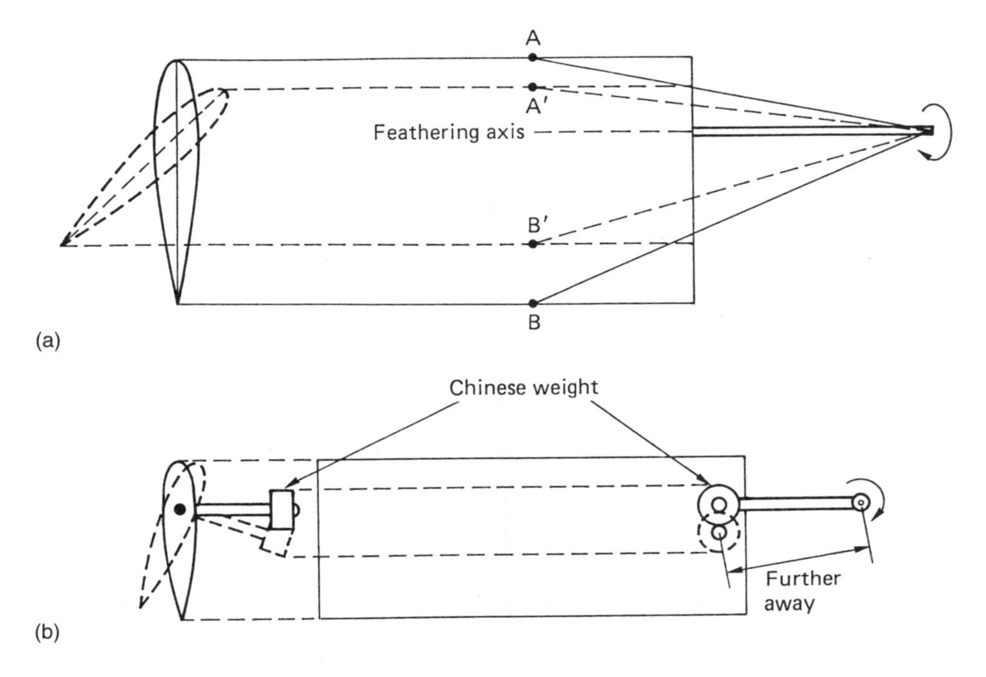

Fig. 3.8 (a) When the blade pitch is increased, parts of the blade are brought closer to the axis of rotation. The blades will tend to be thrown to flat pitch when the rotor is turning. (b) So-called Chinese weights are mounted so that they oppose the effect of (a) by moving away from the axis of rotation as pitch is increased.

down the rotor. Another possibility is for the rotor head to incorporate centrifugal weights that only apply an upload to the collective when the rotor is turning.

3.7 Rotor coning

Anyone who has examined a parked helicopter will immediately notice how droopy the blades are. A considerable deflection can be obtained by lifting the tip with one finger. How, then, can the blades lift the weight of the helicopter? The answer is that the rapid rotation in conjunction with the mass of the blades causes them to be pulled out straight; a phenomenon called *centrifugal stiffening*. Rotating blades are not in static equilibrium and they must be made to accelerate towards the shaft if they are to follow a circular path, and this requires a considerable inward or centripetal force, which will always be an order of magnitude more than the weight of the helicopter.

Figure 3.9(a) shows that if the blade bends upwards, the downward component of the centripetal force will balance the lift at some *coning angle*, and the resultant will be a horizontal force only so no further bending takes place. The reaction of the blades at the rotor head is shown in Figure 3.9(b). The force from the coning blade has an upward component, which is the lift, and an outward component due to the rotation.

In the hover, if the blades are properly balanced and all have the same coning angle, the horizontal forces cancel in the rotor head and only lift results. Adjusting all of the blades to the same angle is achieved using the process of tracking which ensures that the collective control applies exactly the same pitch to each blade. Clearly, if the blades

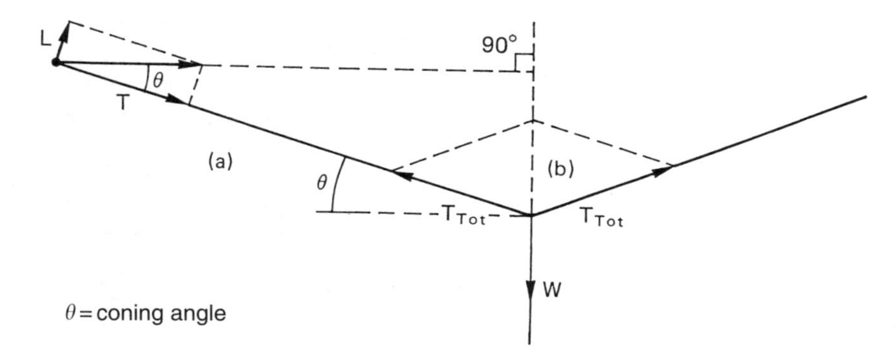

Fig. 3.9 The blades must be accelerated inwards to make them rotate. At (a) the blades cone upwards until the resultant of the lift and the blade tension is perfectly horizontal. At (b) the force balance at the rotor head is shown. The tension in the upwardly coned blades cancels in the horizontal direction leaving only a vertical component to balance the weight of the machine.

are not balanced or not tracking, vibration will result. Accurate blade balancing and tracking is important and the necessary techniques will be considered in section 4.19.

3.8 Torque and thrust in rotors

Whilst resolving the airfoil reaction into lift and drag is useful for fixed wings, it is less useful for helicopters because of the feathering action of the blades. Figure 3.10 shows how the reaction on a rotor blade can be resolved in a more useful fashion. As the only real force on the airfoil is the reaction, resolving the force into components is only taking place in our imagination, so we can resolve into whatever directions we find useful. In helicopters, whatever the mechanical pitch angle of the blade, and whatever angle of attack results, it is more convenient if the reaction is resolved into the rotor thrust acting substantially at right angles to the tip path plane or rotor disc, and the rotor drag acting in the tip path plane.

The angle of attack is the angle by which the pitch angle exceeds the angle of the RAF. The reaction of the airfoil is tilted further back in the presence of inflow, so that a larger component of the reaction is opposing the thrust delivered by the engine, and power is consumed simply driving against it. The rotor thrust is also slightly reduced by the tilt of the reaction. The hover out of ground effect (HOGE) is a particularly power consuming exercise. Inflow tilts back the blade reaction to oppose the engine, so that more torque is required to drive the rotor and more tail rotor power will be needed to balance the torque. HOGE is also one of the worst situations in which to lose power as the high rotor drag will reduce RRPM more before the pilot reacts and lowers the collective lever.

Clearly increasing the collective pitch will increase the induced drag, and so more power will be required to maintain rotor speed. In simple piston engine helicopters, the end of the collective lever is fitted with a twist-grip connected to the throttle. Maintaining RRPM is then the responsibility of the pilot who will need to monitor the rev. counter. In more sophisticated piston engine machines and in all turbine machines automatic rotor speed governing systems or throttle correlators are used. The pilot

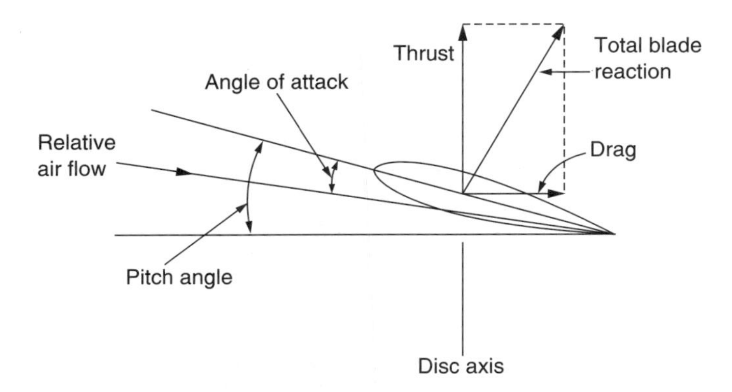

Fig. 3.10 The blade reaction is usefully resolved into components both in and orthogonal to the tip path plane. These allow the rotor thrust and torque to be derived. In the presence of inflow, the angle of attack is the amount by which the pitch angle exceeds the angle of the relative airflow.

just changes the collective pitch and the system computes the engine power for itself. Correlators will be described in Chapter 6.

3.9 The rotor as an actuator

Without considering constructional details, it is possible to conceive of an ideal hovering rotor as an actuator disc that somehow accelerates air downwards over a circular region and develops thrust from the reaction. In this case the operation of an ideal rotor can be analysed theoretically from the change of momentum of the air passing through. The concept of an ideal rotor is useful because it can act as a benchmark with which to compare real designs. Clearly the performance of the ideal rotor can only be approached, but never exceeded.

Figure 3.11(a) shows the actuator concept which makes some simplifying assumptions. One of these is that the air is somehow constrained to flow only within the tapering column shown, known as a stream tube, and doesn't mix with the surrounding air. The mass of air passing any horizontal plane in the column per unit time must be constant. The actuator causes a pressure difference to exist across itself. This pressure must be given by the thrust divided by the disc area. A further assumption is that this pressure is uniform. The pressure difference across the actuator can only be sustained locally. Figure 3.11(b) shows that at a distance, the pressure must be the same as static pressure. To allow the pressure difference across the actuator, the pressure must have a falling gradient along the streamlines except at the actuator itself.

Bernouilli's theorem can be used to predict what happens along a stream tube. Bernouilli's theorem is simply another example of conservation of energy. Air has a static pressure and a dynamic pressure due to its motion. Bernouilli's theorem states that the sum of the two, known as the head, remains constant. As a result when static pressure falls, the dynamic pressure must increase to compensate. As the density doesn't change significantly, the velocity must increase. Clearly Bernouilli's theorem does not hold across the actuator, as energy is put into the air there.

Except for the pressure step at the actuator, the air in the stream tube experiences a falling pressure gradient which causes the stream to accelerate and contract. This is the phenomenon of wake contraction seen in propellers and rotors. Thus the velocity with

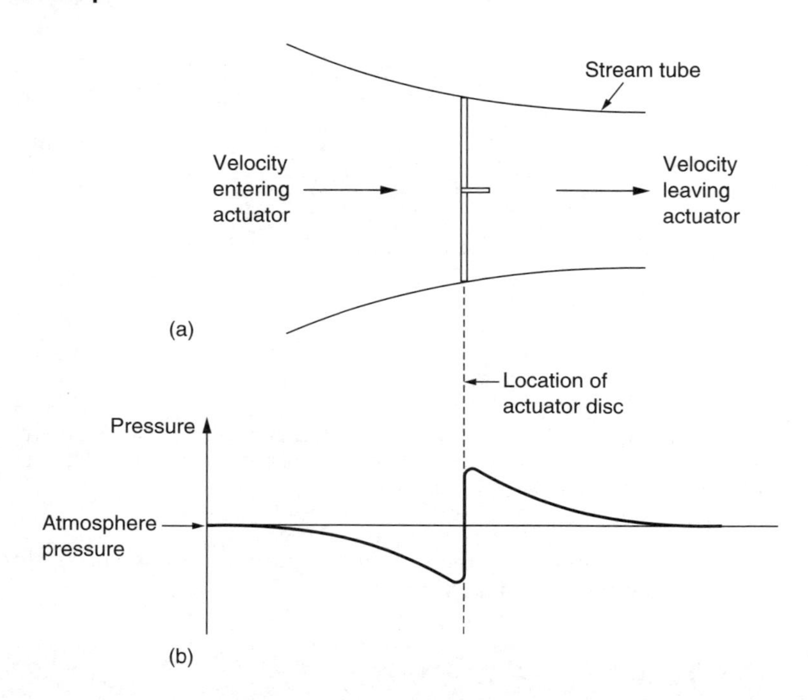

Fig. 3.11 (a) In actuator theory, air passes along a stream tube which is intersected by the actuator. (b) In order to allow a step pressure difference across the rotor, pressure must fall both approaching and leaving. Note the wake contraction to about 0.7 of the rotor diameter.

which air arrives at the actuator is higher than it is some distance before the actuator. This difference is the induced velocity. The induced velocity is important because it determines the velocity the rotor sees which has a large bearing on the RAF and the power needed. It should not be confused with the final velocity which is higher still.

In the case of an actuator that is climbing vertically, the rotor thrust is given by the rate of change of momentum of the air passing through the disc. The power needed must be the product of the thrust and the velocity, where the velocity is the rate of climb plus the induced velocity. This power must be equal to the difference in the kinetic energy well above and well below the disc. Clearly this assumption implies that only the induced drag is being considered. Actuator theory cannot account for profile drag and assumes it to be zero.

The increased velocity applied to the downwash is twice the induced velocity at the disc. In a stationary hover, the total velocity is the induced velocity alone and the downstream velocity will be twice this. In practice the contraction has completed by about one diameter below the rotor. At this point, as the velocity has doubled, the cross-sectional area must have halved in order to have constant mass flow and so the downwash diameter will be about 0.7 of the rotor diameter.

In Figure 3.11, uniform inflow over the disc is assumed. In other words the induced velocity and the pressure are the same all over. This is the most efficient condition. It is easy to show that any other condition requires more power. For a given total thrust, if one area of the disc has greater induced velocity and thrust, another area must have less induced velocity and thrust. The thrust is proportional to the momentum change, which is in turn proportional to the induced velocity, whereas the kinetic energy imparted is proportional to the square of the induced velocity. Thus the power saved in an area of reduced thrust does not compensate for the extra power needed in an area of increased thrust because of the square law. It follows that the minimum power is required when the thrust per unit area is constant.

Helicopter flight is dependent on a pressure difference across the rotor disc that disturbs the air. Such pressure disturbances travel at the speed of sound and can thus easily travel long distances from the helicopter. It should be clear that a helicopter in flight is associated with the induced movement of a significant air mass. Such a moving mass cannot change its velocity in an instant. Consequently a rapid change of blade pitch does not result in an immediate change of inflow velocity. This will result in an angle of attack initially higher than in the steady state and this may result in a thrust overshoot. The same effect is observed in sailing vessels where there is a lag between sheeting the sails and a change in sail reaction. The phenomenon is known as *dynamic inflow* and is a macroscopic version of the dynamic overshoot phenomenon of Figure 3.5. The inertia of the associated air mass will also have effects in translational flight as will be seen in section 3.21.

3.10 Blade element theory

The actuator concept is useful, but it does not consider a number of real world factors, not least profile drag. Blade element theory takes this into account and allows a more accurate result to be obtained. In the hover the forward speed of some point on a blade is proportional to the radius of that point. It is possible to analyse the performance of a rotor by considering it to be made of small elements where the conditions over each element are substantially constant. The overall result is obtained by adding up the contribution from each element.

At each blade element, the inflow must be known so that the RAF and the effective airspeed can be calculated. The characteristics of the blade section employed at that element must be consulted to find the resultant aerodynamic force on the element that will be resolved into an element of rotor thrust and an element of drag. All of the axial thrusts from the elements can simply be added to find the total thrust. The drag of each element is multiplied by the radius of the element concerned to obtain an element of rotor torque. These elements are then added to obtain the total rotor torque.

Blade element analysis is only as accurate as the assumptions made about the inflow. Actuator theory gives some idea about the inflow, but in a real rotor the inflow at the tips and the blade roots will be non-ideal. More accurate results require the use of vortex theory to take into consideration the conditions at the blade root and tip.

3.11 Disc loading

The thrust production mechanism in the helicopter creates a pressure difference, which causes the air to accelerate downwards. To maintain height, the rotor then has to climb up through the air at the same speed as the air is coming down. Work is being done on the air and the power is the product of the weight of the helicopter and the induced velocity. It follows that the power needed to hover can be minimized by reducing the induced velocity. This can be achieved by reducing the disc loading. This is the helicopter's equivalent of span loading in a fixed-wing aircraft. Where power is limited, as in piston engine helicopters, a low disc loading will be needed to extract more lift from the available power. Some of that lift will be lost because the larger rotor will be heavier.

In a turbine helicopter, the engine and fuel supply weigh less for the power the engine produces and so a better result may be obtained by using a smaller rotor and transmission that also weighs less to obtain a greater useful payload. A comparison of, for example, an Enstrom F-28 and a Hughes 500 illustrates this point.

This cannot be taken to extremes, as there is a practical limit to the downwash velocity. This is reached when objects on the ground are blown about and become a hazard. The CH-53E Stallion represents about the limit of acceptable downwash velocity as it is almost impossible to stand in the downwash. A further consideration is that in autorotation a machine with a high disc loading may have a rapid rate of descent, leaving the pilot little margin for error in judging the flare-out in the case of an engine failure. On the other hand a machine with a low disc loading may have very good autorotation performance, the Enstrom being a particularly good glider.

Once a disc area and loading has been decided, some consideration has to be given to the tip speed. Profile power is proportional to the cube of the tip speed, whereas thrust is proportional to the square of the tip speed. The minimum profile power is where the minimum possible tip speed is used to operate the blades at $L/D_{max.}$, which will be just below the stall. Reducing the tip speed will require the blade area to be increased to maintain thrust. Consequently the disc area is chosen to give the desired disc loading, whereas the blade area is chosen to produce sufficient rotor thrust (with an adequate load factor) at the chosen RPM. The ratio of the total blade area to the disc area is known as the solidity. Although low tip speeds and high solidity reduce profile drag and improve hovering performance, a machine built to these criteria would not have an adequate load factor for forward flight. It will be seen in section 3.24 that a low tip speed is also detrimental to forward flight as well as being a liability in the case of a power failure as the rotor stores little energy, giving the pilot little time to react and establish autorotation.

In a conventional helicopter there will always be a compromise between hover and forward flight performance. Possible technical solutions include variable rotor speed and variable diameter rotors.

In the hover a low rotor speed would reduce profile drag by allowing the blades to operate closer to $L/D_{max.}$, whereas in cruise a high rotor speed would improve the load factor. To obtain maximum power from a piston engine at both speeds, a two-speed transmission would be needed, and there would be some weight and cost penalty. Changing the speed would be easy with a free turbine engine. There would also be a problem in detuning the rotor to minimize vibration at two different speeds. These problems are not insuperable.

A variable diameter rotor could have a low disc loading for efficient hover, but a raised disc loading in cruising flight. The technical problems here are enormous but may one day be economically viable.

3.12 Figure of merit

Whatever the disc loading, in a constant height hover, the potential and kinetic energy of the helicopter remain constant, and so no work is being done *on the helicopter*. Thus the mechanical efficiency of all helicopters in the hover is zero. Clearly mechanical efficiency is not a useful metric, because it doesn't allow comparison. A better metric is to compare the power theoretically needed by an ideal actuator with the actual power needed in the real helicopter where the actuator has the same diameter as the rotor. This is the origin of the figure of merit.

3.13 Blade twist and taper

The forward velocity of the rotating blade increases linearly from zero at the shaft to a maximum at the tip. This is not a good starting point from which to obtain uniform inflow, but the rotor is the only practical approximation to an actuator disc that anyone has devised. The figure of merit may be improved by twisting and tapering the rotor blades. Figure 3.12(a) shows that in untwisted, untapered blades, the lift will be produced increasingly toward the tips as a parabolic function, until tip loss dominates.

Given the induced velocity is required to be uniform, it is easy to calculate the pitch required at any radius to make the angle of attack constant along the blade so that the whole blade operates at an efficient L/D ratio. However, in practice the induced velocity will be a function of radius and such a twist will not be optimal. Twist that takes account of induced velocity distribution is known as ideal twist. In fact ideal twist is academic because the amount of twist would have to change as a function of induced velocity and clearly it does not in any practical blade.

Ideal twist does not achieve uniform inflow, because of the reduced effectiveness of the inner slow-moving parts of the blade. Tapering the blade as well as twisting it removes area at the fast moving tips and places it in the slower moving areas where induced velocity is ordinarily lower. Ideally, the rotor chord needs to be inversely proportional to the radius. This results in an impractical rotor for a helicopter, but the technique

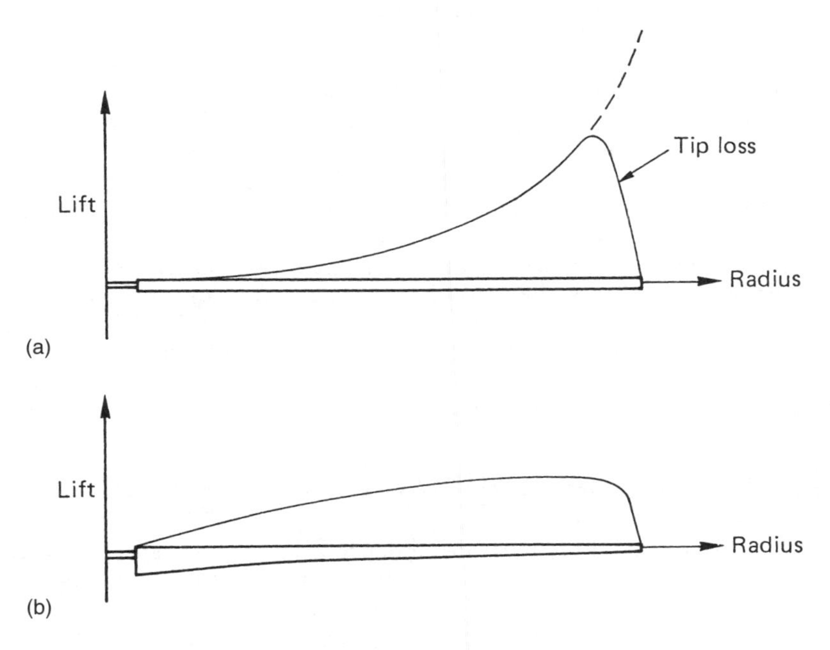

Fig. 3.12 (a) An untwisted, untapered blade has a parabolic lift distribution and far from uniform inflow. (b) Taper and twist improve the uniformity of inflow.

will be found in axial blowers where the root chord may be several times the tip chord. Figure 3.12(b) shows a better lift distribution resulting from a combination of twist and taper. Blade taper has the further advantage that the various vibration modes of the blade will not be harmonically related. This will be considered in section 3.26.

In practical helicopters the blades cannot extend all the way to the mast because of the need to provide a rotor head mechanism. A further consideration is that, in the hover, the extreme roots of the blades simply provide a downwash onto the hull resulting in a download, negating some of the rotor lift. Section 3.20 will show that, in forward flight, the blade root of the retreating blade encounters reverse airflow. As a result blades often have a significant root cut. Ideal twist for hovering will not be ideal for forward flight or autorotation and a compromise is invariably necessary. In the proprotors of a tilt-rotor helicopter more twist will be possible because the rotors work with predominantly axial flow.

3.14 Swirl

In a real rotor, the downwash does not just move downwards, but it also has a rotational component of motion known as swirl. Intuitively it is clear that the turning rotor must turn the air with it to some extent. Figure 3.10 showed that the reaction on a lifting blade is somewhat aft of the rotor shaft axis. It must follow that the direction in which the momentum of the inflow increases is in the opposite direction. The vertical component is the inflow velocity, which produces lift, whereas the horizontal component creates swirl which represents wasted power.

Some of the swirl is due to profile drag and some is due to induced drag. In the absence of profile drag, the blade reaction is still slightly behind the vertical and this must result in a small loss when compared with an ideal actuator that only causes vertical inflow. The main source of swirl is the result of profile drag adding to the rotational momentum of the air. In an ideal actuator the air leaves with increased vertical momentum only. In a real rotor the amount of swirl gives an indication of the figure of merit. The higher this is, the lower the swirl will be. In a twisted blade, swirl will be greatest near the root where the pitch angle is greater.

The torque delivered to the rotor shaft can be divided into two parts, first the torque needed to drive an ideal rotor which only produces an increase in vertical momentum sufficient to create the required thrust, and second an additional torque which is the reaction to creating swirl and tip vortices. With a single rotor, swirl energy is lost forever, but in contra-rotating rotors, the swirl of the second rotor can cancel the swirl of the first. Unfortunately this does not reduce profile drag, it only reduces induced drag.

In the single rotor helicopter, the main effect of swirl is that the downwash on the hull is not vertical, which adds to the general air of asymmetry surrounding the helicopter. The effects will be considered in Chapter 5.

3.15 Vertical autorotation

The direction of the reaction on an airfoil depends upon the angle of attack. When hovering, inflow causes the resultant to be tilted back and opposes the engine thrust. If, at a suitable height the collective pitch is lowered, the machine starts to fall, the inflow reverses and the relative airflow has an upward component. Figure 3.13(a) shows that the principle of the rotor blade in autorotation is no different to powered flight. It continues to accelerate air into a new direction. However, as the air approaches

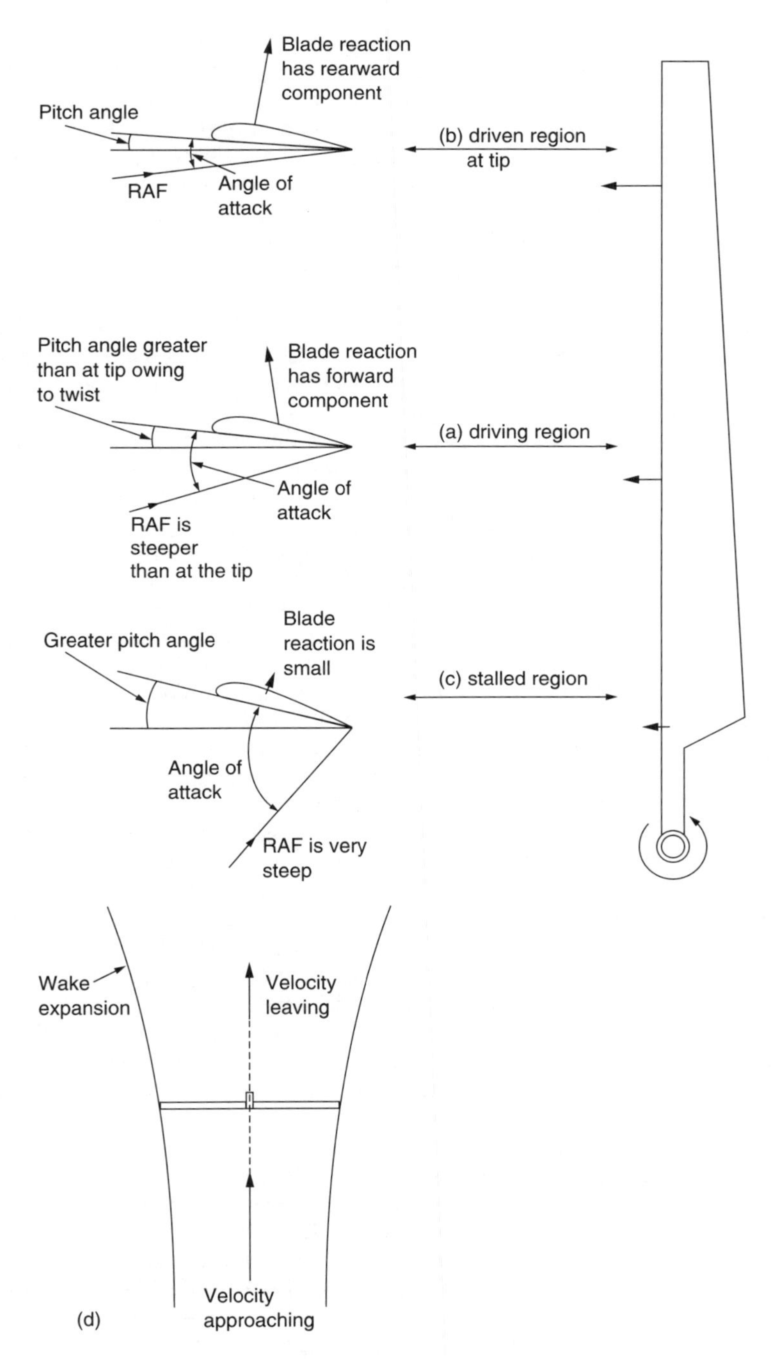

Fig. 3.13 In autorotation, inflow is reversed. Air approaches from below so that the reaction has a forward component. (a) Middle part of blade is driving, taking momentum from the air. These conditions can be obtained with positive blade pitch. (b) Outer part of blade has a lower angle of attack and is being driven as in normal flight. (c) Inner part of blade has high angle of attack and is stalled. (d) The rotor takes momentum from the air and so the wake is slower than the approaching air and expands. This should be compared with Figure 3.11.

upwards, the acceleration has a rearward component. The reaction to this is forward, so the rotor can be driven against profile drag and transmission losses. At a suitable setting of the collective pitch, the overall reaction on the blades is brought to the vertical and so they will continue to turn at the same speed and provide lift. Note that in order to produce lift the blade must have a positive angle of attack with respect to RAF and so it is common for the blade to have positive pitch in autorotation.

The machine is moving steadily downwards using potential energy to overcome profile drag. In fact the rotor is taking momentum from the air so that induced power becomes negative and provides the profile power needed. The strict definition of autorotation is the requirement for zero shaft torque. Where a rotor actually delivers shaft power it is said to be windmilling or in the windmill brake state. A real helicopter in autorotation is technically windmilling because the shaft torque is not zero. In the absence of the engine, some shaft torque is needed to turn the tail rotor and any other important items such as hydraulic pumps.

The speed of a blade element is proportional to radius. A further complication is that most rotor blades have twist to improve the hover figure of merit, but in autorotation the inflow is reversed and the twist is the opposite of what is needed. It can be seen that on moving from tip to hub, the RAF decreases in speed but the angle of attack increases. As a result the central part of the blade, shown at Figure 3.13(a) is being driven by the inflow and is windmilling. Figure 3.13(b) shows that the tips have a small angle of attack and high airspeed. Profile drag dominates and L/D is poor, with the total blade reaction being behind the vertical. This area of the blade is producing lift and induced drag as if it was in the power-on condition. Figure 3.13(c) shows that the innermost parts of the blades are stalled as they have a high angle of attack and a low airspeed. The power needed to drive the tips and the inner parts of the blade comes from the windmilling of the centre parts. Only a small section of each blade at the boundary between the driven and driving regions is technically autorotating with a vertical total blade reaction. Thus what is meant in practice by autorotation is that the entire rotor has zero net torque. The entire rotor can also be considered to have an L/D ratio and in autorotation the slowest rate of descent will be where this is used.

In practice the loss of performance in autorotation due to twisted blades is surprisingly small. One reason for this is that the profile drag is determined only by the rotor speed and is largely unaffected by local angle of attack. Another is that real blades don't have anything like ideal taper and also have root cut-out so the stalled inner part of the rotor doesn't absorb as much power as might be expected.

In autorotation there is still a pressure step across the rotor, but the inflow is now from below. Figure 3.13(d) shows that as the rotor is taking momentum from the air, the slipstream from the rotor is slower than the inflow. The pressure gradient along a streamline is rising and so the wake expands.

In autorotation the best rate of descent will be where the entire rotor (not a blade section) operates at $L/D_{max.}$. The highest RRPM will also be obtained at $L/D_{max.}$, giving the greatest stored energy for landing. Figure 3.14 shows an autorotation diagram in which the vertical axis is drag over lift and the horizontal axis is angle of attack α. It will be seen from the figure that, at an idealized blade radius, the ratio of drag to lift determines the angle φ which must also be the angle of RAF if the blade reaction is to be vertical. If the pitch angle is θ, then $\varphi = \alpha - \theta$. The function $\alpha - \theta$ is shown on the graph to be a straight sloping line. The lowest profile drag will be at the bottom of the drag 'bucket' and the pitch angle to achieve that is where the line of $\alpha - \theta$ intersects the α axis. Most conventional helicopters will be rigged so that the minimum pitch stop on the collective lever is at about this value. Thus operation in autorotation is in the region to the right of the intersection, but not so far as the position of the dotted line where rotor

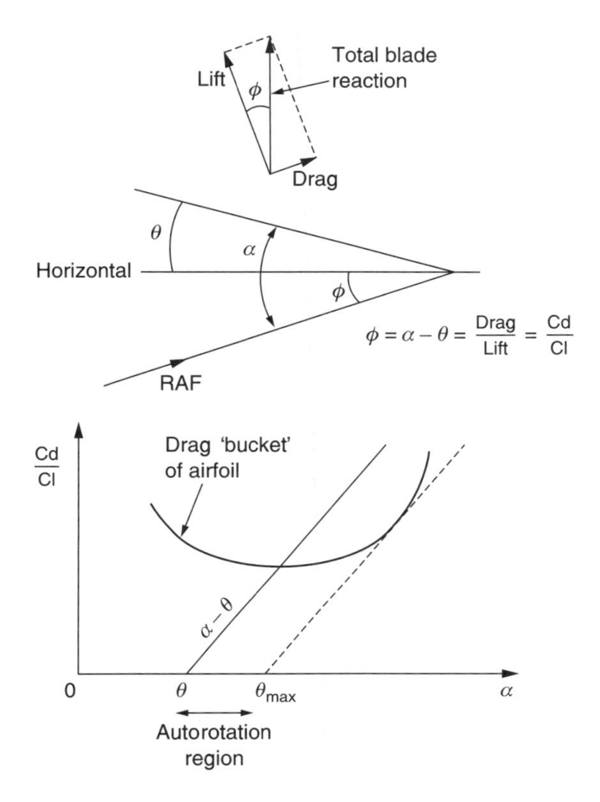

Fig. 3.14 In ideal autorotation, the total blade reaction must be vertical as shown. For a given pitch angle θ, the rotor operates at the intersection of the $\alpha - \theta$ line with the drag bucket. See text.

speed may be irrevocably lost. In this region rotor speed may be changed by alteration of the balance of forces. Reducing the angle of attack causes the blade reaction to move forward from the vertical. Negative induced power exceeds the profile power requirement until the rotor speed has increased. On the other hand, increasing the angle of attack causes the reaction to move behind the vertical, and the rotor speed falls.

3.16 Tip loss and the vortex ring

The pressure differential between the upper and lower surface of the rotor disc causes air to try to flow transversely around the tip in order to equalize the pressure. This rotary flow combines with the blade velocity to leave a corkscrew-like trail shown in Figure 3.15(a) and known as a *vortex*. The loss of pressure differential due to this vortex generation cancels out the lift generated at the blade tip; an effect known as tip loss. Nearly 10% of the blade length is useless. The power needed to overcome induced drag at the tip is wasted generating the vortex. There must also be a vortex at the inboard end of the blade but it is less powerful because the lift gradient is much lower. In the steady hover and in forward flight, the tip vortices are swept away downwards by inflow.

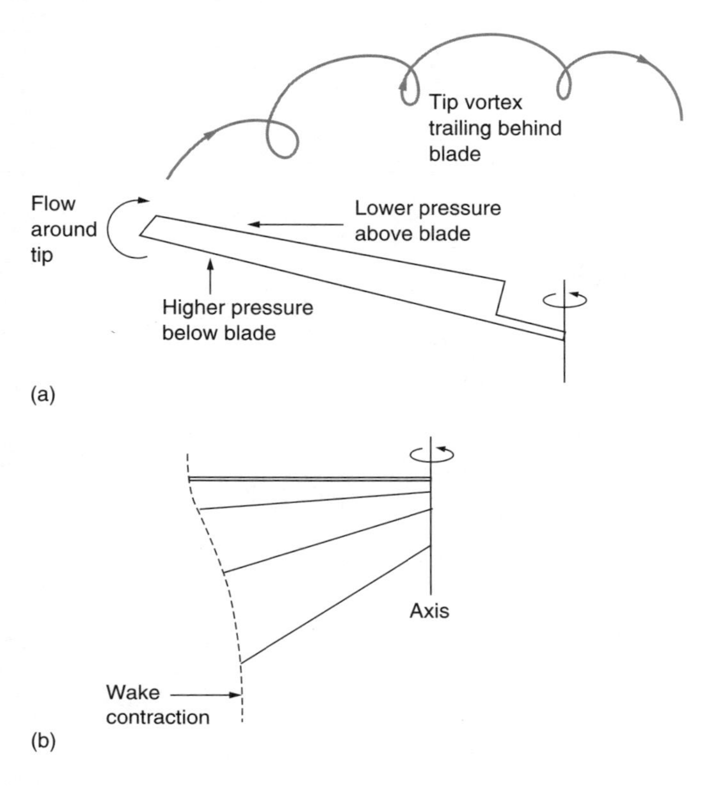

Fig. 3.15 (a) The pressure difference across the blade due to the generation of lift results in a flow from bottom to top around the tip that results in a spiral vortex. (b) In the hover the tip vortices are swept away downwards, but not as fast as the inflow which is generally faster towards the tip.

Tip loss may be reduced by tapering and by twisting the blade near the tip so that the angle of attack is small. The pressure differential is then minimized across the tip. This cannot be taken too far as very small tip chords will operate at poor Reynolds numbers and suffer poor L/D. As a practical matter space will also be needed for tip weights.

It will be seen in Figure 3.15(b) that the result of non-uniform inflow is that the air disturbed by the passage of the blades changes from a disc to a cone and shrinks in diameter due to wake contraction.

In autorotation the vortices are swept away upwards. Thus in climb, hover or autorotation the vortices are always carried off by the slipstream. However, if a vertical powered descent is attempted, the machine descends into its own inflow and the vortices from successive blade sweeps are closer together and begin to reinforce one another. As the vertical rate of descent approaches the induced velocity, the vortices will not be swept away so much, and will begin to augment one another. The airflow will recirculate, in a toroidal motion known as a vortex ring. This is shown in Figure 3.16. The recirculation is increasing the inflow and so more collective pitch is needed to provide the same lift. The blade reaction will tilt back and more torque will be needed.

Paradoxically, a helicopter needs more power to descend in the hover than it does to maintain height. Early helicopters having marginal power used low disc loading

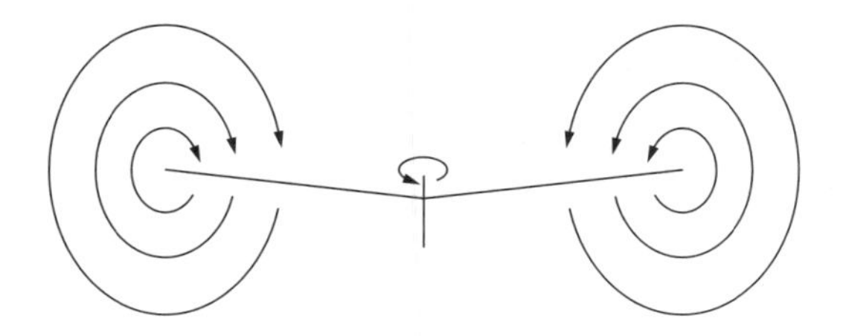

Fig. 3.16 When the helicopter descends vertically, it can catch up with its own wake and the rotor vortices are not swept away, but result in recirculation. If allowed to develop fully, the result is the vortex ring condition.

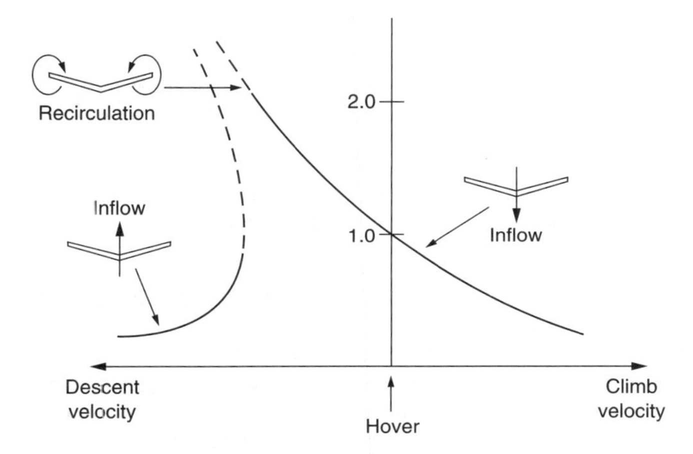

Fig. 3.17 The relationship between induced velocity and vertical climb/descent velocity. Both have been normalized with respect to induced velocity in the hover. Note there is a discontinuity between powered descent and autorotation due to the vortex ring condition.

and so had low induced velocity. Some suffered from power settling where above a certain rate of vertical descent the descent could not be arrested except by moving into forward flight.

Figure 3.17 shows the relationship between the induced velocity needed and the climb/descent velocity. As the effects are proportional to the hover induced velocity, Figure 3.17 has been normalized by dividing both parameters by that factor. As a result the curve passes through unity in the stationary hover. Note that as the rate of descent increases, following the curve to the left, there is a discontinuity where a fully developed vortex ring condition occurs. The recirculation has now enveloped the rotor whose ability to produce lift is considerably reduced. The effectiveness of the cyclic pitch will also be reduced. Increasing collective makes it worse. The machine drops like a stone, pitching and rolling randomly. If this condition is entered close to the ground, recovery may be impossible.

To create lift a rotor must have mass flow whose momentum it can increase. In hover or climb, the inflow is downward and the rotor puts momentum into the inflow by accelerating it. In a vertical autorotation the inflow is upward (effectively making the momentum negative) and the rotor puts momentum into the air by slowing it down. Both of these conditions are stable and the results are predictable from actuator disc

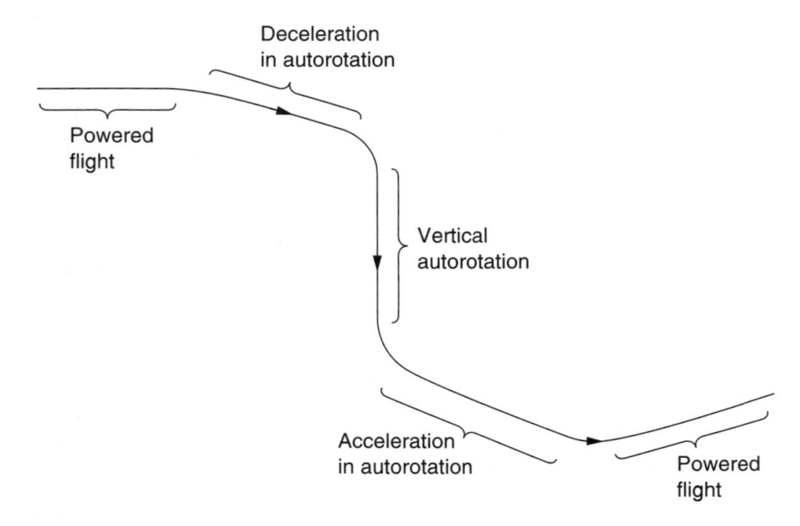

Fig. 3.18 Avoiding the vortex ring. Autorotation is entered in forward flight, and then the forward speed is brought to zero. Prior to resuming powered flight, forward speed must be regained.

theory. The problem is that in vertical flight to get from downward inflow to upward inflow the state of zero inflow must be traversed.

Zero inflow makes nonsense of actuator disc theory, and the real helicopter can't fly in that condition either, hence the discontinuity in the curve of Figure 3.17. The position of the discontinuity is affected by blade twist. A twisted blade produces more thrust at the inner parts of the disc and excites the vortex ring less, so it is possible to descend faster.

It must be accepted that there are certain combinations of vertical rate of descent and induced velocity in which a helicopter will be uncontrollable. Thus an important part of pilot training is to impart a practical knowledge of how to fly outside that region, how to recognize the onset of a vortex ring and how to recover from it.

Using forward speed to guarantee an inflow allows the vortex ring condition to be avoided. Figure 3.18 shows that to enter autorotation from forward flight the collective is simply lowered until the inflow reverses. Once in autorotation the forward speed can be brought to zero for a vertical descent. In order to terminate the vertical autorotation forward speed is regained before raising collective pitch again to obtain normal downward inflow.

3.17 Ground effect

When hovering close to the ground, the air passing downward through the rotor cannot escape as freely as shown in Figure 3.19(a). Air pressure below the machine builds up, and reduces the induced velocity. The RAF is closer to the horizontal and the blade resultant is closer to the vertical. The same lift can be obtained from the rotor with less rotor drag. As Figure 3.19(b) shows, considerably less power is needed to hover close to the ground. As the figure shows, ground effect is noticeable up to a height approximately equal to the rotor radius. In still air, a helicopter will reach equilibrium in ground effect at a height where the lift just balances the weight without any input to the collective pitch

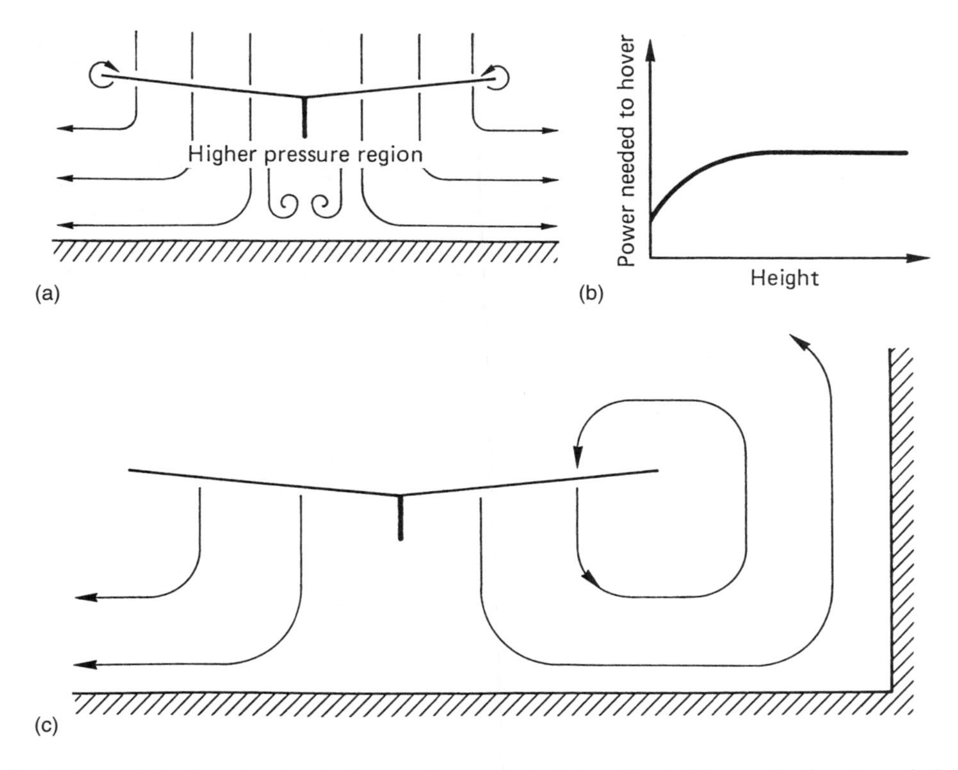

Fig. 3.19 (a) When hovering close to the ground, a pressurized air cushion builds up under the rotor, reducing the induced power needed to hover (b). Hovering close to a building (c) may result in the downwash being returned to the rotor. Recirculation causes loss of lift.

control. If the machine descends, it will obtain more lift from the air cushion; if it climbs, the lift will fall. In many respects the helicopter in ground effect is like a hovercraft.

A helicopter in ground effect over water will displace some of its weight in a saucer-shaped depression in the water surface. This is not often seen because hovering low over water is a recipe for spray in the engines.

The presence of the rotor head and blade root cut-out creates a hole in the middle of the disc through which air can escape upwards in ground effect. This is actually another form of tip loss, which is called the *fountain effect*. In practice the fountain effect can be beneficial because it puts some of the hull into upflow and reduces the hull download. There will be vortices trailing from the inboard end of the blade. As the lift gradient is small here these are relatively weak and their main contribution is that in ground effect they make the hover conditions chaotic especially if they interact with the tail boom or tail rotor.

It is often stated by pilots that hovering over rough ground 'dissipates' the ground effect, when theory would suggest that restricting the ability of the downwash to escape would enhance ground effect. This aerodynamic effect ought to be very small, and the real effect may be psychological. When hovering over a crop, the pilot will estimate his height from the top of the crop, not from ground level. Thus the ground effect may not appear as powerful.

Ground effect normally increases lift so that it offsets the effect of vortex ring generation. The two effects are always opposing one another. However, hovering close to a wall or next to another helicopter can trigger recirculation (Figure 3.19(c)).

3.18 Cyclic control

In order to move around in the hover, the pilot pushes the cyclic stick in the direction he wants to go. This superimposes a cyclic variation of blade pitch on the average or collective setting. The result is that the blades oscillate sinusoidally about the feathering axis as shown in Figure 3.20(a) at the same frequency as the rotor turns. As a result the lift will increase on one side of the rotor and decrease on the other, resulting in a

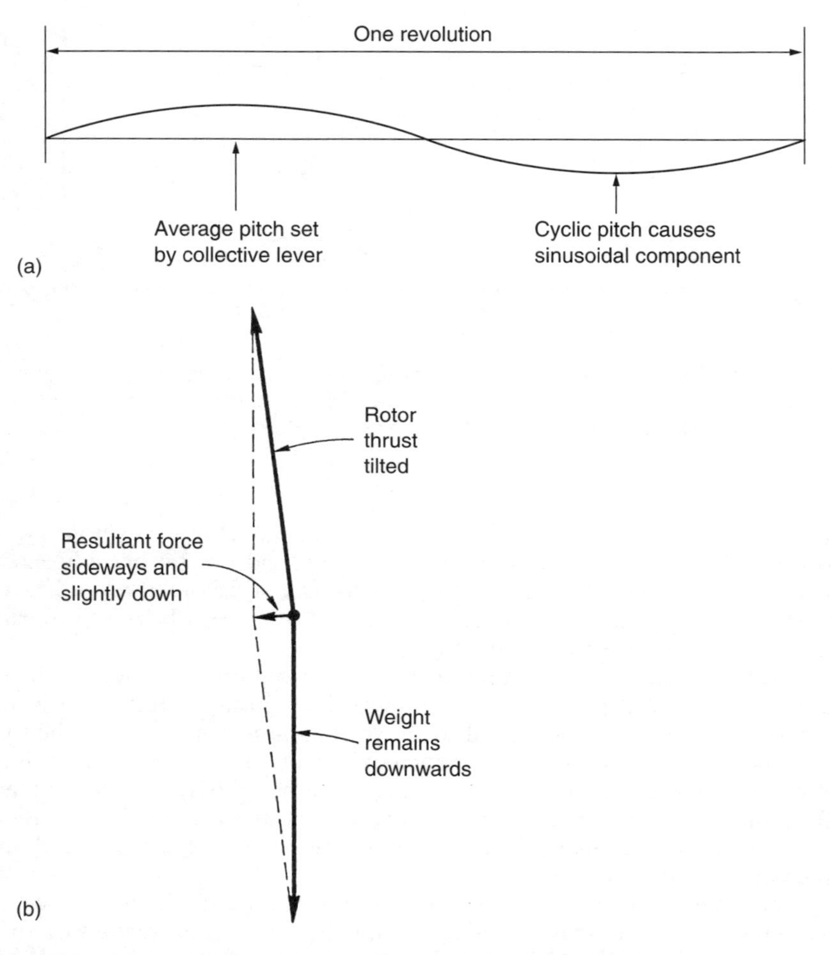

Fig. 3.20 (a) The cyclic control causes the pitch of the blades to change sinusoidally at the rotor frequency. The rotor disc will tilt with a 90° phase lag. (b) Tilting the disc has the effect of reducing the vertical component of rotor thrust. In practice a slight increase in collective is required to maintain height as the cyclic control is used to move off.

rolling couple. As was seen in Chapter 2, the rotor is gyroscopic and so the cyclic pitch change is arranged to occur 90° ahead of the required result to allow for precession in the rotor.

Figure 3.20(b) shows that the application of cyclic control tilts the thrust vector, and that there will be a resultant of the thrust vector and gravity which is primarily horizontal but also has a slightly downward component. This is because the vertical component of the rotor thrust becomes smaller when the disc is tilted. As a result the helicopter will accelerate sideways and lose height. To compensate, the pilot slightly raises the collective lever so that the vertical component of the thrust remains the same.

3.19 Basic manoeuvres

From the hover, the helicopter can be accelerated into translational flight by using cyclic pitch to tilt the disc. The greater the acceleration required, the more extreme will be the disc tilt and the greater the application of collective pitch. Figure 3.21(a) shows a helicopter at maximum horizontal acceleration. If done at low height, the blades can get very near the ground. Figure 3.21(b) shows that at constant speed and height the thrust necessary to balance the drag of the pure helicopter comes from the horizontal component of the inclined rotor thrust. The vertical component of the rotor thrust must balance the weight of the machine. Clearly the rotor thrust must be greater than in the hovering condition. The diagram simplifies the situation in that it assumes the drag of the fuselage acts at the rotor hub, which is not actually the case. In practice the drag acts below the hub to produce a couple which would depress the nose. There are two ways of dealing with that: using a rotor head which can apply an opposing couple from the blades and/or using a tail plane to produce a down thrust.

Note that the tip path plane may be inclined forwards with respect to the hull so that the balance of forces is obtained with the hull level. In this condition the drag will be minimized and the occupants will be most comfortable. A consequence is that when the machine comes to the hover it will sit with a tail down attitude. The length of the undercarriage legs or the angle of the skids is often arranged such that the machine can settle on the ground very nearly in this attitude.

Tilting the disc accelerates the machine along, but bringing the disc back to the horizontal only removes the acceleration, and the machine will continue along, slowing only because of drag. In order to stop, the disc must be tilted the opposite way to obtain acceleration in the opposite sense to reduce the velocity. Figure 3.21(c) shows a helicopter with maximum deceleration in a manoeuvre known as a quickstop. Note that the inflow has reversed so that a quickstop is actually a form of autorotation. Note that the tail boom is very low, and could strike the ground.

Figure 3.21(d) shows a turn at speed. The turn changes the direction but not the speed, so it is a change of velocity or acceleration. The acceleration must be towards the centre of the turn, and it is necessary to obtain a resultant sideways force to cause that acceleration. This is done by banking with lateral movement of the cyclic stick. Collective will need to be increased to maintain the vertical component of thrust equal to the weight. The machine will then fly around as if on the surface of a cone until the desired heading is reached, when the bank will be taken off and the collective reduced again. Note that in a 60° bank the thrust vector has to be twice that needed to hover in order to maintain height. The ratio of manoeuvre thrust to hover thrust is called the load factor. Not all machines can deliver that much power, and not all transmissions can accept it. However, some large transport helicopters can carry their

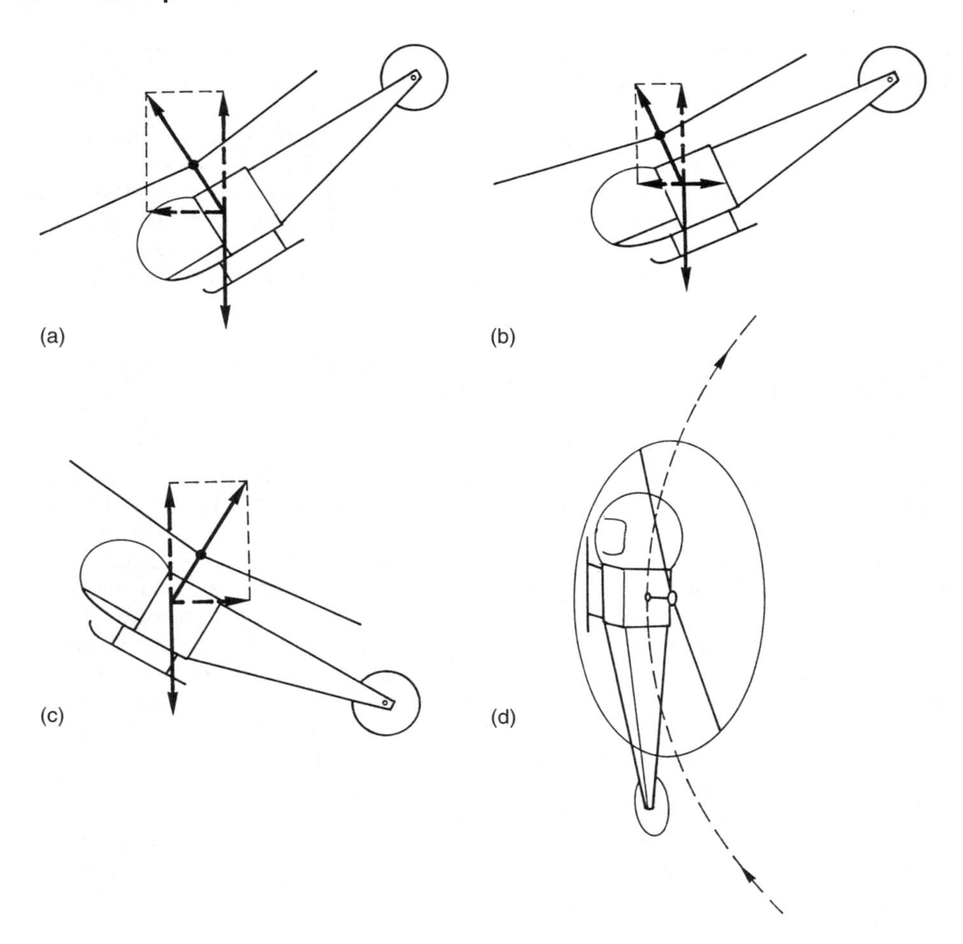

Fig. 3.21 (a) The forces involved during maximum acceleration. Note the front of the disc can be very close to the ground. (b) The balance of forces in forward flight. The forward component of rotor thrust balances hull drag and the vertical component balances the weight. (c) The quickstop is a horizontal autorotation. Note the low position of the tail rotor. (d) A turn at speed is limited by the increased thrust necessary to maintain height whilst accelerating the machine towards the centre of the turn.

own weight in cargo. When unladen, their load factor effectively doubles and they can safely accomplish some quite alarming manoeuvres.

Note that during acceleration, deceleration and a correctly banked turn gravity still appears to act vertically down through the cockpit floor, and it is possible to complete an entire flight without spilling a mug of coffee placed on the floor.

3.20 In translational flight

In translational flight, the forward velocity increases the velocity of the air passing the advancing blade and reduces that seen by the retreating blade. Figure 3.22 shows what happens. In (a) it can be seen that the velocity of a blade element due to rotation

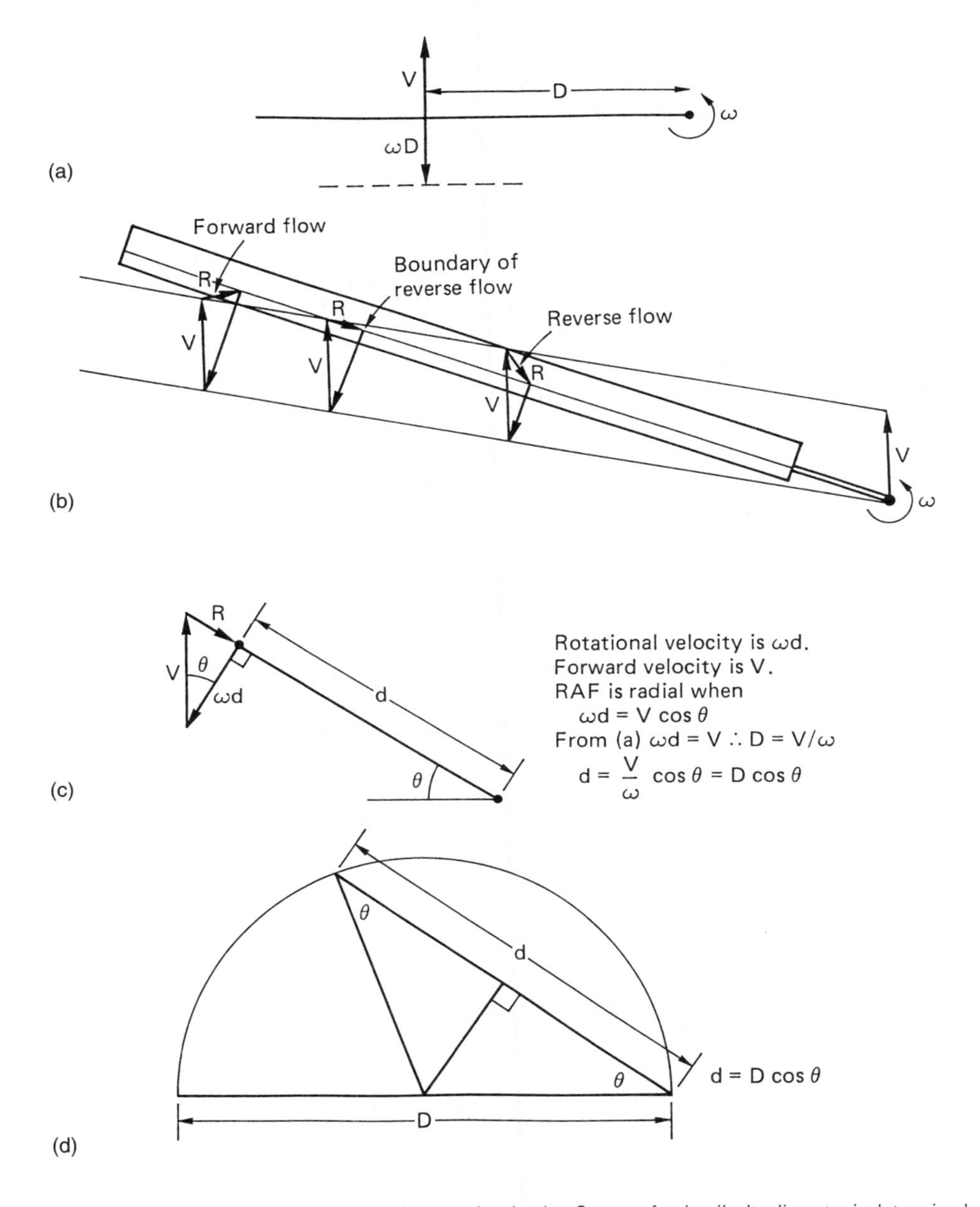

Fig. 3.22 The reverse flow region can be shown to be circular. See text for details. Its diameter is determined by the ratio of the forward speed to the tip speed, which is the advance ratio μ.

is given by the distance D from the shaft multiplied by the angular speed ω. When the blade is at 270°, the edge of the reverse flow region will be where ωD equals the airspeed V. In (b) the method for finding what happens at an arbitrary blade angle is shown. At any distance d from the shaft, the vector resultant R of ωd and V can be found. Figure 3.22(c) shows the derivation of the expression for d, the distance from the shaft to the boundary of the reverse flow region. Figure 3.22(d) shows the expression for d graphically and it will be seen that the reverse flow area is circular and that its diameter increases with speed. Instantaneously the rotor is turning with respect to the outer edge of the reverse flow region. The ratio of forward speed to tip speed is known as the *advance ratio* μ which is the same as the ratio of the diameter of the reverse flow region to the blade radius.

The retreating blade suffers reverse flow near the root; a small downward force may be experienced because the retreating blade effectively has a negative angle of attack due to flow reversal. Here is another reason for root cut-out, to reduce the amount of useless reverse-flowed blade. However, as the area is proportional to the square of the diameter, the amount of disc area lost to the reverse flow region is small. For example, at $\mu = 0.4$, the diameter of the reverse flow region is 0.2 of the rotor diameter but the area lost is only 4%.

Owing to the huge difference between the airspeeds, if nothing were done, the advancing blade would produce more lift and the retreating blade less. This would apply to the rotor a roll couple toward the retreating side. The gyroscopic action of the rotor would result in a response delayed by 90° of rotation corresponding to a rearward pitch effect. This rearward couple is colloquially known as 'flapback' and it tends to reduce the forward tilt of the machine and consequently reduces the forward component of the rotor thrust, slowing the machine down.

Thus sustained forward flight requires a continuous application of forward cyclic control that reduces the angle of attack of the advancing blade and increases that of the retreating blade in order to balance the lift moments on the two sides of the rotor. In cruise the amount of cyclic pitch applied may be of the order of 5°. Flapback results in speed stability because it automatically opposes the cyclic control.

The more forward cyclic is applied, the higher will be the forward speed. Thus in translational flight the fore-and-aft position of the cyclic stick primarily controls the airspeed. If the airspeed is changed, the collective lever will need to be adjusted to maintain height because tilting the rotor thrust vector alters its vertical component.

In forward flight the rotor has access to a greater mass of air. Thus the same momentum increase and the same thrust can be obtained with a smaller induced velocity and this reduces the power needed. Figure 3.23 shows that there is a minimum induced power speed where the inflow velocity is the least.

3.21 Inflow and coning roll

In flight, there is a region of low pressure above the rotor and a region of high pressure below. This pressure difference results in a tendency for air in the plane of the rotor to move upwards. Figure 3.24(a) shows that this happens for some distance from the edge of the disc. Simple inertia keeps the air moving upwards so that in translational flight, the leading edge of the disc encounters upwash. The upwash ceases when the downward impulse of the rotor cancels the upward momentum of the air mass.

In slow forward flight, Figure 3.24(b), the low pressure above the rotor has time to bend the inflow increasingly downwards as it reaches the trailing edge of the disc. The disc can be thought of as a low aspect ratio wing producing tip vortices which roll inward at the top and down through the rear of the disc. The result is a reduction of the angle of attack of the blade at the rear of the disc compared to that at the front. The loss of lift at the rear of the disc is subject to the phase lag, or precession, of the rotor and manifests itself as a roll to the advancing side, which is known as inflow roll.

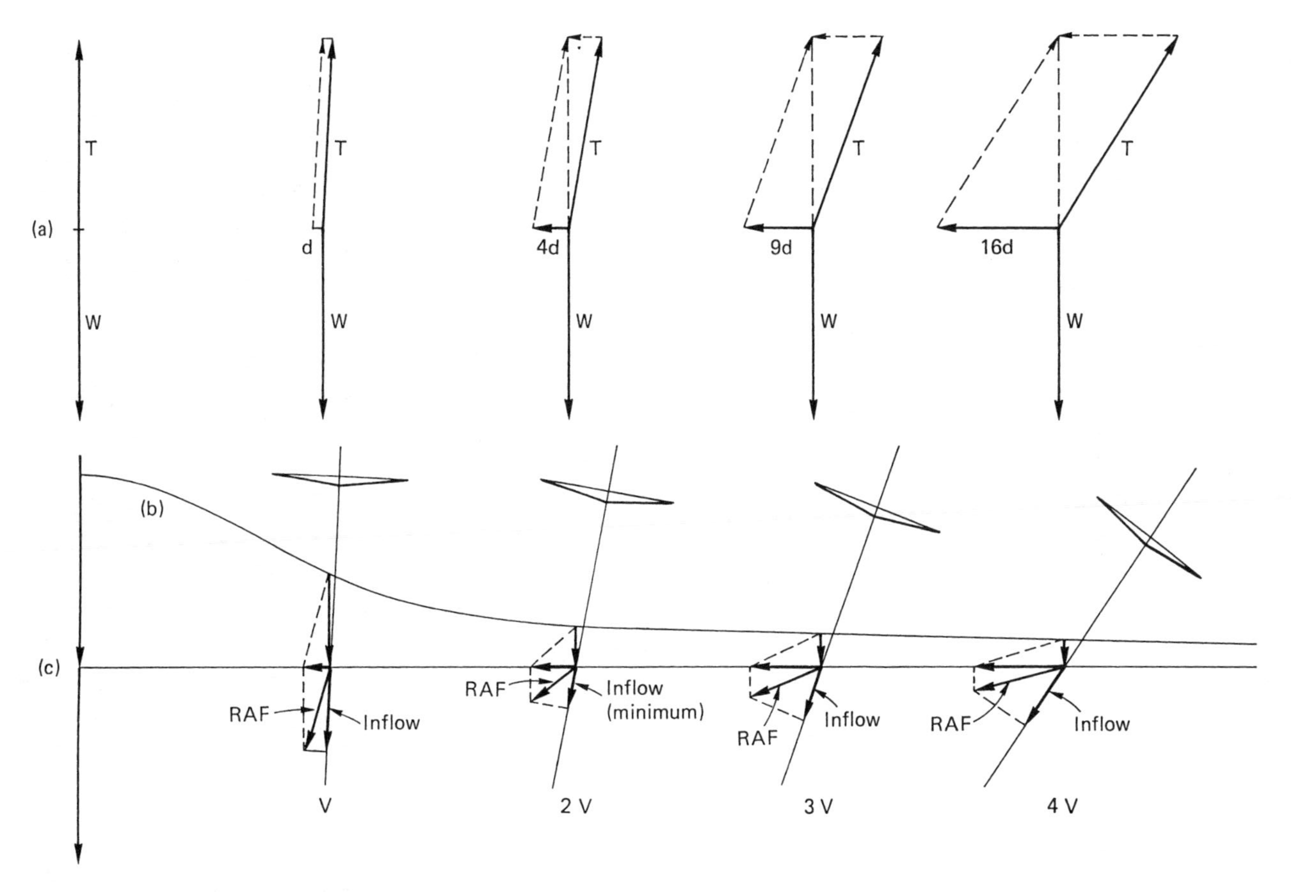

Fig. 3.23 Conditions for hover and four different airspeeds. In (a) note the drag d increasing as the square of the speed and requiring an increasingly forward tilt of the rotor thrust T. Curve (b) shows the vertical component of inflow. In conjunction with the airspeed this determines the direction and magnitude of the RAF (c). The actual inflow is the component of RAF along the rotor shaft. Note there will be a speed at which the minimum power is required to maintain height.

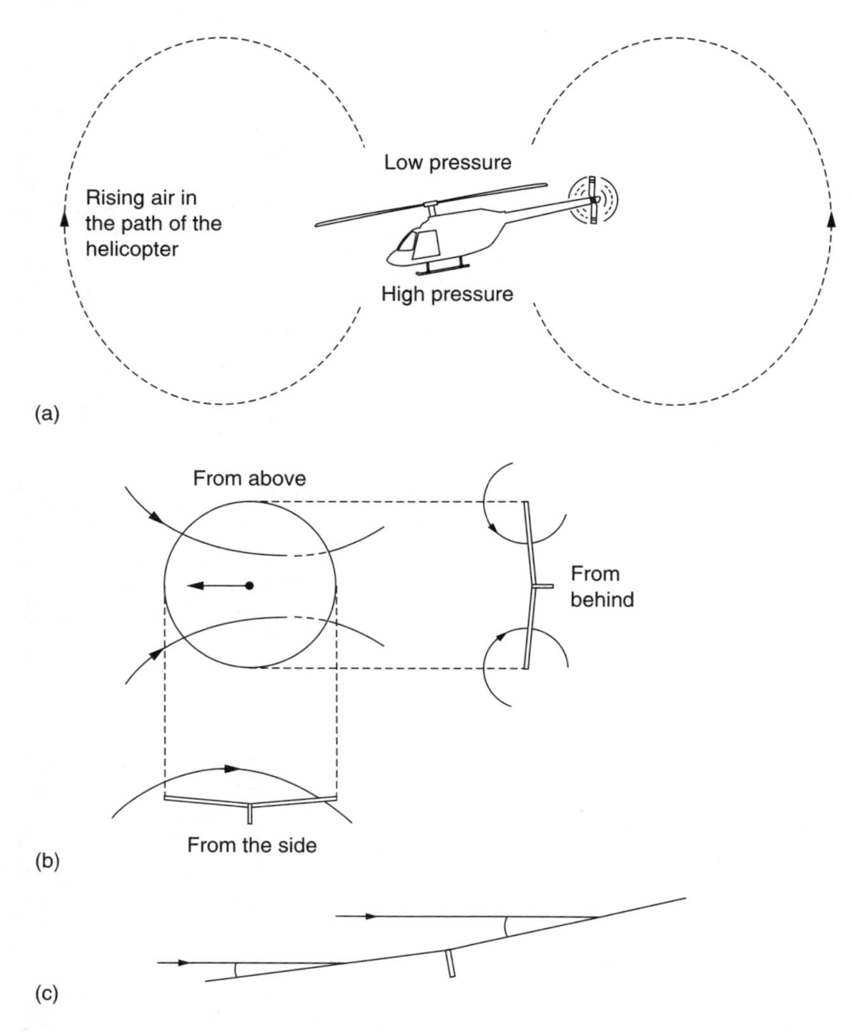

Fig. 3.24 (a) Given the pressure difference across a rotor in the hover, surrounding air will tend to move upwards. In translational flight the edge of the rotor may actually encounter upwash. (b) In forward flight the rotor acts like a low aspect ratio wing. Vortices curl over the tips and down through the rear of the disc, reducing the angle of attack there. Loss of lift at the rear precesses into inflow roll. (c) As the rotor blades are coned, in forward flight the air approaches at a different angle as shown here. The component perpendicular to the blade is greater for blades at the rear, where the angle of attack is reduced.

Whilst coning causes no undue aerodynamic problems to a hovering helicopter, it has an unwanted effect in forward flight. Figure 3.24(c) shows that the effect of coning is to reduce the angle of attack of the blade passing across the tail compared with that of the blade passing across the nose. As for the inflow roll, this results in a tail-down moment applied to the disc, which likewise manifests itself as a roll towards the advancing side because of the phase lag of the rotor. The two effects take place in parallel, but as different functions of airspeed. Inflow roll commences on leaving the hover but reduces significantly as speed increases, whereas coning roll slightly increases

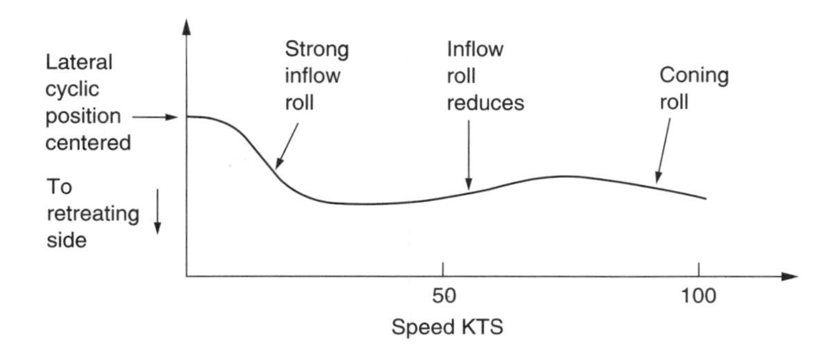

Fig. 3.25 How the cyclic stick might be trimmed at different airspeeds. Inflow roll begins at low airspeed but has less effect as speed rises, whereas coning roll increases with speed as the coning angle may increase to produce more thrust.

with airspeed. Figure 3.25 shows how the lateral cyclic control might have to be moved to the retreating side in order to fly straight at different airspeeds.

Thus in straight flight at constant speed, the cyclic stick must be held slightly towards the retreating side to counteract the inflow and coning induced rolls as well as forwards to counter the advancing/retreating induced pitch-up. In some machines the phasing of the controls is arranged so that forward cyclic automatically creates a degree of lateral cyclic. However, in the hover this results in the machine not responding precisely in the direction the stick is pushed.

Like the collective control, there is no one correct setting for the cyclic control. The cyclic trim control is used to shift the neutral position of the stick, relieving the pilot of the need to produce continuous control forces.

3.22 Rotor H-force

The retreating blade has the airspeed subtracted from the rotational speed and so has to operate with an increased angle of attack. The profile drag will be reduced but the induced drag will increase. Conversely the advancing blade will suffer significantly more profile drag because of the high relative airspeed, but the induced drag will fall because the angle of attack has been reduced. The overall effect of profile drag and induced drag does not balance between the advancing and retreating sides and the resultant is a rearward acting force called the H-force. Figure 3.26 shows that the H-force is the reason why the rotor thrust is not precisely at right angles to the tip path plane. A typical figure is 1°.

Because of inflow and coning, the blades crossing the nose produce more lift than the blades crossing the tail and so the induced drag is not the same. The resultant is the Y-force which acts at right angles to the direction of flight. The H-force is small and the Y-force is very small with the result that the effect on the magnitude of the thrust is insignificant.

3.23 Blade stall and compressibility

Stall occurs when air passing over the cambered upper surface of a blade can no longer accelerate rapidly enough to follow the surface. Airflow breaks away and lift is lost

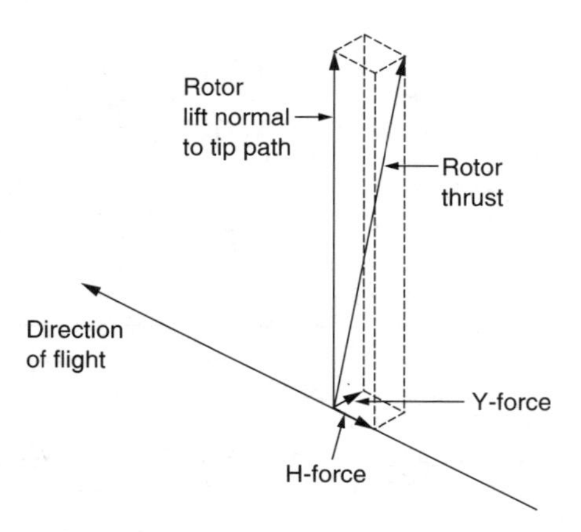

Fig. 3.26 H-force and Y-force. Conditions experienced by the blades in forward flight are asymmetrical and this reflects as cyclic differences in blade drag that produce a force in the plane of the rotor. The rotor thrust is not precisely at right angles to the tip path plane because of this force. H- and Y-forces are the components of this force in the fore-and-aft and transverse directions. The effect is very small and not really noticeable to the pilot.

accompanied by an increase in drag. As acceleration is the limiting factor, at low relative velocities air can negotiate the blade at a larger angle of attack than at high velocity. Figure 3.27(a) shows the stall limit diagram for a typical section. As the speed of sound is approached, the allowable angle of attack becomes very small. Note that the speed of sound falls with falling temperature and in very low temperature conditions helicopters with high tip speed will suffer a performance loss.

In fast translational flight, the tips of the rotor blades will be encountering relative airflow that alternately adds to and subtracts from the rotational speed. The pitch angle will change sinusoidally owing to the use of cyclic feathering but the angle of attack follows a more complex function owing to the effects of coning and inflow. Figure 3.27(b) shows the angle of attack of a part of the blade near the tip plotted against the relative airflow velocity. It will be seen that the plot is an elongated figure of eight. At positions A and C the blade movement is transverse, whereas at B and D the forward speed is subtracted from or added to the rotational velocity.

When the plot is superimposed on the stall limits of the airfoil section, it becomes possible to predict where blade stall will occur. This has been done in Figure 3.28. Figure 3.28(a) shows the conditions near the maximum airspeed. (b) shows that with a heavily loaded machine, stall can occur on the retreating blade where it has very low relative speed. The retreating blade needs to be feathered to a large angle of attack to resist lift asymmetry and inflow roll. The peak angle of attack will be reached at about 285° and the length of stalled blade will be greatest here. The loss of lift around 285° is converted by rotor phase lag to a pitch-up and roll. An attempt to correct the roll with cyclic will accentuate the stall and loss of control will result. Figure 3.28(c) shows that with a lightly loaded machine in straight and level flight or in a dive, advancing blade compressibility stall will occur first at high forward speed.

Partial blade stall can also be provoked at moderate speeds by tightly banked turns and if this is detected, the severity of the manoeuvre should be reduced immediately. Blade stall of either type causes serious vibration and control difficulty before the loss of lift becomes significant. Serious alternating stresses are set up and can cause blade delamination.

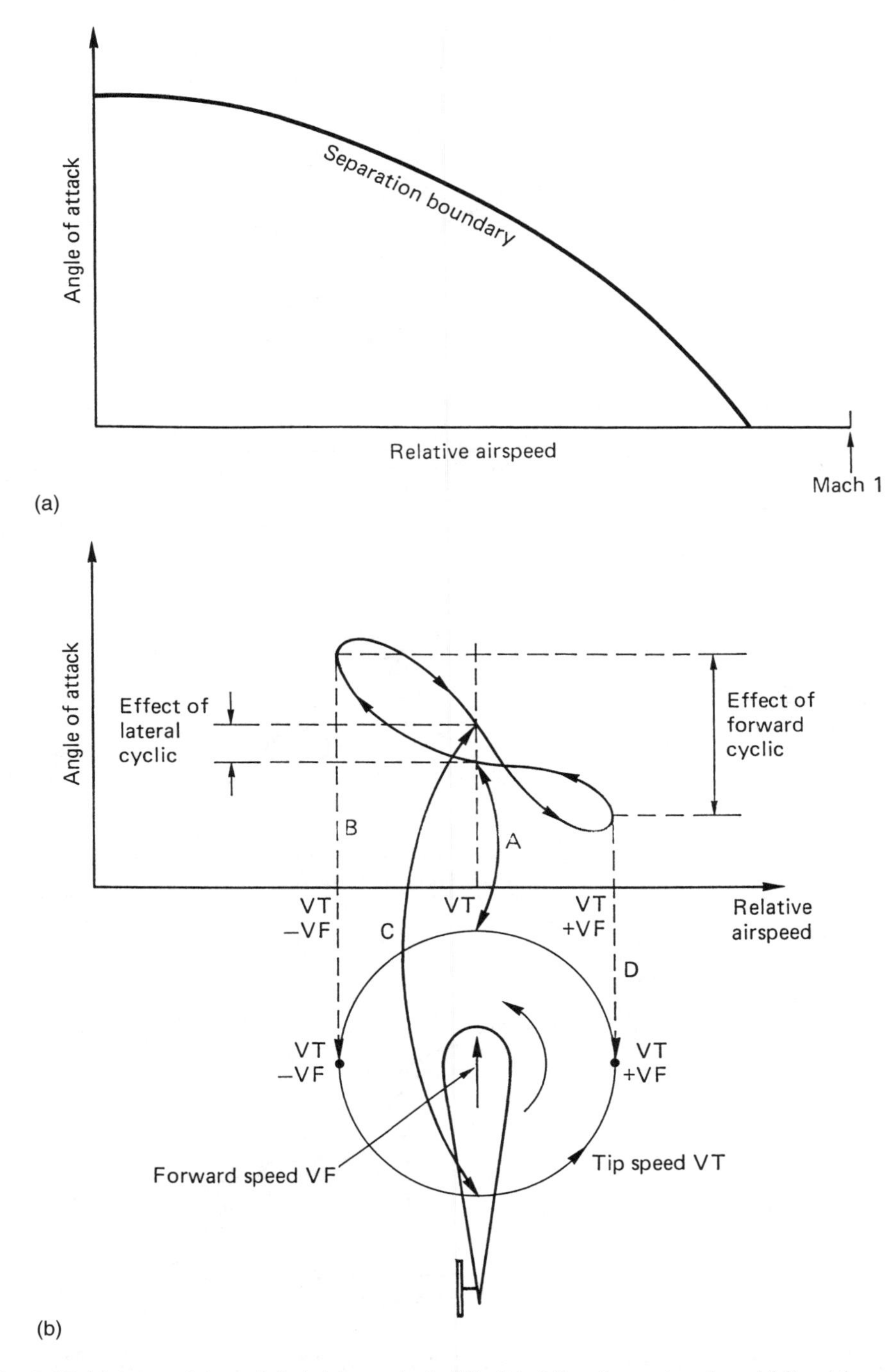

Fig. 3.27 (a) As speed rises, air finds it increasingly difficult to follow the camber of an airfoil and the result is that the angle of attack must be limited as a function of speed to prevent stall. (b) The angle of attack of a blade may vary in the way shown here. The sinusoidal component due to cyclic pitch is modified due to variations in inflow over the disc.

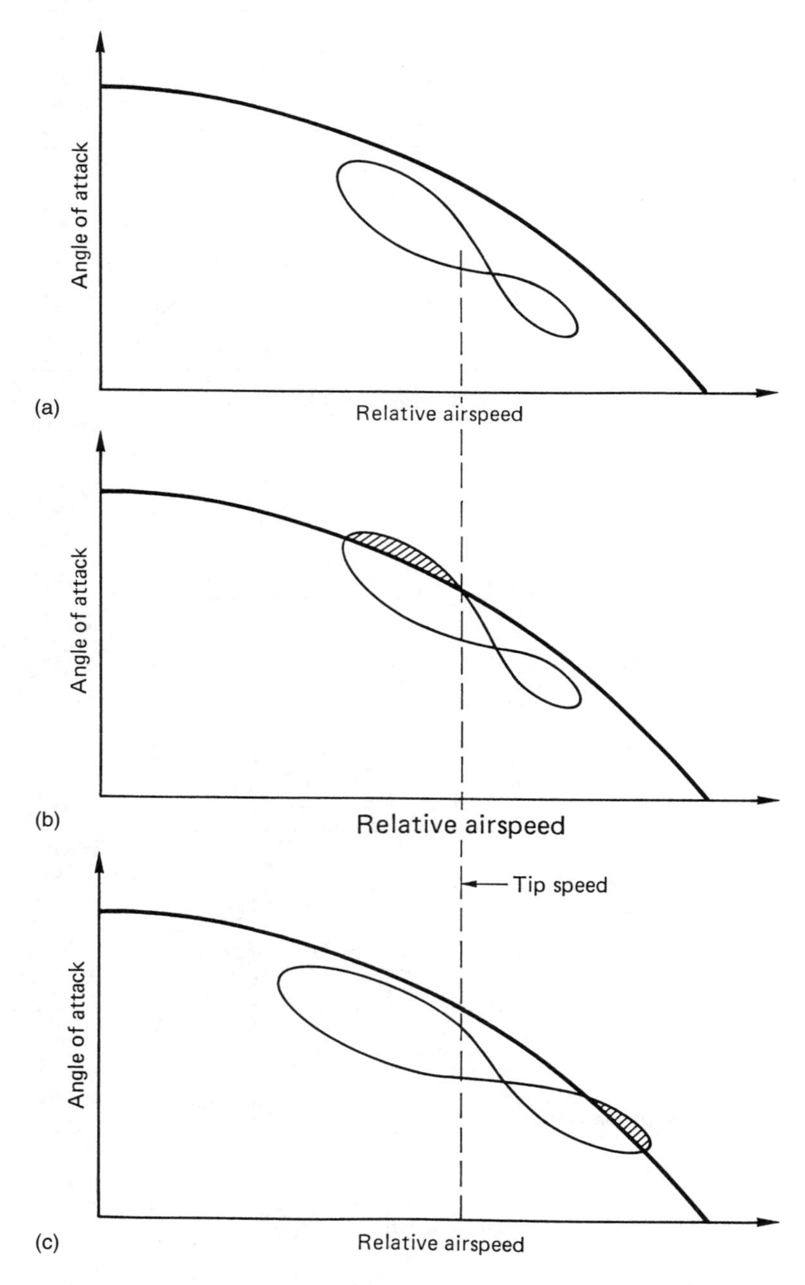

Fig. 3.28 (a) No separation. (b) Retreating blade stall due to high rotor thrust. (c) Advancing blade compressibility at high airspeed.

The never exceed speed for the machine, V_{ne}, will be set at a value which avoids advancing blade compressibility stall.

3.24 The speed limit

The pure helicopter is forever trapped in a forward flight region where it can retain control and contain blade forces. Figure 3.29 shows the constraints. The airspeed is added to the tip speed on the advancing blade, and this must be kept below about 0.92 of the speed of sound to avoid excessive noise and blade forces. On the other hand the airspeed is subtracted from the tip speed on the retreating blade and the advance ratio has to be kept below about 0.5 to avoid a large reverse flow area and retreating blade stall. As a result the pure helicopter with ideal tip speed is unable to exceed about 200 knots at the extreme right of the envelope of Figure 3.29.

If outright speed is not a priority, the possible range of tip speeds is greater. The upper limit is set by noise and the lower limit by the requirement to store enough kinetic energy to handle an engine failure. If a rotor has to provide thrust as well as lift it will be tilted well forward at high speed and the hull will be creating a lot of drag because of its nose down attitude. Higher speeds can be reached if an auxiliary form of forward thrust is available because the rotor thrust can then be vertical and minimized and the hull attitude will be better.

Westland modified the turbine exhausts to produce thrust on the speed record-breaking Lynx, and the Lockheed Cheyenne had a second tail rotor facing rearwards

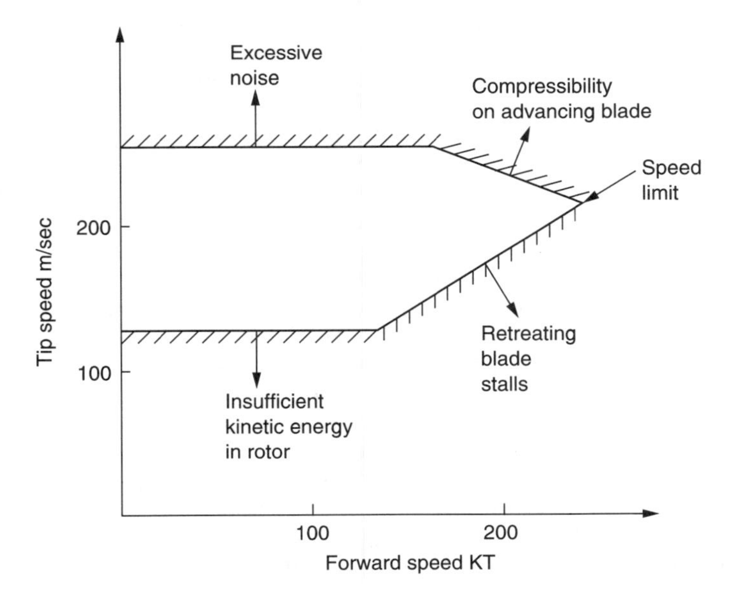

Fig. 3.29 Pure helicopter speed is forever limited by the coincidence of advancing blade compressibility and retreating blade stall at the right of the diagram. Only compound helicopters can break this limit using auxiliary thrust and/or wings to offload the rotor. At lower speeds tip speed is constrained by noise and stored energy limits.

to provide thrust for forward flight as well as a fixed wing to unload the rotor at high airspeeds. These ideas will be considered in Chapter 7.

3.25 Harmonic blade motion

So far the disc has been discussed as an entity since the disc attitude determines the path of the machine. Now the motion of the blades within the disc will be explored.

In trimmed translational flight the application of cyclic feathering opposes the lift asymmetry. Cyclic feathering changes the blade pitch angle sinusoidally. If the lift asymmetry were also sinusoidal, the two effects would be in constant balance. Unfortunately this is not the case. Sinusoidal cyclic feathering is not strictly what is needed, but for practical reasons it is the most widely used solution. Cyclic feathering can only make the *average* lift moment the same on both sides of the disc. It cannot keep the lift of an individual blade constant at all angles of rotation.

Figure 3.30 shows what happens to a blade element as it rotates in forward flight with cyclic pitch applied. The diagram assumes uniform inflow across the disc and neglects coning for simplicity. Figure 3.30(a) shows the relative airspeed experienced by a blade element. The rotational speed produces a constant component, but the forward speed appears as a sinusoidal component to the revolving blade, adding speed to the advancing blade and subtracting it from the retreating blade. The lift generated by a blade is proportional to the square of the airspeed and this parameter is shown in Figure 3.30(b). Note that the squaring process makes the function at 90° very much greater than at 270°.

The coefficient of lift of the blade element is nearly proportional to the angle of attack. The collective pitch setting results in a constant component of the angle of attack and the cyclic pitch control causes this to become a shifted or offset sinusoid as shown in Figure 3.30(c). Figure 3.30(d) shows the lift function which is the angle of attack multiplied by the square of the airspeed. In other words waveform (d) is the product of (b) and (c). The amplitude of the cyclic control in (c) has been chosen to make the lift at 90° the same as the lift at 270°. Note that although the lift on the two sides of the rotor has been made the same by the application of cyclic feathering, the lift is not constant.

Unfortunately a sinusoidal function cannot cancel a sine-squared function. There is a lift trough at about 270° where the square of the relative airspeed has become very low and increasing the angle of attack fails completely to compensate. Lift symmetry is only obtained because the same cyclic input also strongly reduces the angle of attack at 90°. As a result there will be lift troughs at 90° and 270°.

The lift function contains a Fourier series of harmonics, or frequencies at integer multiples of the fundamental, which is the rotor speed. As the lift troughs on the two sides of the rotor are not the same shape, significant levels of odd harmonics will exist, especially the third and fifth harmonics. The lift function excites the blades in the flapping direction and they will respond according to the dynamic characteristics of the blade and its damping. The result is that the blades do not describe a perfect coning motion in forward flight, but as shown in Figure 3.31, they weave in and out of the cone due to the harmonics. Figure 3.31(a) shows second harmonic flapping, completing two cycles per revolution, whereas Figure 3.31(b) shows third harmonic flapping. The flapping action directly modulates the axial thrust delivered to the mast causing a hull motion variously described as ‘hopping’ or ‘plunging’.

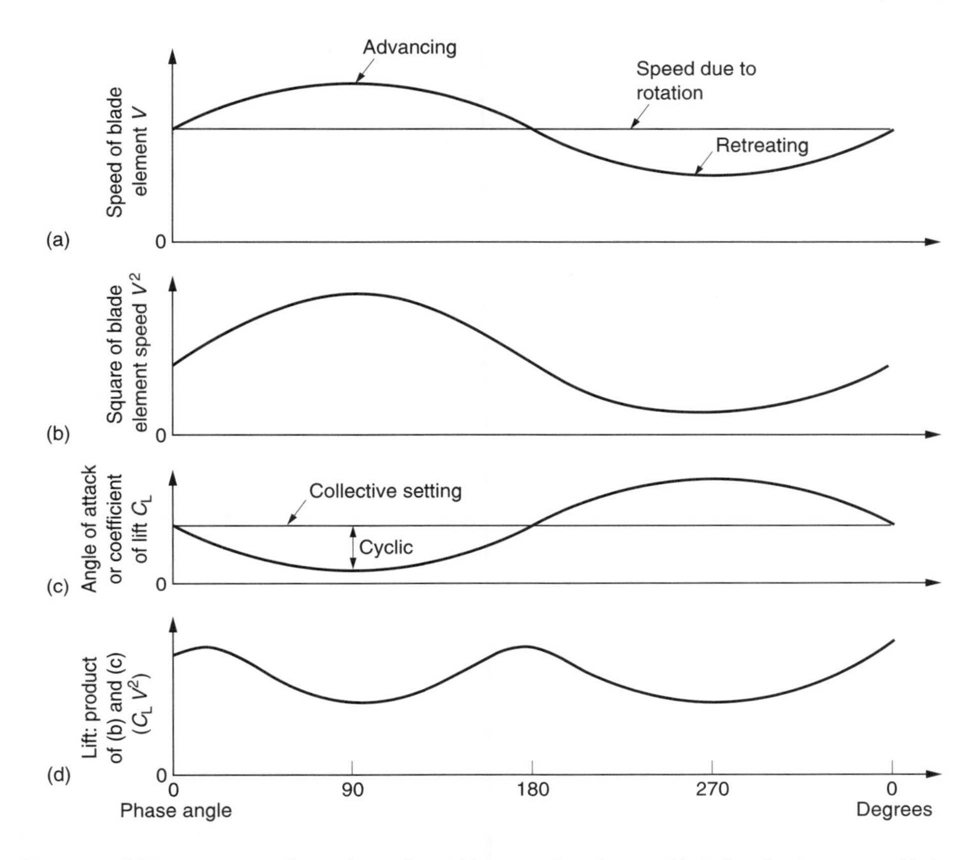

Fig. 3.30 (a) Instantaneous airspeed experienced by a certain point on a blade is a fixed average added to a sinusoidal variation. (b) Lift is proportional to airspeed squared. (c) Sinusoidal control of angle of attack due to application of forward cyclic. (d) Non-constant lift around the disc is a source of vibration.

3.26 Sources of vibration

Whilst vibration in the hover can be minimized by attention to dynamic balance and tracking, vibration in translational flight is inevitable because of the varying conditions caused by the alternate advancing and retreating of each blade. This causes harmonic flapping and dragging of the blades, resulting in forcing functions at the hub which may be axial, i.e. along the mast, lateral, i.e. tending to cause rocking, or torsional, i.e. tending to twist the mast.

On the retreating blade, the lift on the blade is concentrated near the tip with download near the root. This is because of the reverse flow area shown in Figure 3.22. As the same blade moves over the tail boom and into the advancing position, the lift distribution changes so that the lift is now generated more inboard. If the blade is twisted, the angle of attack at the tip may be negative at high forward speeds resulting in a download. The situation is now exactly the reverse of the retreating case. Clearly the blade is being excited by time varying forces and as it is not rigid it will flex.

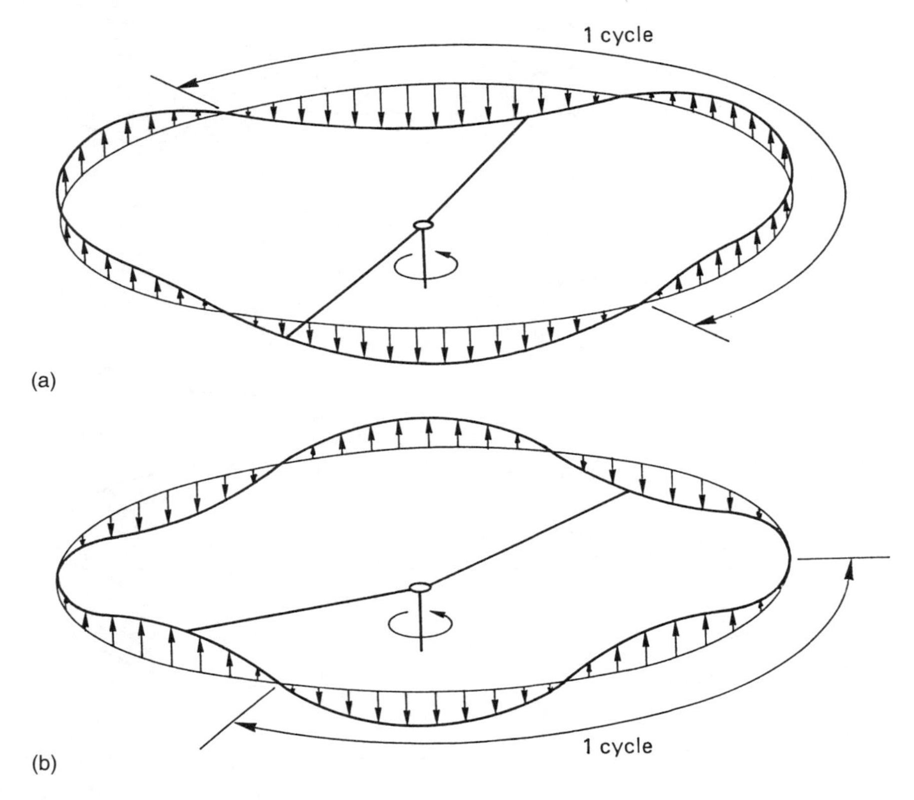

Fig. 3.31 Lift on a given blade is not constant, but a function of azimuth angle. This results in harmonic blade flapping. (a) shows second harmonic flapping. (b) shows third harmonic flapping.

Centrifugal stiffening keeps the blades taut giving them certain of the characteristics of the string of a musical instrument, except that the tension is not constant but is a function of radius. The response to excitation by a Fourier series is shown in Figure 3.32. A variety of harmonic structures will occur. At low frequencies the blade motion can be considered a form of flapping. At some higher frequency it will have to be considered a vibration. Note that blades carried on flapping bearings (a) will have a different harmonic response to that of hingeless blades. Teetering rotors and hingeless rotors have a different harmonic structure (b) from articulated rotors because the blades are fixed together in the centre of the rotor.

When a blade flexes, the CM of the blade elements must be closer to the rotor axis than when the blade is straight. Thus the blade mass is being hauled in and out against the rotational forces, modulating the blade tension. As the root tension is typically measured in tons, it should be clear that these lateral forces are considerable.

In the presence of harmonic flapping, conservation of blade momentum suggests that there will also be harmonic dragging. The azimuthal variation in lift caused by the application of the cyclic control will result in variations in induced drag that excite the blade in the dragging plane. This will be compounded by the azimuthal variation in profile drag. The blade will respond to these dragging excitations as a function of its dynamics and damping. When the blade drags, the effective moment-arm at which the blade outward pull is applied to the hub changes. Thus even if the blade pull were

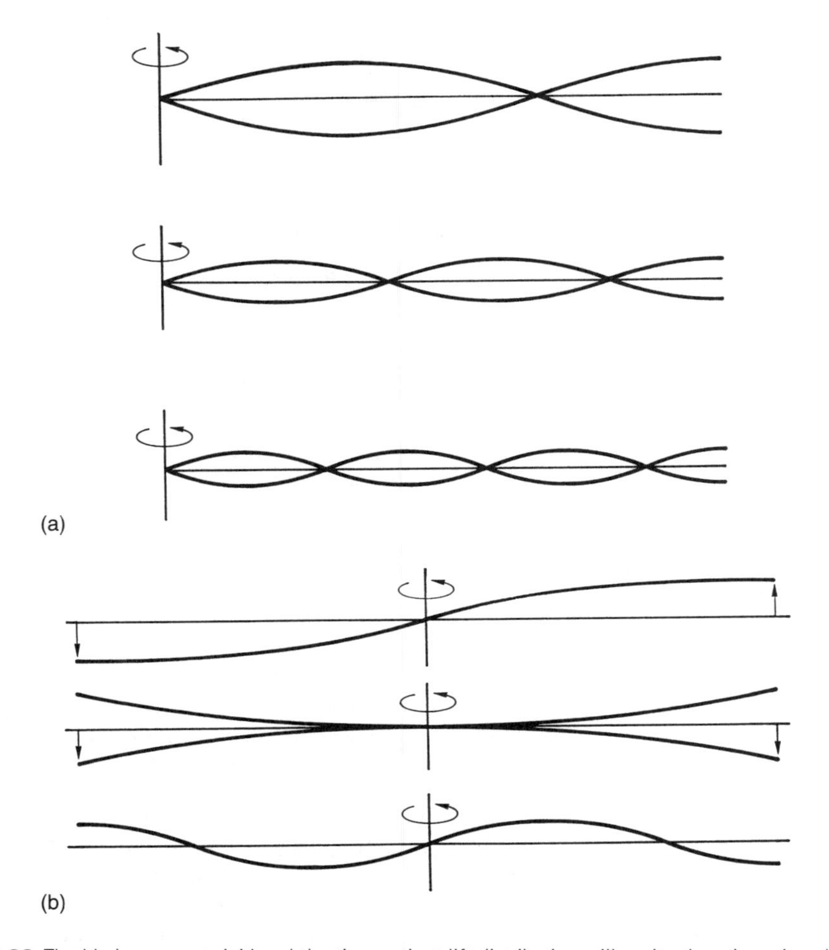

Fig. 3.32 The blades are not rigid and the time variant lift distribution will excite them in various bending modes. (a) Modes for an articulated rotor in which the blades are hinged at the root. (b) Modes for a teetering or stiff hingeless rotor in which moments can pass from one blade to the other.

constant, the moment applied to the mast would be modulated. However, the blade pull is not constant because of blade flexing.

The application of cyclic pitch and the variations in airspeed seen by a blade segment will result in chord-wise movement of the centre of pressure which will excite the blade in torsion, placing alternating stresses on the pitch control mechanism.

A further complexity is that the finite number of blades in a real rotor cannot create a uniform downwash. The downwash velocity changes at the blade passing frequency. Figure 3.33 shows that in forward flight a given blade element traces out a path known as a cycloid. This path will take the blade element across its own path and those of a number of other blades depending on the forward speed and the number of blades. As the blade element crosses a previous path, the higher downwash velocity will reduce the angle of attack, whereas between the previous paths the angle of attack will increase. The result is that the lift of a particular blade element is modulated by the wakes of previous blades. Over most of the rotation the effect is generally not serious because

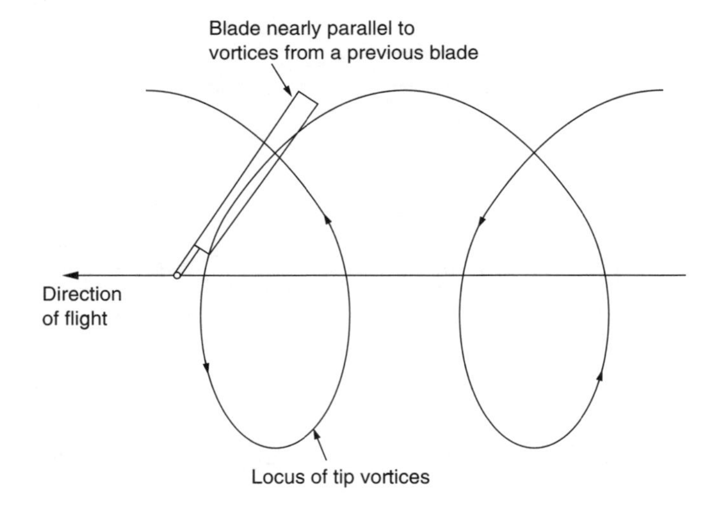

Fig. 3.33 A rotor is not an ideal actuator disc. The downwash velocity is not constant but increases as each blade passes and falls in between. A fixed point on a blade will cross its own path and the paths of other blades in forward flight.

the point on the blade at which the lift dips is constantly changing in radius. However, it can be seen from Figure 3.33 that at about 60° of azimuth the blade is nearly parallel to the cycloid, meaning that a significant length of the blade is experiencing the effect at the same time.

The non-uniformity of the downwash will also directly excite the hull and is commonly observed as flexing of the canopy at the blade passing frequency. In the absence of vibration structurally transmitted from the rotor, this mechanism will still vibrate the hull.

In addition to the vibration sources noted above, in some flight regimes the rotor may ingest its own tip vortices. These are mostly carried away by downwash in powered flight or upwash in autorotation, but in some conditions, such as the end of a landing approach or a roll manoeuvre, the vortex from one blade may hit the next blade head-on. Figure 3.34(a) shows that the effect of a vortex approaching a blade is a local modulation of the RAF direction causing a dip in the angle of attack. Figure 3.34(c) shows that as the vortex leaves the blade the angle of attack is increased. The time taken for the blade to transit the vortex is only a few milliseconds and so the resulting pressure dipulse shown at (c) produces audio frequencies. This is the origin of blade slap.

In order to simplify study and to allow comparison between rotors turning at different speeds, the frequencies of vibration are always divided by the rotor frequency. Thus the units of vibration are not cycles per second but cycles per revolution. An imbalanced rotor will cause a lateral vibration at the rotor frequency having a vibration frequency of one cycle per rev. This is usually abbreviated to 'one-per'.

In establishing the effect on the hull of vibrations created in the rotor, it is vital to appreciate that there are two completely different mechanisms at work. Vibrations in the direction of the mast axis and torsional vibrations about the mast axis are directly transmitted and will be felt in the hull at the same frequency. However, the rotor is turning with a frequency of its own and so vibrations in the plane of the rotor are taking place in a rotating frame of reference. When these vibrations reach the hull, which is in a stationary frame of reference, the same frequency will not be felt. Instead,

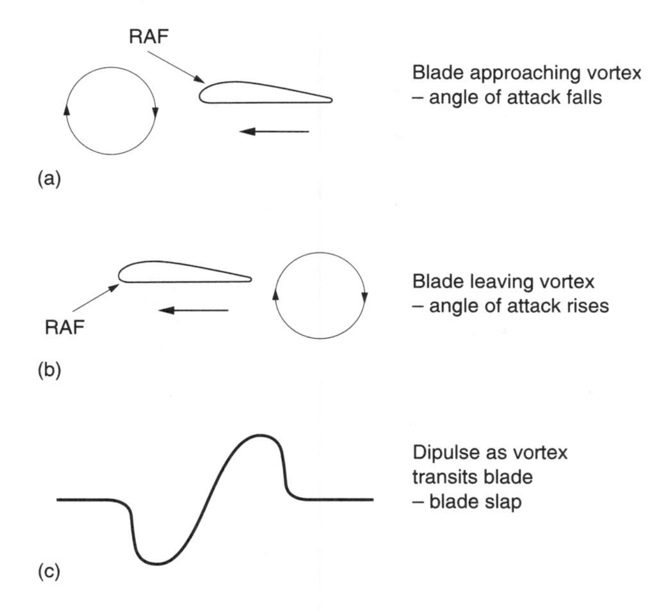

Fig. 3.34 (a) A blade element encountering a vortex from an earlier blade will experience first an increase in angle of attack (a), then a reduction (b). The variations in angle of attack produce a pressure dipulse (c) rapid enough to be audible as blade slap.

in the case of lateral (in-plane) directions the vibration frequencies beat or heterodyne with the rotor frequency to produce sum and difference frequencies.

The concept of sidebands was discussed in section 2.11, but the result can be demonstrated graphically. It is possible conceptually to replace the blade generating in-plane vibrations due to modulation of its root pull with a device generating the same vibration. This consists of a pair of contra-rotating eccentrics called a Lanchester exciter shown in Figure 3.35(a) which produces linear sinusoidal vibration. Figure 3.35(b) shows that if the rotor blades vibrate radially at the second harmonic (two-per) this can be simulated by an exciter which makes two revolutions with respect to the blade in one rotor revolution. The result is that one of the eccentrics makes three revolutions with respect to the hull whilst the other makes only one. Thus a two-per in-plane blade vibration results in three-per and one-per at the hub in stationary co-ordinates. Third harmonic vibration produces four-per and two-per vibration in stationary co-ordinates and so on.

The discussion so far has considered the action of individual blades. However, in practice the hub will vectorially sum all of the forces and moments acting on it. Vectorial summation must consider the phase of the contributions from each blade. Clearly the phase difference between the blades is given by 360° divided by the number of blades. The overall result must then also depend on the number of blades in the rotor. If the blades are identical, and experience the same forces as they turn, they will develop the same vibration, except that the phase will be different for each blade. If, however, one blade has different characteristics to the others, a one-per function may result.

It is interesting to see how the number of blades affects the spectrum of vibration and how this effect is different for in-plane vibrations. Figure 3.36(a) depicts a two-bladed rotor and shows a graph of the fundamental flapping for each blade (one-per) and the second harmonic (two-per). Note that the two blades are 180° apart and so the one-per waveforms are in anti-phase and cancel. As one-per flapping is the result of tilting the disc this is no surprise and remains true for any number of blades. However, note that the second harmonic flapping of the two blades is in-phase and will therefore add.

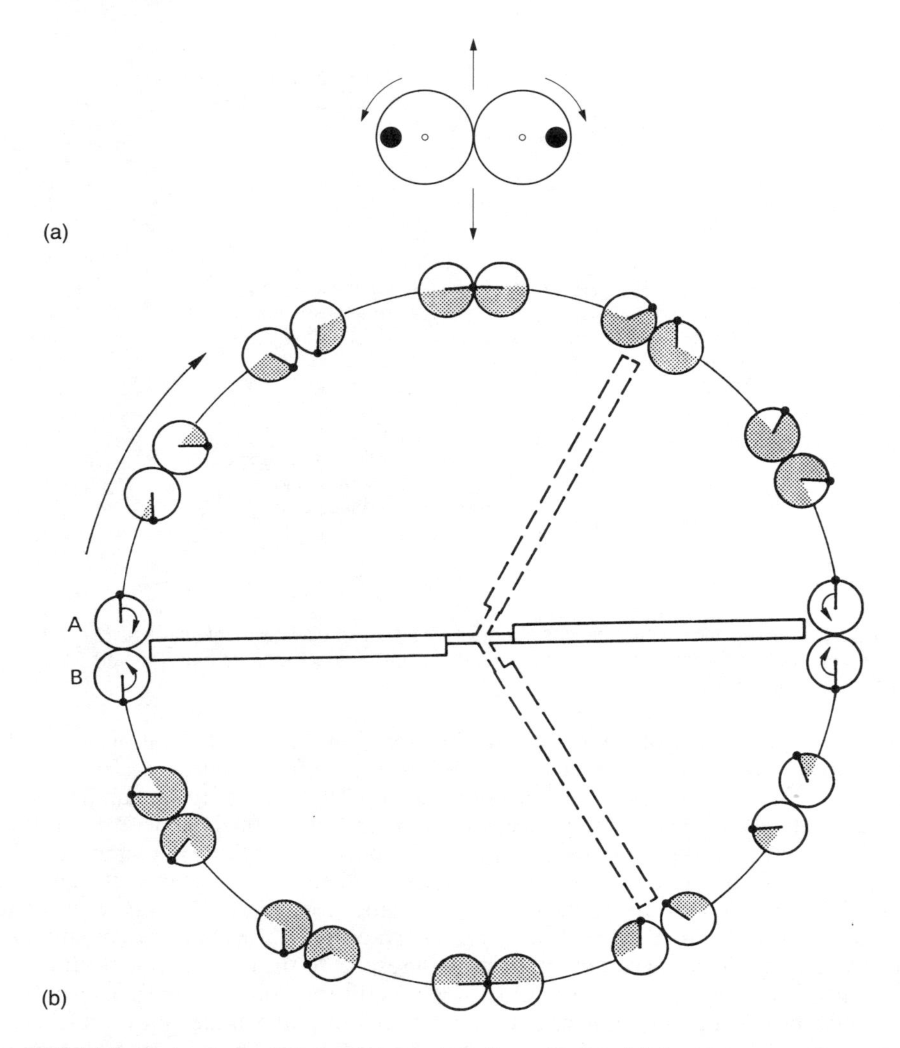

Fig. 3.35 (a) A pair of contra-rotating eccentrics produces linear vibration. (b) If the mechanism of (a) is rotated, one of the eccentrics will turn faster and the other slower. Thus frequencies transmitted to the hull will be the vibration frequency plus or minus the rotor frequency.

It will be clear that a two-bladed rotor is at a disadvantage because both blades will be in the lift trough simultaneously at 90° and 270°, and in a lift peak simultaneously at 0° and 180°. The result at high speed is a vertical vibration known as 2P hop. Figure 3.36(b) shows the situation for the third flapping harmonic. As the third harmonic flapping has one and a half cycles in the time between blade passings, it will be clear that the 3P components will be out of phase between the two blades. Following similar logic it should be clear that in a two-bladed rotor, only even vertical harmonics exist; the odd ones cancel.

Figure 3.37 shows the case for a three-bladed rotor. At (a) it will be seen that the second flapping harmonic waveform of each blade is at 120° to that of the next. The sum of three sine waves at 120° is constant and so there can be no 2P hop in a three-bladed

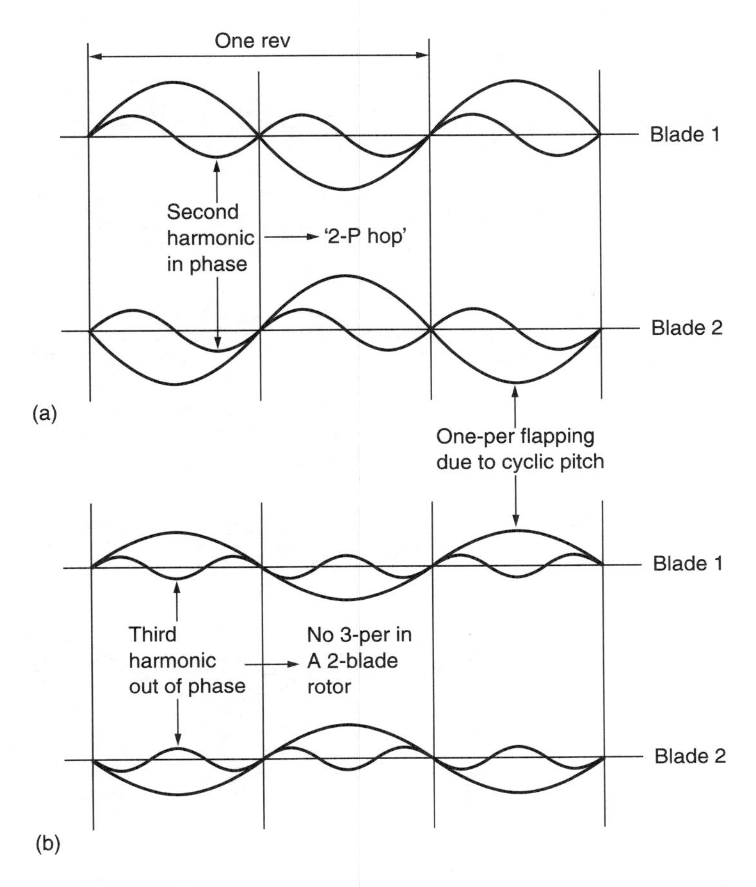

Fig. 3.36 (a) In a two-bladed rotor the blades flap with a 180° relationship. The fundamental flapping cancels, whereas the second harmonic flapping adds. Two bladed rotors are prone to 2P hop. (b) The third harmonic cancels in a two-bladed rotor.

rotor, or indeed for a rotor with a higher number of blades. In Figure 3.37(b) is shown the effect for the third flapping harmonic. Here the third harmonic waveform is in the same phase for each blade and so there will be summation.

Now that the general principle is clear, the result for any number of blades and any harmonic can readily be established. This is summarized in Figure 3.38(a) where it will be seen that the lowest hopping frequency is numerically equal to the number of blades and that these frequencies may be multiplied by integers to find the higher hopping modes.

When in-plane forces are considered, the results are quite different. Figure 3.38(b) shows the result of various harmonics with different numbers of blades. These forces will be heterodyned by rotor speed so that the result on the hull of a rotor frequency nP will be vibration at $(n-1)P$ and $(n+1)P$ whereas the frequency P is absent. Following the principles of vector summation explained above, certain harmonics cancel at the hub. Note that as with hop forces, the lowest frequencies at which rocking forces can exist are numerically the same as the number of blades, with the harmonics obtained by integer multiplication. The difference between Figure 3.38(a) and Figure 3.38(b) is that

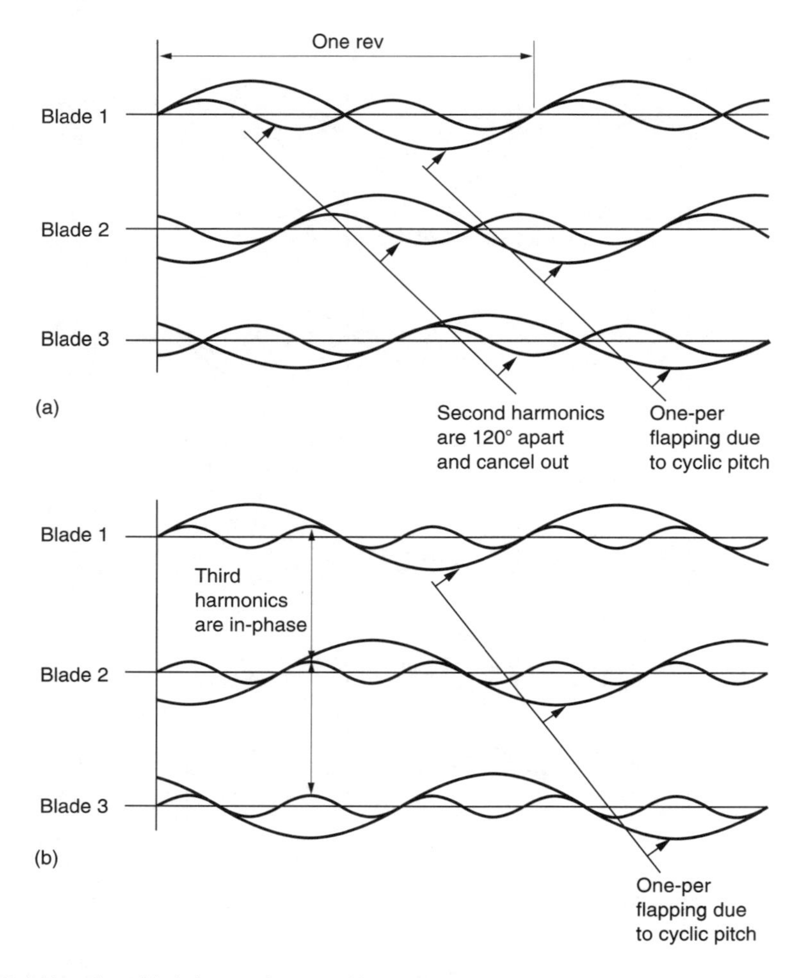

Fig. 3.37 (a) In a three-bladed rotor, the second harmonic flapping is mutually at 120°, the sum of which is zero. (b) Third harmonic flapping is in the same phase in each blade and so will add.

different frequencies at the blades are responsible for the same frequencies at the hull. For example, a five-bladed rotor can suffer 5P hopping and rocking, but the hopping is due to blade flapping at 5P whereas the rocking is due to blade flapping at 4P.

This section has concentrated on the main rotor, but clearly a tail rotor will show many of the same effects as it too is passing edge-on through the air. However, the tail rotor is at a further disadvantage because it is working in airflow that has been disturbed by the main rotor and the hull. This will be considered further in Chapter 5.

3.27 Vibration control

Vibration and helicopters are inseparable. However well balanced the rotor may be, the lift asymmetry in forward flight will cause vibration. Vibration is unwelcome because

Number of blades

Harmonic	2	3	4	5	6	7	8
2P	X						
3P		X					
4P	X		X				
5P				X			
6P	X	X			X		
7P						X	
8P	X		X				X

(a) Hopping or vertical harmonics

Number of blades

Harmonic	2	3	4	5	6	7	8
2P		3P					
3P	2P 4P		4P				
4P		3P		5P			
5P	4P 6P	6P	4P		6P		
6P				5P		7P	
7P	6P 8P	6P	8P		6P		8P
8P		9P				7P	

(b) In-plane or rocking harmonics

Fig. 3.38 (a) Axial or hopping modes as a function of the number of blades. (b) In-plane modes have different characteristics to axial modes because frequencies are heterodyned by the rotor frequency.

it has a number of negative effects. Section 3.26 showed some of the many causes of vibration and it should be clear that many of these may result in severe alternating stresses in the rotor head, the blade roots, the pitch control mechanism and the transmission. This has a large impact on the service life of such parts where a failure cannot be tolerated. Many parts need to have more generous cross-sections, increasing mass and reducing payload. Periodic checking for cracks adds to maintenance costs.

A key problem due to vibration is that it impairs the pilot's vision. The human visual system (HVS) obtains its acuity by averaging visual information over a considerable length of time. In the presence of vibration, the image will be unstable on the retina and will thus appear blurred. There have been cases where damage to rotating components has caused vibration severe enough that the pilot was unable to read the instruments at all. Vibration contributes to pilot stress and workload. One of the earliest approaches therefore was to reduce vibration at the pilot's seat.

Given the low load factor of a helicopter, the hull is not highly stressed and could be very lightly built were it not for fatigue inducing vibration. Increasingly helicopters incorporate electronic systems and sensors used either to control or navigate the machine or as part of its mission equipment. Electronic equipment does not take kindly to vibration, or if specially adapted, must be more expensive.

Vibrating masses are a common origin of sound, so it follows that a vibrating helicopter will be noisy both inside and out. External noise causes disturbance in civil operations and may compromise military missions. A significant part of the design process of the modern helicopter is taken up in minimizing noise and vibration. There are a number of approaches that will be combined in various ways. Study of the aerodynamic origins of vibration may allow a reduction at the source. Once the forcing frequencies are known, the design of the structure should ensure that resonant responses to those frequencies are minimized. This is known as *detuning*. Vibration may be isolated or *decoupled* in some way, *damped* using lossy materials to convert flexing into heat, or opposed or *absorbed* through various cancelling techniques, both passive and active.

The reduction of the vibration at source has to be the technically superior solution. The disadvantage of decoupling the vibration from the passengers is that although their comfort is improved, the vibration experienced by the parts that are not decoupled may actually increase. It is easy to see why. For a given exciting force, the amplitude of a vibration is inversely proportional to the mass being vibrated. In a conventional helicopter the dominant masses are those of the transmission, hull and payload. If the hull and payload are decoupled, the mass seen by the vibration source is now only that of the transmission and the amplitude of the vibration there must be greater and not necessarily beneficial.

An obvious way of reducing vibration at source is to use non-sinusoidal cyclic pitch control that allows the lift moment to be more nearly constant as the blade turns. The means to do this are non-trivial and are described in section 3.28. Another way of reducing vibration is to use more blades. This has several advantages. First, the load on each blade is reduced and with it the magnitude of the forcing function of each blade. Second, the more blades, the higher the frequency of the vibration will be and the lower will be the amplitude for a given mass. Figure 3.37 shows that a three-bladed rotor eliminates the 2P hop of the two-bladed rotor, whereas a four-bladed rotor reintroduces some 2P lateral vibration. It will be seen from Figure 3.38(b) that an even number of blades gives lower rocking frequencies, making an odd number preferable. Consequently the change from four to five blades gives a significant improvement. Large transport helicopters inevitably have a large and generally odd number of blades – seven or nine – simply to contain the large vibrations that must result from generating high rotor thrust. Using many blades will be expensive and result in a large rotor head having high drag.

It is common to design the blades so that their natural frequencies of vibration are not in the spectrum of the excitation. If any of the excitation frequencies coincide with the resonant frequencies there could be a significant response at that frequency. Any coincidence can be seen from an interference diagram as shown in Figure 3.39. This interference diagram is for the Westland Lynx and shows the operating speed (318 RPM) as a vertical line. The excitation frequency may be at any integer multiple of the rotor frequency. Figure 3.39 shows that, as rotor speed increases, the excitation harmonics (thin lines starting at the origin) fan out. The resonant frequencies of the blades also change with rotor frequency because of the effect of centrifugal stiffening. Figure 3.39 shows how flapping, ωf, and lagging, ωl, resonant frequencies may change with rotor speed.

The only coincidence between excitation and response is the close correspondence between the fundamental flapping resonant frequency and the rotor fundamental frequency. This is the characteristic that results in rotor precession. Note that as the Lynx

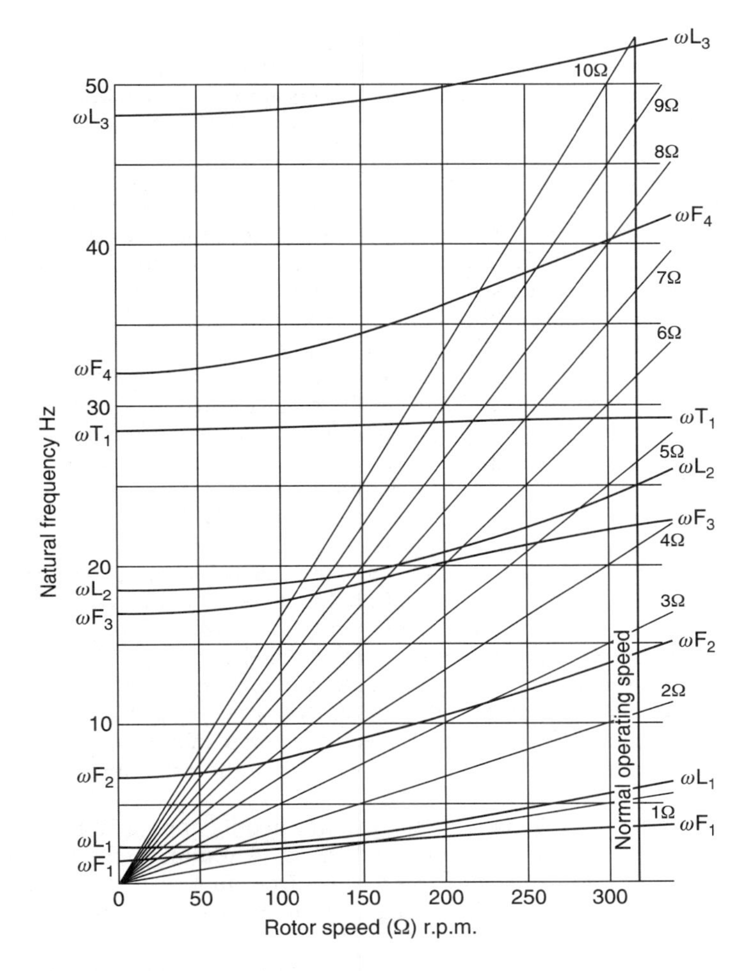

Fig. 3.39 The excitation spectrum in a rotor fans out at RRPM increases. The natural frequencies of flapping and dragging of the blade will also change with RRPM due to centrifugal stiffening. If these two effects are superimposed, it is possible to see where the excitation frequencies coincide with response frequencies. This is known as an interference diagram. (AugustaWestland)

has a hingeless rotor, the flapping resonance is above the rotor frequency and so the rotor phase lag will be less than 90°.

In all other cases, coincidence is avoided so that at normal RRPM no resonances are excited. This result was not an accident. During the design process the blade was designed with a tapering spar resulting in the stiffness and the mass per unit length being non-uniform so that the flapping resonant frequencies are not integer multiples of the fundamental. Where constructional techniques such as extrusion do not allow the blade or spar to taper, the blades can also be detuned by placing weights at critical points along the span so that the effective mass per unit length is no longer uniform. Note that the technique of detuning only works at one rotor speed, at other speeds the excitation frequency is different and the response will also differ. Detuning will also be employed in the hull so that hull resonance does not magnify rotor-induced vibrations.

Although vibration may occur at a number of frequencies, often one of these will dominate and cause a problem. Vibration absorbers are devices intended to reduce vibration at a single specific frequency. The term is actually a misnomer because what they actually do is to create an opposing or near-anti-phase vibration. These devices work using mechanical resonance.

As was seen in section 2.9, a mass supported on a spring displays resonance. If the damping is very low, the phase response just above the resonant frequency becomes nearly 180°. If the resonant frequency is set just below the frequency of vibration, the mass will automatically resonate in inverse phase and oppose the vibration. In helicopters the battery may be used as a suitable mass and this is supported on carefully calibrated springs. A mass absorber of this kind can only reduce vibration in the vicinity of the device itself. Flexibility of the hull will allow vibration elsewhere; however, such devices can oppose the hull vibration due to non-uniform downwash. Mass absorbers only work at a fixed frequency and require the rotor to turn at a precisely controlled RPM. The alternative is to use a system in which the spring stiffness can be modified by a servo so that the absorber can be tuned to the actual RRPM in use. Such a system is fitted in the Chinook. In some early helicopters vibrations fed into the control rods were opposed in this way. Figure 3.40(a) shows a resonator consisting of a bob weight on a leaf spring attached to a control bell crank.

A pendulum is a kind of resonator. The type used in a clock uses the earth's gravitational field to provide the restoring force. In helicopters, pendular vibration absorbers fitted on rotors use the acceleration caused by rotation. Figure 3.40(b) shows that pendular absorbers used to handle flapping vibration consist of weights that can swing on the end of cranked shafts pivoted in the blade roots. When the rotor is turning, the weights fly out horizontally and attempt to stay in the same plane. Vertical movement of the blade excites the pendulum which is tuned to resonate in anti-phase. These devices can be seen on the Kamov Ka-32 and on the Bo-105 and are known colloquially as bull's balls.

As these devices employ resonance, a single unit can only oppose vibration at a single frequency. However, it is possible to fit more than one pendulum to the same mounting. The Hughes 500 has a four-bladed rotor and so according to Figure 3.38(b) will be prone to 4P rocking forces due to flapping at 3P and 5P. The two pendula are tuned to the third and fifth harmonics of blade flapping.

Torsional vibration of the rotor shaft may be due to blade excitation in the dragging plane. This is an application for the bifilar pendulum. A pendulum can be made to swing in one axis only by attaching it with two strings, hence the name. Restoration forces are much smaller for in-plane motion, suggesting that a pendulum would need a very short arm in order to resonate at a harmonic of rotor frequency. Thus in practice the arms are replaced by a pair of pins which operate in oversized holes as in Figure 3.40(c). As the weight moves to one side, the geometry of the pins moves the weight inwards. As the weight tends to fly out as the rotor turns, there will be a restoring force if the weight is displaced. This is all that is necessary to create a resonant system.

The restoring force that determines the resonant frequency in pendulum and bifilar absorbers comes from the acceleration due to rotation. This is a function of RRPM, as is the frequency to be absorbed. Consequently these devices are self-tuning.

Another popular and obvious way of isolating vibration is to use flexible mountings between the transmission and the hull. It is a fundamental concept of such mounts that there will be a resonance whose frequency is determined by the sprung mass and the stiffness of the mounts. The degree of isolation is proportional to the difference between that resonant frequency and the frequency to be filtered out. Consequently the lowest possible resonant frequency is required. This frequency may be lowered by

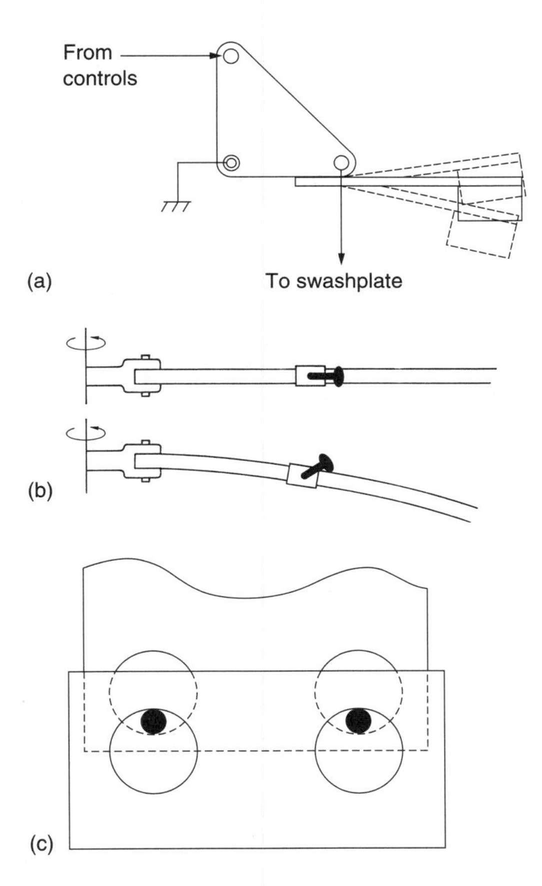

Fig. 3.40 (a) A bob weight on a leaf spring attached to a bell crank will oppose vibrations in a control run. (b) The pendular absorber is designed to oppose vibrations in the axial direction. (c) A bifilar pendulum absorber achieves the effect of very short arms by using rollers as shown here.

softening the mounts, or by increasing the supported mass. Soft mounts may cause other problems, such as handling difficulties because the hull would tend not to follow the rotor in a manoeuvre. As a result, the best solution is to increase the mass to be isolated as much as possible.

This is the principle of the machinery raft, which is a rigid structure carrying the engines, transmission, control actuators, oil coolers, generators, etc. as a single unit. This raft is then mounted to the hull by widely spaced resilient supports. The large mass of the raft opposes the forcing function of the rotor and makes the resonant frequency as low as possible, as well as reducing the amplitude of vibration seen by the components on the raft.

The problem with resilient supports is that for good isolation they have to be soft, and this will always result in a static deflection. Some techniques have been developed for helicopters to circumvent the problem. Figure 3.41(a) shows a system developed by Kaman known as DAVI, or dynamic anti-resonant vibration isolator. This works using a mechanical impedance convertor to raise the moving mass artificially. Between the transmission and the hull is a lever carrying a bob weight. The mechanical advantage of

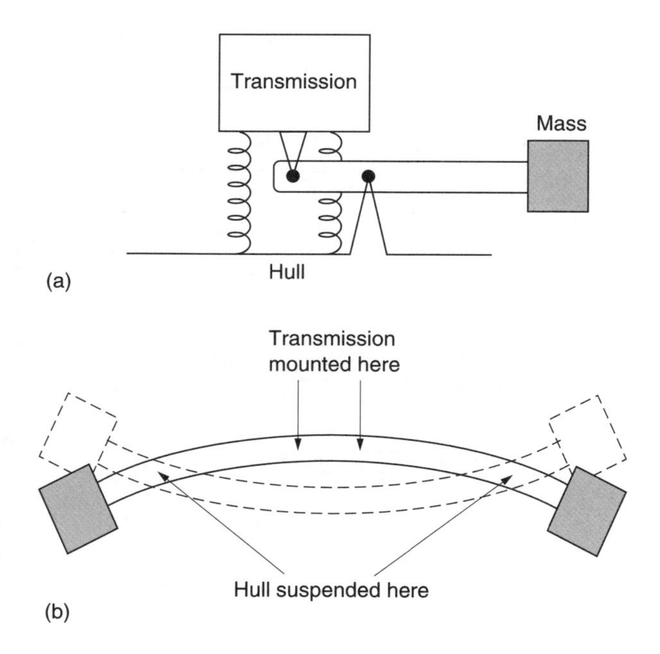

Fig. 3.41 (a) In a dynamic anti-vibration isolator, a lever is used to amplify the motion of a weight thus giving it enormously increased effective mass. (b) The nodal beam system. A suitably tuned flexing beam will have nodes where the hull is attached.

the lever causes any relative motion to be amplified. The lever changes the mechanical impedance of the bob weight by a factor equal to the square of the mechanical advantage. Thus a 1 kg weight on a 10 : 1 lever would appear to weigh 100 kg. Mounting the transmission on, for example, four such mounts would raise the effective weight by 400 kg. Given the increase in apparent weight of the transmission, the supporting spring can be made much stiffer than would be possible with a conventional resilient mount, without increasing the resonant frequency.

Figure 3.41(b) shows the nodal beam system developed by Bell. At one time most Bell helicopters had two blades and suffered 2P hop. The nodal beam was developed to isolate the hop. Between the hull and the transmission is a flexible beam made from a sandwich of elastomer and steel. The transmission mounts are closer together on the beam than the hull mounts. As relative movement takes place between the hull and the transmission, the beam must bend. The proximity of the hull and transmission mounts makes a virtual lever system resulting in larger amplitude of movement at the centre of the beam. The effective mass of the beam is amplified by the mechanical advantage of the virtual levers, making the system mathematically equivalent to a pair of DAVI isolators in parallel. The geometry of the system is tailored so that at the bending resonant frequency of the beam the hull attachment points are at nodes and so do not move.

In active vibration cancelling systems, Figure 3.42, a number of high speed actuators, typically hydraulically operated, are used to attach the hull to the transmission. The actuators are controlled by signals from processors and accelerometers in the hull. The signal processors are needed to adapt the waveforms fed to the actuators to changing conditions such as all up weight and airspeed. An increase in all up weight would

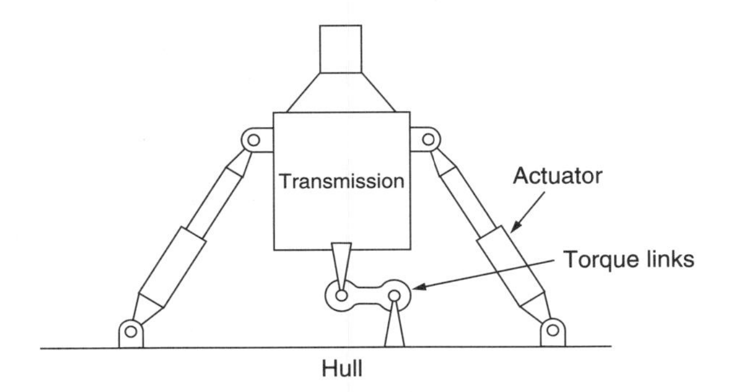

Fig. 3.42 In an active anti-vibration system, the transmission is mounted on actuators that are driven by a suitable forcing function to cancel the vibration.

require increased static pressure in the actuators to prevent them extending. In practice a feedback mechanism measures the average extension of the actuator and centralizes each one with a relatively long time constant.

No practical hull can be rigid, and so there will be various modes of vibration excited in it. One of the sources of vibration will be due to the non-uniform downwash impinging on the hull surface. Conventional isolators between hull and transmission can do nothing about this because they only oppose structure-borne vibration. However, an active system can use the transmission as a reaction mass to create forces opposing vibration from any source. Unfortunately the amplitude of vibration in the transmission will be increased. Better results may be obtained by applying active control to a raft carrying the engines, etc. as described above.

3.28 Harmonic pitch control

It was seen in Figure 3.30 that the use of sinusoidal feathering in a cyclic pitch control is incorrect. The airspeed seen by a blade element changes sinusoidally in forward flight, but the lift is proportional to the square of the airspeed. This non-linear function generates flapping harmonics that cannot be cancelled by sinusoidal feathering. It follows immediately that if there are to be no flapping harmonics, the feathering function must contain harmonics.

A harmonic cyclic pitch waveform is approximately a square root function of a raised sine wave although some modification will be necessary because of inflow effects. When multiplied by the sine-squared function due to the speed, the product, namely the lift, is constant. The pitch function above will need further refinement to account for upwash at the leading edge of the disc. Figure 3.43 shows how an actuator between the swashplate and the blade control arm could add the harmonic control.

In practice perfect constant lift can only be achieved at one radius, but it should be possible to find a pitch function that makes the average blade lift moment essentially constant. A more advanced system would have means to control the angle of attack of different parts of the blade independently. This may be possible using, for example, piezo-electric actuators. It is impossible to obtain constant lift over the whole blade because of the reverse flow area, and so some harmonic flapping will always remain.

Whilst theoretically obvious, harmonic blade control, also known as higher harmonic control, is rarely found. One of the difficulties is that the cyclic control becomes a

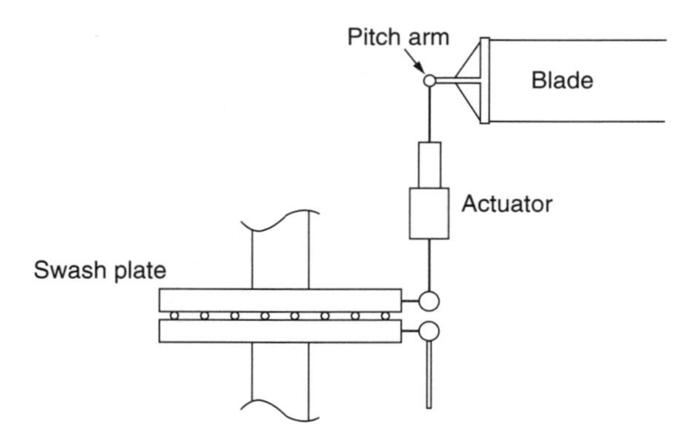

Fig. 3.43 In harmonic pitch control, the blade pitch is not sinusoidal, but has a waveform calculated to make the lift more uniform around the disc. This may be controlled by an actuator between the swashplate and the blade pitch control arm.

function of airspeed, returning to a near sinusoidal function in the hover. This is probably beyond what might be achieved mechanically but some kind of electronic system would easily generate the required functions. High speed actuators are necessary to drive the blade feathering, and the blades must be torsionally stiff in order to respond to the rapid pitch variations. Another consideration is that as the number of blades increases the effects of lift troughs are averaged out more and there is less of a problem.

However, the elegant result of constant lift moment around the disc is one that is most desirable. The reduction in vibration makes the flight more comfortable for passengers and prolongs the life of the machine. A further consequence of constant lift is a reduction in power needed in forward flight. In a conventional rotor the increased power needed where the lift is greatest is not balanced by a reduction in power where the lift troughs are. Consequently constant lift needs the least power.

It would seem clear that the greatest cost benefit of harmonic blade control would be in the two-bladed helicopter where the 2P hop could be eliminated. The improvement in ride would be greatest but the number of actuators needed is least. In some light helicopters a two-bladed harmonically controlled rotor could replace a conventional three-bladed rotor and the overall cost might be the same. At some point designers will find ways of economically achieving the goal of constant lift. Some experiments have been done with systems which rock a conventional swashplate, but this cannot give the required result as each blade needs to be individually controlled.

3.29 Blade design

The helicopter blade is a complex subject because of the number of interacting variables involved. These will be considered here. It might be thought that the aerodynamic characteristics of the blade would be determined only by shape. In general this is untrue. Figure 3.44 shows some of the issues. The long thin structure of a practical blade can never be rigid and flight loads will cause the blade to flex. However, it is not enough to know the stiffness of the blade to calculate the result. In forward flight the loads have a powerful alternating component and this generates harmonics. At the frequencies of the higher harmonics the motion may be mass controlled. A further complexity is that the flexing of the blade will affect the aerodynamics; a phenomenon

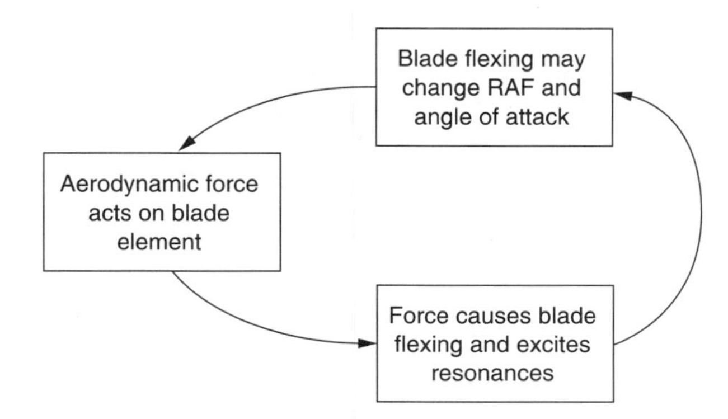

Fig. 3.44 The aeroelasticity effect loop in a rotor blade. The blades are not rigid and can flex. This will affect their local angle of attack, changing the lift distribution that in turn causes flexing.

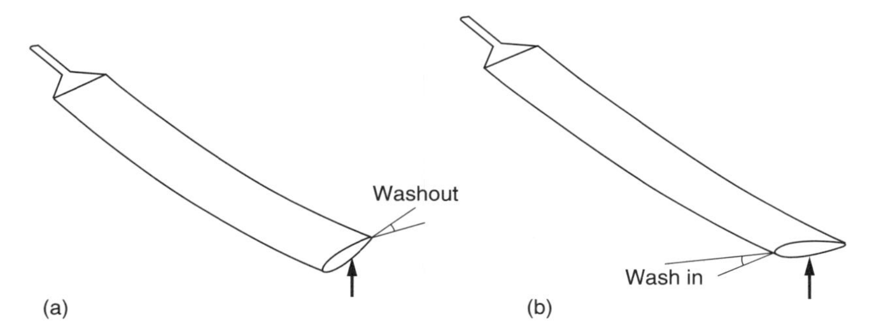

Fig. 3.45 (a) If the blade has greater bending stiffness at the leading edge, it will wash out as lift increases. This is a stable condition as wash-out reduces the lift. (b) If the trailing edge is stiffer than the leading edge, lift causes wash-in that will further increase the lift. Such a blade would suffer serious flutter and probably disintegrate.

called *aeroelasticity*. As it is clear that the aerodynamics affect the flexing, then there is a loop. This loop may be stable or unstable. Part of the design process must be to ensure that aeroelastic effects always result in blade stability.

The shape of the blade is completely defined by the root cut-out, the tip shape, the degree of twist and taper and by the blade section used, which may change with radius. Assuming first a rigid blade, the blade shape will allow the time variant aerodynamic forces to be predicted in a given flight regime. The way the blade responds to this excitation is complicated. The torsional forces depend on the locus of the centre of pressure with respect to the mass centroid. The effect of flight loads on blade twisting is significant because twist will alter the local angle of attack as well as feeding loads back into the control system. The degree of twisting depends on the moment of inertia, the stiffness, the damping and the frequency.

The chord-wise distribution of stiffness needs to be known accurately. Figure 3.45(a) shows a blade in which the leading edge is stiffer than the trailing edge. The application of lift at the centre of pressure causes the blade to wash out; a stable condition. However, Figure 3.45(b) shows a blade in which the trailing edge is stiffer than the leading edge. Here lift causes the blade to wash in; leading to flutter. As a result blades are generally constructed in such a way that the mass centroid and the centre of bending stiffness is well forward on the chord. The result is a section where the structural mass is concentrated in a D-shaped structure at the leading edge, with the trailing part of the section being comprised of a low density structure.

Blade taper is not always employed because in practice the penalty of a constant chord blade is only a few per cent. Taper prevents constructional techniques such as extrusion that requires a constant section. Blade twist is advantageous in hover, but the ideal amount for hover will be detrimental in translational flight because the application of cyclic pitch may cause the finely pitched blade tips to go to a negative angle of attack. The exception to this restriction is in the tilt rotor aircraft that both hovers and cruises with axial inflow. This gives the designer more freedom to use blade twist and this is clearly visible in rotors of the Bell-Boeing Osprey.

4

Rotors in practice

4.1 Introduction

The loads on the rotor blades are large and time variant and the articulated head was developed to minimize them. Articulation also reduces the rate at which the aircraft responds to the controls. In this chapter the dynamics and control of the articulated rotor will be introduced, along with later developments such as hingeless rotors that became possible with the development of modern materials.

The essential functions of the rotor head are to transfer shaft power to the blades, to transfer the resulting rotor thrust to the mast, to resist the blade tension and yet to allow the blades to move on their feathering axes so that cyclic and collective pitch control is possible. This simple and elegant arrangement was initially not possible because it requires extremely strong materials. Thus practical rotor heads are often much more complex.

Figure 4.1 demonstrates that the rotor blade of a typical helicopter pulls outwards with a force roughly equal to the weight of a bus. It's no wonder those floppy blades straighten out. It requires little imagination to predict what would happen if a blade attachment failed. Clearly the rotor head has to be extremely strong and reliable and as a result it is one of the heaviest individual components in the machine.

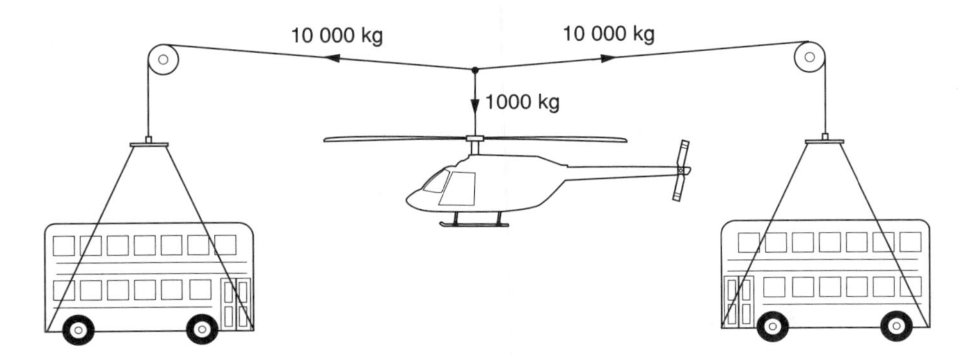

Fig. 4.1 Given the coning angle and the weight of the helicopter, the blade root tension follows. Here it will be seen that the same coning angle could be obtained by suspending the helicopter on cables with a bus on the other end. Blade attachments have to withstand tremendous loads yet allow the blades to feather.

4.2 Why articulated rotors are used

The variation in lift, lift distribution and drag as the blades turn, especially in forward flight, produces alternating stresses which can fatigue materials. The adoption of articulation was first suggested by Renard in 1904 and was essential in early rotorcraft to reduce the stresses involved, particularly the alternating stresses in the blades and the moments applied to the mast when rolling and pitching manoeuvres are performed. Some texts claim that articulation is necessary to handle the lift asymmetry in translational flight, but this is quite incorrect.

In addition to the feathering bearing, the fully articulated rotor head carries each blade on freely turning bearings which allow flapping, or movement above or below the plane of the rotor head, and dragging, a swinging movement in the plane of the blades which is also called lead/lag. The presence of the flapping and dragging bearings means that moments about their axes cannot be transferred from the rotor head to the blades, and so bending stresses in the blade roots are dramatically reduced.

Figure 4.2(a) shows one arrangement which has been widely used by, for example, Sikorsky and Enstrom. The feathering bearing is outboard of the flapping bearing. This arrangement has the advantage that the loads fed into the control system when the blades flap and drag are minimized. The flapping and dragging hinges can be displaced or coincident in various designs and this is considered in section 4.7.

If the axes of the flapping hinges pass through the shaft axis, the result is called a zero-offset rotor head and these will be considered in section 4.10. In a conventional articulated head, the flapping bearings are horizontally displaced, or offset, typically by a few per cent of the rotor diameter, and it will be seen from Figure 4.2(b) that blade tension can produce a control moment on the rotor head if the shaft is not at right angles to the tip path plane. Consequently the hull tends to follow the disc better when offset is employed and so the machine becomes more manoeuvrable, although a stronger mast is needed to withstand the moments.

4.3 Axes galore

Whether by flexing or by movement of a bearing, each blade can flap and feather as it rotates. A consequence of these degrees of freedom is that three rotational axes need to be considered when studying the behaviour of a turning rotor. The way in which these axes interrelate is fascinating and allows an insight into the behaviour of a blade in flight.

Figure 4.3(a) shows a helicopter with an articulated rotor in forward flight. The cyclic stick will need to be trimmed forward to maintain airspeed, since this will reduce the angle of attack of the advancing blade and increase that of the retreating blade, so that they generate equal lift moments. The stick will also need to be trimmed towards the retreating side to oppose the inflow roll. The tip path plane is tilted forward to obtain a forward component of rotor thrust to balance drag.

4.3.1 The shaft axis

The most obvious, and least useful, of these axes is the shaft axis on which the rotor head turns. An observer riding on the shaft axis could see flapping, dragging and feathering taking place simultaneously, but some of these motions could disappear at certain relationships between the shaft axis and the other two axes.

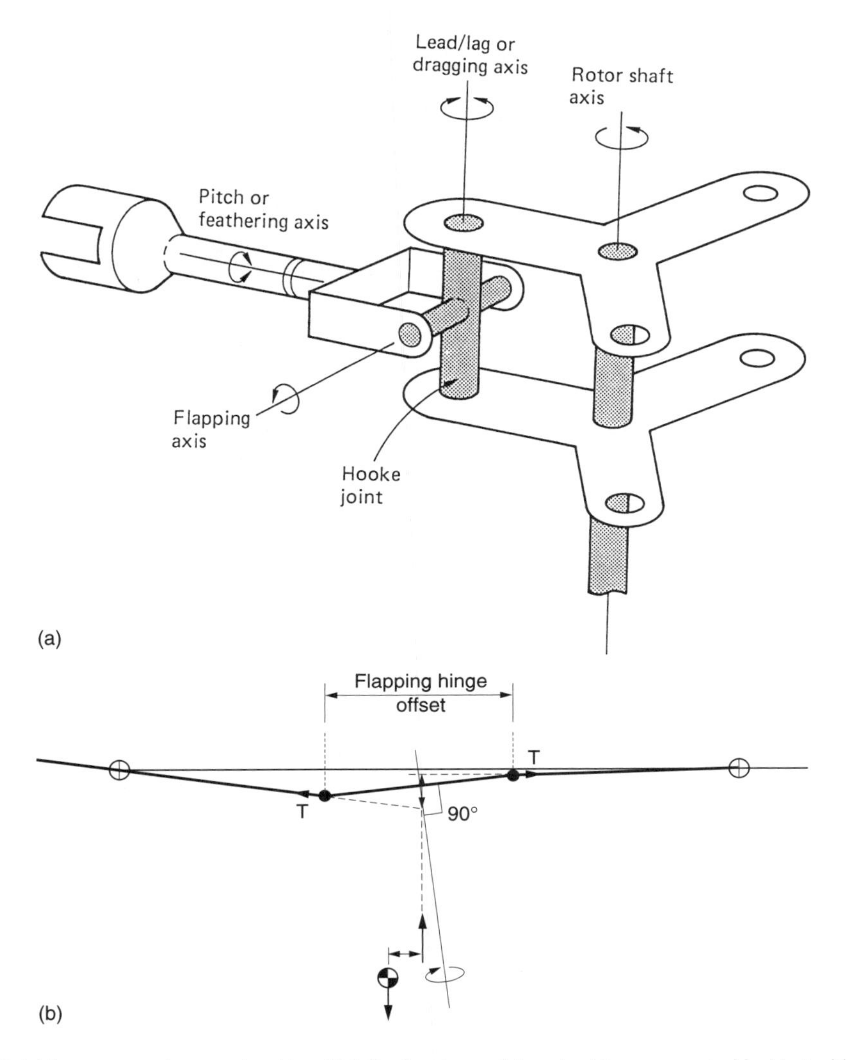

Fig. 4.2 (a) A representative rotor head in which the flapping and dragging hinges are coincident but with the feathering hinge situated outboard. (b) Offset flapping hinges allow a couple to be transferred to the head when the tip path plane is not orthogonal to the shaft axis.

4.3.2 The tip path axis

The axis at right angles to the tip path plane is generally called the tip path axis or disc axis. The resultant thrust of the rotor aligns within about a degree of the tip path axis, and for most purposes they can be considered to be coincident. The tip path axis is also called the axis-of-no-flapping in some texts, although this term is not strictly correct because the existence of harmonics in the blade motion cause the blades to flap slightly with respect to the tip path axis. By the same token the tip path plane is not strictly a plane.

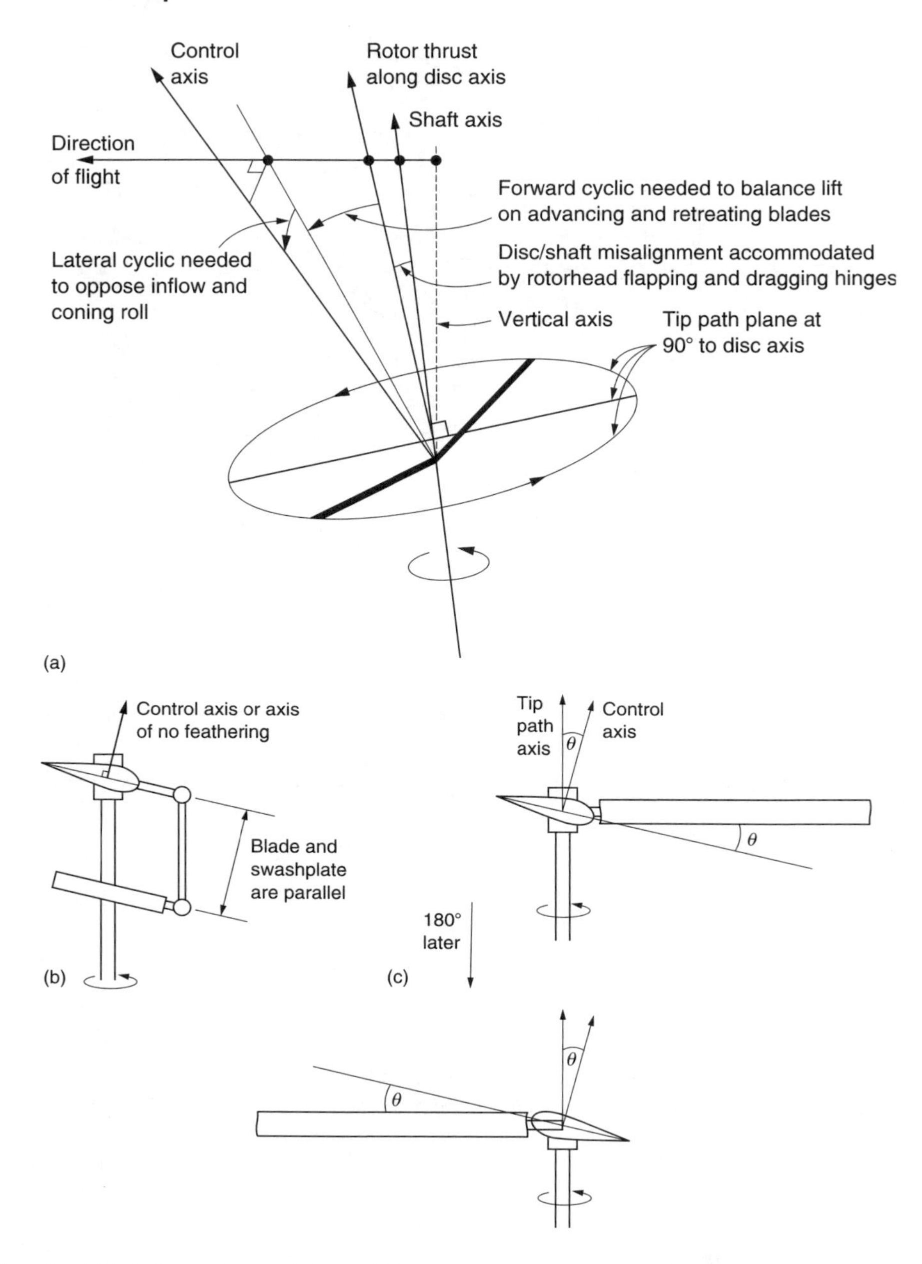

Fig. 4.3 (a) A helicopter in steady forward flight requires a combination of forward and lateral cyclic inputs to neutralize flapback and inflow/coning roll. Note the tip path axis that is very nearly aligned with the rotor thrust, the control axis, and the shaft axis that are generally not coincident. When observed with respect to the control axis, the blades do not feather, but instead appear to flap. Note that even a hingeless rotor with rigid blades can flap with respect to the control axis. (b) If the pitch control rods are parallel to the shaft and 90° from the feathering axes, then the axis of the swashplate is identical to that of the control axis. (c) The angular difference between the tip path and control axes is equal to the amount of flapping observed with respect to the control axis and to the amount of feathering observed with respect to the tip path axis.

Designers often try to trim the machine so that in cruise the action of the tail plane opposes couples due to drag on the hull in order to align the shaft axis with the tip path axis because this minimizes the amount of flapping and hence the amount of wear and vibration. However, this will often be compromised because of variations in the longitudinal position of the CM. In tandem helicopters it is easier to align the tip path axis with the shaft axis. During a roll manoeuvre the tip path axis and the shaft axis must diverge in order to roll the hull.

An observer turning with the tip path axis would see the feathering action changing the pitch of a given blade sinusoidally about the collective pitch setting at one cycle per revolution but as he is turning in the tip path plane he would not see any flapping (except for harmonics). In the steady hover no dragging would be observed, but in forward flight dragging would be seen due to the increased drag on the advancing blade compared with that on the retreating blade.

4.3.3 The control axis

There exists a third axis, not parallel to the disc axis in translational flight, about which no blade feathering takes place. This is called the control axis, and it is the hardest one to visualize. An observer turning with the control axis in forward flight would note that the pitch of the blades was constant, but that the blades flapped up at the advancing side of the nose of the machine and down at the retreating side of the tail. A given blade would appear to drag back when advancing, and then forward when retreating. The dragging would appear much greater than in the case of the observer on the disc axis.

As Figure 4.3(b) shows, in the special case of a two-bladed rotor, if the pitch control rods stay parallel to the mast and are at 90° to the blades, the swashplate also rotates on the control axis. The blade stays parallel to the swashplate and so cannot feather with respect to it. In the general case the control axis and the swashplate axes are not the same as will be seen in section 4.13.

There is an exact equivalence between the degree of flapping seen by an observer on the control axis and the degree of feathering seen by an observer on the tip path axis. Figure 4.3(c) shows that when the control and tip path axes differ by $\theta°$, if the blades reach a peak feathering angle of $\theta°$ with respect to the tip path axis, they will flap to a peak deflection of $\theta°$ with respect to the control axis.

The reason for the difference between the tip path axis and the control axis should now be clear. In the hover, there would be no permanent difference as the blades follow the cyclic stick. However, in translational flight, the blades encounter highly asymmetrical conditions that pull the tip path axis away from the control axis. In order to fly straight and level, the pilot has to find a position for the cyclic stick in which the asymmetrical conditions on the blades are precisely opposed by cyclic input so that the rotor disc attitude is no longer affected by the asymmetry. For example, if at a certain airspeed the advancing/retreating effect would result in three degrees of flapback, the application of three degrees of forward cyclic will put the rotor disc back in the attitude it would have had in the absence of flapback.

As the shaft axis and the tip path axes can be non-parallel, clearly some joints or bearings are necessary or something must be designed to flex. The flapping and lagging bearings in an articulated head approximate to a constant velocity joint between the shaft axis and the tip path axis. The spherical bearing in the centre of the swashplate and the feathering bearings in the rotor head allow the control axis to be different from the shaft axis.

The motion of the blades can be studied with respect to any one of the above axes. Each has its own merits. The use of the control axis is instructive when studying the helicopter in detail. For hingeless rotors use of the shaft axis has some advantages but the tip path axis excels when discussing articulated and teetering heads because it is easier to visualize what is taking place and the direction of the resultant rotor thrust is clear. It is best used for flying training because it is most tangible to pilots. In practice it is useful to be able to switch conceptually from one axis to another to obtain a proper understanding.

The control axis was naturally used in the study of the autogyro and as the theory of the helicopter was in many ways derived from work done on the autogyro much early helicopter analysis also uses the control axis, including the classic work by Gessow and Myers.[1] Although the authors make it perfectly clear that the control axis is being used as a reference, not every reader understands what that means and there is scope for misunderstanding if the axes become confused. One example is the occurrence of numerous explanations that are seen claiming that flapping hinges prevent asymmetry of lift in translational flight in gyroplanes and helicopters. This is complete nonsense and comes about because the tip path and control axes have been confused. The blades can flap with respect to the control axis without any flapping hinges or flexibility at all simply because the control axis is not parallel to the tip path axis.

It is important to appreciate that all of these axes are hypothetical and the rotor doesn't care how we choose to observe it. Thus whatever axis is used for analysis the result must always be consistent with the result of an analysis on another axis. To give an example, as the blade pitch appears constant to an observer on the control axis, no force is necessary to make the blades perform cyclic pitch changes other than to overcome friction in the feathering bearings. If the same analysis is carried out on the tip path axis, the blades appear to be oscillating along the feathering axis. The effect shown in Figure 3.8 produces a restoring action tending to return the blades to flat pitch. This causes the blades to have a torsional resonant frequency. It can be shown that this is the same as the rotational frequency of the rotor. As a result the feathering blades are in torsional resonance and so no force is required to operate them other than to overcome friction. This is consistent with the analysis on the control axis.

If harmonic pitch control is used, then strictly there is no axis about which the blades do not feather. However, the control axis could be redefined as the axis about which there is no first harmonic feathering. In other words only the higher harmonics of feathering would be seen with respect to the control axis. This is a reasonable approach because the tip path plane is not really a plane because of flapping harmonics and yet the concept is still useful. Interestingly with the correct application of harmonic pitch control the feathering will see more harmonics so that the tip path may see fewer.

4.4 Flapping

Flapping is the movement of the blade CM along the shaft axis. It was shown in section 2.14 that the fundamental resonant frequency of the flapping blade is almost the same as the rotational frequency. With a zero-offset rotor the frequencies are identical and this results in a 90° phase lag in the rotor. When flapping hinge offset is used, the resonant flapping frequency rises slightly because the blades become a little shorter without the diameter changing and the result is that the phase lag of the rotor becomes a little less than 90°. This is compensated for in the design of the controls.

Flapping is self-damping because of the effect it has on relative airflow. When a blade flaps down, its vertical velocity adds vectorially to the airflow to increase the angle of attack, providing more lift to oppose the flapping. When a blade flaps up, the opposite occurs. As a result no additional damping is required about the flapping axis, aerodynamic damping being perfectly adequate.

4.5 Droop stops

When a rotor is turning at flight RPM, centrifugal stiffening tends to hold the blades straight and they will cone to whatever angle is necessary for a force balance in the rotor head. With flapping hinge offset, if the tip path plane tilts, a couple will be created which tilts the hull to follow. As a result the freely flapping blades will not strike the tail boom. There have been some spectacular incidents in wind tunnels where the hull of the helicopter was fixed and this mechanism could not operate.

As a rotor slows down, the blades will droop and stops are needed to restrict the travel. These can be fixed if a tall mast is used. However, Figure 4.4 shows that at flight RPM the blades are nearly straight whereas at rest they droop. Thus the actual angular motion allowable at the head is greater when the blades are turning. This requirement can be met by using moving droop stops which are thrown out of engagement as the rotor speeds up, and which return under the influence of springs as it slows down. Stops are also required to restrict the upward motion of the blade if it 'sails' whilst starting up in a wind.

4.6 Introduction to dragging

In the absence of coning, the flapping hinges would be sufficient for the shaft axis to tilt with respect to the tip path axis, but greater freedom is needed in the presence of coning. Figure 4.5 shows an example of a machine hovering with the CM not directly below the rotor head so that the mast is not vertical. When viewed along the tip path axis at (a), the blades are evenly spaced, but when viewed along the shaft axis (b), the blades are not evenly spaced, and dragging hinges are necessary to allow this angular change with respect to the rotor head so that the blades can travel at constant velocity to conserve momentum. If the blades were not allowed to drag, they would experience alternating bending moments due to the geometrical conflict.

The dragging motion in the hover is due to the rotor head turning on the shaft axis whilst the blades turn on the tip path axis. If a flexible joint were installed in the rotor shaft, the rotor head could turn on the same axis as the blades and then there would be no dragging movement due to coning and no need for dragging hinges. A simple flexible joint such as the Hooke or universal joint used in automotive propeller shafts

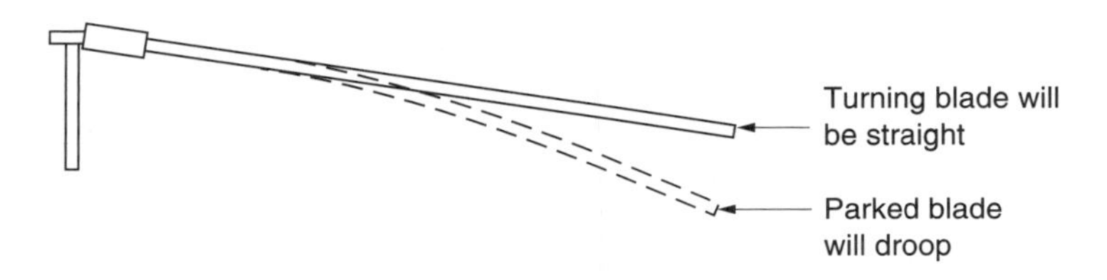

Fig. 4.4 In flight the blades are held straight by centrifugal stiffening. At rest the blades will droop further. Retractable stops may be used to prevent excessive droop without limiting flapping in flight.

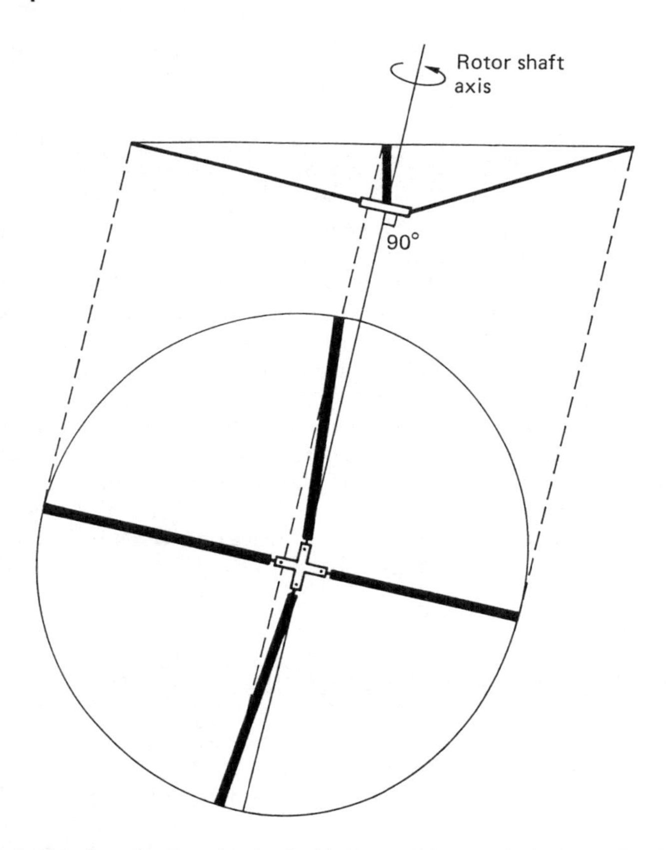

Fig. 4.5 When viewed along the tip path axis, the blades are 90° apart in the hover, but if the CM is offset, the shaft axis and the tip path axis are not aligned. With respect to the shaft axis the blades are not evenly spaced. Dragging hinges allow this geometric conflict to be resolved.

is shown in Figure 4.6(a). This, however, does not give constant angular velocity when it is deflected and this would cause torsional stress in the transmission. What is needed is a true constant velocity joint of the type used in front-wheel-drive cars. In the LZ-5 helicopter designed by Glidden Doman a true constant velocity joint was used (b).

A rotor head having only a constant velocity joint would not be able to transmit any moments to the mast and so would be classified as a zero-offset head. Such heads are only suitable for lightweight machines. However, it would be possible to fit springs to the head to give it some ability to transmit moments.

Clearly the blades are being driven through the air by torque supplied to the rotor head. In an articulated head the dragging bearings prevent torque being transmitted to the blades. The blades actually drag back until the blade tension applied to the rotor head is no longer truly radial but is displaced with respect to the mast axis. Figure 4.7(a) shows that this results in the torque from the transmission being balanced in the rotor head. The rotor head will often be constructed with a typical offset built in as shown in (b).

Note that the blade grip attaches to the blade on the axis of the mass centroid of the blade so that no bending stress is caused at the mounting. As the mass centroid is

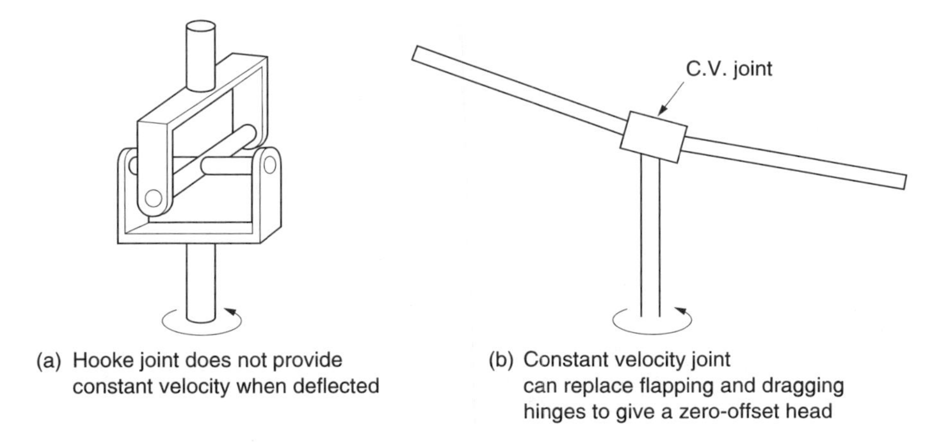

Fig. 4.6 (a) The simple Hooke joint does not transmit constant velocity when deflected. (b) A constant velocity joint may be used instead of hinges in the rotor head.

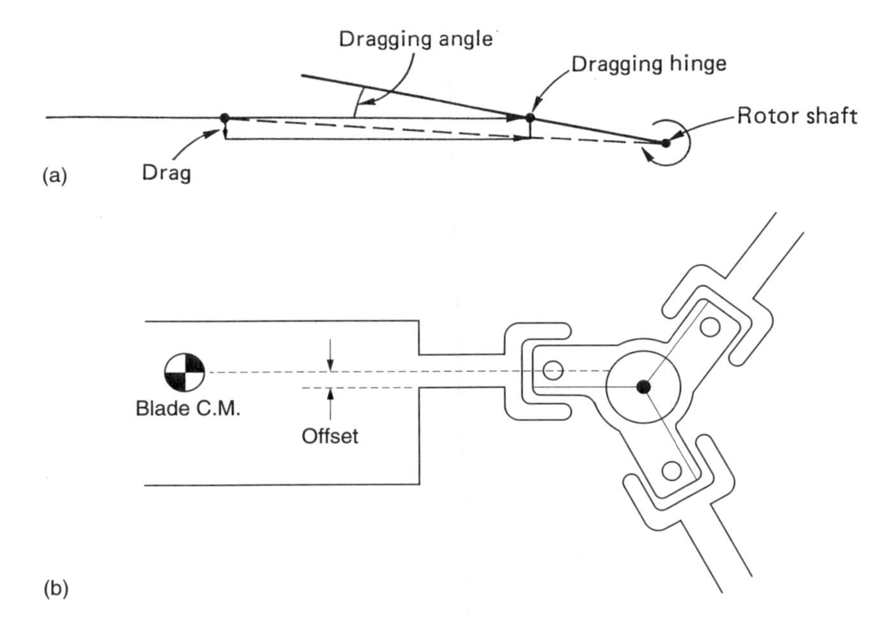

Fig. 4.7 (a) Blades on dragging hinges will drag back until the root tension is offset from the shaft axis to allow torque to be delivered. (b) The head may be constructed with the necessary offset built in.

set forward to about 25% of the chord for stability, the attachment point will also be forward of the mid-chord point.

When a blade flaps up, perhaps in response to a cyclic input, the centre of mass of the blade moves towards the shaft and conservation of momentum causes the blade to accelerate in the disc plane. This explanation is for a stationary frame of reference. However, in a rotating frame of reference, the blade appears to be stationary and has no momentum. In order to explain how it appears to move forward a virtual force,

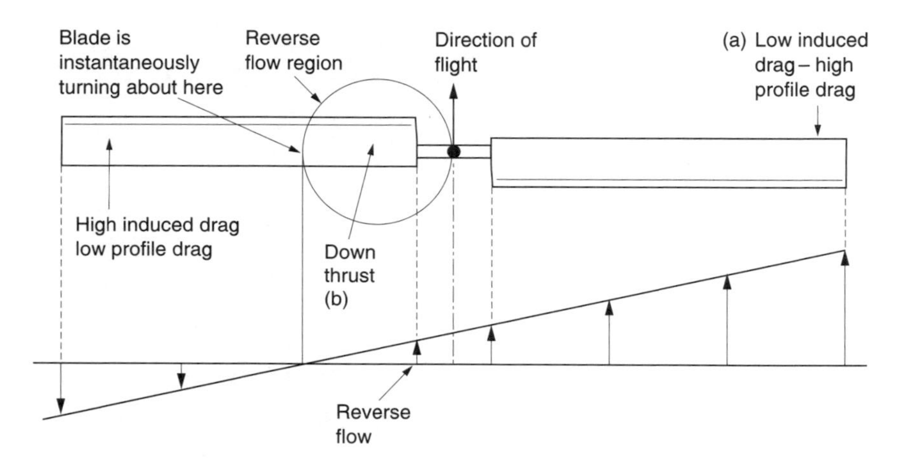

Fig. 4.8 (a) The application of cyclic pitch in translational flight results in reduced induced drag on the advancing blade, but increased profile drag. (b) The root of the retreating blade experiences down thrust due to reverse flow.

called a Coriolis force, has to be imagined to act. If the head itself is completely rigid, the in-plane blade movement due to conservation of momentum results in bending in the blade. The resultant forces are often called Coriolis forces. In fact the forces arise because the rigid head tends to prevent the blade conserving its momentum whereas drag hinges allow momentum to be conserved.

In translational flight there is considerable asymmetry in the conditions experienced by the advancing and retreating blades. The most obvious consequence of this asymmetry is the application of cyclic pitch in order to produce equal lift moments on each side of the rotor. This will result in the induced drag and the profile drag being functions of blade phase angle and radius. Note that these two functions are quite different. Figure 4.8 shows that the induced drag falls on the advancing blade because the angle of attack is reduced, whereas the profile drag increases because the velocity of the RAF is higher. Figure 4.8 shows that, at an inboard radius, reverse flow causes the root end of the blade to develop an undesirable down force.

The result of these induced and profile drag variations will be in-plane bending moments within the blade as well as overall in-plane moments at the blade root. Dragging hinges are intended to relieve in-plane root moments. However, the presence of dragging hinges allows in-plane blade motion and it is important that this motion is stable. Unlike flapping, the restoring force when the blade drags away from its neutral position is quite small and so the resonant frequency is much lower than the rotational frequency. For in-plane motion, the degree of aerodynamic damping is also very small. It was seen in Chapter 1 that additional damping is often necessary on the dragging axis to prevent ground resonance. This phenomenon will be discussed in sections 4.16 and 4.17.

4.7 Order of hinges

The order in which the flapping, dragging and feathering hinges are disposed is subject to a certain amount of variation from one design authority to the next. The use of coincident flapping and dragging axes is common because it allows a compact bearing assembly that is particularly important on multi-blade heads, where the head can become a serious source of drag if it is larger than necessary. Figure 4.2 showed an example of a head in which the flapping and dragging hinges are combined in a single Hooke joint having a cruciform central member. The logical place for the feathering bearing is outboard.

Figure 4.9(a) shows that the three hinge axes are not necessarily orthogonal. This results in coupling between the axes, sometimes deliberate, sometimes an unavoidable geometric compromise. A hinge has an axis and this can only uniquely be defined in three dimensions. In Figure 4.9(a) O is the rotor head, OZ is the rotor shaft axis and OX is the blade span axis. OXY is the plane of the rotor. In an orthogonal head, OY would be the flapping axis, OZ would be the dragging axis and OX the feathering axis.

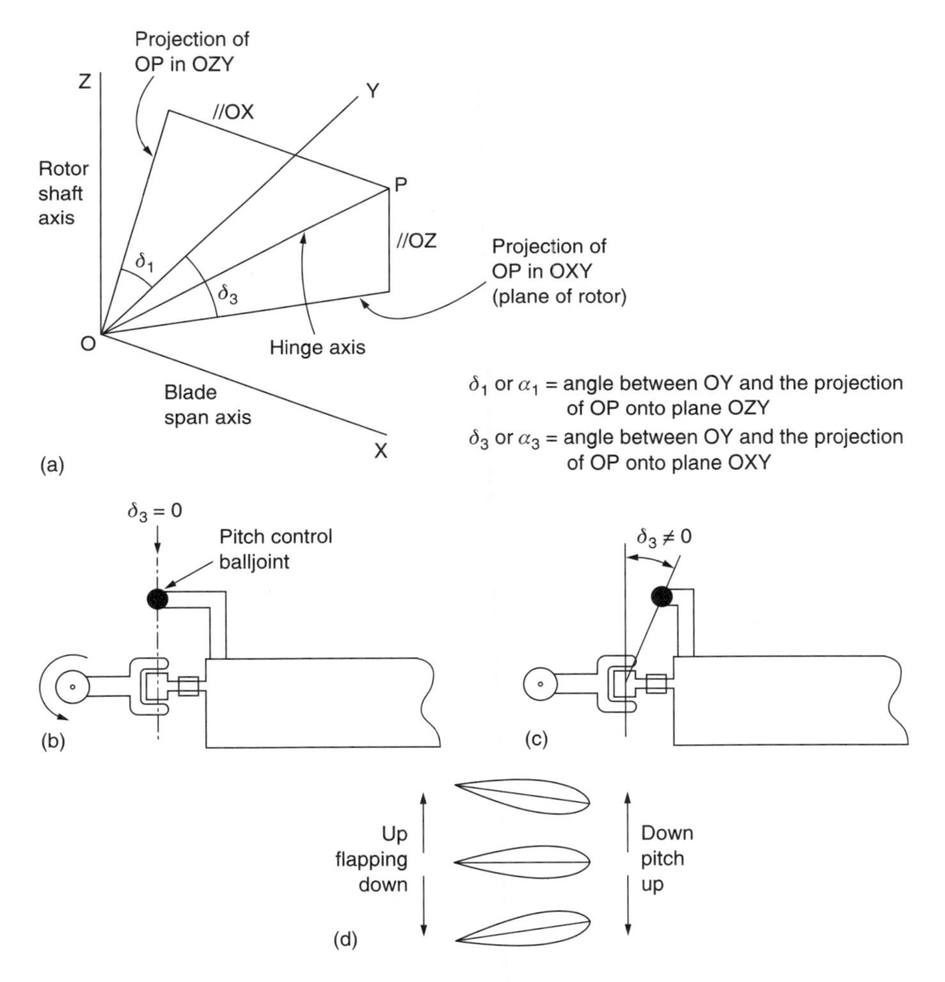

Fig. 4.9 (a) The axes of hinges in the head are not always orthogonal. Two of the axes may be turned as shown by amounts delta-1 and delta-3. (b) To avoid coupling between flapping and feathering, the pitch change rod must terminate on the flapping hinge axis. (c) The pitch change rod may terminate outboard of the flapping axis. This results in a delta-3 effect (d).

OP is an arbitrary hinge axis and could be a flapping or a lagging hinge. In the literature, the flapping hinge is called the δ hinge and the dragging hinge is called the α hinge. These hinges can be turned away from the mutually orthogonal in three axes, 1, 2 and 3. Only two of these axes are shown in Figure 4.9(a); axis 2 is redundant. If OP is a flapping hinge and is projected into the plane of the rotor, OXY, the angle between the projection and OY is called δ_3 (delta-three). If OP is projected onto the plane OZY, the angle between the projection and OY is δ_1 (delta-one). In an orthogonal head these angles are both zero.

Figure 4.9(b) shows that to avoid coupling between flapping and feathering, the spherical bearing between the pitch control rod and the pitch control arm must be on the axis of the flapping hinge which must also be at 90° to the feathering axis.

The pitch control arm is generally on the leading edge of the blade. If the pitch control arm is made slightly shorter as in Figure 4.9(c), there will be an interaction between flapping and feathering. Figure 4.9(d) reveals that upward flapping will reduce the pitch of the blade, tending to lower the blade, whereas downward flapping would have the reverse effect. Effectively a positive δ_3 hinge has been created. This is a stable condition, whereas the reverse relationship between the control rod joint and the flapping axis to give a negative δ_3 hinge would be unstable.

In some autogyros the dragging hinge was inclined to produce a δ_1 hinge. This had the effect of coupling the dragging and the blade pitch to allow jump take-off. The blades would stay in fine pitch whilst being driven, but upon the drive being disconnected the blades would swing forward, increasing the pitch.

In some toy free-flying helicopters a reverse δ_1 hinge is used which sets the machine automatically into autorotation when the fuel runs out.

4.8 Types of rotor head

For many years the designer was faced with a choice between a teetering two-bladed head and an articulated multi-bladed head. These techniques were developed because they greatly reduce bending loads on the blades. The teetering head imposes the least stress on the mast, but has some drawbacks as will be seen in section 4.10. In the early years of helicopter design, available blade materials and designs precluded the use of other types of rotor head and the drawbacks had to be accepted. Although it is less demanding on the blades and the feathering bearings, the articulated rotor head contains a mass of bearings subject to oscillating motion that causes wear. These bearings require frequent replacement. Large machines may require continuous oil feed to the bearings, whereas smaller machines require the periodic application of a grease gun. The hingeless rotor head is a desirable goal if only because it reduces the amount of maintenance required.

The effect of blade flapping in articulated and teetering heads is also used to prevent excessively rapid response to cyclic control inputs. This will be made clear in section 4.11.

The stress due to coning can be relieved by fitting the blades at a preset coning angle. In practice, the disc and shaft axes will be slightly different since the hull blowback will never be perfectly balanced by the tail plane, particularly if the CM is displaced. As the hinges are only resolving a small geometric problem, it is clear that a rotor strong enough or flexible enough to accommodate the geometric conflict can dispense with actual flapping and dragging hinges. The result is a hingeless rotor. The term 'rigid rotor' is sometimes used to describe such a system, but clearly it is a misnomer and the term hingeless is to be preferred.

Pitch change or feathering is, of course, still necessary, and this may be achieved by an actual bearing or by a further degree of flexural freedom. The feathering bearing must be able to withstand the flapping and dragging moments in addition to the axial pull.

Although the elimination of hinges reduces maintenance, it is not necessarily a goal to reduce flapping. An articulated rotor can flap to decouple rapid variations in thrust from the hull. There is thus a good argument for a rotor head that combines the mechanical simplicity of the hingeless head with the ride quality of the articulated head. As will be seen in section 4.11, the designer can make different compromises between ride quality and manoeuvrability by changing the stiffness of the flexures.

4.9 Zero-offset heads

The main characteristic of the zero-offset head is that the blades cannot transfer moments to the mast, and the helicopter hull hangs from the rotor like a pendulum. The mast and gearbox do not have to be as strong and so can be lighter, an important consideration in early machines where power was limited. The Sikorsky R-4 had a three-blade head with zero-offset as shown in Figure 4.10.

Figure 2.9 showed the result in the hover if the centre of gravity of the hull does not coincide with the mast. The tip path plane remains horizontal, as the lift vector must be vertical, but the articulation must act like a universal joint so that the mast can tilt with respect to the tip path plane until the centre of gravity is below the centre of the rotor head.

The zero offset rotor head needs to be mounted on top of a relatively tall mast to obtain a reasonable range of CM positions without undue changes in hull attitude. It may be necessary to shift balance weights if the number of passengers is changed, and the fuel tanks will have to be located on the CM.

Figure 4.11(a) shows that the teetering rotor is a special case of a two-bladed zero-offset articulated rotor where the two blades are rigidly attached to one another (except for the feathering bearings) and share a common articulation bearing. Both Bell and Hiller have produced a number of machines using teetering rotors. Bending stresses due to coning are reduced in the teetering rotor by attaching the blades at the typical coning angle. As the coning of the blades raises the centre of mass of the rotor, the teetering

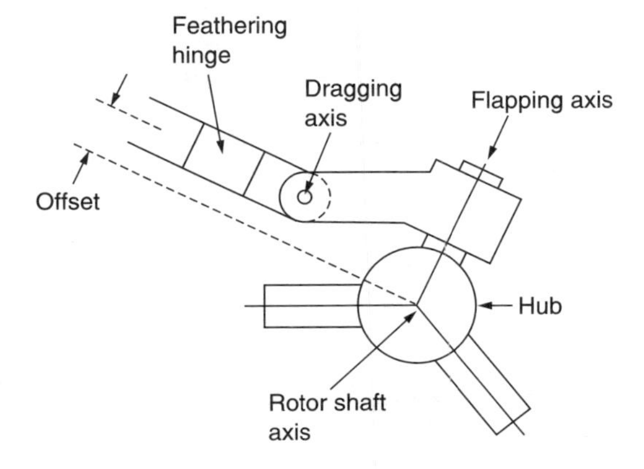

Fig. 4.10 A zero-offset head. The axes of the flapping hinges are coincident with the shaft axis.

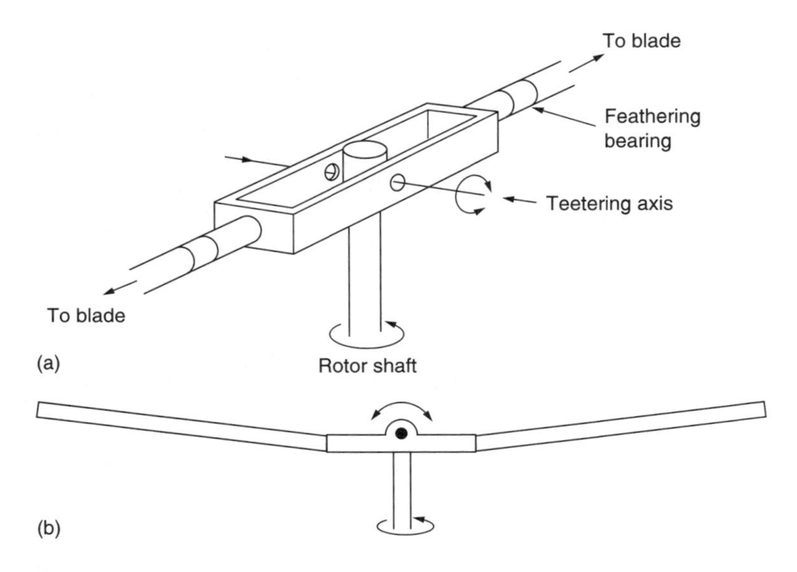

Fig. 4.11 (a) A teetering rotor has a single flapping hinge. The blades are rigidly connected together in the flapping and dragging axes and may be mounted at a typical coning angle to reduce stress. (b) In an underslung teetering head, the flapping hinge is elevated so that it approximately coincides with the CM of the rotor.

bearing is then raised above the feathering axes at the hub so that it is essentially at the rotor CM. There will then be less vibration when the tip path axis is not parallel with the shaft axis. The result is called an underslung rotor shown in Figure 4.11(b). As the rotor head itself tilts to stay in the plane of the rotor disc, the effect of Figure 4.5 cannot occur and so dragging hinges are not needed to overcome it. Forces due to varying drag in translational flight are resisted by blade stiffness.

The rigid connection between the blades means that their dragging frequency will be very high. It will be seen in section 4.17 that this means no drag dampers are required. The undercarriage needs no damping to prevent ground resonance and can be a simple tubular skid arrangement. This further reduces the weight of the machine.

Two-bladed rotors have a disadvantage that both of the blades simultaneously enter a lift trough in forward flight at 90° and 270° of rotation. This causes a twice-per-revolution hop (two-per) at the hull. Later teetering rotors incorporated a degree of flapping flexibility in the rotor head to decouple the hop from the hull. This is shown in section 4.20.

At speed the hull of a zero-offset machine tends to blow back because the drag D is applied a moment arm c below the suspension point as shown in Figure 4.12. At speed, the tail plane will have to apply a download T at moment arm b to counter the blowback. As a result the shaft and tip path axes nearly coincide. In a machine with an articulated or flexural head the hinge offset will generate a restoring couple if the hull is blown back so a less powerful tail plane is sufficient.

4.10 Dangers of zero-offset heads – negative *g*

The main advantage of the zero-offset rotor head is that no moments can be transferred to the mast from the blades and so the mast can be very light. The absence of dragging

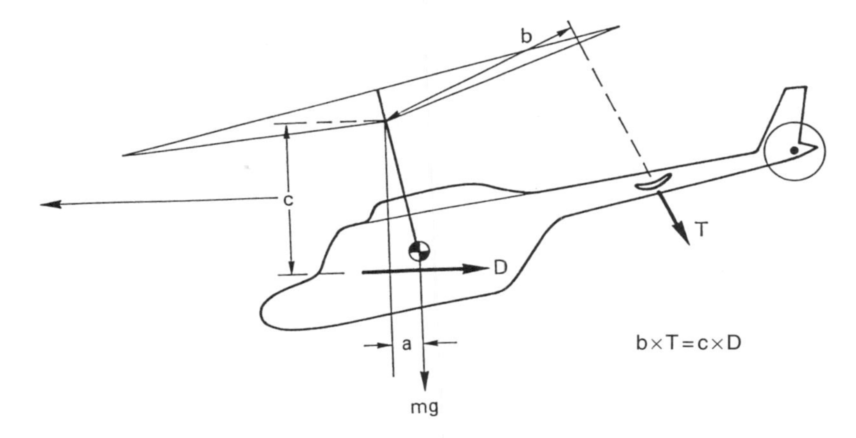

Fig. 4.12 With a zero-offset head, blowback in forward flight must be opposed by a down force from the tail plane.

and undercarriage dampers allows a further weight saving. However, it is vital that the machine is never placed in a flight condition where the rotor thrust is downwards, because in this negative-*g* situation the hull attitude is unstable, like a pencil balanced on its point.

In the worst case the hull can roll so that the limit of travel of the teetering or flapping hinge is reached. The side thrust from the tail rotor can initiate such a roll. The resultant pounding of the limit stops can break the mast. In zero or negative *g* the rotor will cone flat or downwards raising the possibility that if the hull pitches the tail boom can swing into the rotor disc.

A number of Robinson helicopters have been lost due to these effects, although the manufacturer stresses the avoidance of manoeuvres such as pushovers that can induce the negative-*g* condition. In the event of getting a zero-offset helicopter into negative *g*, there is only one solution and that is to return lift to the blades. This can be achieved by the application of back cyclic or raised collective or both.

In some zero-offset helicopters, such as the Bell 222, a strong spring is fitted to the teetering axis that minimizes the negative-*g* instability, but it does require a stronger mast and gearbox to withstand the spring forces.

Zero-offset heads will not tolerate a wide range of CM movement in any direction and are more prone to blade sailing when starting in high winds.

4.11 Rotor response

Different rotor head designs respond to cyclic control inputs in different ways contrasted in Figure 4.13.

Consider a two-bladed rotor where the pitch control rods are at 90° to the blades. For example, if the swashplate is tilted down at the front, the pitch angle of the blade moving forwards will be reduced, and the pitch angle of the opposite blade will be increased. This will cause a lift moment difference that results in a rolling couple. However, the gyroscope-like phase lag of the rotor will convert this into a pitch forward, so the front of the rotor disc will drop. In the hover, if the rotor is articulated, it will continue to pitch forward until it is parallel with the swashplate. At this point there will be no more

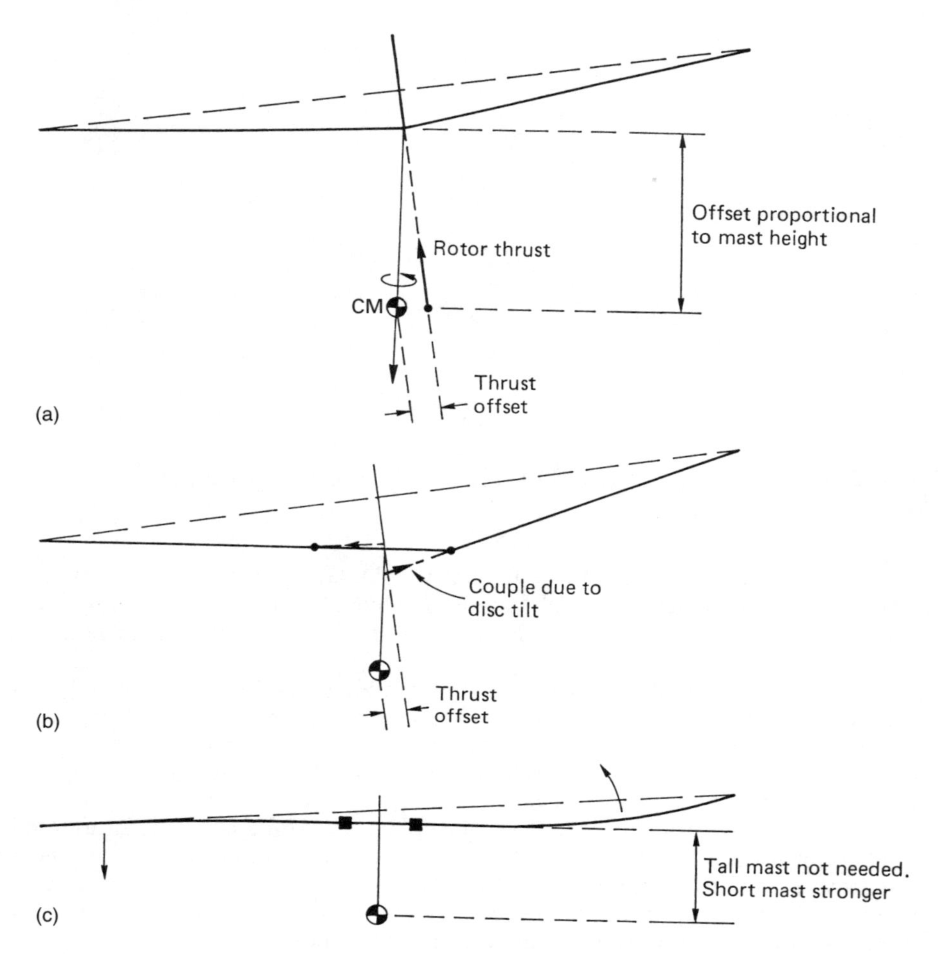

Fig. 4.13 Comparison of the response of zero-offset, offset and hingeless rotors to a cyclic input. (a) In the zero-offset head, a moment is produced because the rotor thrust no longer passes through the CM. (b) In the offset head, the blade root tension is not symmetrical and a couple is produced in addition to the effect of (a). In a hingeless rotor head, (c) couples can be applied to the mast via the stiffness of the blades.

cyclic variation of the blade pitch angle, and so the rotor will remain at this angle. The tip path plane tries to stay parallel to the swashplate. This is known as following, and the speed with which the rotor follows the swashplate is known as the following rate.

The following rate depends on the relationship between the aerodynamic forces trying to tilt the disc and the inertial forces trying to oppose the tilt. The aeromechanical parameter that is used to quantify the phenomenon is called the Lock number.

For a given rotor, the following rate is proportional to the RPM and the air density. Operation at high density-altitude will result in the cyclic response reducing, and changing in phase.

In practice the following rate is generally high enough that the rotor will respond to a change of control axis within about a revolution. This does not mean that the whole machine will respond in this way, which is just as well. There is a considerable difference between the following rate and the response rate of the machine.

In a real helicopter, the rotor is not rigidly attached to the hull because of the presence of flapping hinges or flexures. As the rotor disc tilts, the hull will lag behind so that the tip path axis and the shaft axis do not coincide. This non-coincidence causes moments and it is these that govern the tilt of the hull. As the hull tilts, it causes a further tilt of the swashplate. This causes a further tilt of the rotor and so on. This is an example of positive feedback. Consequently in practice a steady application of cyclic control causes the rotor disc to tilt at constant speed. The ratio between the two is the response rate. The stiffer the connection between the rotor and the mast, the more the rotor is able to tilt the hull and swashplate, and so the greater is the positive feedback and the greater the response rate for a given cyclic application.

In general the stiffer the rotor head, the more rapid the response rate will be. Before the mechanism was understood, some experimental helicopters had response rates beyond the ability of the pilot and were uncontrollable. The use of hinged rotor heads in early helicopters had the advantage of reducing the response rate to what the pilot could manage.

The case of a machine that is initially hovering in equilibrium will be considered. In the zero-offset head, the application of a cyclic input causes the tip path plane to tilt with respect to the mast. As a result the thrust vector no longer passes through the CM, and a roll couple is created as shown in Figure 4.13(a). This will be proportional to the height of the rotor above the CM, and is a further reason for tall masts on teetering machines. The tip path plane initially tilts without the mast following it. Since the cyclic pitch control operates with respect to the mast axis, tip path tilt opposes the cyclic pitch control, so the further the disc tilts with respect to the mast, the less cyclic pitch change the blades get. Excessive following rate is thus prevented. As the teetering head cannot exert a couple on the mast, it can only pitch or roll the hull if the rotor thrust is tilted. If the rotor thrust is reduced for any reason, the cyclic control becomes less effective. This may happen in a pushover manoeuvre. The rotor disc will follow the cyclic control, but the hull doesn't follow the disc. Such a situation may result in mast or droop stop pounding and all pilots who want to live will avoid it.

In the case of an articulated rotor head with offset flapping bearings, the tip path plane tilts as before, and moves the thrust vector with respect to the CM as before, but in addition the blade tensions no longer align and this results in a roll couple being applied to the rotor head as shown in Figure 4.13(b). The further apart the flapping bearings are, the more powerful this effect will be, speeding up the following rate.

Consider a fully articulated head with torsion springs fitted to the flapping and dragging axes. If the springs are stiff, there will be little deflection of the bearings. Deflection will be transferred to the blades and they will flex. If the springs are weak, there will be little to prevent deflection of the bearings, and bending loads on the blades are minimized. Clearly the flexural rotor head (c) can behave more or less like an articulated head depending on the stiffness of the flexures. The stiffness of the flexures may be different in the flapping and dragging directions. Sections 4.16 and 4.17 will explore the effects of dragging stiffness.

In the flexural rotor head, a single structural element often acts both as a bearing and as a spring. It will bend, but not without an applied moment, and so flapping and dragging are allowed by a combination of rotor head and blade flexing. The distribution of bending is clearly a function of the relative stiffness. All other things being equal, in an articulated head the flapping hinge offset determines the magnitude of the couple applied to the mast when the tip path axis deviates from the mast axis, whereas in a hingeless head this couple is determined by the stiffness of the system. In order to simplify analysis of the dynamics of a helicopter having a hingeless head, it is possible

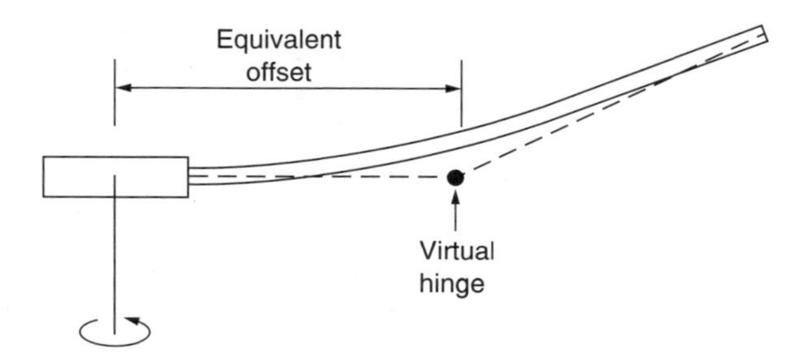

Fig. 4.14 In a hingeless rotor, the same couple could be produced by a rotor with flapping hinges having an equivalent offset.

to describe the rotor by its effective or equivalent offset. Figure 4.14 shows that this is the radius of a virtual flapping hinge that would produce the same couple for the same disc tilt.

In the hingeless rotor head the tip path plane will attempt to tilt as the result of a cyclic control application, but the roll couple applied to the rotor head will depend on the flapping stiffness. Hingeless rotor machines with stiff flapping flexures can be highly manoeuvrable, as the Westland Lynx regularly demonstrates. In fact a hingeless rotor helicopter can be rigged to hover upside down as many model helicopter pilots have demonstrated.

Whilst the stiff hingeless rotor head leads to an agile machine, it will impose heavy bending loads on the blades and possibly also on the feathering bearings, which must be designed accordingly. Unless carefully designed, such rotors can also respond so rapidly to cyclic inputs that control is difficult. A permanently operating gyrostabilizer may be needed.

In hingeless rotors with stiff flapping flexures, the thrust vector offset with respect to the CM is much less important, and the mast does not need to be so tall, although machines with very short masts have suffered from powerful canopy vibration due to the non-uniform rotor downwash. Stiff rotors can withstand a wider range of CM travel because a rotor head couple can be used to keep the hull in the correct attitude, being limited by stresses in the mast and its bearings.

The stiff hingeless rotor is the goal for the ultimate in manoeuvrability, at the cost of a rougher ride and the need for a stronger mast and airframe. A more compliant hingeless rotor head allows a more cost effective solution and a better ride. The bearings of the articulated head are replaced by elements that can flex and thus need no lubrication or maintenance. It is not inconceivable that flexural heads will be developed which will last the life of the machine. Given the high maintenance costs of helicopters, any development that tangibly reduces wear and maintenance is highly significant.

4.12 Feathering

Despite the enormous pull of the blades, it must be possible to rotate them on their feathering axes in order to control the pitch angle. There are several approaches to the problem of the feathering bearing and these are contrasted in Figure 4.15.

Raoul Hafner first used a torsion bar shown at (a) to eliminate the axial thrust in the feathering bearing so that the remaining bearing became essentially a locating sleeve. It is a characteristic of bending and torsion that the greatest stress is caused at the surface

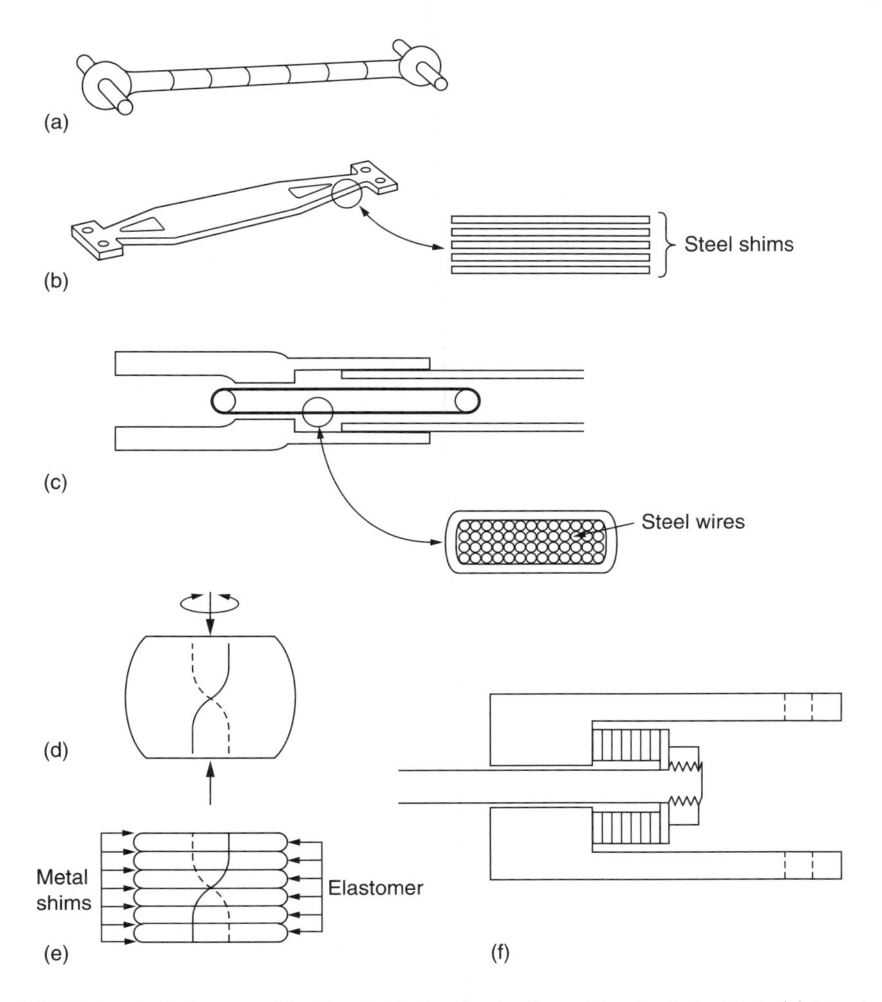

Fig. 4.15 Various techniques used to allow feathering despite the root tension in the blade. (a) A torsion bar resists the axial pull. (b) Thin steel shims may be laminated to produce a flexible structure. (c) Strap made from strands of wire. (d) Elastomer under compression allows rotation in shear, but bulges under load. (e) Metal shims between layers of elastomer prevent bulging. (f) A practical elastomeric thrust bearing which is failsafe because the blade cannot come off if the bearing fails.

so the interior of thick components is unstressed. This is why tubes are used instead of solid bars. A given cross-section of material will be more rugged if it is divided up into many slim elements in parallel and in addition will be easier to twist. This division into many elements also gives a degree of protection against failure because some elements can fail without the whole assembly failing.

The blade may be attached by laminated steel shims shown at (b) which can twist as the pitch is changed. PTFE tape may be placed between the shims to allow one to slide over the next. A variation in the idea is used by Bell on the JetRanger. Here the tensile element is a strap made from hundreds of strands of steel wire (c) like the cable in a suspension bridge. The blade is guided by a sleeve bearing that only takes bending loads.

Enstrom use an elastomer in compression to retain the blade. Elastomers are virtually incompressible but flexible synthetic compounds. Figure 4.15(d) shows that if a block of elastomer is placed under compressive stress, it will bulge at the sides. If the bulging is serious, the strength of the material will be exceeded and it will tear. Bulging would also allow the blades to move away from the mast, causing balance problems. Figure 4.15(e) shows that this problem is overcome by interleaving thin shims of metal with layers of elastomer. The metal shims prevent bulging under compression, but have little effect on the ability of the unit to twist so that pitch changes can be made. As the diagram shows at (f), the blade is guided by sleeve bearings, but the elastomer takes the thrust. Note that the system is failsafe because the bearing is in compression. Even if the bearing begins to disintegrate, the blade cannot come off.

In the Enstrom only the feathering axis is elastomeric. Aerospatiale pioneered a spherical elastomeric laminated bearing which allows full articulation in addition to feathering.

4.13 Pitch control

There are a number of mechanisms suitable for pitch control. No two designs have exactly the same control linkages, but the diagrams here are representative of actual practice. One of the best ways of appreciating how these systems work is to have someone move the controls of a helicopter one at a time whilst observing how the linkages move.

The most common control system is the sliding swashplate shown in Figure 4.16(a). The swashplate is made in two halves. The lower part is fitted with a spherical bearing so that the swashplate can tilt to any angle in any direction. The sphere can also slide up and down, often on the outside of the gearbox. A large diameter ball race is fitted between the two halves of the swashplate. A pair of jointed scissors links is fitted. One of these ensures that the lower half of the swashplate cannot revolve and the other ensures that the upper part turns with the rotor shaft. Each rotor blade grip is fitted with a pitch-operating arm. The end of each arm is connected to the periphery of the swashplate by a pushrod fitted with spherical bearings.

Movement of the collective lever raises or lowers the swashplate, and this causes all of the pitch operating arms to rotate all of the blades about their feathering axes by the same amount. In addition, movement of the cyclic stick in two axes causes the swashplate to tilt in two axes. Simultaneous lifting and tilting of the swashplate by two controls requires a device called a mixer, an example of which is shown in Figure 4.16(b). One end of the mixing lever is pivoted on the helicopter frame or on the gearbox and is connected to the collective lever by a rod. The opposite end is fitted with a bell crank operated by one axis of the cyclic control. The cyclic pitch changes take place differentially around the average pitch determined by the collective setting.

Figure 4.16(b) shows only one dimension of the cyclic mixing for clarity. In practice the cyclic control works in two axes and two bell cranks are fitted to the mixing lever. These bell cranks are connected to the swashplate by vertical rods and to the cyclic stick by horizontal rods. Moving the collective lever moves both bell cranks bodily up and down together, and so the swashplate rises and falls. Moving the cyclic stick rocks the bell cranks and tilts the swashplate about the collective setting in any desired direction.

Figure 4.17 shows an alternative arrangement known as a spider. The spider has radial spokes, one for each blade. Pushrods connect the ends of the spokes to the pitch operating arms. The hub of the spider is spherical so it can tilt and slide within the shaft and it is fitted with a vertical post passing down through the centre of the shaft

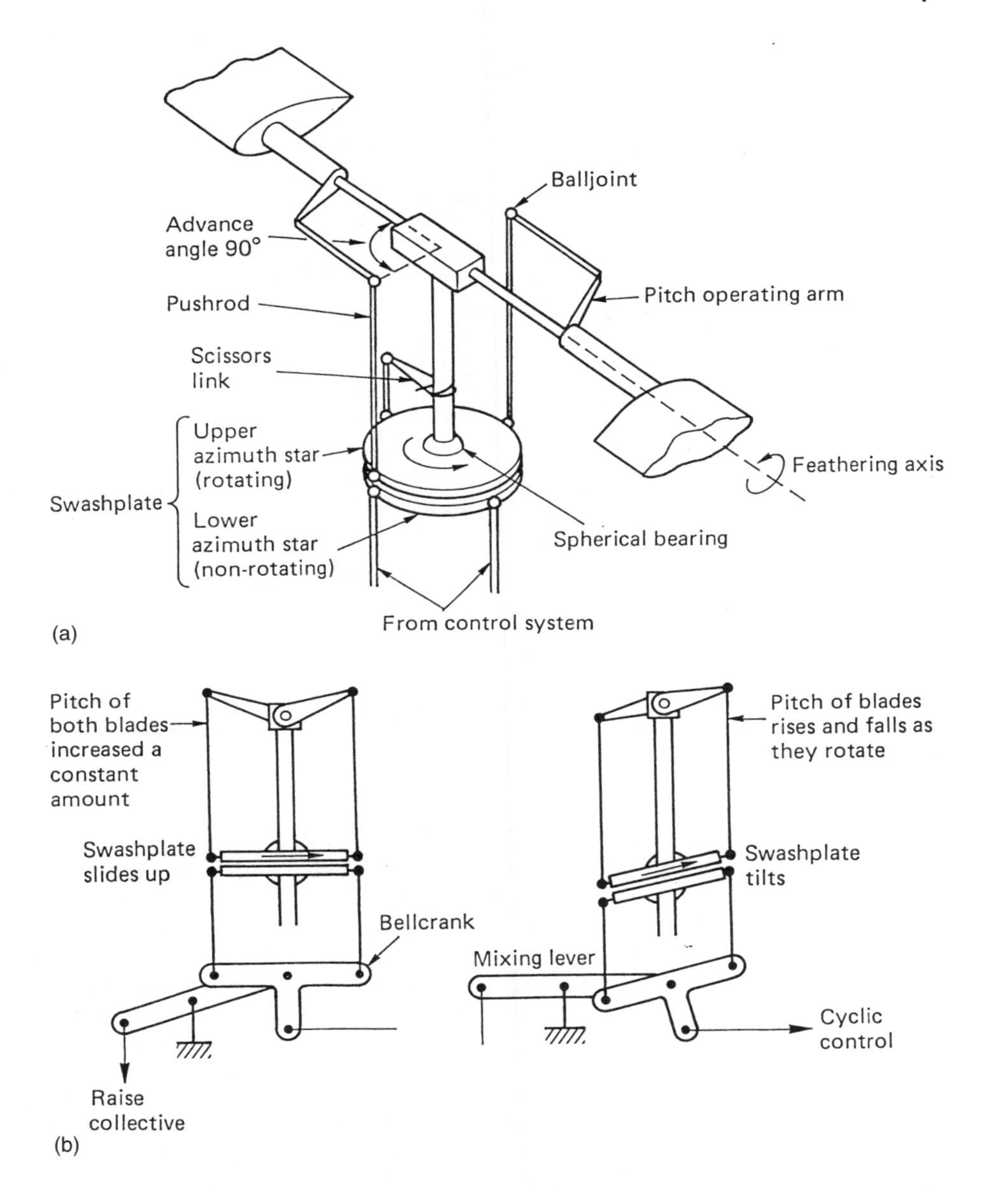

Fig. 4.16 The swashplate is a common method of controlling a rotor. (a) The two halves of the swashplate are connected by a ball bearing. (b) A mixer allows the swashplate to rise and fall with collective input whilst tilting with fore-and-aft cyclic inputs.

and protruding below the gearbox. The lower end of the post is moved up and down to obtain collective pitch changes and from side to side in two axes to obtain cyclic pitch control. Figure 4.17 shows an installation in which the radial arms pass through slots in the rotor shaft. Such a system is used in the Lynx where a significant part of the control system is protected inside the rotor head. In the spider system, jointed rods approach the bottom end of the spider post in three axes, and tilt and lift the spider.

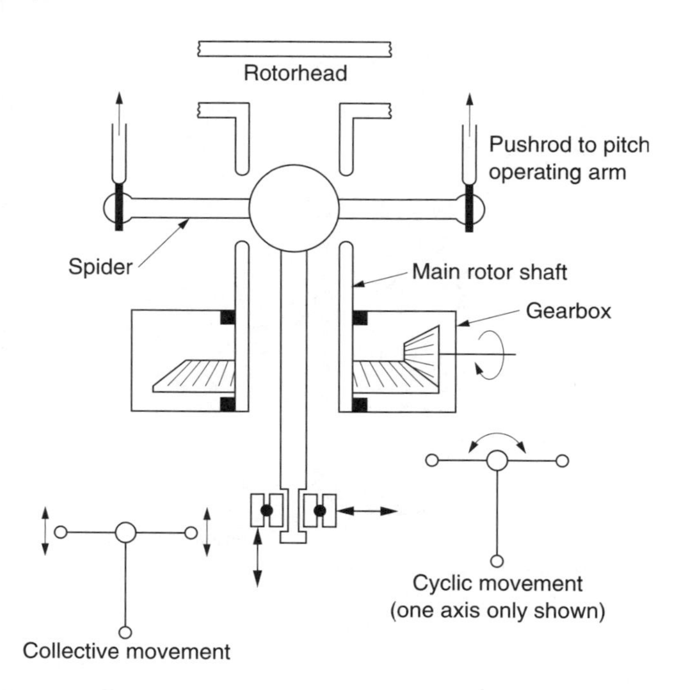

Fig. 4.17 A spider installation in which vertical movement of the spider controls collective pitch, whereas tilting the spider applies cyclic pitch. In this example the arms of the spider emerge through slots in the rotor shaft. In some cases the spider emerges from the top of the rotor shaft.

The Enstrom has a control linkage in which the pitch control rods pass up the centre of the rotor shaft to give an extremely neat external appearance and reduced drag. Figure 4.18 shows that the rotor head is fitted with rocker arms to transfer pushrod motion to the pitch operating arms. The pushrods are connected to a miniature spider located below the gearbox. The spider itself swivels on a universal joint so it can tilt in two axes whilst turning with the rotor. The spider support slides up and down in the rotor shaft under the control of a fork on the collective crank. The cyclic control rods tilt the spider to obtain cyclic pitch variation about the average collective setting.

Figure 4.19 shows the principle of servo tabs. Instead of attempting to change the blade pitch at the root, the blade is fitted with a small aerodynamic control surface rather like a miniature tail plane, typically at about 75% of rotor radius. Pushrods inside the blade allow the angle of the servo tab to be controlled. Servo tab systems have the advantage that the pilot only has to move the tab; the blade is moved by aerodynamic force on the tab. As a result power assistance is not necessary. The blade shank is made flexible to allow the tab to twist it. As the pitch control force is provided directly to the blade at an outboard location, there are structural and flutter resistance advantages. Servo tabs have been used extensively in the various models produced by Kaman, and are extremely effective. It is not clear why they have not been more popular elsewhere.

In a zero-offset rotor head such as the one shown in Figure 4.16, the rotor has a phase lag of 90° and to compensate, the pitch operating arms have an advance angle of 90°. In this case, the swashplate tilts in two axes exactly as the cyclic stick does and

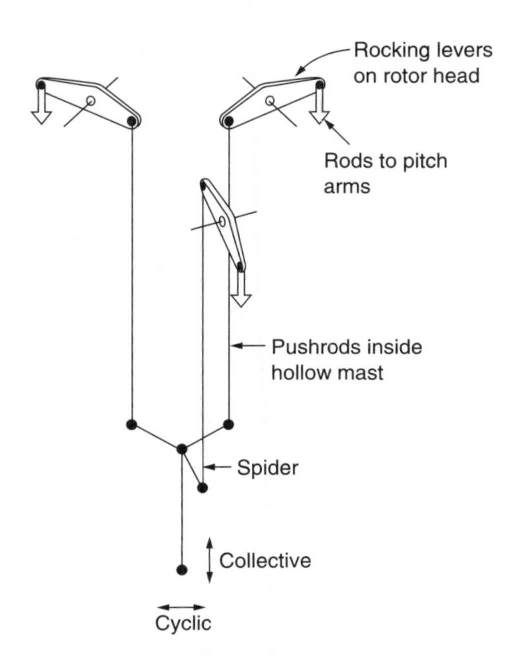

Fig. 4.18 The rotor control arrangement used by Enstrom. The spider is below the transmission and pushrods travel inside the shaft to rockers on the top of the rotor head.

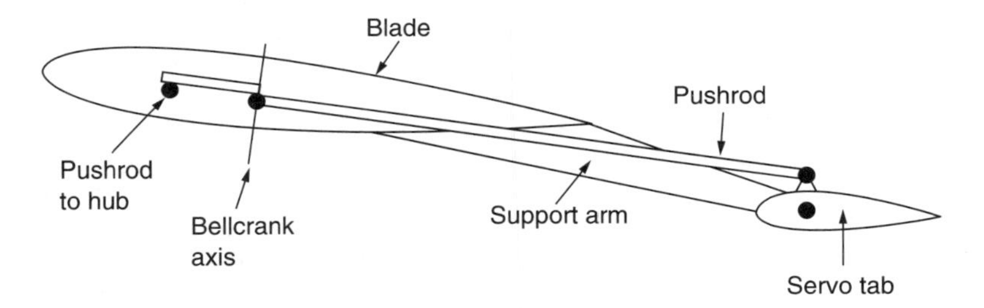

Fig. 4.19 Servo tabs are small control surfaces hinged on the trailing edge of the blade operated by pushrods running inside.

the axis about which the swashplate turns is the control axis. In fact the control cranks could be designed so that the swashplate stays parallel to the cyclic stick.

In multi-bladed heads this situation is seldom found. Such heads will be articulated or flexural and their flapping frequency will not be identical to the rotational frequency. As a result their phase lag will not be precisely 90° and the control system will need to compensate by arranging to tilt the swashplate in a slightly different direction to the cyclic stick.

Although an advance angle of 90° is easily obtained in a two-bladed teetering rotor, with articulated heads it becomes impossible because the end of the pitch-operating arm must terminate on the flapping axis to avoid the δ_3 effect (see Figure 4.9(b)). In multi-blade heads the end of the pitch-operating arm must remain clear of the next blade root. When the advance angle of the pitch operating arms is less than 90°, this is corrected by advancing the stationary part of the swashplate. Figure 4.20 shows that

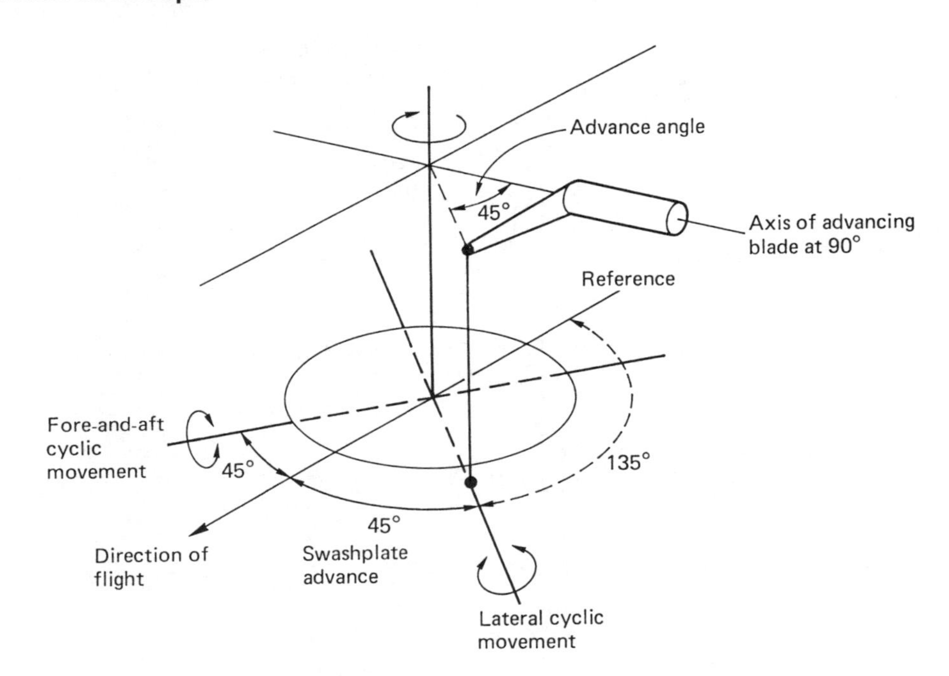

Fig. 4.20 Where the advance angle is less than 90°, the stationary part of the swashplate is advanced by rotation of the points at which the pushrods from the cockpit are attached. In this example, a four-blade head has an advance angle of 45°. The pushrods from the cyclic stick approach the swashplate with a further 45° advance. In a head with a phase lag of less than 90°, the swashplate advance could be reduced to compensate.

the pushrods from the cockpit must now connect at different angles. By consensus, all rotational angles are measured in degrees from the tail boom in the direction of rotation as shown.

The control axis can stay parallel to the cyclic stick, but the axis of the swashplate will tilt with a phase shift with respect to both. For example, a four-bladed head may have an advance angle of only 45°. The swashplate tilt will need to be advanced a further 45° with respect to the cyclic stick. Pushing the stick forward would cause the swashplate to tilt down midway between the front and the advancing side, i.e. at a phase angle of 135°, so the lowest point on the control orbit is reached 45° before the straight ahead position. At this point the pushrod going straight up to the pitch-operating arm will set the advancing blade at 90° to minimum pitch. The same blade will reach maximum pitch 180° later, and so there will be a roll couple towards the advancing side. The phase lag of the rotor will convert this to a forward tilt.

In other words the swashplate linkage has advanced the stick movement with respect to the control axis by 45°, the pitch-operating arms have advanced it a further 45°, and the 90° rotor lag cancels both, making the rotor tilt to follow the cyclic stick. Inspection of the cyclic pushrods in the Enstrom or the Lynx will show that they approach the spider at 45° either side of the centreline. The pitch-operating arms have an advance of a further 45°. The Chinook, having three-bladed heads, has an advance angle of 60°. In contrast the JetRanger, having only two blades, can have an advance angle of 90° in the pitch operating arms, and the swashplate moves parallel to the control axis. Since the advance angle of the pitch-operating arms and the phase advance of the swashplate always add up to the phase lag of the rotor, they can generally be ignored for the purpose of studying rotor response, and all that is of interest is the control axis.

The direction of main rotor rotation is quite unimportant. On most American machines the blades rotate anticlockwise when viewed from above, whereas most French and Russian machines use the opposite rotation.

4.14 Cyclic trim

The controls of a fixed-wing aircraft tend to be blown to the neutral position by the slipstream, and this results in the pilot feeling resistance roughly proportional to control deflection. This feel is important so the pilot knows how much control power he is using. In the helicopter, there is no equivalent of feel in the cyclic control. Forces fed back to the cyclic stick from the rotor head bear little relationship to the deflection, so feel must be provided artificially. This is done by springs that tend to return the cyclic control to a central position.

The cyclic displacement needed in translational flight to counteract lift dissymmetry and inflow roll would be tiring to maintain against spring pressure, so a trim system is used to alter the neutral position of the springs. Figure 4.21 shows how a trim servo works. The trim servo has telescopic cylinders containing strong springs that normally

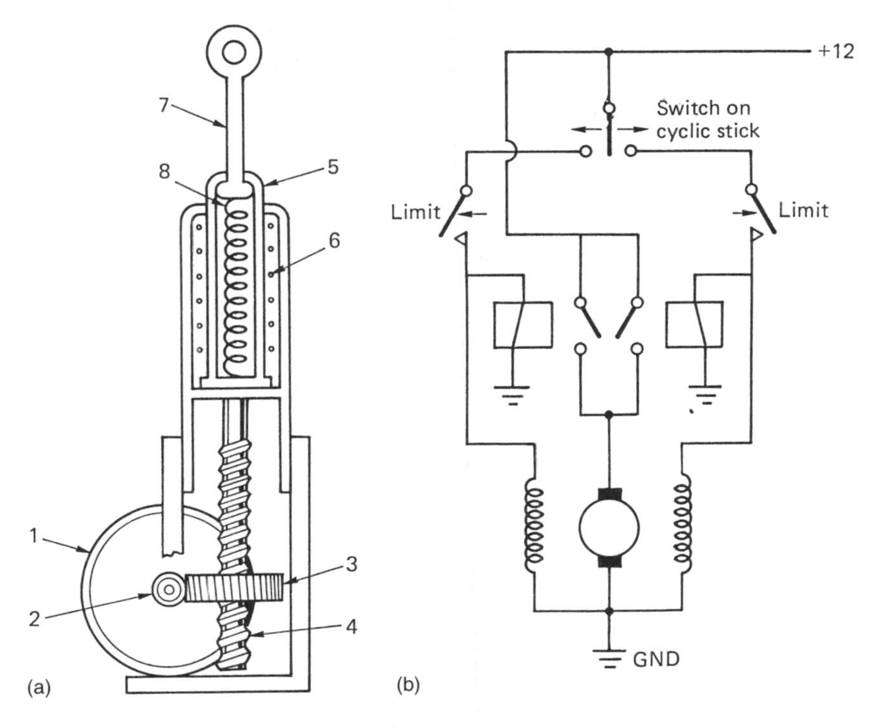

Fig. 4.21 A cyclic trim servo. (a) Springs 6 and 8 bias the stick to a neutral point which can be moved by an electric motor 1 which drives a worm 2, a wheel 3 and a leadscrew 4. If the pilot moves the stick away from the trim servo, sleeve 5 and rod 7 move together, compressing spring 6. In the opposite direction, spring 8 compresses and sleeve 5 remains stationary. (b) Circuit of trim motor drive. The motor direction is selected by choice of field coils which are wound in opposite senses.

hold the telescopic elements at their end stops. Control movement will compress one or other of the springs to give feel. The neutral position of the springs can be adjusted by an electric motor operating a screw jack. One trim motor is fitted to each cyclic axis. The motors are controlled using a small four-way switch mounted on the cyclic stick where it can easily be reached by the pilot's thumb. The switch is simply moved in the same direction the pilot is pulling the stick to hold the desired attitude. The force needed to hold the stick will be felt to reduce as the trim motor runs. Limit switches prevent the trim motor running too far and jamming.

As an alternative to trim motors, the centring springs may be attached to the hull reference by magnetic clutches. These are normally engaged by permanent magnets and the application of current in a coil can cause them to release. If the pilot holds the stick in the desired position and briefly applies the trim release current, the springs will extend to their neutral position and when the clutches grip again, the cyclic stick will be held in the trimmed attitude. The force trim release button will be on the handle of the cyclic stick.

4.15 Tilting heads

The equivalence of flapping and feathering was introduced in section 4.3. A rotor flapping with respect to the control axis is feathering with respect to the tip path axis. Figure 4.22 shows that this principle can be used for cyclic control. If the head has flapping hinges, it can be tilted with respect to the tip path axis. The result is that the blades are now cyclically feathered and the rotor will try to roll to a new attitude where the flapping and feathering cease. In the hover it will be able to do so, but in

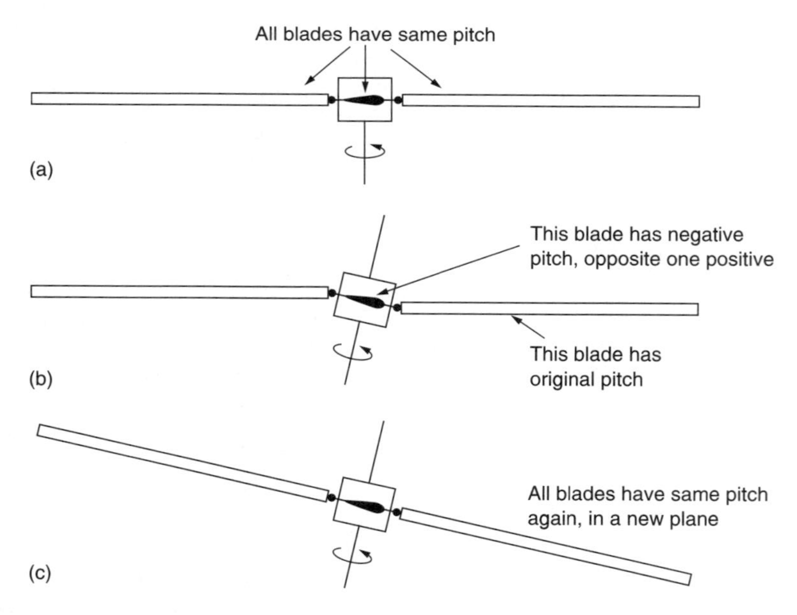

Fig. 4.22 Tilting head control. If the blades have flapping hinges (a), they can remain in-plane whilst the head tilts, (b). Head tilt applies cyclic pitch causing the rotor to follow the attitude of the head (c) (which is the control axis).

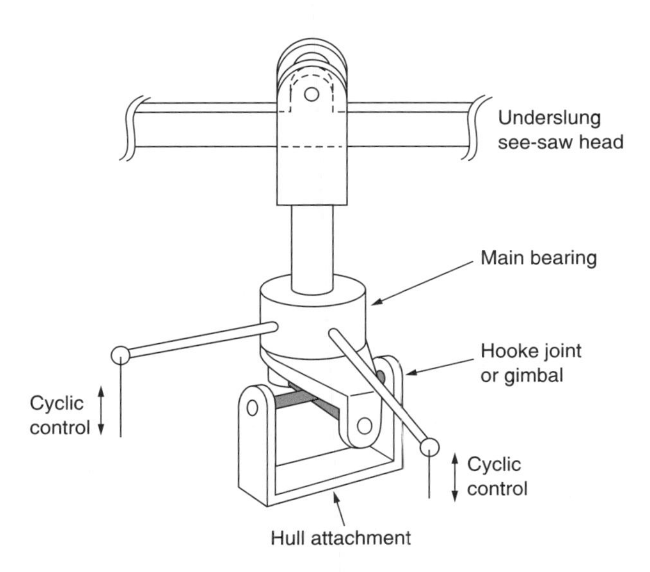

Fig. 4.23 A fixed-pitch tilting head needs no more than a gimbal and control rods leading to the cockpit.

translational flight a steady application of cyclic feathering is required to overcome lift asymmetry.

Early de la Cierva autogyros had aircraft-style control surfaces and no rotor control at all. The rotor was equipped only with flapping hinges. In forward flight, lift asymmetry would result in extra lift on the advancing side, but this would precess to an aft tilt of the disc. Figure 1.10 showed that the thrust of the airscrew overcomes the rearward component of rotor thrust due to the aft tilt. As the attitude of the hull does not change when the rotor tilts aft, the mast is now tilted forwards with respect to the tip path plane. The result is that the rotor has cyclic pitch applied. In an autogyro of this kind the rotor will flap back until the correct amount of cyclic pitch is obtained for the airspeed. Whilst the autogyro flies in this way, it should be emphasized that the helicopter cannot and does not.

In later de la Cierva machines, the aircraft control surfaces were dispensed with and the pilot could tilt the rotor head for pitch and roll. This approach is used to this day for control of lightweight fixed-pitch autogyros. Where there is no shaft drive or collective control, the rotor head can be very simple, as shown in Figure 4.23. The rotor requires just a teetering hinge and the head is mounted on a gimbal so the pilot can tilt it. The tilting head can only apply cyclic pitch. Where collective control is required, a feathering bearing will be needed and so the tilting head has no advantage. Hafner was the first to build an autogyro with cyclic and collective control and the mechanical solutions he devised would be used in his subsequent and successful Sycamore helicopter.

4.16 Dragging dynamics

In-plane blade motion is variously called 'lead lag', 'lag' or 'dragging'. This author prefers the term dragging because lead lag is cumbersome and lag can be confused with phase lag; another thing altogether. Section 4.6 illustrated the mechanisms responsible for dragging. In many rotor designs, dragging hinges or flexible members are provided

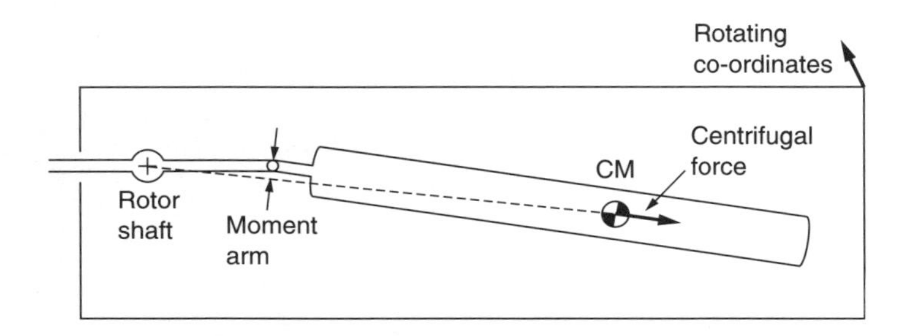

Fig. 4.24 When a blade drags, the CM moves closer to the rotor axis. In rotating co-ordinates, centrifugal force creates a restoring moment proportional to deflection such that the blade has a resonant response to excitation in dragging. Unlike flapping, the aerodynamic damping is low.

to reduce stress. In other designs, the rotor is designed to be extremely stiff, but as stiffness is always finite, there will still be some dragging motion. In all cases the dynamics of the system must be considered in order to avoid resonances.

In a traditional articulated rotor head, the dragging hinge cannot apply a restoring moment to the blade. The only restoring force for dragging motion in this case is due to rotation. Figure 4.24 shows that when a blade drags on a hinge having an offset from the rotor axis, the CM of the blade moves closer to that axis. When the rotor is turning, the CM will tend to be at the greatest possible radius and so if the blade drags forward or back, there will be a restoring force. Figure 4.24 is drawn in a rotating frame of reference in order to arrest the rotation. The acceleration of the blade is now zero and has been replaced by an equivalent centrifugal force. It will be seen that the centrifugal force, which must pass through the rotor axis, creates a small moment due to the drag hinge offset.

The restoring moment is proportional to the deflection and the result is that a blade which suffers an in-plane disturbance will tend to execute simple harmonic motion at its resonant frequency. In an articulated rotor, resonant frequency is proportional to the square root of the stiffness. The stiffness is due to centrifugal force and is proportional to the square of the RRPM. Thus the dragging resonant frequency will be proportional to RRPM as was the case for articulated flapping. However, unlike flapping, the only aerodynamic damping available to the dragging motion is due to changes of profile drag. Damping will often need to be augmented by mechanical means.

In hingeless rotors there will be some degree of structural stiffness that will provide a restoring force for the blade even when the rotor is not turning. The dragging resonant behaviour of a rotor can be characterized by the Southwell coefficients. Figure 4.25 shows how the ratio of dragging resonant frequency to rotor frequency is determined by $K1$ and $K2$. $K1$ reflects the resonant frequency when the rotor is not turning, as determined by the blade inertia and the stiffness of the rotor/head system in the dragging plane and is called the structural stiffening component. For an articulated head, $K1$ is zero because the dragging hinge has no stiffness. $K2$ is determined by the effective or actual dragging hinge offset and represents what is usually called the centrifugal stiffening component.

Figure 4.25(a) shows that in the articulated rotor, the hinge offset and consequent restoring forces are generally small and so the dragging resonant frequency is a small proportion of the rotor frequency, typically between 20 and 30%.

In a hingeless rotor, shown in Figure 4.25(b), at low RRPM the structural stiffness will dominate, whereas at high RRPM rotational forces dominate and the dragging

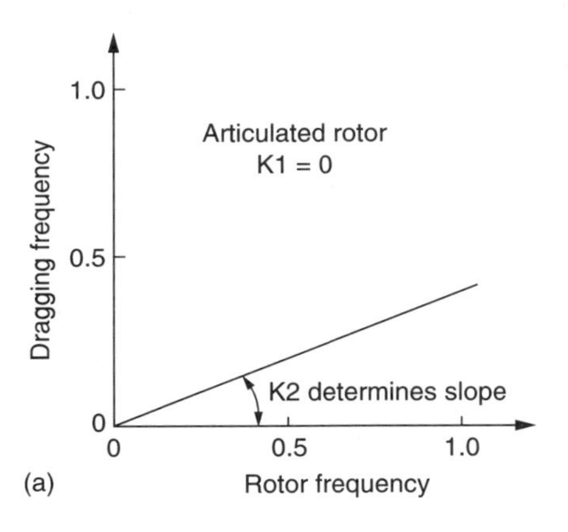

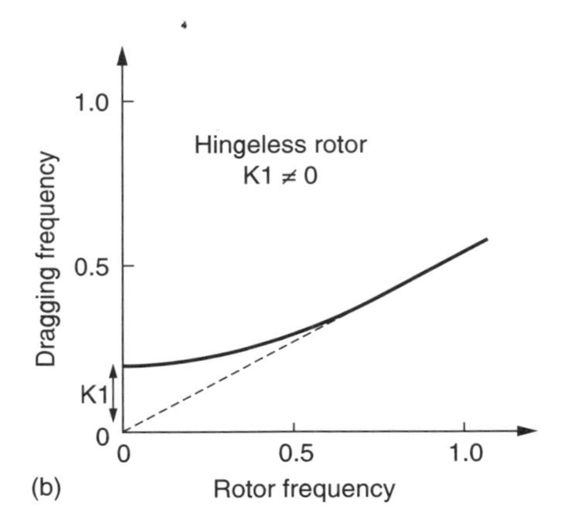

Fig. 4.25 Blade dragging is characterized by Southwell's coefficients which describe the static and centrifugal components of the restoring force. (a) In the articulated rotor, there is no static stiffness. The dragging frequency is a fraction of the rotor frequency. (b) In the hingeless rotor there is static stiffness so there is a finite resonant frequency even at zero RRPM. As RRPM increases, centrifugal stiffening dominates and the frequency becomes asymptotic to what it would have been for a hinged blade with the same mass distribution.

frequency becomes asymptotic to what it would have been with no structural stiffness. As a result the dragging frequency is no longer proportional to RRPM.

It should be appreciated that in the above discussion and in Figures 4.24 and 4.25 the dragging frequency is with respect to the turning rotor, not with respect to stationary co-ordinates. Frequencies experienced at the hull will be in stationary co-ordinates and will be different from the frequency discussed here.

4.17 Ground resonance

Given the flexible nature of rotors, it is possible for the centres of mass of the blades to get 'out-of-pattern' as shown in Figure 4.26. This is the term describing the condition where the blades are not evenly spaced around the disc. This may be due to an external influence such as the impact of a run-on landing or starting the rotor on a slope. It may also be self-excited.

Section 4.16 considered the resonant dragging motion of a single blade, whereas in practice it is necessary to consider the phase relationship from blade to blade. Figure 4.27 shows some possibilities for a three-bladed rotor. Rotors with more blades will also behave in the same way except that there are more phases of blade motion. Two-bladed rotors have different characteristics, but because such rotors seldom employ dragging hinges these differences will not be enlarged here.

It must be stressed that throughout Figure 4.27 the diagrams are in rotating co-ordinates so that the rotor appears to be stopped. Figure 4.27(a) shows the case where, for example, a sudden increase in drive torque has caused all of the blades to drag back by the same amount. They will all resonate in the same phase and the result is that the CM of the rotor is undisturbed. Figure 4.27(b) shows the case where there is a 120° phase difference between the motion of each blade. Now the centre of the rotor hub is moving in a circle. The motion may be visualized by considering the operation of a three-cylinder radial engine.

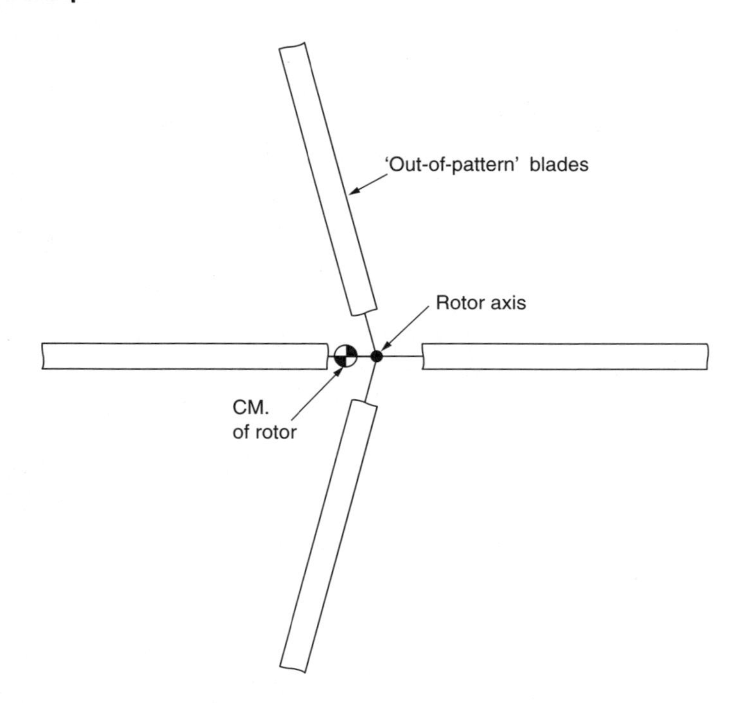

Fig. 4.26 Out-of-pattern blades shift the CM of the rotor and whirl the top of the mast.

Figure 4.27(c) shows the same situation except that the phase relationship is $-120°$. The hub is still moving in a circle, but in the opposite direction. There is a strong analogy here with the three-phase electric motor that can be reversed by changing any two connections. This circular motion is known as whirling and it may be forward, in the same direction as the rotor, or backward, in the opposite sense to rotor direction. Figure 4.27(d) shows the case where one blade is not dragging, but the other two are dragging at 180°. Now the hub motion is linear.

If the effects of Figure 4.27 are considered in a non-rotating frame of reference, (a) will have no effect. Effect (b) causes circular whirling at a frequency which is the sum of the rotor frequency and the dragging resonant frequency, whereas (c) will result in circular whirling at a frequency which is given by the rotor frequency minus the dragging frequency. The whirling frequency seen by the hull has been modified by the addition of the rotor frequency. If the whirling experienced by the hull is in the same direction as the rotor turns, it is said to be *progressive*. If it is in the opposite sense it is said to be *regressive*. The whirling orbit is reminiscent of some types of food mixer. A mild form of such an orbit is usually experienced during rotor starting with an articulated rotor and is called 'padding' by pilots.

In a real helicopter these whirling forces are applied to the top of the mast. The CM of the hull is well below the rotor head, and the result is a combination of pitching and lateral rocking. The moment of inertia of a helicopter about the pitching axis is quite high owing to the long tail boom, but there is less inertia about the roll axis, so generally lateral movement will dominate. The hull will not be infinitely stiff and may also be supported on a sprung undercarriage, and so it will have natural resonant frequencies of rocking.

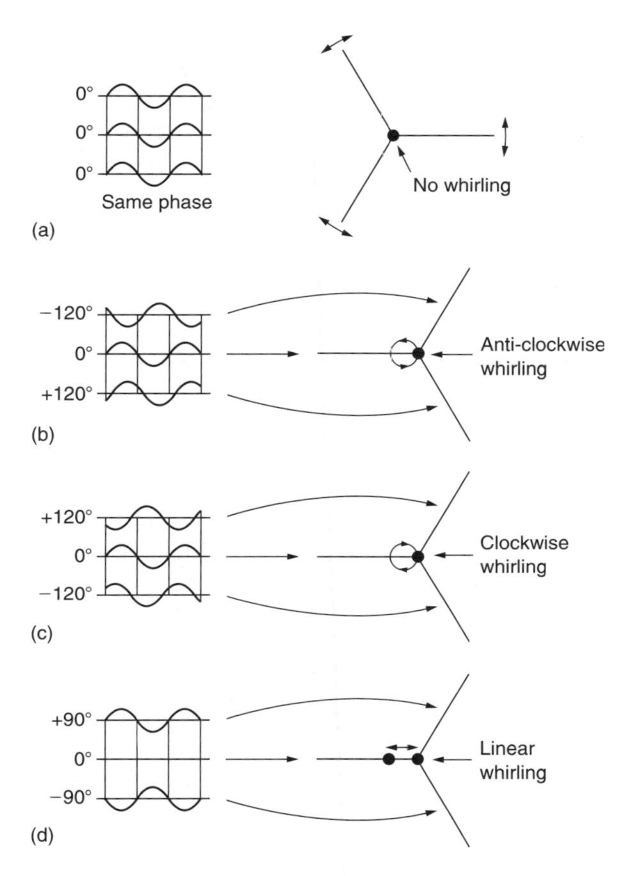

Fig. 4.27 Effects of dragging blade motion. (a) A sudden application of drive torque may result in all three blades dragging in the same phase. (b) When the dragging of each blade is at 120° phase to the next, the hub moves in a circle. (c) If the phase relationship of (b) is changed, the direction of whirling is reversed. (d) Where two blades drag in anti-phase and one blade is stationary, the hub is driven in a line. Note that all of these examples are in rotating co-ordinates.

In the absence of preventive measures, if a resonant frequency in the hull is the same as that of a whirling frequency, one of two things can happen. Either the hull will act as a vibration absorber as was described in section 3.27, and the whirling will be opposed, or the rocking will become increasingly violent until the machine either comes apart or turns on its side. The phenomenon was first observed when a taxiing autogyro struck a rock with a wheel and literally disintegrated.

Needless to say this Jekyll and Hyde behaviour in seemingly similar circumstances means that the mechanism involved must be extremely subtle. Whilst the results of ground resonance are painfully obvious, the mechanism causing it is not obvious at all. Until a mathematical basis for the phenomenon was found, designers proceeded empirically. For example, Frank Piasecki obtained an improvement on his first machine by filling the tyres with cork. It was concluded early that ground resonance is purely a mechanical phenomenon and that aerodynamic forces are not significant. There are some parallels with whirling phenomena observed in other disciplines such as steam turbines.

The first full mathematical treatment of ground resonance was due to Robert Coleman and Arnold Feingold who were working at NACA (the forerunner of NASA). The mathematics turned out to be so complicated that the authors had to present their results in the form of charts intended to be practically useful without the reader needing to have advanced mathematical skills. Using these charts, designers were able to tame ground resonance, but this is not the same thing as understanding it. In order to understand ground resonance it is necessary to consider the geometry of whirling. Initially this will be considered for the case of the rotor alone.

The rotor is assumed to be isolated, and turning without translating. There is no external force on an isolated system so, according to Newton's laws, the overall centre of mass of the rotor cannot move. If the centre of mass of the hub is whirling, this must mean that the blades together must have an effective centre of mass that is whirling in the opposite direction. Figure 4.28, which is in rotating co-ordinates, shows the orbits of various points on a blade and hub in a forwards whirling system. There will be a null point on the blade between the drag hinge and the blade CM that is oscillating radially but not tangentially. Note that the motion for a backwards whirling system can be seen by reversing all of the circles and ellipses.

In whirling, the KE of the hub is constantly changing because the circular motion requires a constant change of velocity. It follows that the kinetic energy of the blades must also be changing constantly. The blade KE variation is due to motion of the blade CM plus that due to in-plane rotation of the moment of inertia of the blade about the null point. Essentially whirling is a continuous interplay of blade and hub energy and in the absence of friction at the hinges and any aerodynamic effect it could continue indefinitely.

Figure 4.29 is in stationary co-ordinates. Figure 4.29(a) shows an articulated rotor turning anticlockwise and whirling forwards whereas (b) shows the same rotor which is still turning anticlockwise but which is whirling backwards. Let these rotors be fitted to a helicopter having a rocking hull resonance. In both cases the figure shows the blade at

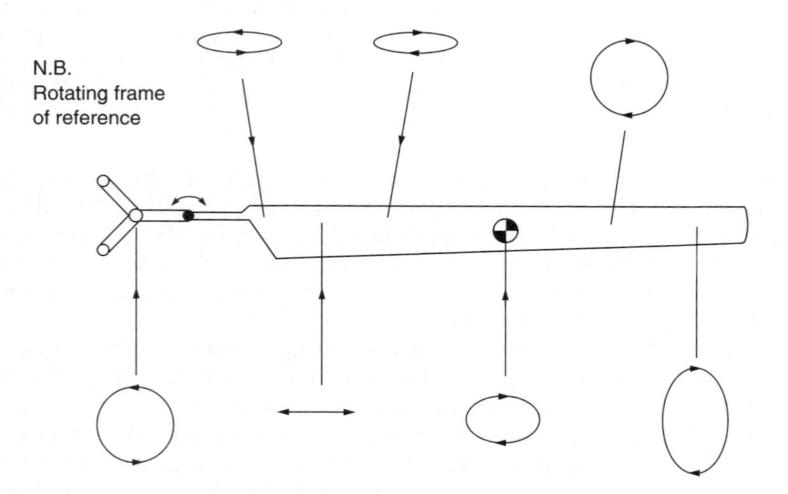

Fig. 4.28 In an isolated whirling rotor, the overall kinetic energy must remain constant. Consequently there must be an energy interchange between the blades and the hub. If the hub is whirling in one direction, the blade CMs must be whirling in the opposite direction. The loci of several points in the whirling system are shown.

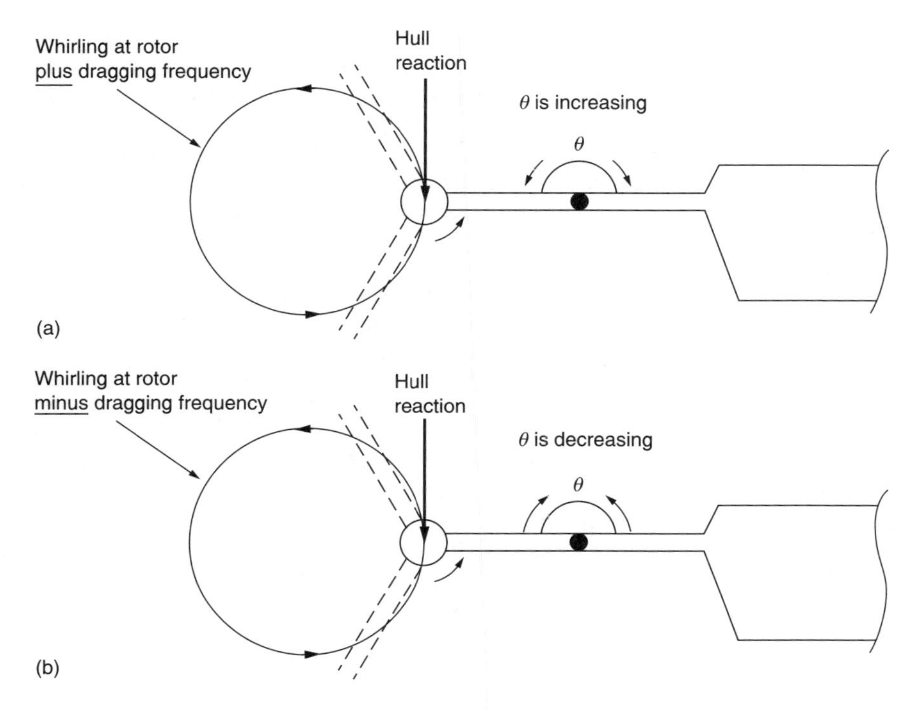

Fig. 4.29 (a) A rotor turning anticlockwise with forward whirling. The reaction from the hull opposes the whirling. (b) A rotor turning anticlockwise with backward whirling. The reaction from the hull which is the same as in (a) is now in the same direction as the whirling and amplifies it. This is the mechanism of ground resonance.

180° passing over the nose. In the case of (a), to obtain the rocking frequency seen by the hull, the forward whirling frequency must be *added to* the rotor frequency. Clearly the whirling is progressive. If the hull/undercarriage system is resonant at this frequency, then it will produce a force opposing the rocking excitation because the response of the hull at resonance must be anti-phase to the exciting motion as was shown in section 3.27. This force is applied at the mast. It will be clear from Figure 4.29(a) that at the instant depicted by the figure, the angle θ between hub and blade shank is *increasing*. The force applied by the mast tends to *reduce* θ, thereby opposing the whirling. As a result, the system is stable. Ground resonance *cannot* occur due to the higher whirling frequency resulting from forward whirling.

In the case of Figure 4.29(b) the whirling is backward. To obtain the rocking frequency seen by the hull, the forward whirling frequency must be *subtracted from* the rotor frequency. However, the whirling frequency is lower than the rotor frequency, so the whirling seen by the hull is still progressive. If the hull has a rocking resonance at this frequency, it will produce a force in anti-phase to the whirling as before. However, it will be clear from Figure 4.29(b) that at the instant depicted by the figure, the angle θ between hub and blade shank is *decreasing*. The force applied by the mast tends *to decrease* θ *further*, thereby augmenting the whirling. As a result, the system is unstable. Ground resonance occurs due to the lower whirling frequency resulting from backward whirling.

In mechanical terms, the mechanical impedance of the system has become negative, so that it can gain energy from forces that would otherwise oppose motion. Negative impedance is used in electronics to construct oscillators. The criterion for this to happen is that the whirling must be both progressive and backwards. Given that the system is unstable in this mode, an undamped rotor can spontaneously display ground resonance.

In the absence of preventive measures, a helicopter on the ground with progressive backward rotor whirling is a mechanical oscillator. Given the huge amount of energy stored in the rotor, once started the whirling will increase in amplitude until something breaks. Hull rocking resonance can only occur if there is a reaction from the ground, hence the name of the phenomenon. This also explains why a machine can fly safely but disintegrate on landing, as has happened on a number of occasions.

There are a number of solutions to ground resonance which will be explored. It will be seen from a consideration of Figure 4.29 that damping any changes in the angle θ will be highly effective hence the use of dragging dampers in the traditional fully articulated rotor head. In many cases damping is provided in the undercarriage to dissipate landing impacts and this damping can augment but not replace the damping in the head.

In modern helicopters employing damping, ground resonance is virtually unknown provided the dampers are kept in good order. These dampers may be hydraulic, similar to automobile dampers, which work by forcing oil through a small orifice, or elastomeric, which work by dissipating heat in hysteretic flexing. The latter have the advantage of needing no maintenance. Oil filled dampers will lose effectiveness if the oil leaks.

Given the destructive nature of ground resonance, it is a good idea to examine the dampers as part of the pre-flight inspection. By moving the blade on its dragging hinge, the resistance of the damper can be felt and the oil can be heard rushing through the damping orifice. All of the blades should feel and sound the same. If one blade feels different the damper may have some air in it. As the air is forced through the damping orifice the sound will change. Whilst one weak damper may not cause ground resonance, it may result in an increase in vibration in forward flight. It is also useful to learn the characteristics of the machine's padding on start-up. If the rotor dampers are satisfactory, but there is unusual padding, the undercarriage oleos may need attention. A smoother rotor start may result if all of the blades are first moved to their rearward damper travel limit. Unusual padding may also result if the machine is parked on a slope when gravity will tend to take the blades out of pattern during the early stages of starting.

Figure 4.30 shows an interference diagram or Coleman diagram for an articulated rotor. $K1 = 0$ and the dragging frequency is small in relation to the rotor speed. Thus the rotor speed always overcomes the backward whirling and so all of the frequencies

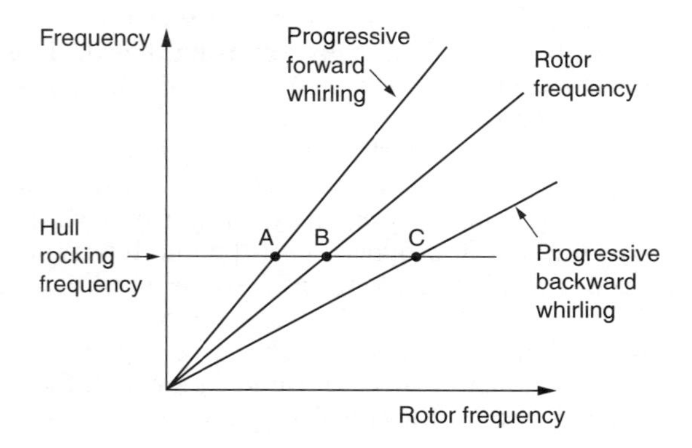

Fig. 4.30 An interference diagram for an articulated rotor. The dragging frequencies fan out above and below the rotor frequency. Both are progressive but one is backward.

concerned are positive. It was seen above that the dragging resonant frequency is proportional to rotor frequency. Thus as rotor frequency increases, the upper and lower whirling frequencies will be found symmetrically above and below rotor frequency. The horizontal line in the figure is the rocking resonant frequency of the hull. As such a rotor is started from rest, there will first be random padding due to the initial blade orientation. At a rotor frequency corresponding to point A, the upper whirling frequency coincides with the hull frequency. As the upper whirling coincidences are always stable, the hull acts as a vibration absorber and this will be noted as a lull in the padding. If the padding motion is noted, it will be seen to reverse phase after point A because the system has passed through a resonance.

Further increase in speed makes the rotor frequency coincide with the hull frequency at B. Only if there is any mechanical imbalance in the rotor (a defect) will there be any response. Further increase in rotor speed results in coincidence between the hull frequency and the lower (backwards) whirling frequency. This is the coincidence responsible for ground resonance and the dampers will be working to prevent it. The padding may be worse at this rotor speed. Flight RRPM will be above C where rotation should be smooth.

A 'soft-in-plane' rotor has no dragging hinges, but a relatively compliant blade root or hub so that there is a small restoring force even when the rotor is stationary. Consequently the dragging resonant frequency starts at a minimum value and increases with RRPM. Figure 4.31 shows that when this characteristic is added to the rotor frequency, the forward whirling is always progressive and the frequency increases from the static value to become asymptotic to the value determined by centrifugal stiffening. Points A and B are stable as for the articulated rotor. However, the backwards whirling frequency is initially higher than the rotor frequency. The whirling is actually in the opposite direction to the rotor and so is regressive. If this regressive whirling coincides with a hull frequency, the result is stable because the criterion for ground resonance is not met. In regressive backwards whirling the whirling is backwards in both rotating and stationary co-ordinates and the phase reversal needed for instability is absent.

However, when the rotor frequency in Figure 4.31 becomes equal to the backwards whirling frequency at C, the whirling frequency becomes zero. At any higher rotor frequency the rotor frequency exceeds the whirling frequency and the backwards whirling

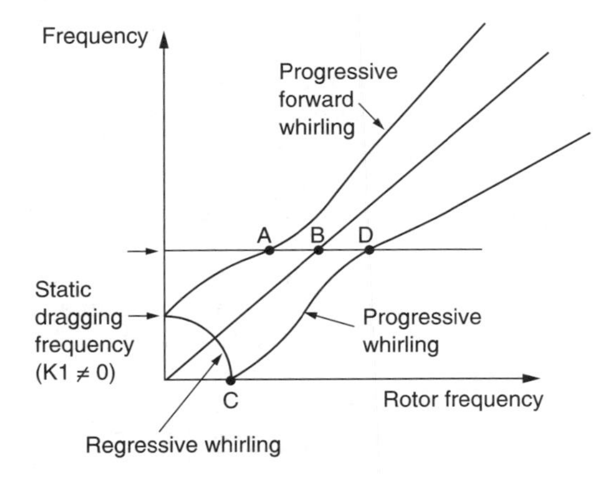

Fig. 4.31 A soft-in-plane rotor has a regressive dragging region as far as point C at low RRPM, but this is not unstable.

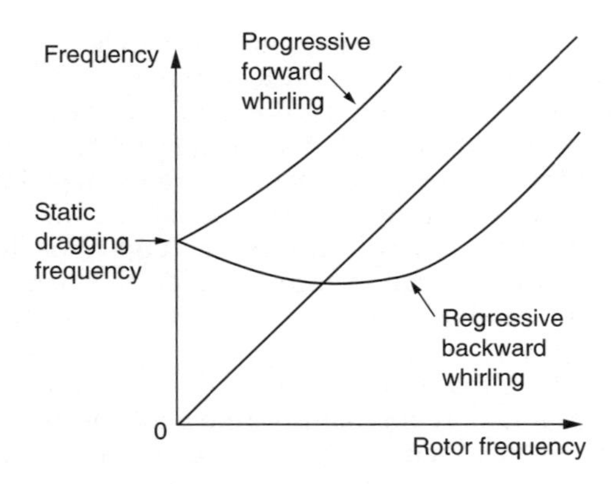

Fig. 4.32 A supercritical rotor has such a high dragging frequency that the rotor can never reverse it. No damping is then needed.

becomes progressive again. The frequency then rises to become asymptotic to the value determined by centrifugal stiffening. Point D in Figure 4.31 corresponds to a coincidence between a hull frequency and a progressive backwards whirling frequency, at which ground resonance can occur. Consequently soft-in-plane rotors still require damping.

A 'stiff-in-plane' or 'supercritical' rotor has a powerful dragging restoring force due to its construction and so has a significant dragging resonant frequency, even at rest. When the rotor is turning, the dragging frequency increases even further due to centrifugal stiffening. Figure 4.32 shows that, by definition, in the case of a 'stiff-in-plane' rotor, the whirling frequency is always higher than the rotor frequency. The behaviour of the forward whirling is benign as before. However, as the backwards whirling frequency is so high, the rotor frequency can never reverse it and so backwards whirling is always regressive in the case of a stiff-in-plane rotor. In that case the unstable combination of progressive backwards whirling can never occur and such a rotor can function with no damping at all. Two-bladed teetering rotors take advantage of this concept. They are built without dragging hinges to obtain supercritical dragging behaviour.

There is a small possibility of a damper failure and so it is as well to know how to handle it. There are two ways to recover from incipient ground resonance, depending on the rotor speed at the time it happens. If sufficient revs are available to permit flight, the answer is to take-off immediately. With no ground reaction against the undercarriage, the hull rocking resonance cannot occur and the shaking stops immediately. An attempt at a gentle landing may succeed, but if not then a landing should be attempted on a soft or high friction surface which is more likely to absorb energy than tarmac. It is important that there is no groundspeed whatsoever at the moment of touchdown. If insufficient rotor revs exist for flight, the solution is to change the frequency of the excitation away from the hull/undercarriage resonance. This may be done by reducing rotor speed by cutting the throttle and using the rotor brake if fitted.

4.18 Air resonance

Air resonance is a condition primarily relevant to helicopters having hingeless rotors. In articulated rotors the natural frequency of the blade about the lagging hinge is low and lag dampers are in any case present to prevent ground resonance. Furthermore the motion of the hull is decoupled to some extent by the articulation. In a hingeless rotor the blades lag by flexing and the stiffness is higher, leading to a higher natural frequency of lagging motion. The relatively stiff connection between the rotor and the hull means that large rotor moments can excite hull flexing. If such hull flexing is resonant then an unstable system could exist. A bending mode of the hull could result in a lateral motion at the rotor head that is similar to the rocking experienced in ground resonance.

It should be appreciated that resonance may also be excited by the tail rotor. This will be discussed in Chapter 5.

4.19 Dynamic rollover

Dynamic rollover is a hazard that can only occur when a helicopter is in contact with the ground in some way. Normally the cyclic control has complete authority over the attitude of the helicopter, but interference from the ground or from external loads can reduce or even overcome the cyclic authority. Section 4.15 introduced the tilting hub mechanism for cyclic control. If the hull of a helicopter is forcibly tilted, the rotor will follow. Dynamic rollover occurs when external tilting forces overpower the cyclic control. Teetering or zero-offset rotors are more prone than offset or hingeless rotors because their cyclic authority is small to begin with.

Dynamic rollover becomes a possibility when the conditions on one side of the hull are different from those on the other side. This may be due to an attempt to land or take-off from a slope, an underslung load snagging a skid or a skid being stuck in mud or the surface tension of wet sand. Most winches for personnel recovery are fitted at one side of the machine and if a winch rope snags on the ground it can cause difficulty. A side wind and the thrust of the tail rotor complicate the matter.

Figure 4.33(a) shows what can happen with a stuck skid. The pilot raises the collective lever to lift off, but the stuck skid combines with the rotor thrust to produce a roll couple that could exceed the power of the cyclic control. In many cases the machine will be on its side before the pilot can react and the cyclic authority becomes academic. The roll accelerates because as the hull tips it increases the cyclic input to the rotor and causes further roll. A variation of this hazard is where the stuck skid suddenly comes free and the machine ends up on its other side because the pilot does not remove the lateral cyclic quickly enough.

Figure 4.33(b) shows a helicopter about to lift a slung load. Unknown to the pilot the load strap has snagged a skid. As the machine rises, once more the rotor thrust and the load strap tension create a couple on the hull and the dynamic rollover takes place. Most helicopter winches are installed with a view to bringing loads alongside one of the hull doors. The load is applied with a considerable moment arm from the machine CM. Moments due to normal loads can be resisted by the rotor, but in the event that the winch line became snagged an attempt to climb would result in dynamic rollover. This is one of the reasons why the pilot is provided with means to jettison loads in an emergency. Underslung loads can be jettisoned by releasing the load hook that may be electrically operated. Winches are fitted with cable cutters that may be explosively actuated.

Figure 4.33(c) shows an attempted landing on a slope. After the first skid has touched, the machine rotates about that skid. As it comes down, the control axis must be kept vertical by the application of cyclic control. The question is, what happens when the cyclic reaches the end of its travel and the other skid still is not in contact? The rotor disc cannot be kept horizontal and there is a danger of dynamic rollover. If the landing

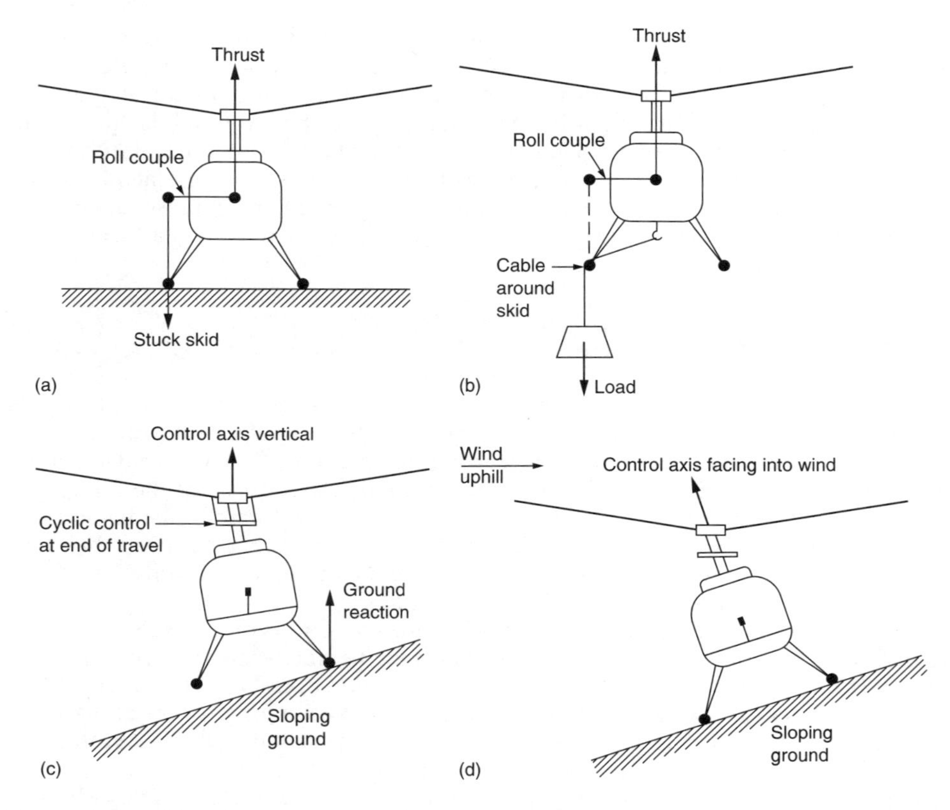

Fig. 4.33 The origins of dynamic rollover. (a) A skid sticks in mud and applies a roll moment as the helicopter attempts to take-off. (b) A load strap snags around a skid and applies a roll moment as the tension is taken up. (c) An attempted slope landing in which the cyclic control has reached the end of its travel with the lower skid still not in ground contact. (d) A landing made with an upslope wind is easier, but if the wind changes to downslope, there may not be enough cyclic travel to level the disc in the presence of flapback.

is continued, the hull will roll and carry the rotor with it due to the tilting head effect. When the downhill skid contacts, there may be enough roll momentum that the roll continues, with the uphill skid lifting. Once the CM of the machine is outside the skid base there is no recovery. This explanation is the origin of the slope landing limit laid down for each type of machine. These figures should be ignored at one's peril.

Slope landings can be made harder or easier by wind. Figure 4.33(d) shows the wind blowing upslope. This is an advantage because more cyclic travel is available. However, if the wind subsequently reverses, it may become impossible to take-off because insufficient cyclic travel exists to tilt the disc into wind when the mast is tilted downwind. An experienced pilot might perform a 'hop' take-off which consists of bringing the rotor to flight RPM with neutral cyclic and then pulling collective rapidly to exit at a right angle to the slope. Once the gear is clear of the ground full cyclic authority becomes available.

The preferred technique in all external load handling or slope landing situations is to proceed very slowly, and if the cyclic travel approaches the limits the manoeuvre will have to be halted. This means in the event of attempting a take-off, rotor thrust should be applied very gradually so that any roll does not proceed too far. If anything goes wrong, the collective should be lowered. With the rotor thrust gone, the couple disappears and the machine should fall back onto its gear. In the event of a sloping landing, if the cyclic stick travel is becoming extreme, the landing should be rejected. The collective lever is pulled up to get the uphill skid off the ground. Again the couple is eliminated because there is no ground reaction. Even if the roll has started, a smart and powerful application of collective will recover cyclic control as soon as the skid leaves the ground.

Rollover is also possible in a hingeless rotor machine if the cyclic control is not trimmed to neutral before collective is raised to take-off, although the mechanism is not the same as that of dynamic rollover. Pilots like to tilt the disc around on the ground to check the cyclic controls before flight. Hingeless rotor machines can have extremely high cyclic power without much rotor tilt being present. The amount of tilt that would be safe in an articulated machine might roll over a hingeless machine. Sometimes the undercarriage squat or ground contact switch may be used to limit cyclic authority when the machine is on the ground.

4.20 Some rotor head examples

The number of approaches used in rotor head design over the years has been subject to wide variation. In the earliest days the theory of rotor dynamics was not well established and most designs mixed theory with empiricism. As understanding increased, attention turned first to optimization of design for a particular purpose and subsequently to such refinements as the reduction of vibration and of build and maintenance cost.

Figure 4.34 shows the evolution of zero-offset heads. These are usually underslung so that the hinge pin passes through the CM of the coned rotor. Supporting the rotor at its CM allows the blades to conserve momentum as the disc tilts and dragging hinges can be eliminated. This means that the rotor can be made supercritical in dragging. Ground resonance cannot occur and no damping is needed in the head or in the landing gear. This allows a cost and weight saving in small machines.

At (a) is the fully gimballed teetering head which essentially contains a Hooke joint so that the rotor can tilt in any direction with respect to the mast. The blade grips are mounted on feathering hinges. The feathering axis is parallel to one of the gimbal axes and the attitude of the hub in that axis would be indeterminate were it not for an equalizing linkage that makes the pitch of each blade equal with respect to the hub. This type of head was used on early Bell and Hiller machines.

At (b) is the later type of teetering head where only a flapping hinge is fitted and tilting in the other axis is accommodated by the feathering hinges. This type of head can be seen on the Bell 206 JetRanger. It is not obvious how this arrangement can act as a universal joint to allow the disc to tilt in any direction relative to the mast. The answer is that the feathering bearings are used to provide the additional degree of freedom. Figure 4.34(c) shows a helicopter with a teetering rotor hovering with an offset CM. The control axis and the tip path axis are vertical, but the shaft axis is not. As the rotor turns, the blades remain at a constant pitch relative to the tip path axis and the control axis, but need to turn in the feathering bearings to do so. Thus in this case it is the rotor head that is oscillating about the blade root and not vice versa.

The feathering bearings and the underslung teetering bearing act as a Hooke joint to allow shaft axis to be different to the tip path axis. The Hooke joint is not ideal because of the inclination of the feathering bearings at the coning angle, but for typical shaft/tip path relationships the discrepancy is small. The Hooke joint is not a constant velocity joint, and some torsional flexibility will be required in the transmission.

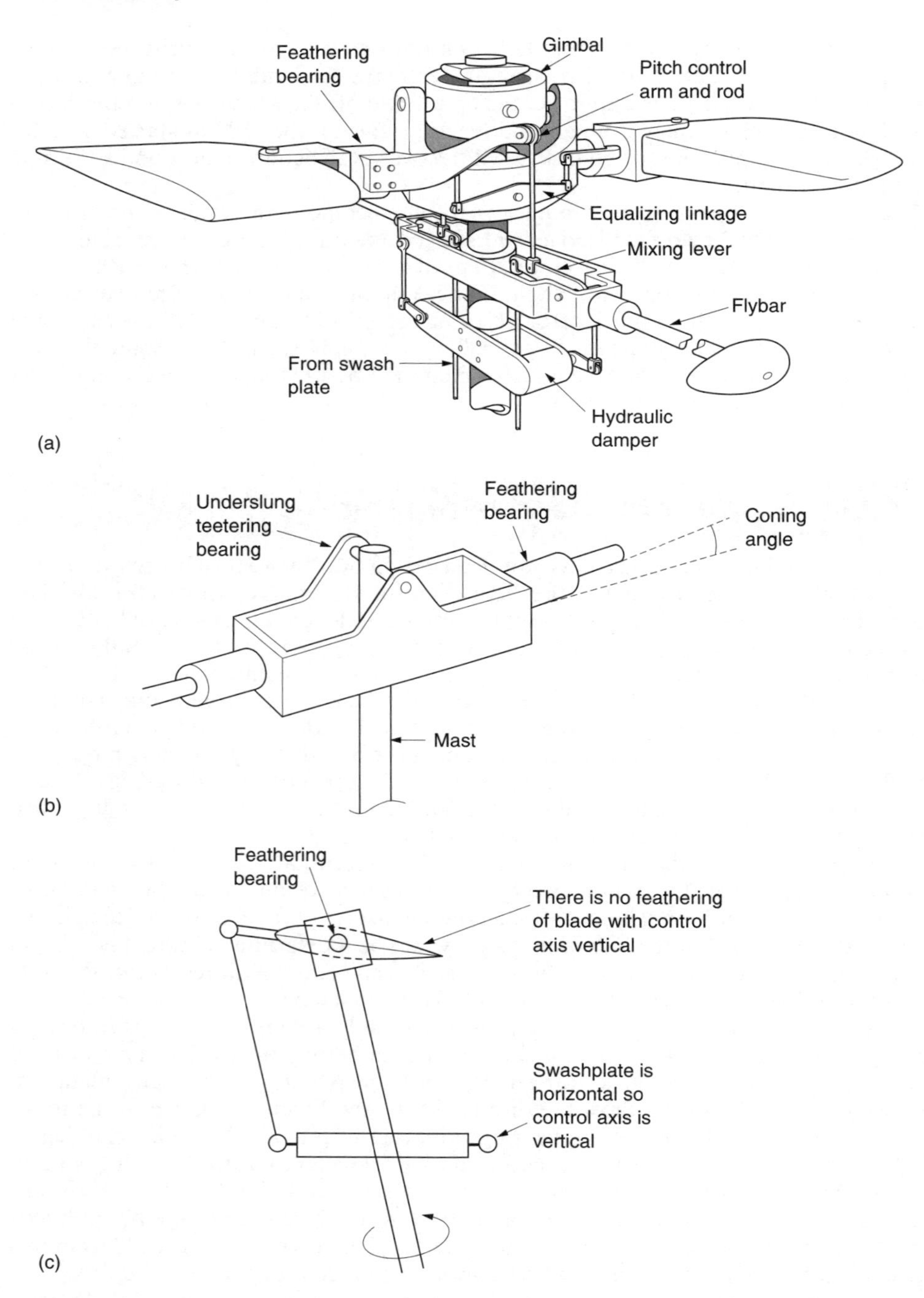

Fig. 4.34 Teetering heads. (a) Early Bell fully teetering head has a two-axis gimbal and an equalizing linkage to keep the pitch of the two blades with respect to the hub the same. (b) Later teetering hub uses the feathering bearings as one axis of the gimbal and only needs a single transverse hinge. (c) Hub of (b) hovering with offset CM. Note level disc and swashplate and feathering bearing allowing the mast to be non vertical.

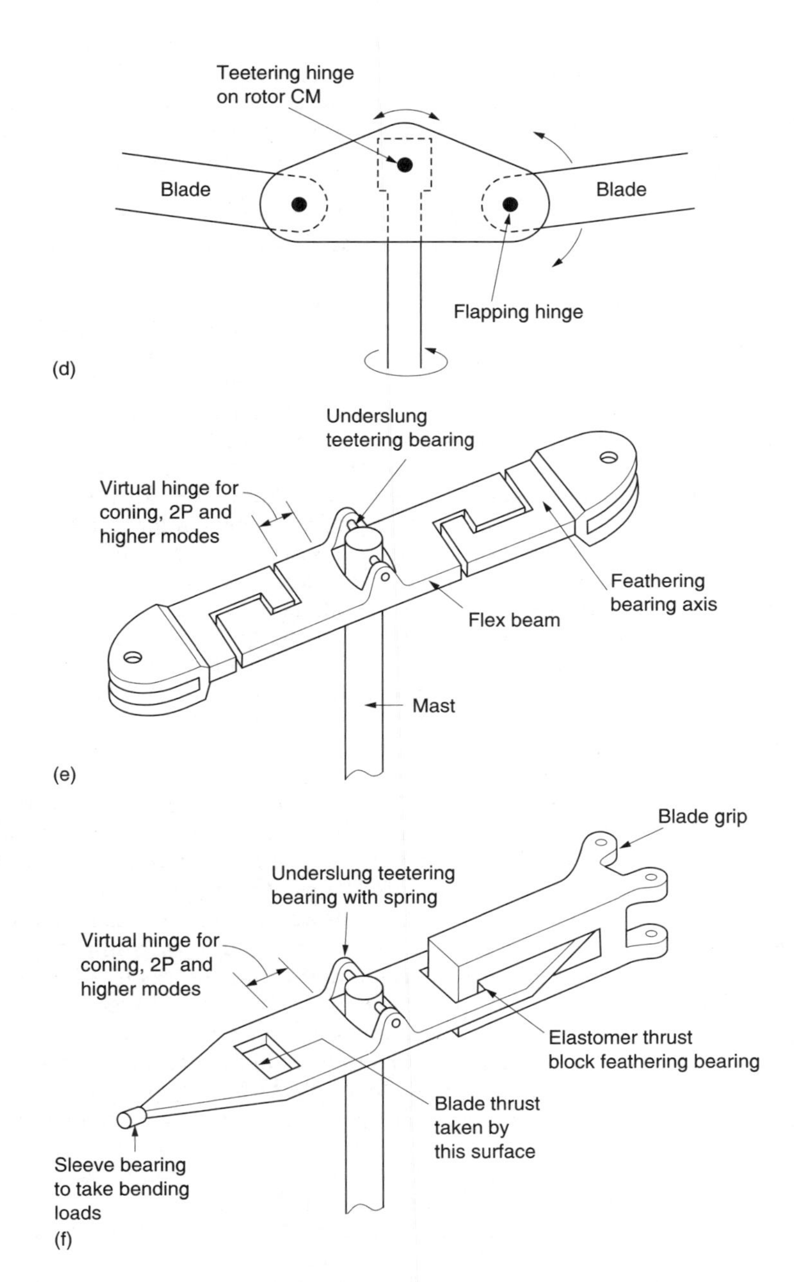

Fig. 4.34 (Continued) (d) Underslung tri-hinge head has flapping hinges but no dragging hinges. (e) Door-hinge teetering head has a flexible section to allow some flapping. (f) Rotor head having elastomeric element in compression for feathering bearing.

At (d) a variant of the teetering rotor is one in which the blades have individual flapping hinges so that no bending moments occur due to coning. This allows the blades to be more lightly built than in a conventional teetering system, and also reduces 2P hop. The effect of the offset between the flapping hinges is eliminated by the action of a third underslung bearing. This approach is known as a tri-hinge head and is used in the Robinson R-22.

At (e) is the so-called door-hinge hub. This hub is relatively thin in section to reduce drag. The thin section also introduces some flexibility to reduce the effect of two-per hop. This type of head was used on the Bell AH-1G.

At (f) is the semi-teetering head of the Bell 222. This contains elastomeric feathering bearings in compression, a thin flexural section for accommodating 2P hop and a centring spring on the teeter bearing to give stability in low-*g* manoeuvres.

Zero-offset heads are relatively intolerant of the CM location of the helicopter and cause handling difficulties in low-*g* conditions. The teetering hinge also allows blade sailing and starting the rotors may be impossible above a certain wind velocity because of the danger of a boom strike. This was acceptable in early machines, and possibly remains acceptable in today's light machines, given the weight advantage due to the elimination of dampers. However, it is not acceptable in helicopters intended for military or emergency service. Other types of head can overcome these difficulties.

Although most zero-offset heads have two blades, more blades can be fitted. Figure 4.10 showed the zero-offset head of the Sikorsky R-5 that has three blades. As the flapping hinge axes pass through the mast axis no flapping moments can be communicated to the mast. However, the blades are individually hinged and no underslinging is possible. This means that the CM of the coned rotor is above the flapping bearings. When the disc tilts relative to the mast this arrangement cannot allow the blades to conserve momentum and dragging hinges must be fitted. This then requires drag dampers and landing gear damping. Such a head is a poor combination because it needs the drag hinge/damper complexity of the articulated head but does not offer the advantages.

In the classical articulated head the flapping bearings are offset from the mast axis so that a couple is exerted on the mast when the disc tilts, tending to make the hull follow the disc attitude. This couple may also be used to allow a non-ideal hull CM position. Any number of blades can be fitted in this way, usually between three and nine. When coned, the disc does not flap about its CM and so dragging hinges are needed. It was realized early that making the flapping and dragging hinges intersect allowed a conventional Hooke joint to act as flapping and dragging hinge. The feathering bearing was then placed outboard. This arrangement became relatively standard for a long time and appeared on rotor heads from numerous manufacturers including Sikorsky, Westland and Enstrom. The larger Mil machines place the dragging axis outboard of the flapping axis, probably to allow space for dragging dampers when a large number of blades are required.

The flapping blades need droop and sailing stops and sometimes these are centrifugally operated so that they only operate as the rotor slows down.

The articulated head results in a rotor that can start in higher winds than can a teetering rotor, and which can retain some control at zero *g*. On the other hand the articulated rotor is extremely complicated and needs a lot of maintenance. As materials have improved, it has become possible to build structures that will withstand stress without excessive weight or that will flex without fatigue. Thus instead of using a mechanical hinge to relieve the stress, the designer has the choice of withstanding the stress with a stiff structure or relieving it by flexing. With modern materials such as Kevlar and carbon fibre it is possible to make a rotor in which flexing is minimal. However, this is not desirable as a great deal of vibration would be transmitted to the hull and a very uncomfortable ride would result. Flapping blades act much like the suspension of an automobile and decouple the effects of gusts and alternating components of lift in forward flight. Another problem is that a perfectly rigid rotor would respond extremely quickly to cyclic inputs, possibly too quickly for a human pilot.

As a result the extremely stiff rotor is little used. Instead the goal is to make hingeless or bearingless rotors where vibration decoupling and stress relief are achieved not by an identifiable hinge, but by a low maintenance structure which is designed to flex. The term *virtual hinge* will be met to describe the existence of an axis about which flexing effectively takes place. Virtual flapping hinges decouple the non-constant lift in forward flight and virtual dragging hinges allow the blades to conserve momentum. The presence of the flexibility reduces bending stress in the blades and reduces the cyclic response rate of the rotor. It is also possible to incorporate the feathering hinge into the flexible structure.

The designer can choose an appropriate degree of flexibility depending on the role of the machine. For some military purposes, such as nap-of-the-earth flying, extreme manoeuvrability is required along with the requirement to sustain zero or negative g when cresting a hill at speed. This would suggest a rotor that is relatively stiff in flapping, which would also be able to start in almost any wind. Conversely a civil machine would benefit from a softer flapping flexure.

Dragging flexures are also subject to variations. If the dragging virtual hinge is made very stiff, the rotor can be made supercritical so that dragging dampers are not needed. However, stiff-in-plane rotors tend to produce more lateral vibration because the blades transmit the drag changes as they move between the advancing and retreating sides. As an alternative, the soft-in-plane rotor reduces vibration by allowing the blades to drag, but requiring drag damping to avoid ground resonance. The use of elastomeric damping will still allow a low maintenance structure. Clearly the dragging and flapping stiffness can advantageously be different and some ingenuity is required to provide suitable geometry.

Figure 4.35(a) shows the hingeless rotor head of the Bo-105. The only real bearings are for blade feathering. Flapping and dragging movements are accommodated through flexing of the blade shank. The dragging stiffness is supercritical and no dampers are needed. This is also a relatively stiff head in flapping, and pendulum vibration dampers are needed.

Figure 4.35(b) shows another stiff-in-plane head, that of the Lockheed AH-56 Cheyenne. This has door-hinge feathering bearings for low drag and a virtual flapping hinge. The flapping hinge is very stiff indeed and the following rate of the rotor is very high, requiring full-time gyro stabilization. This will be considered in detail in Chapter 7.

Figure 4.35(c) shows a soft-in-plane hingeless rotor from the Westland Lynx. Here the flapping flexure is a massive piece of titanium that is only flexible in the context of the enormous forces set up in a rotor head. The flexures are relatively stiff in flapping. Since the rotor can exert large moments on the hull, the Lynx does not need a tall mast; in fact it was a design goal to keep the height down to allow the machine to fit into transport aircraft, and a special low profile gearbox was designed to go with the rotor. With such a low rotor, the Lynx needs the tail boom to be angled down to give blade clearance before turning up to mount the tail rotor.

Figure 4.35(d) shows the head of the Bell 412. This has outboard dragging hinges and the drag damping is obtained by an elastomeric block acting on a inward extension of the blade grip which also mounts the pitch arms. The dragging hinges are mounted on flex beams that can twist to act as feathering bearings and bend to allow flapping. Two such flex beams are stacked to make a four-blade rotor.

Figure 4.35(e) shows a head developed by Aerospatiale. This uses a spherical laminated elastomeric bearing which takes the axial blade thrust whilst allowing feathering, flapping and dragging. Elastomeric blocks are fitted inside the blade grip and these couple with extension arms on the hub. The arms are stiff in the dragging plane, but the elastomeric block is in shear for dragging movements, giving a soft-in-plane characteristic and providing the drag damping. In flapping the blocks are in compression where they are quite stiff. Flapping flexibility is provided through a virtual hinge in the arms. The head arms also twist to allow feathering.

Figure 4.35(f) shows the head of the Bell 680. Pairs of blades are joined by long composite yokes that divide to pass clear of the mast. The yoke bends to allow flapping and dragging and twists to allow feathering. The yoke is long so that the strain due to feathering is moderate. The blade is attached at the end of the yoke, at some considerable radius from the mast. The blade is hollow at the root forming a cuff so that much of the yoke is inside the cuff. The inboard end of the cuff carries the pitch arm connected

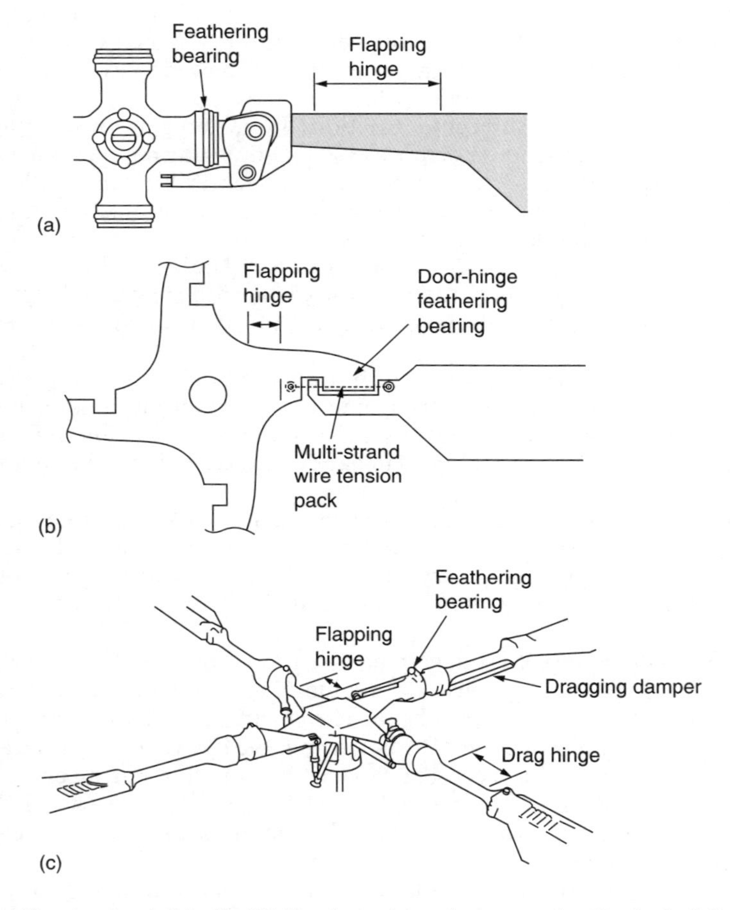

Fig. 4.35 (a) Hingeless head of the Bo-105. Flapping and dragging are permitted by the flexibility of the blade shank. (b) Lockheed Cheyenne head has virtual flapping hinge but is very stiff in dragging. (c) Westland Lynx head has flexible members to allow flapping and dragging. Virtual flapping hinge is stiff. Dragging hinge is soft and drag dampers are needed.

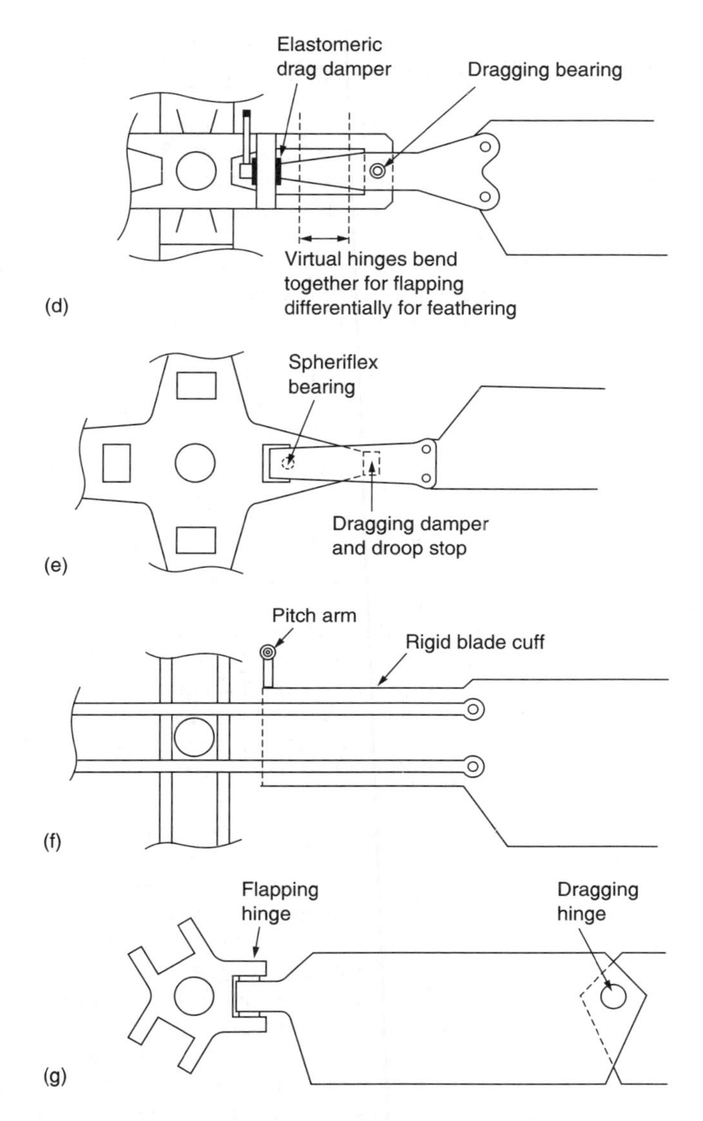

Fig. 4.35 (Continued) (d) Bell 412 head having elastomeric drag dampers. Flexural members bend to allow flapping and twist to allow pitch change. (e) Aerospatiale spheriflex head uses spherical elastomeric bearing for flapping, dragging and pitch change. (f) Bell 680 head using flexible elements inside a stiff blade cuff that transmits pitch control and connects to the drag damper. (g) Brantly has conventional flapping and feathering bearings, but dragging bearing is outboard making dragging supercritical.

to the swashplate. When the blade drags back, the inboard end of the cuff will move forwards. Thus the drag damper is placed between the inboard end of the cuff and an arm attached to the mast. The hollow cuff approach gives a very clean head.

Figure 4.35(g) shows the rotor of the Brantly (now Hynes) helicopter. This has conventional offset flapping hinges, but the dragging hinges are placed part way down the blade. This large drag hinge offset makes the rotor supercritical and no dampers are required.

4.21 Blade construction

The rotor blade has an arduous life being subject to considerable and time-variant forces which in the early days made that life all too short. The construction of rotor blades has been the subject of numerous advances directed towards increasing blade life.

Early blades were built up rather like the wings of aircraft. Blades must have a certain weight both to store energy in the case of engine failure and to prevent an excessive coning angle. High torsional stiffness is also required. These requirements were initially met by building the blade around a steel tube. Collars were attached to the tube and these supported wooden ribs. A wooden leading edge was attached to the ribs and this would contain a metal insert to bring the mass centroid of the blade forward. The leading edge of the blade as far back as the spar might be covered with thin plywood. The whole was then covered with canvas and doped. Such construction lent itself readily to the incorporation of blade taper and twist. Machines having such blades had to be hangared because the blades were porous and moisture absorption could cause them to go out of balance. Such blades were of no use in tropical countries where the glues used could not withstand heat, humidity and bacterial/insect attack.

The solution was to make all-metal blades. Typically the leading third of the chord was a heavy D-shaped extrusion with the remainder of the chord made up of thin sheet metal, perhaps supported by aluminium honeycomb. Initially riveting was used; later advances in adhesives allowed bonded structures. Metal structures are subject to fatigue, particularly given the alternating loads experienced by the helicopter blade. In some blades a warning of cracks was given by pressurization of the blade and fitting a pressure indicator. In other cases the blade or spar was evacuated and a vacuum indicator was used instead. If the indicator reading was incorrect this might indicate an incipient crack or bonding failure. The use of extrusions made it difficult to incorporate blade taper.

As the military helicopter developed, consideration was given to resisting battle damage by designing redundant structures. In civilian applications a redundant structure can have a longer service life because a failure is not catastrophic. Redundant structures work by providing parallel paths for forces. In a rotor blade the single spar can be replaced by a series of smaller parallel box spars. The box sections may be of glass fibre which give support against buckling to steel spars between them. In some cases the spars can be made entirely of glass fibre, but it will usually be necessary to incorporate a metal strip bonded into the leading edge to give the correct mass centroid.

The trailing edge of the blade is relatively lightly loaded and it is advantageous if it is not too stiff in bending in comparison with the leading edge because this combination results in the blade washing out as it bends up; a stable condition. Generally a thin outer skin is used which is supported against buckling by a low density filling material such as alloy or composite honeycomb, foam plastic or end grain balsawood. In some machines the trailing edge skin is deliberately weakened by chord-wise slits that allow the blade stiffness to be defined by the spar. The trailing edge of the blade is a relatively non-critical structure and many machines have returned safely following substantial trailing edge damage.

As the conditions in which machines can operate are extended, it becomes necessary to protect against lightning strikes. The result of a strike is massive energy dissipation in any material presenting electrical resistance to the current path. The means of protection is to make the outside surface of the blades electrically conductive and to bond the blades together electrically. In this way potential differences across the rotor are minimized along with the damage.

Abrasion is also a fact of life for rotor blades. Hovering stirs up dust and grit and this can impact the leading edge of the blade. Composite blades may incorporate a thin metal anti-abrasion skin wrapped around the leading edge. In many cases a sacrificial coating made of self-adhesive tape is applied. This can be peeled off without damage to the blade and replaced.

4.22 Blade tracking

Blade tracking is the process of adjustment that makes each blade fly identically. In the hover with the CM of the hull at the mast the blades should all have the same coning angle and the tips should all be in a single plane. If this is not the case there will not be a force balance at the hub and the result will be vibration. The tracking adjustment is performed by change to the length of the pushrod going to each blade pitch arm. The ball joints are attached to the pushrods with a screw thread to allow length adjustment.

It should be pointed out that a tracking adjustment can only succeed if the blades concerned are correctly statically and dynamically balanced. There may also be vibration if a piece of blade abrasion tape has come off or if the tape application on the blades is not identical.

The traditional way of measuring the tracking was to apply a different colour of chalk or greasepaint to each blade tip. A sheet of stout cloth on a frame was then brought carefully up to the edge of the spinning rotor so that coloured witness marks would be left on the cloth. If one blade were out of track it would leave a witness mark above or below the others. Clearly this method was not for the faint hearted and could only be done on the ground with the helicopter supporting some of its weight with the rotor.

Subsequently a system was devised in which inward facing coloured reflectors were temporarily fixed to the blade tips. These were illuminated by a strobe light triggered by a sensor on the mast so that as each blade passed the same azimuth angle the strobe light would flash. The machine could be flown with the apparatus attached and if the tracking was incorrect the reflectors would appear at different heights. Clearly the machine would have to be shut down each time an adjustment was required. Some helicopters have motorized adjustments that allow the pitch rod to change its length in flight. The optimum adjustment may be obtained by adjusting to minimize vibrations at some specified operating condition.

In some machines trim tabs are fitted to the trailing edge of the blades, which can be used to compensate for the effect of slight differences between the blades in forward flight.

4.23 Blade folding

The blades of a helicopter may take up a great deal of room and in some applications, such as operation from ships, folding blades will be needed as a practical matter. In a small machine this may be achieved entirely manually after first removing the locking pins. However, in large helicopters manual folding is not viable and a power-operated system will be needed. Figure 4.36 shows the blade folding system of the Sea King. In this five-bladed machine, one blade is positioned over the tail boom and does not fold. Blade folding can only take place if the rotor head is turned to exactly the correct angle. After the rotor has stopped, an hydraulic motor will turn the rotor to the correct angle and the shaft will then be locked.

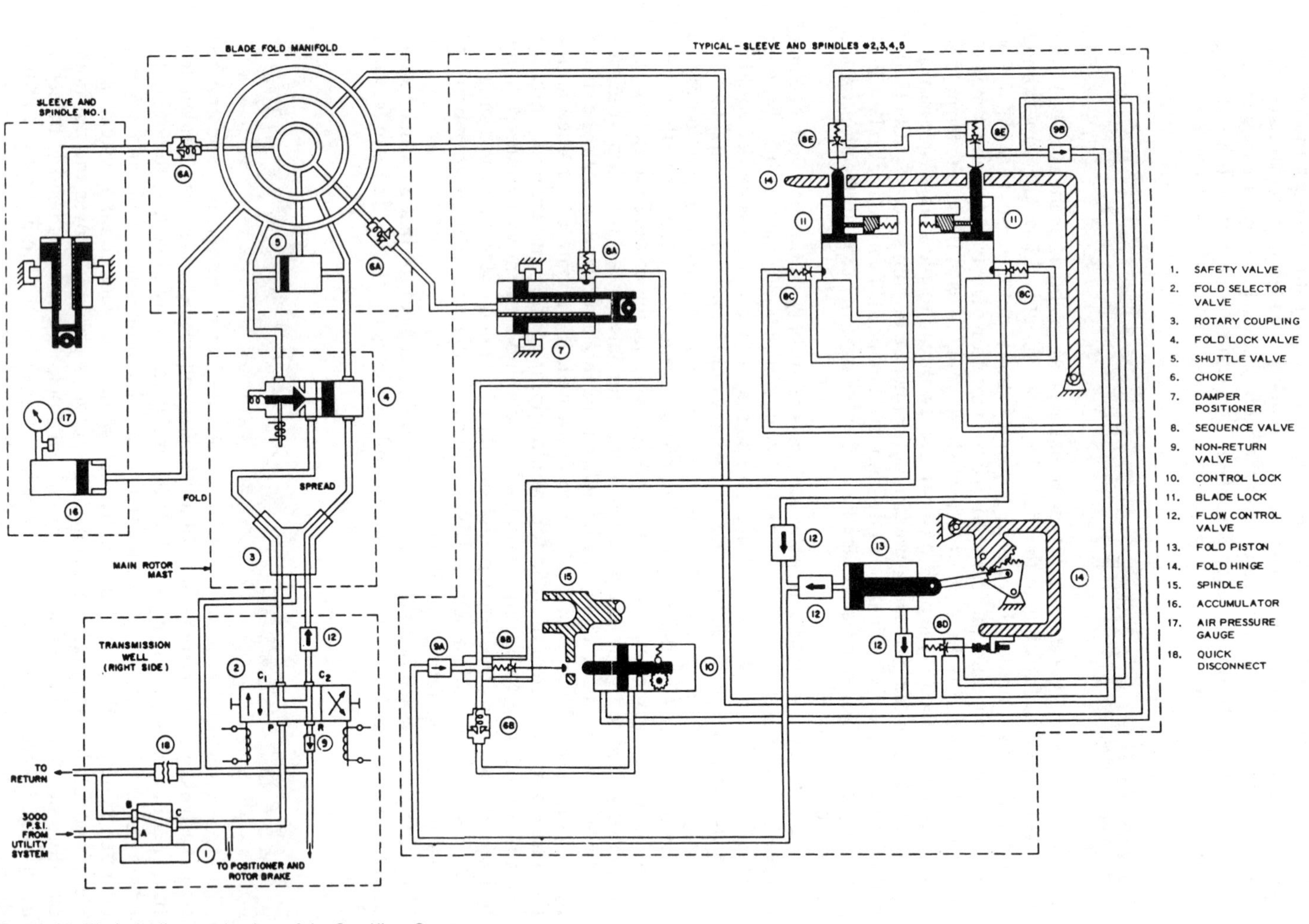

Fig. 4.36 Blade folding mechanism of the Sea King. See text.

There are some additional complexities involved in blade folding. First, folding the blade will put the blade centre of mass a long way from the feathering axis and the resulting couple could damage the swashplate and controls. This is prevented by a mechanism that locks the feathering bearings. Second, motion of the lead-lag hinges can interfere with the folding and it is necessary to position the lag hinges correctly. The lag dampers are also hydraulic rams.

Hydraulic pressure is supplied to the rotor head through a rotating seal. Initially, the lag hinges are positioned by applying pressure to the dampers. The dampers carry limit valves to allow fluid to flow once they have reached the end of their travel. This fluid flow operates the feathering bearing locks. The feathering bearing lock pins operate further limit valves to allow fluid to flow to the folding mechanism proper.

Locking pins in the hinges will then be withdrawn by hydraulic power. The retraction of the lock pins operates sequence valves which then apply fluid pressure to the blade-folding rams inside the rotor head. These drive sector gears in order to turn the hinges through the large angles needed. The hinge axes are not vertical, so that the blades fold back and down alongside the hull.

As the folding mechanism is self-sequencing, unfolding the blades is achieved simply by reversing the direction of hydraulic pressure on the rams. The folding rams retract to swing the blades out, and limit valves then apply pressure to the lock pin rams, which lock the folding hinges. Limit valves on the lock pins then allow pressure to the feathering bearing locks in order to release the pins. When the feathering lock pins retract, further limit valves allow flow to reposition the lag dampers.

For obvious safety reasons, the hydraulic interlocks ensure that the sequence cannot complete if any stage fails to operate. As a further precaution, the positions of the major elements are monitored by switch to operate a cockpit warning if any part is out of sequence.

Blade folding mechanisms are inevitably complex. In certain versions of the Osprey, not only do the blades fold parallel to the wing, but the wing also rotates 90° to align with the hull. Blade folding is expensive, adds weight and increases maintenance and will only be used where it is absolutely essential.

Reference

1 Gessow, A. and Myers, G. (1951) *Aerodynamics of the Helicopter*. Ungar, New York.

5

The tail

5.1 Introduction

Helicopters with a single main rotor must have some means of balancing the torque reaction due to driving the rotor. Whilst the anti-torque rotor could be mounted anywhere for hovering, in forward flight the most stable location is aft, supported on some kind of structure called a boom. In addition to the anti-torque function helicopters need some means of yaw control and the tail rotor also serves that purpose. In order to balance the weight of machinery at the tail, the helicopter cabin usually extends some way forward of the mast. This large forward side area is unstable in yaw and generally some fin area is needed to give directional stability in forward flight. A further consideration is that the main rotor on its own is unstable in pitch in forward flight and a tail plane is usually required.

As a result the tail of the conventional helicopter will be a structure supporting a variable pitch tail rotor, its transmission and controls, some fin area and a tail plane. In reality tail booms are also encrusted with antennae, flare launchers, navigation lights, strobe beacons, static vents, registration letters and warning notices, not to mention tail skids and occasionally part of the undercarriage.

In a helicopter with more than one rotor the torque may be cancelled in a different way and a great variety of yaw control mechanisms will be found. These multi-rotor yaw mechanisms are considered in Chapter 9. In this chapter the conventional type of tail rotor will be considered, along with a number of alternatives having various advantages and drawbacks. The conventional tail rotor is well understood and relatively inexpensive owing to its wide use.

The far aft location of the tail rotor assembly means that it must be lightly built in order to avoid the machine becoming tail heavy. The thrust needed from the tail rotor is much smaller than that needed in the main rotor. These two features conspire to ensure that the tail rotor is considerably more fragile than the main rotor. Unfortunately the fragile tail rotor is in a more exposed location where the pilot cannot see it. During certain manoeuvres, such as quick stops and rearward flight it can get very close to the ground, and it may come off worse in any encounter with vegetation the pilot has not seen. Tail rotor blades are usually fitted with soft aluminium 'telltales' at the tips which will be distorted by any impact and indicate that a close inspection for damage is required.

Generally the tail boom will be fitted with some kind of skid intended to impact the ground before the tail blades. In some cases this is a tubular structure partially encircling the rotor and known as a D-ring. This gives better protection as well as being more visible to ground personnel than the spinning rotor. The Enstrom F-28 series has a conspicuous D-ring.

The fast spinning tail rotor is actually very hard to see under many lighting conditions and a significant number of ground personnel have literally walked into one, often with fatal results. The problem is that the human visual system cannot respond to light changes above the critical flicker frequency which is at about 50 Hz. Unfortunately the blade-passing frequency of most tail rotors is beyond this. The result is that when looking at a spinning tail rotor, there is literally nothing to focus on and the eye tends to see only the background beyond the rotor, especially if this is more brightly illuminated than the rotor. A further problem is that when a hovering helicopter yaws, the tail rotor may move laterally at some speed, too fast for someone on the ground to move clear. Unfortunately most helicopters don't have rear view mirrors and the pilot may be unable to see a person near the tail.

There have been too many tragedies due to these effects, and these can be avoided by some simple rules. Ground personnel should never approach a hovering helicopter or one on the ground with turning blades unless the captain has indicated that it is allowed. If a conventional helicopter must be approached, it should only be from directly ahead, in full view of the captain. Helicopter pilots should avoid initiating rapid yaws in a low hover, as this gives an unnoticed person on the ground no chance of escape and increases the risk of striking the tail rotor on obstacles. It is good practice to paint both dark and bright patches on the tips of the tail rotor blades so that some contrast will be available whatever the background. Painting the patches at a different radius on each blade causes a spiralling or flickering effect that is more noticeable. In the case of a multi-bladed tail rotor, the flicker frequency can be lowered into the visible range by painting all of the blades a dark colour except for one which should be as brightly painted as possible. Another useful safety feature is to have a tail plane-mounted light to illuminate the tail rotor.

An increasingly relevant drawback of the conventional tail rotor is that it seems to generate a lot of noise. Although the tip speed is typically about the same as that of the main rotor, the tail rotor turns at higher RPM and so the blade-passing frequency is higher. This in itself doesn't make more noise, but human hearing is more sensitive to the increased frequency and so it seems louder. Another problem is that the tail rotor often works in the disturbed wake of the main rotor and impulse noise will be created when a main rotor vortex passes through the tail because this causes rapid variations in local angle of attack.

From some directions the tail rotor may be the noisiest part of the machine. Society is quite reasonably becoming less tolerant of noise, and there is no reason for the aviation community to expect special treatment. In military applications helicopter noise may also be an issue. The helicopter excels at inserting and retrieving special forces, but covert missions are likely to be compromised by excessive noise.

Helicopter designers have explored various ways of countering the main rotor torque in a way that reduces or eliminates one or more of these problems. These techniques give a noise reduction and a safety advantage, but currently at an increased cost. The *fenestron* system uses a fan enclosed in a short duct in much the same location as a conventional tail rotor. The NOTAR system (NO TAil Rotor) uses a combination of a sideways-lifting tail boom and air jets at the end of the boom. There is a fan inside the hull providing air for boundary layer control over the boom and for the tip jets. These systems will be considered in later sections of this chapter.

The tail rotor needs power and control. The power is generally delivered from the main gearbox by a light shaft supported by regularly spaced bearings to prevent whirling. Flight loads can cause the tail boom to flex and the drive shaft must be fitted with couplings to accommodate small angular errors and plunging. This avoids putting unnecessary stress on bearings and on the shaft itself. Often the shaft will be run on the outside of the tail boom to allow easy inspection. The shaft may be exposed, or covered by a D-shaped detachable cowling as in, for example, the Bell 206.

The tail rotor gearbox is relatively simple since it has only to turn the drive through 90°. The output shaft bearings will be designed to support the tail rotor and withstand flight loads. An oil level sight glass will be provided, and generally a chip detector. In large machines an oil temperature gauge may be fitted.

The tail rotor only requires collective pitch control and so it will have a swashplate which moves along the shaft axis but without tilting. There are various ways of moving the swashplate via the foot pedals. The pedals usually communicate with the tail using a pair of stranded steel wires, since these are light in weight. Figure 5.1 shows that there is some variety in the mechanisms used to convert the wire motion into movement of the swashplate. Figure 5.1(a) shows an example where the swashplate runs on the gearbox casing between the rotor and the gearbox. A simple bell crank and fork arrangement will move the swashplate. Figure 5.1(b) shows a system in which a pitch control rod runs through the hollow tail rotor shaft to a swashplate which is outboard of the rotor. This may be operated by a bell crank as before. The pitch control rod may be terminated in a coarse screw thread. The wires from the foot pedals are wound round a drum that

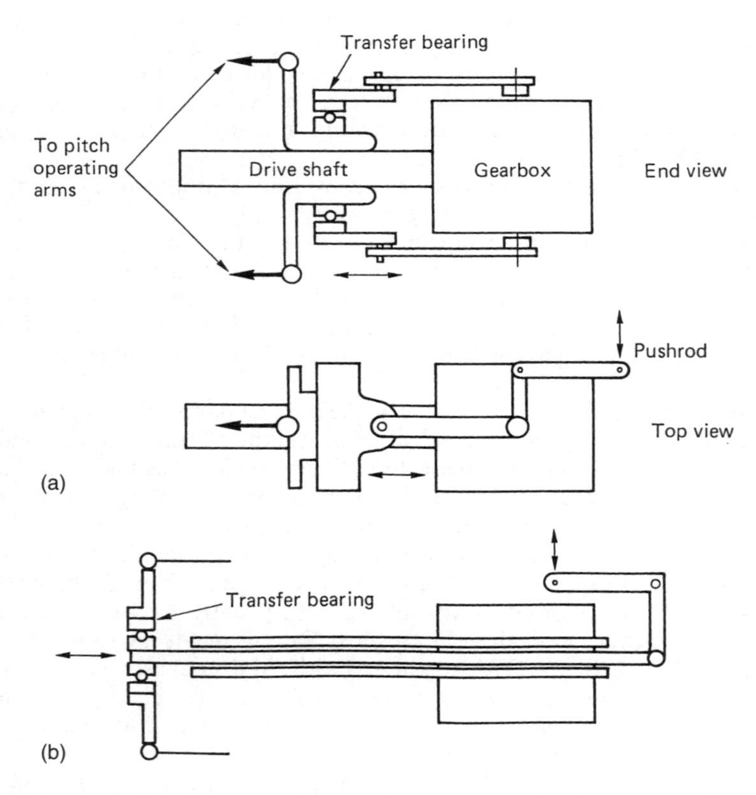

Fig. 5.1 (a) The tail rotor pitch may be controlled by a swashplate between the gearbox and the rotor. (b) The swashplate may also be found outboard of the rotor.

rotates the rod in the screw thread. Alternatively the drum may be replaced with a sprocket engaging a length of roller chain inserted in the wire.

An elegant improvement is to couple the collective pitch setting into the tail rotor pitch controls so that to some extent torque variations are automatically compensated. The Westland Lynx has such a system.

In larger machines a power assistance system may be necessary and this will be considered in Chapter 7.

5.2 Balancing the torque

The torque reaction of the main rotor is a pure couple, which is to say it has only a turning effect, and no resultant force in any direction. The tail rotor is mounted pointing sideways at the tail, and so the thrust it produces results in a moment about the centre of gravity of the machine. Unfortunately a moment cannot cancel a couple. The thrust of the tail rotor can be adjusted completely to cancel the rotation of the machine, but if the main rotor thrust is vertical, the tail rotor thrust will cause the machine to move sideways, a phenomenon called tail rotor drift. Tail rotor drift is prevented in the hover by tilting the main rotor disc in the opposite direction. Figure 5.2(a) shows that the main rotor thrust can be resolved into a vertical component opposing the weight of the machine, and a horizontal component. When the latter is equal to the tail rotor thrust, a perfect couple has been produced that opposes the main rotor torque reaction completely. The pitch of the tail rotor blades is controlled using the foot pedals. By disturbing the balance of the couples, yaw control is obtained.

A very slight increase in rotor thrust is necessary so that the vertical component still balances the weight. The tilted attitude of the disc during hover can readily be seen. The hull may or may not tilt due to the disc tilt, depending on the design of the rotor head, the way the transmission is installed and the location of the tail rotor. If the hull tilts, the CM of the helicopter, being some way below the rotor hub, will develop a rolling couple that can only be zero when the CM is directly below the rotor head. The final attitude of the hull will be at whatever angle this restoring couple balances the tilting couples from the rotor head and the tail rotor.

Figure 5.2(b) shows that if the tail rotor thrust is not in the plane of the main rotor, there will be a couple about an axis through the rotor head running in a fore-and-aft direction. The figure shows the situation viewed from the rear of a helicopter having a clockwise-from-the-top rotor and the tail rotor below the plane of the disc. In this case, the side force from the main rotor is above the side force from the tail rotor and this causes a clockwise roll couple called tail rotor roll.

If the tail rotor is above the plane of the rotor (seldom seen in practice), the side force results in an anticlockwise couple. Clearly it is only when the tail rotor shaft is in the plane of the main rotor hub that this couple will be zero, and helicopters which mount the tail rotor high on the top of the fin can achieve this elegant result, although there are more important reasons for such a location, which will be explored in section 5.5.

The reader is cautioned that some texts assert that the tail rotor must be mounted at the same height as the centre of mass to eliminate tail rotor roll, but this is incorrect. In fact if the tail rotor is mounted at the vertical centre of mass of the hull, the couple due to the different heights of the main and tail rotor side thrusts makes the hull tilt until it is balanced by the couple due to the laterally displaced CM. This is shown in Figure 5.2(c). The result is that the shaft axis and the tip path axis align and so there is no lateral flapping and the type of rotor head is irrelevant because, with no flapping,

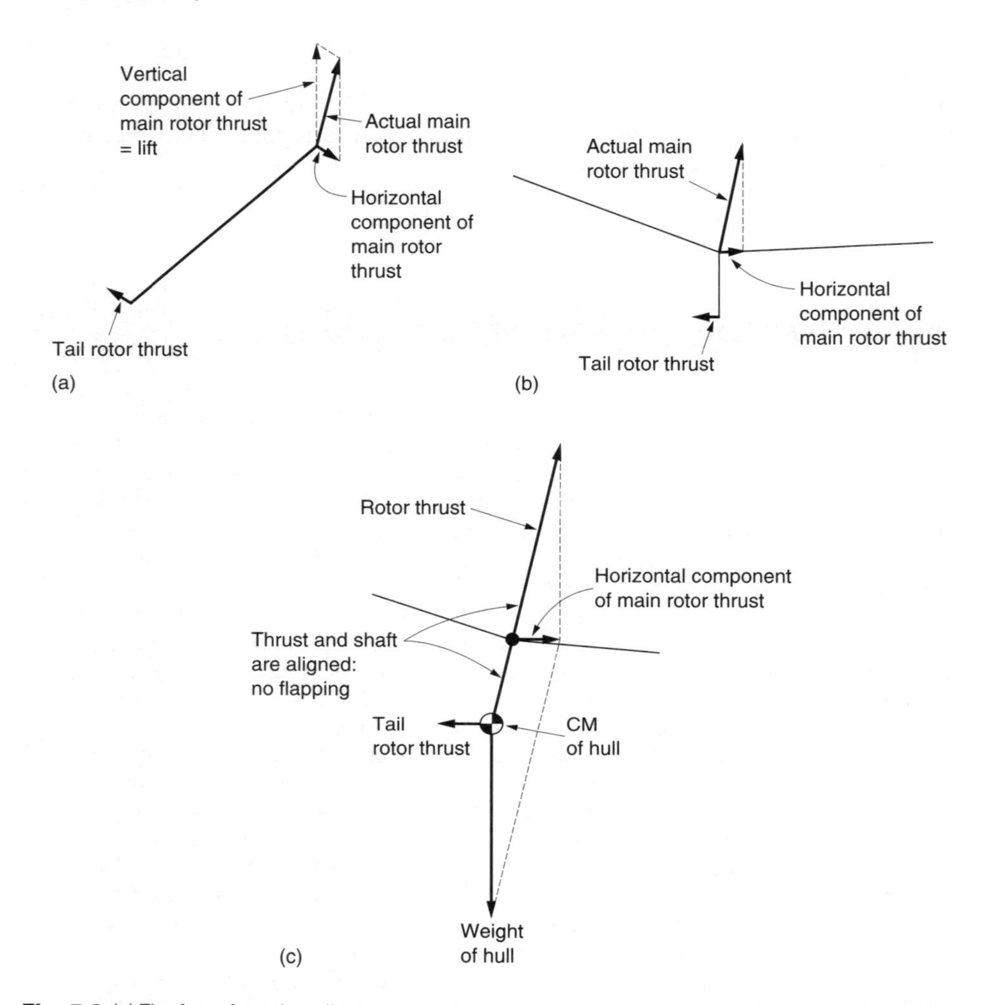

Fig. 5.2 (a) The force from the tail rotor cannot cancel a couple. In practice the main rotor is tilted so that a horizontal component of the main rotor thrust exists. When this is exactly equal to the tail rotor thrust a pure couple has been created. If this is equal to the torque reaction due to driving the main rotor the helicopter will not yaw. (b) If the tail rotor is not at the same height as the rotor head there will be a rolling couple. (c) If the tail rotor is aligned with the vertical location of the helicopter's CM, the tail rotor roll couple is exactly balanced by the horizontally offset CM so that the main rotor does not tilt with respect to the hull. With all other configurations the hull and the main rotor will have different attitudes.

there can be no rolling couple from the head. In all other cases the shaft and tip path axes are different and so the type of rotor head becomes significant.

As was seen in Chapter 4, the zero-offset rotor head is effectively a universal joint that allows the hull to hang straight down. Other rotor heads are stiffer and the hull will hang at whatever angle causes all of the couples to cancel one another.

If the helicopter has a zero-offset rotor, and the tail rotor is in the plane of the rotor hub, the fuselage will hang so that the CM is vertically below the rotor head. This will result in a laterally level cabin, assuming there is no payload asymmetry.

In the case of a rotor having offset hinges, a teetering spring or a hingeless head, the rotor head can apply a couple to the fuselage trying to roll the shaft axis in line with the tip path axis. The angle at which the fuselage settles in the hover will now be a function of the rotor head flapping stiffness, the vertical position of the hull CM and the tail rotor roll.

Finally, the designer may mount the main transmission at a suitably slight angle so that the hull stays more or less level with the disc and shaft tilted by the same amount. This means there will be little or no flapping in the hover. The Sikorsky Skycrane is an example of this approach. In the Mi-24 Crocodile (NATO code name Hind), the hull and transmission are tilted with respect to the undercarriage so that disc is tilted at approximately the correct angle. To compensate, the hull is twisted ahead of the main rotor so the cockpit remains level.

In forward flight the hull is moving through the air and so can develop a side thrust if it is set to a suitable angle of attack. If the machine is flown with no sideslip, the side thrust must come from the rotor tilt, whereas if the rotor disc is level, the side thrust must come from side slipping the hull. Clearly the machine can be flown with any combination of these two effects.

The least drag will be suffered if the hull is aligned with the direction of travel, and this may be significant under marginal power conditions. Zero-slip trim has the further advantage that the compass or direction indicator is actually displaying the helicopter's heading, making navigation easier. Flying at zero slip is aided by an airflow-sensing device showing the direction from which the air is approaching the hull.

Unfortunately most helicopters are fitted with an instrument inherited from fixed-wing aviation, where it is more useful. It is called a slip indicator because in fixed-wing aircraft that is what it does. It is not commonly appreciated that in helicopters the same instrument does not indicate true slip. This will be discussed in detail in Chapter 7.

5.3 The conventional tail rotor

The conventional tail rotor is mechanically a small main rotor. The term small being relative because, for example, the tail rotor of the Mi-26 is about the same size as the main rotor of an MD-500 and produces a similar thrust. The tail rotor needs no cyclic pitch control, only a collective mechanism actuated by the pedals. Aerodynamically it is also a scaled down main rotor, having much the same physics in the hover, and suffering the same indignity of being thrust through the air edge-on in translational flight.

As the tail rotor is expected to produce thrust in either direction, the blade section will generally be symmetrical and blade twist is only occasionally used. Blade taper can still be usefully employed to make the inflow more even, and this has been seen in practice, although it is not common because constant chord blades are cheaper to make from metal. When blades were made of wood, taper was relatively easy to adopt and as the use of moulded composites grows there is a possibility that taper will make a return.

The teetering rotor has many advantages for use at the tail, and the disadvantages it has as a main rotor are not relevant. As a teetering tail rotor is supercritical (see section 4.15) it cannot suffer from ground or air resonance. The two-bladed teetering tail rotor is simple and therefore light and has become extremely common. Teetering can still be used with four-blade rotors. Two independently teetering rotors can be fitted to a common shaft with a small offset between the disc planes. It is not necessary to mount the two rotors 90° apart; in fact it is advantageous not to do so. Mounting the blades in an X or scissors configuration produces less noise because there is no longer a dominant blade-passing frequency. Additionally with a carefully chosen skew angle and spacing between the two rotors a small gain in aerodynamic efficiency can be obtained because the axially spaced pairs of blades act to some extent like staggered biplane wings.

As with the main rotor, airspeed alternately adds to and subtracts from the blade speed due to rotation. The resulting roll couple is subject to precession and the result will be flapback and dragging. As the tail rotor tip path axis is tilted backwards due to the flapping, the thrust has a rearward component acting as a drag which the main rotor has to overcome. However, the inflow has a small component along the tip path axis in forward flight reducing the angle of attack and the shaft power needed. The power saved is then available to the main rotor and is precisely the correct amount the main rotor needs to overcome the drag. As a result tail rotor flapping is not detrimental to efficiency but it could lead to stress and/or wear.

In very large helicopters the tail rotor will need both flapping and dragging articulation to contain the stresses and the dragging axis will need damping. In smaller machines it is enough to have flapping hinges. The hinge will often incorporate some delta-three effect (see section 4.7). The delta-three hinge has the effect that as the rotor flaps, some cyclic pitch change is applied. The rotor finds equilibrium with a smaller amount of flapping.

The tail rotor counters the torque of the main rotor in the hover, but it also aids the directional stability of the machine in forward flight by acting as a fin. Figure 5.3 shows how this happens. If the tail swings to one side or the other, a component of the airspeed will change the inflow and with it the angle of attack. The result will be a change of thrust in such a sense as to return the tail to its original position. This is a highly desirable characteristic, except in rearward flight where the effect makes the machine unstable in yaw. An attempt to fly backwards at speed results in the tail swinging round, a phenomenon known as weathercocking.

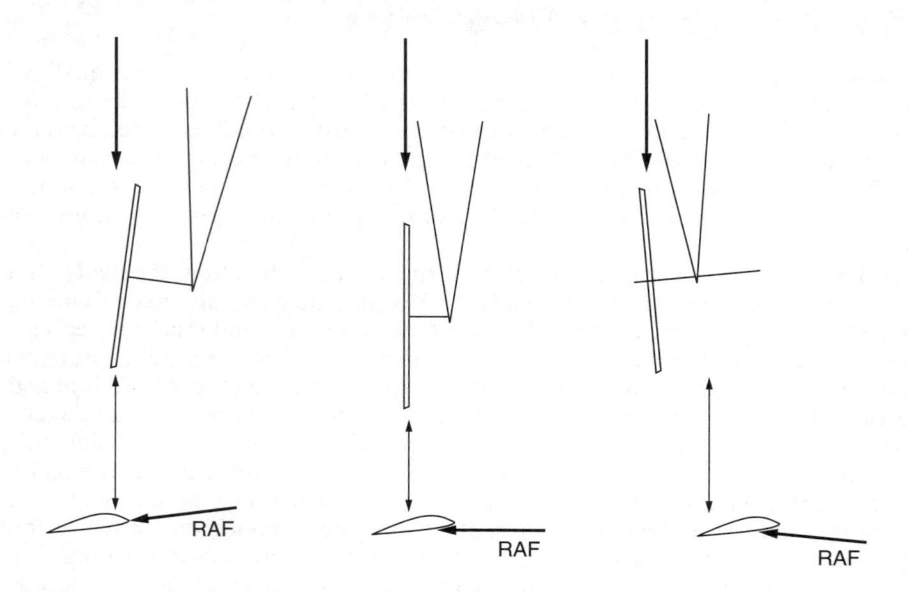

Fig. 5.3 The tail rotor acts like a fin in forward flight because if the machine yaws, the angle of attack of the tail rotor blades will change in the sense that opposes the yaw.

The tail rotor designer is faced with permanent compromise. Chapter 3 showed that the least power is needed if a rotor has a large diameter and a low, uniform, induced velocity. Profile power is reduced if the tip speed is low. Unfortunately the solidity will have to go up and the resulting tail rotor will be very heavy. To make matters worse, the large tail rotor will have to be mounted further aft to maintain clearance with the main rotor and this will require a longer and heavier tail boom. Ground clearance may also be an issue. There are further problems with large, slow, tail rotors. With a very low tip speed the translational speed will be limited by the growth of the reverse flow area. In the hover if the induced velocity is too low the tail rotor can enter a vortex ring condition during a yaw.

In fact minimizing tail rotor power is not the goal. A more useful goal is to minimize the ratio of the total power absorbed by both rotors to the overall machinery weight. A highly efficient but heavy tail rotor may need more total power as the main rotor has to work harder to lift it. As a result the designer will settle for a tail rotor diameter that contains the weight and clearance problems, and a tip speed similar to that of the main rotor so that the same advance ratio is obtained. As a result the only room for variation is in the solidity and taper. As a design evolves, the solidity can be adapted by changing the number of blades. For example, the Super Puma has four tail blades, whereas the original Puma had five. In some helicopters the high tip speed tail rotor was replaced in a later model with a slower version to reduce noise. This may need more blades to increase the solidity.

5.4 Tail rotor location

The tail rotor and the main rotor are both actuator discs and both produce thrust by virtue of the induced velocity they impart to their respective inflows. The main and tail rotors cannot be considered independently because of their proximity. They can and do affect each other by interaction of inflow to a degree that varies considerably from one regime of flight to another.

In the hover a low mounted tail rotor will actually draw air in from the edge of the main rotor. This will result in an increase in the induced velocity experienced by the main rotor and it will need more power. This might amount to 10–20 kW in a medium-sized helicopter. If the tail rotor is mounted higher the effect is largely eliminated and some power can be saved.

Figure 5.4(a) shows that the increased pressure below the main rotor disc and the reduced pressure above causes air to flow in a toroidal path around the edge of the disc. This is the mechanism of tip loss. The tail rotor operates in this region and there is a strong interaction. When the tail rotor turns in the same direction as the main rotor vortices (b) the relative airspeed of the tail blades is reduced and the available thrust is limited. When the tail rotor turns against the main rotor vortex (c) the performance is considerably enhanced because of the square-law connection between thrust and speed. There appears to be no detrimental aerodynamic effect of this direction, and so it is now considered to be the only direction to employ.

This phenomenon was understood relatively late in the history of the helicopter with the result that many machines were designed with the tail rotor going the 'wrong' way. In many cases in later models the tail rotor direction was reversed to universal acclaim. Oddly enough many model helicopters are still designed (if that is the word) with the wrong tail rotor direction. The author has modified a number of models to have the 'right' rotation and can vouch for the improvement in these cases also.

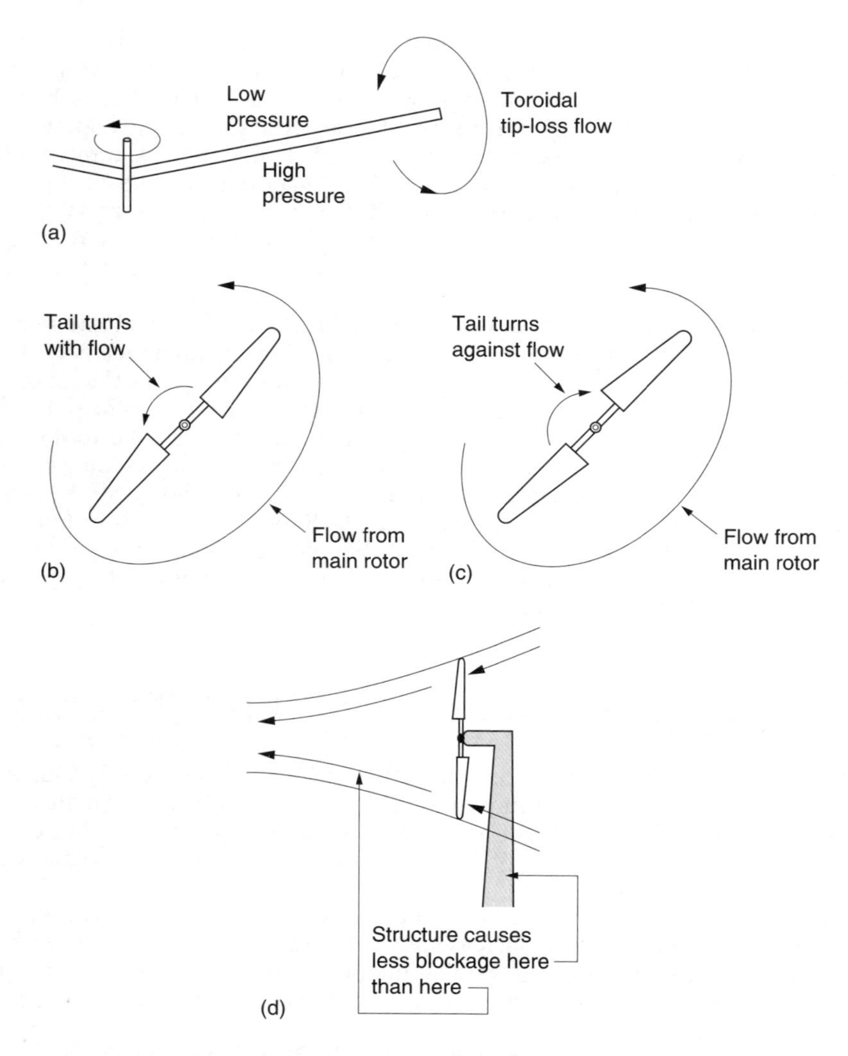

Fig. 5.4 (a) In the hover there is a toroidal flow around the edge of the main rotor in which the tail rotor operates. (b) A wrong-way tail rotor has its effective speed of rotation reduced by the main rotor flow. (c) A right-way tail rotor has its effective speed of rotation increased by the main rotor flow and will thus have greater authority. (d) Owing to wake contraction, it is more efficient to use a 'blower' tail rotor that takes in air over the mounting structure. If the tail rotor is on the wrong side the structure blocks a larger proportion of the wake.

As was seen in Chapter 3, with axial flow in the hover the wake of the tail rotor contracts and speeds up just as it does in the main rotor. This means that there is a right way and a wrong way to mount the tail rotor with respect to its supporting structure, which inevitably causes some blockage. The supporting structure may be a slim tube or a substantial tail fin. Figure 5.4(d) shows that if the tail rotor slipstream is directed away from the fin, a so-called 'blower', the proportion of the inflow which is blocked is reduced. The relative airspeed at the blockage is also lower. If instead the thrust is directed away from the fin, this blocks a larger proportion of the now

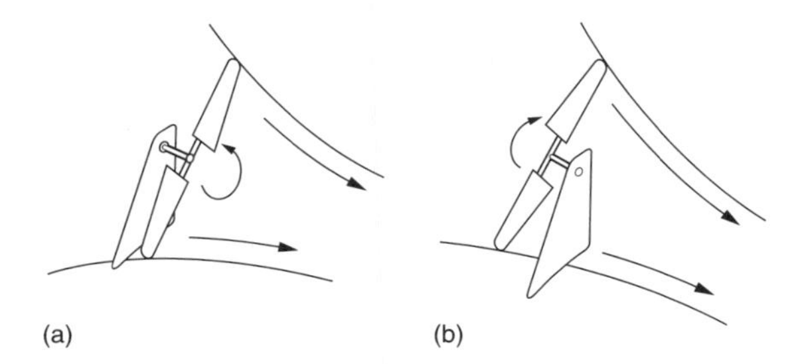

Fig. 5.5 A right-side wrong-direction tail rotor (a) can be made into a wrong-side right-direction tail rotor by turning the gearbox over (b). The improvement due to the right direction exceeds the loss due to being on the wrong side.

contracted slipstream which also has a higher relative airspeed. The thrust loss due to blockage will be significantly higher. There is thus a 'right' side and a 'wrong' side to mount a tail rotor; the blower installation will always be more efficient in the hover. In an *ab-initio* design, the right side, right direction tail rotor is an automatic choice.

The existence of the right side was appreciated well before that of the right direction, and many machines appeared with this configuration as shown in Figure 5.5(a). When it was realized that the tail rotor should be reversed, cost considerations often dictated the simple expedient of turning the tail rotor gearbox through 180° about its input shaft. This had the desired effect on the rotation, but then put the tail rotor on the 'wrong' side as shown in Figure 5.5(b). The improvement obtained by turning the right way exceeds the loss experienced by the increased blockage. As a result there are quite a few right-direction wrong-side machines. The Westland Lynx and the Mil Mi-24 are both in this category. In the Enstrom F-28A, the tail rotor originally went the 'wrong' way, but by turning over the gearbox as described the blockage increase was negligible because the tail rotor is mounted on a slim tube. The result was a significant improvement in tail rotor authority.

Figure 3.24(b) showed that in forward flight the main rotor produces a trailing vortex structure not unlike that of a wing. The streamlines converge above the plane of the rotor and diverge below it. The tail rotor operates in this airflow and the yaw-stabilizing action is affected by it. When the tail is mounted high, as in Figure 5.6(a) the amount of inflow change for a given yaw is increased by the converging flow, increasing the yaw-stabilizing action. However, a low mounted tail rotor may be operating in the diverging flow (b) where the yaw-stabilizing action is actually reduced.

There are thus many good reasons to mount the tail rotor high on a cranked boom. In addition to power saving in the hover, a high tail rotor is less likely to strike the ground in a low level quick stop, and it enables a safe landing to be made in scrub. The blades are less likely to suffer leading edge erosion from grit blown up by the downwash during hovering. In forward flight a raised tail rotor will produce greater directional stability, and encounters a cleaner airflow. The elimination of tail rotor roll as shown in section 5.2 is an elegant but not an essential bonus.

Unfortunately to site the tail rotor on a cranked tail boom requires an extra gearbox in the tail rotor shaft, and a stiffer tail boom. This adds expense that is unwelcome in smaller machines. In this case the requirement for a little extra power is not a problem because the turbine engine is usually heavily derated. In larger machines, the power loss becomes significant and in very large machines a correspondingly large tail rotor will be required and the issue of ground clearance also arises. Thus a cranked boom is

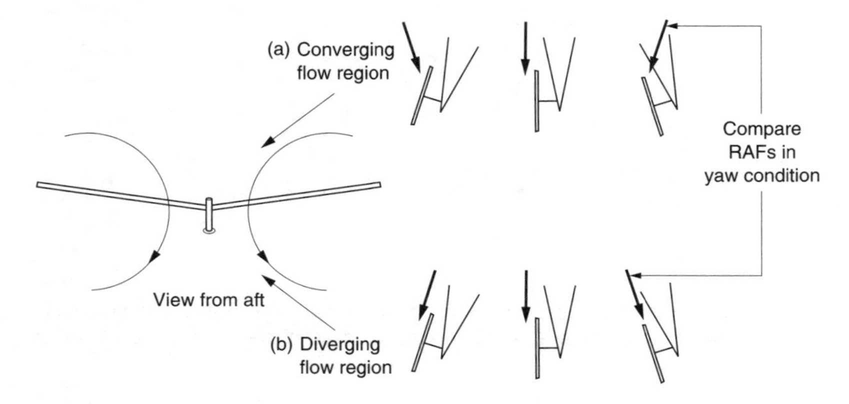

Fig. 5.6 The pressure drop across a main rotor in forward flight causes a vortex system as shown. A high mounted tail rotor (a) operates in the converging area of the vortex system and will have good weather-cocking properties. A low-mounted tail rotor (b) may operate in a diverging area of the vortex system and be less effective.

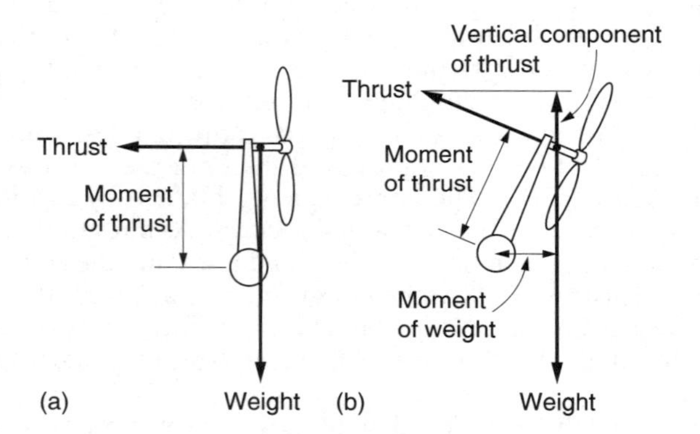

Fig. 5.7 (a) A high-mounted tail rotor applies a moment to the tail boom. (b) By canting the tail rotor, the weight of the rotor and gearbox produces an opposing moment.

a better option on a large machine, good examples being the Mi-26 and the Sikorsky CH-54 Skycrane.

In some helicopters the tail rotor is mounted in an unusual way. The Sikorsky series S-65/Sea Stallion/Sea Dragon is one example. This is a very powerful machine having a high disc loading, three engines and a seven-bladed main rotor needing an equally powerful tail rotor on a cranked boom. Figure 5.7(a) shows that in a conventional installation the side thrust from the high mounted tail rotor puts the tail boom under torsional loading. By tilting over the fin and tail rotor as in (b), the weight of the tail rotor assembly applies a torsional load in the opposite direction and the vertical component of the tail rotor thrust helps to carry the weight of the tail so that stress in the tail boom is reduced. The extra power consumed by the tail rotor is offset by a slight reduction in main rotor power as the latter is no longer carrying the whole weight of the machine.

The Sikorsky Blackhawk also has a canted tail rotor, but for a different reason. In solving various contradictory requirements, this machine turned out with a rearward centre of mass. The solution was to cant the tail rotor so that it produces some lift as well as side thrust. Canted tail rotors solve some problems but they do introduce interaction between controls. For example, pedal inputs will also cause some pitching. This is usually cancelled out to some extent by interconnections in the control system.

5.5 Tail rotor performance

In a still air hover, the slightest imbalance in the two couples will allow a slow yaw. If the pilot applies pedal input, the helicopter will accelerate about the yaw axis, but as it does so the tail rotor will then move through the air at a velocity determined by the yaw rate and the boom length. This velocity will change the inflow conditions and alter the RAF seen by the blades.

Figure 5.8 shows that the hardest yaw direction is against main rotor torque. The tail rotor has to produce a higher than normal side thrust and the pitch of the blades will increase to provide it. However, as the yaw proceeds, the tail rotor accelerates towards the inflow, so that the inflow velocity increases. This has the effect of reducing the angle of attack of the tail blades. This result is not surprising because in section 4.2 it was seen that the collective control determines the vertical velocity in the main rotor because of the same inflow effect. The yaw rate will come to equilibrium when the change in RAF reduces the angle of attack to the point where the tail rotor thrust imbalance is equal to the drag of the yawing tail boom.

If the yaw is required to be with the main rotor torque, the tail rotor needs to produce less thrust and the blades will be set to a reduced pitch. As the tail accelerates, the inflow velocity falls and this has the effect of increasing the angle of attack, again bringing the yaw rate to a constant value. If a greater yaw rate is required, the blades will need to be set to a negative pitch. This will also be a requirement of yaw control in autorotation where the main rotor torque reaction becomes very much smaller and changes direction.

When in a fast power-on yaw with the direction of main rotor torque, the tail rotor may be in negative pitch and the inflow may reverse. The tail rotor can enter a windmilling state where it is actually being driven by the airflow. At a critical yaw rate the inflow becomes zero. As was seen in section 3.16, zero inflow is an undesirable condition because it causes the rotor to enter a vortex ring or recirculation condition.

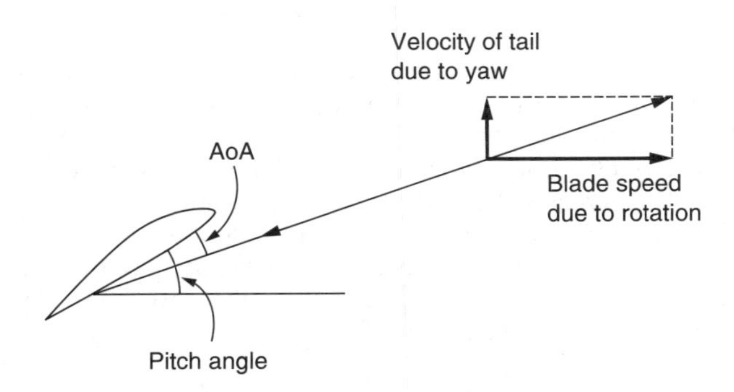

Fig. 5.8 When yawing against main rotor torque the tail rotor needs a large pitch angle in order to maintain the angle of attack with the increased inflow velocity.

This represents a worst case in tail rotor performance. The result of a fast with-torque yaw may be that when the pilot attempts to arrest it there is little response and the required heading may be seriously overshot. This goes down badly in the military as it spoils the weapon aiming.

It should be clear from Figure 5.8 that the tail rotor must have a wide pitch range to allow rapid yaws in either direction and then even more to maintain that degree of control in autorotation. A large positive pitch value is essential to permit a positive angle of attack to be reached despite the high inflow of an against-torque yaw. This means that when the with-torque yaw rate is high the rapid application of full opposite pedal can result in an enormous angle of attack that may be enough to stall the blades.

This is doubly bad, first because the yaw obtained will be in the opposite direction to that intended and second because a stalled tail rotor puts the tail transmission under an enormous torque loading. Everything at the tail of a helicopter is designed to be light in weight and this includes the transmission. Stalling the tail rotor may do damage. Consequently it is considered bad practice to stamp on the tail control pedals. A better result will be obtained if the pedals are operated gradually. In some helicopters tail drives actually did suffer damage and the designers responded by putting viscous dampers on the pedals.

In addition to the yaw function, effective tail power is also needed for sideways flight. It is not obvious why anyone would want to fly sideways, but there are plenty of examples. Large film cameras are often mounted in the main cabin and only have a clear view to the side. Flying the machine sideways allows the camera to shoot forwards. Flying sideways allows the pilot of an attack helicopter to dodge fire whilst keeping his rockets aimed at the target. Pilots who wouldn't dream of doing the above also fly sideways as a matter of course, because this is exactly what happens when hovering in a side wind. The helicopter only appears to be hovering; it's actually flying along sideways at the same speed as the wind but in the opposite direction.

In the case of a clockwise-from-the-top helicopter, the wind coming from the starboard side is undesirable as it increases the tail rotor inflow and so requires more power. The worst case will then be where the pilot wishes to make a maximum speed yaw to port in a strong wind from the starboard side. The tail rotor now has to overcome main rotor torque, boom drag due to the side wind and the yaw under conditions of greatest inflow where its angle of attack is reduced. Needless to say this is how the military test tail rotor performance.

Again assuming a clockwise-from-the-top helicopter, there will be a critical airspeed in sideways flight to starboard where the tail rotor is moving at the same speed as its induced velocity and a vortex ring is again a possibility when the pilot tries to arrest the manoeuvre.

Sufficient shaft power is not generally a problem in a turbine helicopter as the main rotor will be in translational lift in a side wind hover and there will be plenty of power available for the tail rotor. The problem is whether the tail rotor can use the power available. The blade pitch cannot be increased indefinitely or the tail rotor may stall. The RPM cannot be increased because of compressibility in forward flight and the diameter can't be increased because this will need the boom to be lengthened and will reduce ground clearance. One option is to increase the solidity of the tail rotor, typically by adding blades. This is less efficient, but it does at least solve the tail power problem. Other options may include an assessment of the degree of blockage and/or the amount of rear side area.

In a piston engine helicopter the tail rotor may not be powerful enough to allow a crosswind hover above a certain windspeed. In some light helicopters this is not a problem because that windspeed may also exceed the maximum speed at which it is safe to start the rotors. However, some piston engine helicopters, such as the Enstrom, can safely start the rotors in a gale and on moving to the hover it will be found that the machine simply weathercocks into wind with the pedals only allowing a certain yaw angle either way. The tail rotor is defeated by inflow and cannot obtain an angle of attack.

In any tail rotor operation that increases the inflow, the blade reaction will tilt back and the tail rotor will require more torque to drive it. This torque is provided by the engine, and delivered by the long tail drive shaft. If the tail rotor absorbs more power and nothing else happens, there will be less power for the main rotor. In a simple piston engine machine, where there is seldom any sophisticated engine governing, yawing in the same direction as the rotor turns will result in the machine also tending to descend.

The pilot has to compensate by raising the collective lever slightly, which will then require more engine power. The tail rotor takes roughly 15% of the total power in the still hover and more in the circumstances described above. Many piston-engine helicopters don't have a lot of surplus power and the pilot soon learns that large pedal inputs in the hover in one direction are to be avoided. Given a choice, the pilot will always prefer to yaw in the direction requiring the least tail power, using smooth and gradual pedal applications. With limited power, yawing must be done slowly as there is no point initiating a yaw in the 'easy' direction if it cannot be stopped.

In a turbine-powered helicopter the RRPM is accurately governed and so if a yaw in the direction of main rotor rotation is initiated, the extra torque needed by the tail rotor is automatically provided by action of the governor. As a result the RRPM relative to the helicopter does not change. However, the whole helicopter is now turning with the yaw and so the RRPM relative to the air has slightly increased. One might not expect this to have much effect, but the yaw RPM can raise the effective RRPM by a few percent and as lift is proportional to the square of the RRPM, a significant increase in lift can occur. Thus the turbine helicopter will climb under the same conditions that caused the piston engine helicopter to descend.

5.6 The tail plane

The tail plane is a necessity in forward flight for two reasons. First, because the main rotor alone doesn't have stability in pitch; and second, because the drag of the hull acts some way below the rotor head, a moment results which tends to pitch the hull down. The nose-down attitude of the hull will result in higher drag than if it is aligned with the RAF. Figure 4.12 showed that an aft-mounted tail plane developing a down force produces a moment in the opposite direction to the hull drag moment.

In a machine with a teetering or zero-offset head there can be no moments from the rotor and so this mechanism determines the hull attitude. The hull will adopt a pitch angle where the drag moment, the tail plane moment and any moment due to an offset CM all sum to zero. The system is stable because if the hull pitches down the (negative) angle of attack increases and produces a restoring moment.

In Figure 5.9 the machine has rotor head offset and the hull drag moment can be balanced by any combination of the tail plane moment and a couple from the main rotor due to the tip path axis being tilted with respect to the shaft axis. In practice a couple from the main rotor will be obtained with the penalty of increased mast stress and some vibration and so it is beneficial to trim the hull attitude in cruise with the tail plane so that a minimal rotor couple exists.

In the hover the tail plane is an unnecessary weight to be lifted and may actually produce significantly more down force than its weight if it is in the rotor downwash.

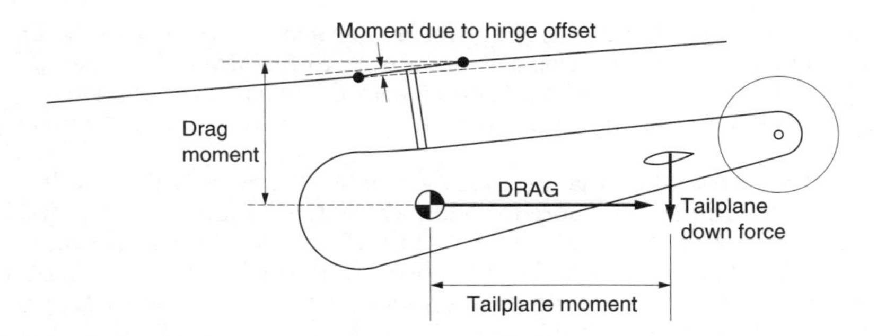

Fig. 5.9 In rotor heads having flapping stiffness, a couple can come from the main rotor to help balance the hull drag.

This is another example of requirements in hover and translation being at odds. In practice it is not the hover download on the tail plane that is the major problem. Of more concern is the serious fore-and-aft trim change that results should the downwash move clear of or back onto the tail plane. In a teetering or zero-offset machine with limited cyclic power this could cause control problems. In machines having higher cyclic authority, a greater tail plane load in climb and autorotation can be handled. In this case the trim shifts can be handled, but may present an excessive pilot workload, especially if instrument flight is contemplated.

There are three possible solutions to avoiding trim shifts of this kind. One is to ensure that the tail plane is always in the downwash, another to ensure it is never in the downwash and a third is to use a variable incidence tail plane or *stabilator*. Stabilators are considered in the next section.

Helicopters often go through a surprising number of tail plane modifications and locations as development proceeds and designers try to find the right combination of characteristics. Figure 5.10 shows some possible locations for the tail plane. At (a) the tail plane is mounted well forward so that it is in the downwash at all but the highest forward speeds. This is the approach used on, for example, the Bell 206 (JetRanger) which has a teetering head. At (b) the tail plane is mounted high on the fin so that it is never in the downwash. This gives the least down force in the hover and a long moment arm for stability at speed, but it does require a stiff and strong fin and tail boom. The MD-500 has a tail plane of this type with the tail rotor below and mounted in-line with the tail boom to simplify the transmission.

Where the tail rotor shaft is mounted in the plane of the main rotor, an out-of-downwash location can be obtained with an asymmetrical tail plane fitted on the opposite side of the fin to the tail rotor. Many Sikorsky machines use this approach, the S-65 being a good example, shown in Figure 5.10(c).

The tail plane can also be put below the tail rotor as in (d). This is almost out of downwash in the hover, but as the machine moves forward from the hover the downwash impinges on the tail plane from above causing a strong trim shift. In this location, however, the tail plane does at least prevent ground personnel approaching the tail rotor.

The section used on the tail plane is subject to some variation. In machines with zero-offset heads, there is no couple from the rotor to prevent hull blowback in forward flight and this must all be balanced by tail plane downthrust. Such machines have limited cyclic control authority and in autorotation the tail plane would produce an undesirable upthrust. One solution is to use an upside-down cambered section that

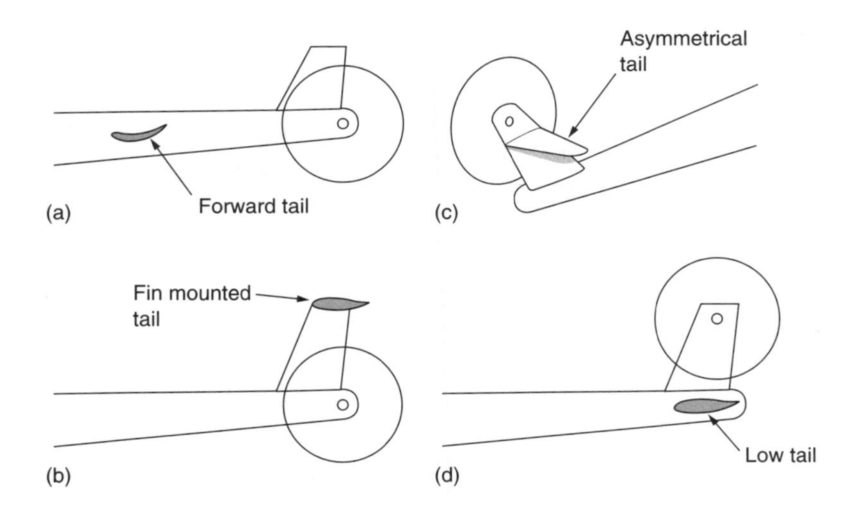

Fig. 5.10 Tail plane locations: (a) forward location almost always in downwash. (b) fin-top location is never in the downwash. (c) Asymmetrical tail plane opposite the tail rotor. (d) Low aft-mounted tail plane causes strong trim shifts but protects tail rotor.

efficiently produces downthrust in forward flight but which stalls in autorotation. The Bell 206 JetRanger uses this approach.

Another concern is avoiding tail plane stall in the climb where the (negative) angle of attack becomes large. Sometimes leading edge slats will be used. Another solution is to use a device called a Gurney flap. This non-intuitive device consists of a thin tab set across the trailing edge mounted at 90° to the chord line. The effect is to delay stalling up to extremely large positive and negative angles of attack, with a small amount of increased drag at low angles of attack. The Sikorsky S-76B uses such a device.

5.7 The stabilator

The stabilator is basically a variable-incidence tail plane that is moved to different angles, generally automatically, according to the flight regime. The stabilator and its actuator will be more costly than a fixed tail plane, so there must be a good reason to employ it. The advantage of the stabilator is that it can be aligned with the downwash so that it can give the stabilizing effect of a tail plane without the unwanted down- or uploads in climb or autorotation. A beneficial reduction in pilot workload will result, along with improved hover performance.

Figure 5.11(a) shows that in forward translational flight the stabilator may assume a slightly negative angle of attack to counter the hull drag moment. In the hover and in slow forward flight the stabilator will move to a large positive angle (b) in order to minimize the download. In a steep climb, the stabilator may also adopt a small positive angle (c). This not only reduces download and aids the climb, but it also levels the hull. In a tandem seat attack helicopter this may give the pilot in the rear seat a better view forward in a climb. Figure 5.11(d) shows that in autorotation the stabilator should move to a negative pitch to avoid an upload which would pitch the machine nose down. In addition to these deterministic movements, the stabilator may also be linked to an artificial stability system based on gyroscopic pitch sensors in the hull.

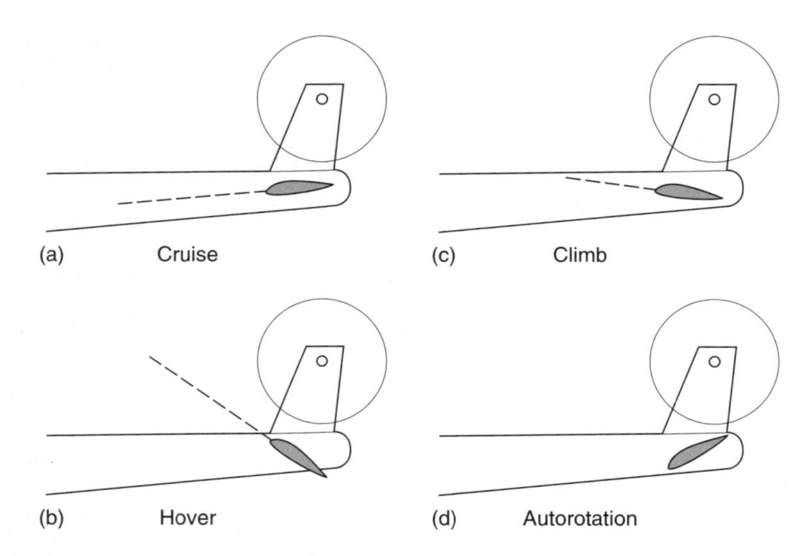

Fig. 5.11 The stabilator is a variable incidence tail plane that aligns itself with the direction of airflow in different flight regimes to improve performance. In forward flight, (a) a slight negative angle of incidence produces a downthrust to counter hull drag which acts below the rotor head. At (b) in the hover the stabilator adopts a large positive angle of incidence to prevent the creation of a download. At (c) in the climb a smaller positive angle of attack is beneficial. In autorotation (d) a negative angle of incidence prevents an unwanted upthrust.

Although stabilators are generally considered a good thing, some thought has to be given during the design process to the results of an actuator or control system failure leaving the stabilator in an inappropriate attitude. In a machine with a zero-offset head this would probably not be acceptable. In a machine with higher control power the incorrect tail loads following an actuator failure could be overcome by the cyclic control at least at low airspeed.

In light helicopters the actuator could be a pneumatic device driven by a combination of dynamic pressure from a pitot so that it responds to airspeed and manifold pressure so it responds to power level. It may be advantageous to mount the tail plane on a viscous damper with a suitable time constant that would stabilize brief perturbations whilst aligning to long-term trends.

5.8 Fins

Fins may be applied to helicopters for the same reason as on fixed-wing aircraft: to enhance yaw stability in forward flight. Secondary purposes of fins may be to allow the machine to fly on following a tail rotor failure or to enhance control in autorotation, especially in helicopters not having a tail rotor.

Fins are seen in a variety of locations as designers seek various compromises. In the case of the high-mounted tail rotor, the tail rotor pylon is naturally extended and streamlined to form the fin. Low-mounted tail rotors with a single fin tend to suffer blockage as the tail rotor and the fin have to be in the same place to clear the main rotor. Blockage can be avoided by using end plates on the tail plane as fins. These also make the tail plane more efficient as tip loss is reduced.

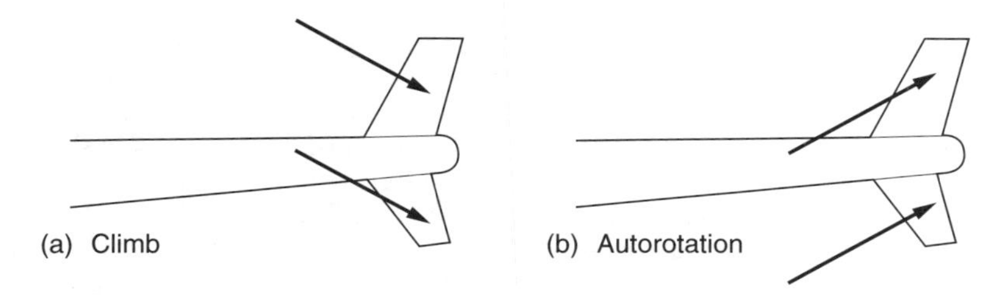

Fig. 5.12 (a) A top-mounted fin is efficient in climb, especially if it has some sweepback. A fin under the tail boom is in turbulent air in the climb. (b) In autorotation the fin under the boom becomes more efficient.

A centrally mounted fin is not in an ideal location as it will be in the turbulent wake of the rotor head and mast. The result may be tail shake as vortices pass the fin. Mounting the fin area at the ends of the tail plane will beneficially move the fins into cleaner airflow.

The helicopter differs from fixed-wing aircraft in that it doesn't have to move the way it is pointing. A helicopter with a near-level hull may be flying level, climbing or autorotating. Figure 5.12 shows that the relative airflow can approach the fin from a wide range of directions. At (a) in a climb, the RAF is angled downwards, and a top-mounted swept-back fin will present a high aspect ratio and be efficient whereas a bottom-mounted fin is at a disadvantage and may also induce tail shake or 'squirreling' because of turbulence around the tail boom.

Figure 5.12(b) shows that in autorotation the position is reversed. The swept top-mounted fin now has a low aspect ratio and will be inefficient. Thus in practice fins are frequently found with upper and lower sections both having sweepback as shown so that good performance can be obtained both in climb and autorotation. The fin of the Bell 206 JetRanger is a good example. The same sweep configuration may be found on end-plate fins such as the BK-117.

A large fin area is beneficial in forward flight and if the tail rotor fails, but it will be a drawback when hovering crosswind as it will create a large source of drag which must be overcome by the tail rotor. As a result fin area must be a compromise. In civilian machines larger fins may be found than those used in military machines because the crosswind hovering requirement is not so important.

Fins are generally, but not always, fixed. In some cases the fin may be cambered and/or fitted at a small angle of attack to offload the tail rotor in cruise. Sometimes the fin may be fitted with a rudder or tab connected either to the pedals or to a yaw-stabilizing system. Pedal operated rudders are more common on coaxial helicopters and synchropters.

Unlike the fins of fixed-wing aircraft, the helicopter fin has to behave predictably over a wide range of possible horizontal RAF directions because of the crosswind hover requirement. One possibility is the adoption of the blunt trailing edge, as if the rear of a conventional airfoil had simply been cut off. This technique allows the fin to operate up to a larger angle of attack without stalling. The penalty in forward flight is remarkably small because the increase in turbulence is balanced by a reduction in wetted area. Blunt trailing edges were adopted on all Chinooks after the CH-47A, and can also be used as a neat way of finishing off the rather fat fin needed to contain a fenestron.

5.9 The tail boom

The tail boom must provide structural support for the tail rotor, the fin and the tail plane, as well as having some aerodynamic characteristics. Figure 5.13(a) shows that a tail boom which is a smooth continuation of the hull will have lower drag in forward flight than the 'pod and boom' construction shown at (b). However, a wide tail boom will suffer a greater download in the hover. Figure 5.13(c) shows that the tail boom often has a rounded rectangular shape to allow reasonable depth without excessive hover download.

A streamlined tail boom may be practical on an executive transport, but for other purposes rear loading ramps or clamshell doors may be needed and these always result in a hull having higher drag. The boom has to be high set and slim to give clearance for rear loading.

The tail boom effectively couples two masses together; the main rotor and the tail rotor. These will each experience different forces, some static and some alternating. The main rotor will apply vibrations to the hull, but the tail rotor will tend to lag behind because of its mass and the result will be stress on the tail boom. Figure 5.14(a) shows that when starting, whirling forces from the main rotor will rock the hull from side to side. A tail rotor mounted atop the tail fin will resist the rocking and place the tail boom in torsion. The mass of the tail and the torsional stiffness of the boom will create a resonant system and if the resonant frequency coincides with an exciting frequency the tail assembly will oscillate in torsion.

If torsional oscillation results from a flight frequency, the resonant frequency will have to be changed. Intuitively, stiffening the tail boom would do this, but weakening it would also change the resonant frequency to a lower value. A lengthwise slot in the

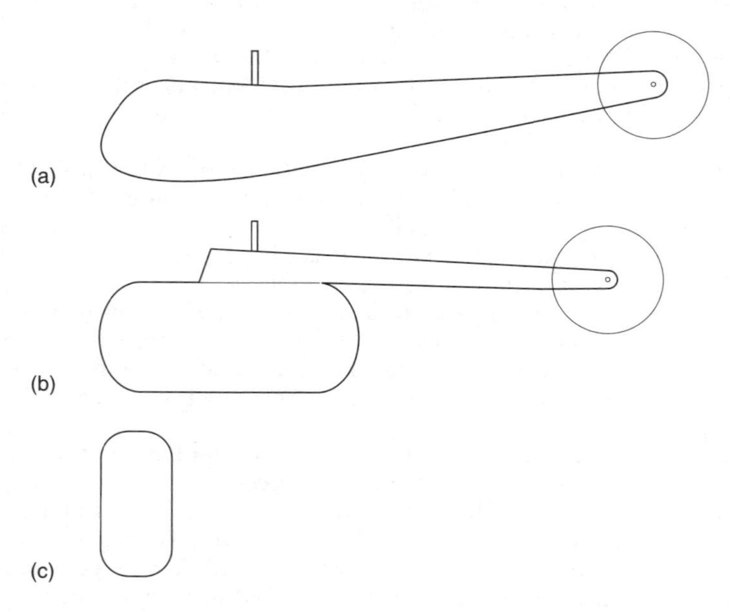

Fig. 5.13 (a) A hull in which the tail boom is smoothly faired into the cabin will cause less drag in forward flight. (b) A pod-and-boom structure suffers from turbulence at the rear of the pod. (c) Tail boom cross-section which is deep but narrow gives strength without excessive download in the hover.

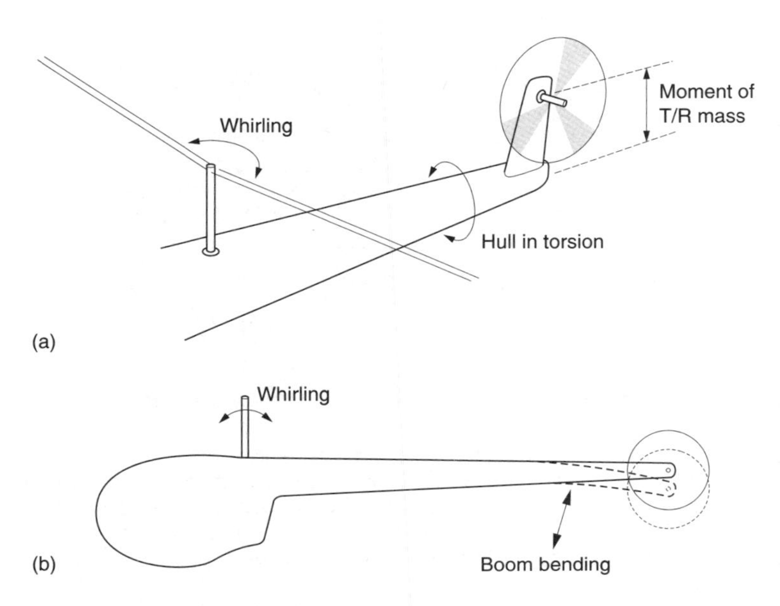

Fig. 5.14 (a) Main rotor lateral rocking during start-up places the tail boom under a torsional load. The inertia of the tail rotor and gearbox may combine with the torsional stiffness of the boom to produce a resonance. (b) Fore-and-aft rocking causes bending loads in the tail boom with a potential for resonance.

skin of a tail boom will reduce the torsional oscillation frequency without materially altering the bending stiffness. Damping material may be placed across the slot where it will be highly effective. Figure 5.14(b) shows that main rotor whirling in the fore-and-aft direction subjects the tail boom to bending loads as does tail rotor imbalance. Again a resonance may be present. The Robinson R-22 passes through some quite noticeable tail boom bending resonances during rotor starting and the rotors must not be run continuously at the RPM that excites them.

As the tail boom is in the downwash of the main rotor, it will create aerodynamic forces. Figure 5.15(a) shows that in still air the force will be primarily downwards, possibly with a slight, and beneficial, sidethrust due to swirl. However, if there is a crosswind, the resultant of the crosswind and the downwash will be a non-vertical RAF seen by the boom. This may result in a significant amount of boom lift, shown at (b) which is not necessarily in a useful direction. If the boom lift opposes the tail rotor thrust, this may restrict the crosswind hovering performance. A possible solution is the addition of a boom strake mounted high on the boom. This has the effect of causing early separation and a reduction in the amount of lift developed.

A cylindrical shape may be great for the pressurized hull of an airliner that goes the way it is pointing, but it is generally less than ideal for transverse flow. This is because transverse airflow across a cylindrical shape is unstable. Separation takes place from alternate sides causing lift in alternate directions. Those spiral strakes fitted to factory chimneys are designed to cause flow separation at defined points in order to prevent shaking in a wind. A rounded helicopter hull or boom will also suffer from an indeterminate separation point and this may result in chaotic shaking in the hover. In this case a strake fitted low on the boom will produce a defined separation point and reduce the shaking. Boom strakes generally align reasonably well with airflow in cruise and so have little detrimental effect.

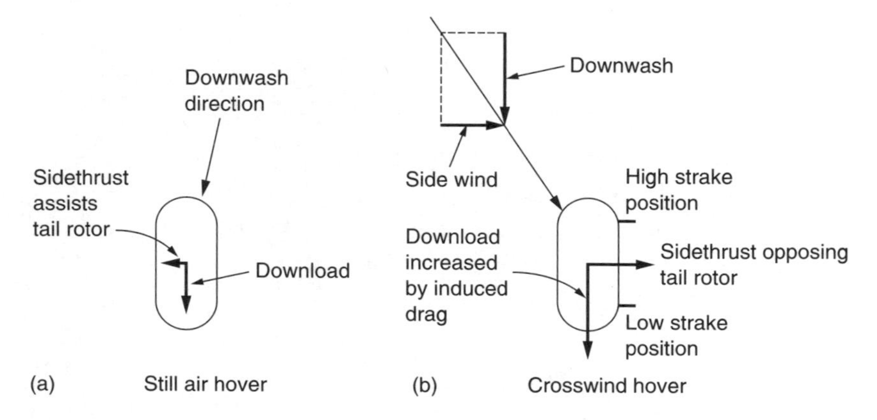

Fig. 5.15 (a) In still air hover, downwash on the tail boom is vertical except for a swirl component. (b) In crosswind hover, the resultant airflow direction due to downwash and wind may cause significant boom lift. If this opposes the tail rotor thrust the crosswind hover performance may be impaired. A high-mounted boom strake may be used to suppress boom lift in crosswinds, whereas a low-mounted boom strake may be used to define a separation point for downwash in the hover and thereby reduce shaking.

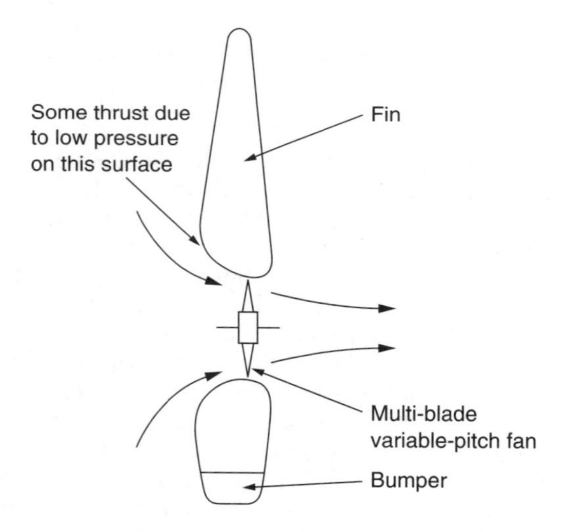

Fig. 5.16 The fenestron is more than a ducted fan because some of the thrust is produced by the low pressure on the intake side acting on the side of the fin.

5.10 The fenestron

The fenestron is basically a ducted fan built into an opening in the tail fin and was developed by Aerospatiale. The name comes from the French fenêtre meaning a window. It is commonly thought that the fenestron works like a small tail rotor, but this is not quite the case. In fact the fin structure is responsible for a significant amount of the system thrust. Figure 5.16 shows that the fan in the fenestron is a multi-bladed variable pitch device of high solidity. However, the fan is set in a duct that is relatively long. The cross-section of the duct matches the stream tube contraction as the fan is approached and so the reduced pressure caused by the fan is applied to the duct. As a result the effective area of the fenestron is the area of the duct mouth. The close fitting duct effectively prevents blade tip loss. The direction of rotation of the fenestron should be chosen on the same basis as that of a conventional tail rotor, namely the bottom blade going towards the nose.

An advantage of the fenestron is that as it is set in the fin, there can be no fin blockage as happens with a conventional tail rotor. In fact the fin is augmenting the effect of the fan. The fenestron presents a much reduced hazard to ground personnel and is quieter. However, as the fenestron is inseparable from its duct assembly, a powerful fenestron can become quite large and ground clearance may be an issue. Some machines equipped with fenestrons have a sacrificial collapsing bumper intended to absorb the energy of an unintended ground contact after which it must be replaced.

5.11 NOTAR

A helicopter without a tail rotor has a number of advantages. It can land in scrub without damage, ground personnel are less likely to be injured and there may be a reduction in noise. The first attempt at a tail rotorless machine was the experimental Weir W-9, shown in Figure 5.17, that flew in the UK in 1944. This machine had a piston engine that, in addition to driving the main rotor, drove a variable pitch fan to blow air

Fig. 5.17 The Weir W-9 had no tail rotor but instead ducted cooling air and exhaust gases down the tail boom to provide an anti-torque jet reaction. Twenty-five per cent of engine power was lost in the anti-torque system, which was not pursued.

down the tail boom, cooling the engine in the process. The exhaust gases were mixed in to provide further energy. At the rear of the hollow boom the gases were ejected from a sideways-facing nozzle at about 150 feet/sec (45 m/sec). The Weir was also unusual in having a constant velocity joint in its zero-offset rotor head. The Weir proved that the tail rotor could be eliminated, but it also proved that a simple fan/nozzle arrangement is very inefficient: around 25% of the engine power was consumed in the anti-torque system.

Stanley Hiller built a similar machine that flew in 1946. At around the same time Antoine Gazda built the Helicospeeder. These had piston engines. In France in 1954, Nord flew a machine called the Norelfe that used the exhaust from its turbine engine to oppose torque. All were abandoned. Today it is clear that these results follow from momentum theory. It is less efficient to derive thrust by ejecting a small cross-section of air at high speed than it is to eject a large cross-section slowly.

These experiments were enough to convince the industry that the tail rotor would not quickly be replaced; not in fact for about 30 years. In the 1980s, McDonnell Douglas developed a viable system which they called NOTAR and which does not require any more power than the conventional tail rotor. The NOTAR system has two components, a unidirectional system that provides about 60% of the anti-torque requirement, and a bi-directional system that provides the remainder of the thrust as well as providing yaw control.

The unidirectional system uses the tail boom as a wing operating in the rotor downwash. The tail boom is cylindrical and boundary layer control is used. Figure 5.18 shows that on one side of the boom are two downward facing slots fed with air from the interior and pressurized to about 1 psi by a fan. The air emerging from the slots energizes the boundary layer and delays separation significantly so that the downwash flows around the boom and exits with a sideways component. Effectively the downwash has been accelerated to the side and the reaction to this is a thrust on the boom. The thrust has a lateral component and a downward component corresponding to the induced

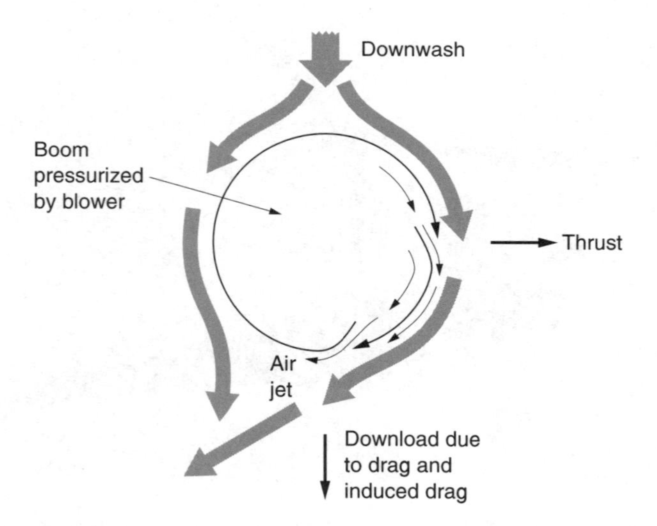

Fig. 5.18 In the NOTAR system, the boom is pressurized by fan and fitted with boundary layer control slots that energize the flow of the downwash to turn the tail boom into an airfoil.

drag. However, the slots are emitting downward jets and this causes some upthrust that nearly cancels the induced drag leaving a side thrust to counteract the torque.

About 60% of the anti-torque thrust comes from the boom. The remainder comes from a nozzle at the extreme end of the boom. This has a stationary inner part and an outer rotating sleeve. The inner part has a port on both sides, whereas the outer part has a single port. The pilot's foot pedals rotate the sleeve so that it will uncover one or other port to allow yaw control. The degree to which the sleeve turns determines the amount of port opening and so controls the thrust.

The boom is pressurized by a variable pitch fan. The pitch required is not a simple function of the pedal position and never becomes negative. Instead it is a function of how much thrust is needed. In other words the application of right or left pedal will cause the fan pitch to increase to provide extra airflow to the yaw control ports. In the case of autorotation the boom will be in upwash and the boundary layer control is no longer effective, but nor is it needed. The fan is driven from the main rotor and yaw control is by rotating the tail sleeve as usual.

The NOTAR system is remarkably quiet and safe and demonstration pilots have made their point by backing the tail boom into bushes and using back cyclic so that the machine rests on the rear of the skids and the end of the boom.

In the absence of a tail rotor to weathercock, the NOTAR machines, not surprisingly, need a substantial amount of fin area. In the production machine, minor yaw instability is handled with a gyroscopically controlled rudder tab on one of the fins. It is possible that this is not a fundamental problem and further development may eliminate it.

5.12 Tail rotor failure

It is often overheard that loss of tail rotor drive will cause the machine to go out of control, but this is not necessarily the case. If the tail rotor drive fails without causing other damage and the machine is put promptly into autorotation, the torque reaction of the main rotor will cease, and the tail rotor will be unnecessary. In some machines the tail boom and fin area are large enough to counteract engine torque by yawing into the slipstream at reasonable translational speed and then autorotation will only be necessary for landing. Nevertheless most pilots would rather lose the engine than the tail rotor drive.

Unfortunately tail rotor failures can result in more than loss of drive. A failed component in the rotor can cause serious imbalance and this can tear the whole tail gearbox out of the boom. A broken drive shaft may flail and destroy hydraulics and wiring or even sever the boom. In these cases the machine will suffer a serious forward CM shift and a loss of rear side area. A machine with a zero-offset rotor head probably wouldn't retain control. Figure 5.19 shows a zero-offset helicopter that has developed a serious nose down trim having lost its tail rotor in a collision with a ship. A machine with a substantial tail boom and a rotor head having a wide CM tolerance would have a much better chance.

Despite all the theory, retaining control requires the pilot to take the correct actions and in order to take good decisions the pilot has to know the situation. In the case of tail rotor problems the pilot cannot see what is wrong and has to deduce the problem from whatever symptoms of vibration, noise, yaw and pitch he experiences. When this happens suddenly it is asking a lot of the pilot to do exactly the right thing. Nevertheless many pilots have walked away from tail rotor failure, in some cases with no further damage to the machine.

Fig. 5.19 Following a collision with a ship, this helicopter has lost its tail rotor and gearbox. Having a zero-offset rotor head, there is insufficient cyclic authority to oppose the serious trim shift and the nose down attitude that results is apparent.

Others were not so fortunate and consequently regular inspection of all components associated with the tail rotor is to be recommended.

One point sometimes overlooked is that as the tail rotor is at the back, anything that comes off or out of the helicopter in forward flight can strike it. The designer may try to think of every eventuality, which is why the fuel filler cap is often retained by a length of wire, but there have been cases of engine access covers coming loose in flight and inflicting grievous harm to the tail rotor.

Helicopters are often flown with one or more doors open or removed, for film and television shooting, for example. In this case it is essential that everything within the machine is fixed, stowed or tethered. The helicopter with a high set tail rotor is at an advantage here, as debris is more likely to pass below it.

6

Engines and transmissions

The power plant is an important component in all aircraft and no less so in helicopters. However, helicopters necessarily have a more complicated power transmission system than aeroplanes. In this chapter the choice of power plant is considered, along with the operating principles of piston and turbine engines and associated transmissions. Power plant control and instrumentation is also treated.

6.1 Introduction

The power source of a conventional helicopter must be able to drive a shaft, and have reasonable weight and fuel consumption in relation to the power delivered. This generally means an internal combustion engine: piston, rotary or turbine. Real engines always turn too fast for real rotors and some reduction gearing will be needed to transmit engine power to the various rotors and accessories. There have been some exceptions to convention. Numerous efforts have been made to fit rockets, ramjets or turbojets directly to the blades, or even to duct gases into the blades from the hull. The goal is to eliminate torque reaction and reduce weight. These alternatives are considered in section 6.28.

6.2 Choice of engine

All internal combustion engines work by burning fuel in air. This raises the temperature, causing expansion which can do work. In practice more power can be obtained if the air is compressed before the fuel burns. The turbine engine has dominated the helicopter power plant market for some time. The main advantage of the turbine is a very high power to weight ratio. The lightness comes from simplicity; the compression of air and the extraction of shaft power are both done by rapidly spinning blades. There is, however, a penalty for that simplicity which is that the fast moving blades suffer from profile drag. Just keeping the blades moving consumes power. When used at high power, the internal power loss is a small part of the overall power produced and the turbine is efficient. When used at lower power, the power lost in the blades does not reduce in proportion and so the efficiency falls. Small turbine engines tend to be inefficient because the Reynolds numbers at which small blades must operate will be less favourable. As a result there are very few small turbine engines and these tend to be used as APUs (auxiliary power units) that are only used intermittently. Although simple in concept, the turbine engine uses highly stressed parts and the initial cost is high.

Statically compressing air with a piston in a cylinder suffers much less power loss than dynamic means such as a turbine. This makes piston engines more efficient at partial throttle settings, hence their popularity in automotive applications. The energy released on each power stroke is finite and so a good way of increasing the power is to increase the RPM. The piston engine must handle the stresses raised by reciprocating that increase with the square of the RPM.

This limits the RPM available as a function of piston size. Very small piston engines used in models can run at over 25 000 RPM and produce an astonishing power to weight ratio, but this cannot be scaled up because of the forces involved. Another fundamental of piston engines is that the power available is a function of the charge that can be admitted to the cylinder. The cylinder volume is proportional to the cube of the bore whereas the cylinder head area, where the valves are, is only proportional to the square of the bore. As a result the small cylinder has the advantage because not only is the piston lighter, but more valve area is available in relation to the displacement. Thus in piston engines there is a power to weight advantage in using a lot of small cylinders rather than a few big ones. There is an optimum cylinder capacity for a given RPM. The difficulty is that in order to produce a lot of power, a lot of cylinders are needed and this causes practical difficulties in induction, exhaust, cooling and reliability, to say nothing of cost. During World War II aircraft engines having as many as 28 cylinders were built and these were immensely complicated.

At 400–500 kW the turbine and piston engine are about equal. For higher powers the turbine is to be preferred, whereas for lower powers the piston engine would be chosen. The power to weight ratio may not be as high, but if the fuel efficiency is good, the saving in fuel load may more than compensate. As a result, helicopter design has diverged into two camps. Large machines use two or three turbine engines and tend to use high disc loading because plenty of power is available. Small machines use piston engines and lower disc loadings.

In between the extremes there is some variability. The small single-engine turbine helicopter cannot use high disc loading because it must be designed to autorotate well. As a result it will not need much power and so the turbine will be heavily derated and have poor fuel efficiency. Unfortunately derating an engine does not lower the cost. Well-engineered piston machines can compete in this market simply by costing less to purchase and less to run. Frank Robinson has demonstrated this amply with the R-44, despite its elderly engine technology.

Unfortunately the aviation piston engine went into a period of stagnation for several decades after the arrival of the turbine, with the result that progress in piston engine development moved to the automotive sphere. This situation is now being remedied and a new generation of aviation piston engines will give helicopter designers some better solutions in light machines. In the automotive sphere fuel economy does not drastically improve performance because the fuel load is a relatively small proportion of the vehicle weight. In helicopters the fuel weight at take-off is limited by available rotor thrust. Thus an improvement in fuel efficiency in the helicopter will translate directly into either an increase in range for the same payload, or an increase in payload for the same range.

The power of the gasoline engine is controlled by an induction throttle, and when this is used, the effective compression ratio of the engine falls and with it the efficiency. The induction throttle reduces the manifold pressure and this opposes the motion of the piston during the induction stroke causing pumping loss. The power of the Diesel engine is not controlled with a throttle. There is negligible pumping loss and the compression ratio is always at its highest value. As only air is compressed, there is no risk of detonation and the compression ratio can be very high indeed. This makes the Diesel fundamentally more fuel efficient than the gasoline engine as will be seen in section 6.14. Traditionally, however, the Diesel engine has been neglected in aviation although some Diesel powered aircraft flew in World War II. The automotive industry has dramatically improved Diesel performance in recent years and has shown that the Diesel engine lends itself very well to turbocharging. With modern technology the turbo-Diesel engine can equal the power of a gasoline engine for a given time between overhaul (TBO), but with better fuel economy so that the weight of engine plus mission fuel is actually less. Given that Diesel engines can run on AVTUR, which is significantly cheaper and less volatile than AVGAS, the economic and safety advantage of a Diesel engine is significant.

6.3 A piston-engine installation

Figure 6.1 shows the layout of the engine and transmission in a typical light piston-engine machine. Many light helicopters use an air- and oil-cooled engine originally designed for aeroplane use in which the crankshaft will clearly be horizontal. These typically have four cylinders arranged with a pair on each side. This arrangement is called a 'flat four' also known as a 'boxer' because the pistons on opposite sides appear to be sparring with each other. The propeller is replaced by the cooling fan and possibly a belt drive pulley. The cooling fan directs air through a series of baffles so that it passes round the cylinders. Some of the air is directed through the oil cooler. The centrifugal cooling fan of the Robinson R-22 is a conspicuous feature of the machine. The Enstrom has an axial fan.

Piston engines do not provide uniform torque because of the power impulses when each cylinder fires, and it is necessary to have some flexibility in the transmission to even out the power delivery. Even so, the transmission of a piston-engine helicopter has to be stronger to withstand the fluctuating torque. The use of anti-vibration technology in the drive train to smooth out the power impulses of piston engines with torsional tuned filters is in its infancy. Currently, flexibility in the drive train may be provided by using a belt drive and by mounting the engine on rubber blocks. The belts used are reinforced with polyester chords and have a tapering cross-section intended to run in vee-shaped grooves in the pulleys. Belt tension wedges the belt into the groove so that

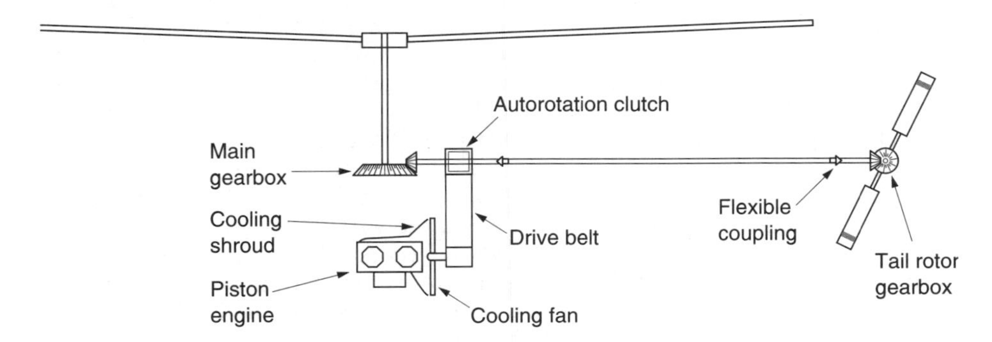

Fig. 6.1 The drive train arrangement used in a number of light piston-engine helicopters including Robinson, Enstrom and Schweizer models. The piston engine drives a cooling fan and a belt pulley. The belt drives a shaft between the main and tail gearboxes. The upper pulley contains a one-way clutch to allow the rotors to turn if the engine fails.

slipping will not occur. Machines vary in detail; the Robinson uses two double-vee belts, the Schweizer uses eight single belts, and the Enstrom uses a single Poly-vee belt about six inches wide. Belts can last as long as five years, but flying in dusty conditions causes more rapid wear.

The belt drive commonly does duty as the clutch. By slackening the belt tension, the engine can be started and warmed up without the rotors turning. When the engine is warm the belt tension is applied gradually so that the belt can slip in order to spin up the rotors smoothly. Once the rotors are up to engine speed the full flight tension can be applied. Starting the engine with the clutch engaged should never be attempted. In the Robinson the starter is powerful enough to turn engine and rotors, but when the engine fires the sudden speed increase could damage the main rotor blades. In an Enstrom the rotor blades are heavy and articulated and instead the result would probably be a burned-out starter.

The belt tensioning is performed in various ways. In the Robinson the shaft between the main and tail rotors is supported near the pulleys by a bearing that can be raised and lowered by a small electric gear motor. The flexible couplings in the shaft allow this movement. The belt tension is measured and the motor can only run until the correct tension is achieved. Some care is needed when operating the motor switch; rapid engagement will stall the engine. In the Enstrom the belt is tensioned by bringing in a jockey pulley against the flat outer side of the belt. This is mechanically operated by cable from a lever in the cockpit. When the drive is fully engaged a system of links passes over centre and maintains belt tension with no tension in the cable. In this system the pilot has better control of the clutch engagement force. In the Schweizer the belts are tightened by pulling out an idler pulley with a steel wire driven by a servo motor. The wire is permanently in tension. In all of these machines, the upper belt pulley drives the long shaft joining the main rotor gearbox and the tail rotor gearbox.

The pulley on the gearbox shaft has a one-way clutch inside it so that the engine can drive the shaft but the shaft cannot drive the engine. If the engine fails, or, worse still, siezes, the rotors can still turn freely so that an autorotation can be performed.

6.4 A turbine installation

As will be seen in section 6.19, turbine engines are found in two types. In the first there is only one rotating assembly and this is integral with the output shaft. No starting clutch is necessary because the starter motor spins the turbine and the rotor. In the preferred type of turbine engine there are two power turbines. The first generates shaft power to turn the compressor. The second turbine develops shaft power to drive the helicopter. Engines of this type are called free turbine engines because there is no mechanical connection between the first part of the engine, which is essentially a gas generator, and the second part that converts gas energy to shaft power. The gas generator may be spun with the starter motor even though the free turbine is locked by a rotor brake. In this way the engine can be started without the rotors turning. If the brake is released and the power is increased, the rotors will run up smoothly. The free turbine engine also has the advantage that the rotors can be stopped without stopping the engine. The turbine is set to idle power and produces minimal torque at the free turbine. The rotor can be stopped with the rotor brake.

The turbine itself is light and often mounted on or in the roof of the hull. Figure 6.2(a) shows the arrangement in the JetRanger. The power turbine gear train, through the autorotation clutch, drives the tail rotor shaft to the rear and the main rotor shaft to the front. The main rotor gearbox is resiliently mounted and the drive shaft is fitted with

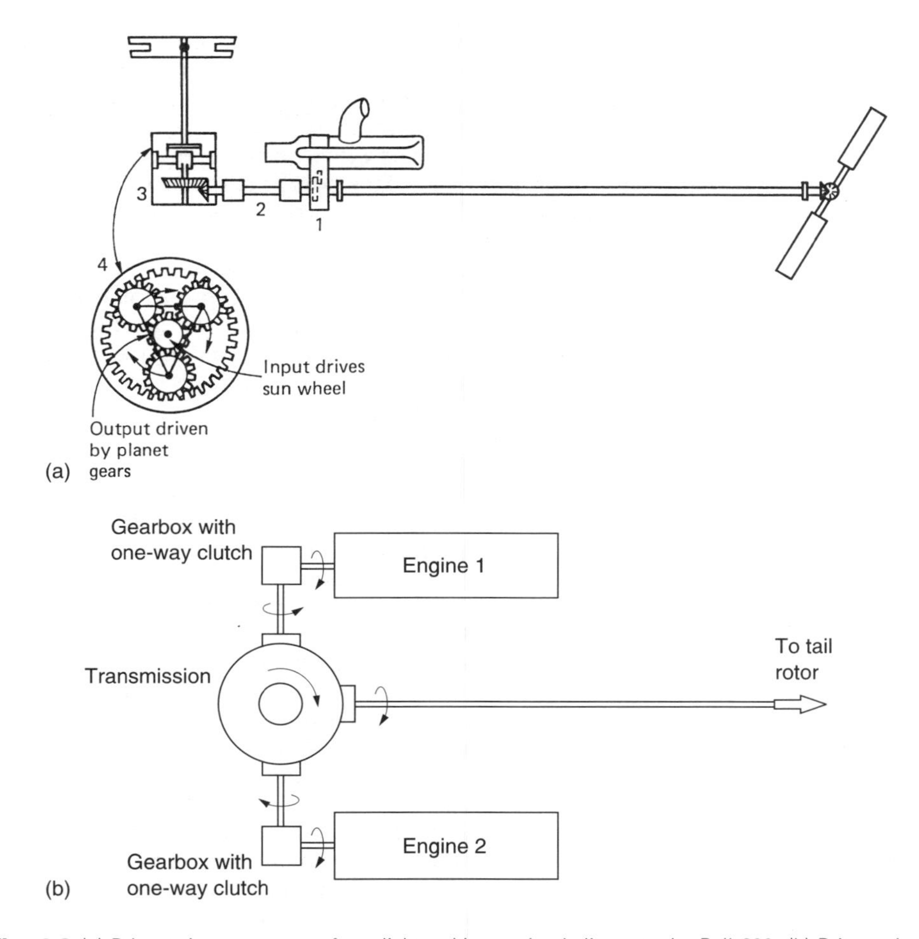

Fig. 6.2 (a) Drive train arrangement for a light turbine-engine helicopter; the Bell 206. (b) Drive train for twin-turbine helicopter. Each engine has its own one-way clutch so one can fail leaving the other to power the machine. The tail rotor will be driven from the main rotor transmission.

two constant velocity joints to allow relative motion of the gearbox and the engine. In some machines the turbine drives only the main transmission as in Figure 6.2(b). In many machines twin engines are installed. In this case a pair of autorotation clutches is needed, one for each engine. Then either engine can fail and the remaining engine drives the transmission. The tail rotor drive must also come from the transmission.

6.5 Correlators and governors

The RRPM determines the rate of response to the controls and must be kept constant. It was shown in Chapter 3 that the power needed to drive the rotors at constant speed varies considerably with the attitude and flight regime. Governors and correlators are devices intended to make control of the throttle more or less automatic so that the pilot has less to do. The governor is a system using negative feedback to adapt the power setting to the load placed on the engine. In contrast the correlator is a device using feedforward to predict the power setting necessary from the position of the collective lever. This may be done mechanically with a shaped cam on the collective linkage. The cam shape is obtained during tests of the prototype. Feedforward may also be applied electrically, in which case the collective lever is fitted with a position sensor. A correlator only gives an approximate control of RRPM since the position of the collective lever is only one factor affecting the power required. There is, for example, no compensation for air density or the degree of inflow through the rotor.

The gasoline engine is literally throttled to control its power. The induction passage is restricted and this is a stable process. If the engine speeds up, the pressure drop across the constriction will increase, dropping the induction pressure. If the engine slows down, the pressure drop across the constriction will decrease, increasing the induction pressure. Thus on a piston-engine helicopter governors are not essential and most machines just have a correlator. One school of thought suggests that it is better to learn to fly an uncorrelated machine because by manipulating the throttle a more thorough understanding of the power needed in different conditions is obtained.

Neither the turbine nor the Diesel has a throttle because the induction system is fully open at all times. As a result these types of engine will not hold their RPM unaided and a governor is essential. The governor is a system that compares the desired speed with the actual speed and uses the speed error to operate the power control to apply a correction. This is an example of negative feedback as described in Chapter 2.

The principle of the governor goes back to the first steam engines. In the centrifugal governor shown in Figure 6.3 a pair of weights are attached to a shaft driven by the engine such that they will fly out as speed increases. This is arranged to reduce the power so a set speed cannot be exceeded. The speed is set by a change to the spring tension applied to the linkage. If the spring tension is increased this tends to increase the power and the weights have to spin faster before they can produce enough force to reduce power.

In practice the engine revs will not be held perfectly stable because if more power is demanded the power control must be operated to provide it. The engine RPM must reduce slightly in order to create an error. Figure 6.4(a) shows that this is known as *static droop*. Static droop can be reduced by increasing the amount of power control movement for a given speed error. This effectively changes the gain applied to the speed

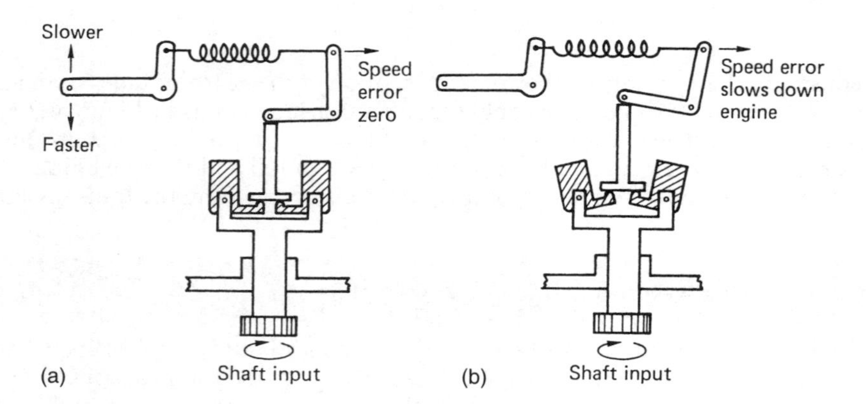

Fig. 6.3 The centrifugal governor is a very simple device as it relies on rotating weights being thrown outwards as speed increases.

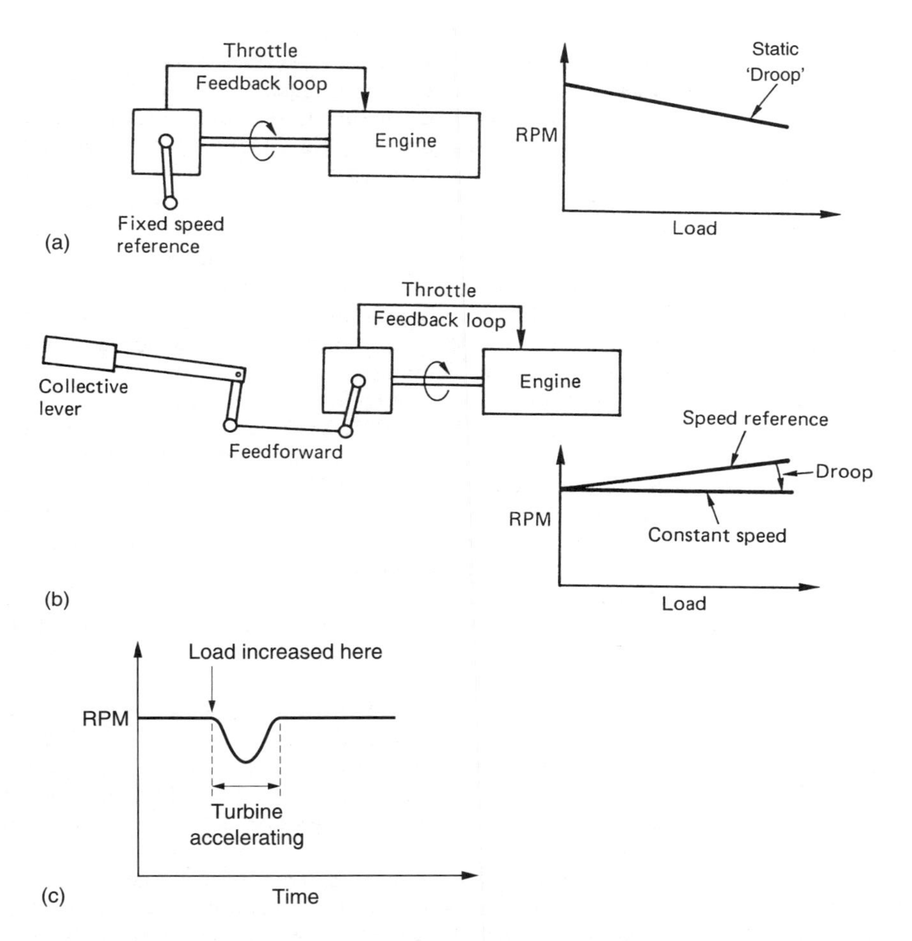

Fig. 6.4 (a) Following a fixed increase in power demand due to, for example, an increase in collective pitch, the RRPM may fall, because the feedback loop needs an error to increase power. This static droop can be eliminated using feedforward. (b) An input from the collective lever is used to anticipate approximately the amount of power needed. The feedback system then has to make smaller power corrections that it can do with a smaller RRPM error. (c) Transient droop due to the time needed for the gas generator to increase speed.

difference of the system. Increasing the gain will allow tighter control of speed but it also increases the chance that the system will become unstable so that it swings between too fast and too slow without stabilizing. A better solution is to combine governing and correlating systems so that there is both feedback and feedforward. Figure 6.4(b) shows how this is done. The power setting is determined by the sum of two signals. One of these is the feedforward signal from the collective lever. This provides approximately the right power setting. The second signal is the RRPM error feedback from the governor that slightly adapts the power setting to minimize the speed error. As most of the power setting is provided by the feedforward signal, the static droop will be minimal.

Piston engines are geared to the rotor and so run at near-constant RPM. The power delivered is changed with a change of torque and with a naturally aspirated engine this can be done instantly by adjusting the throttle or, in the case of a Diesel, changing the amount of fuelling. This does not happen in a free turbine engine where only the power turbine is geared to the rotor. Instead, greater power will be supplied by an increase in the gas generator speed (N_1) that delivers greater torque at the free turbine. N_1 cannot increase immediately. It takes time for the spool to accelerate. As a result there will be a temporary loss of RRPM during the time taken to spool up. This is known as *transient droop*. Figure 6.4(c) shows transient droop following a step change in power demand. Turbocharged piston engines will also display transient droop as the turbocharger accelerates.

Transient droop may be reduced through sensing the collective lever velocity. During a landing, collective will be lowered and this will cause the gas generator to spool down. If the landing is to be rejected, the pilot will pull the collective lever up. A system sensing collective velocity will be able to start spooling up sooner than one which simply waits for the lever to come up to provide power demand.

If the collective lever is raised too far on a turbine machine, the result will be an overtorque condition as the governor maintains RRPM. However, on a piston-engine helicopter once the throttle is fully open, further application of collective will cause *overpitching*. Induced drag causes a reduction in RRPM. In a machine with a manual throttle the correct recovery from overpitching is to maintain full throttle and to reduce collective pitch momentarily to regain rotor speed. If such a machine were fitted with a correlator the act of lowering the collective pitch lever could reduce the throttle setting and delay the RRPM recovery. The operation of a governor would not be impaired because a reduction in RRPM would result in a large speed error that would fully open the throttle. In the optional governor system of the Robinson R-22 the throttle is controlled by a governor mechanism responding to RRPM. However, an additional system is fitted which prevents overpitching. If the throttle is wide open to recover RRPM but the rotor speed fails to respond, an additional servo motor reduces the collective pitch setting until the speed error has been reduced. The collective servo linkage is fitted with a slipping clutch so that in the case of a failure the pilot can override it. The first use of a mechanism to maintain rotor speed by lowering collective was by Flettner in the 282 Kolibri.

In practice a free turbine engine needs two governors, one for the gas generator spool and one for the power turbine. These will be part of the fuel control system. Either of them can limit the turbine fuelling to prevent overspeeding. The gas generator governor is used for starting, idling and shutdown, and is controlled by the throttle lever in the cockpit. For flight, the throttle lever is set to maximum so that gas generator RPM is no longer able to restrict fuelling. This results in acceleration of the gas generator spool and increased gas flow that brings the rotors up to flight RPM. The rotor speed is then controlled using the power turbine governor which limits fuelling as correct RRPM is reached.

6.6 The gasoline engine

The piston engine is now over 100 years old but modern designs still work in the same way as the machine built by von Otto.

Figure 6.5 shows a section through a typical horizontally opposed piston engine. The pistons slide to and fro inside the cylinders and they are joined to the rotating crankshaft by connecting rods. The connecting rods, generally abbreviated to con rods, must be able to swivel in the pistons, and this is the function of the gudgeon pin, which passes through the little end of the con rod. The other end of the con rod must have a larger bearing in it to fit around the strong crankpin. In the boxer engine the crankshaft

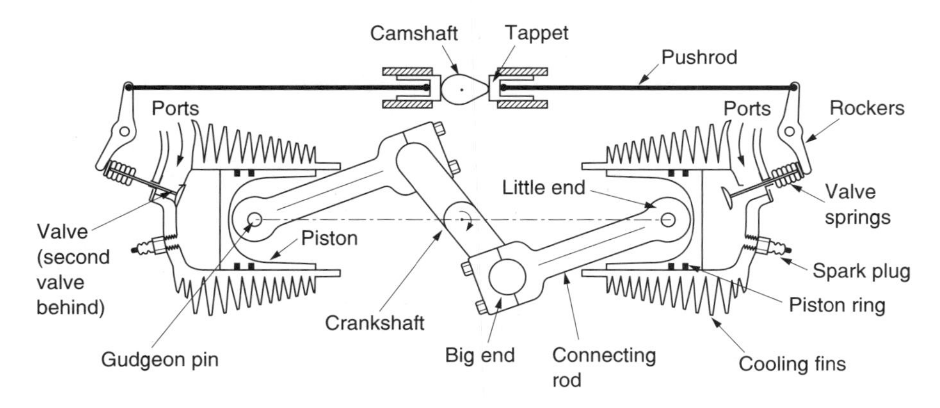

Fig. 6.5 A typical horizontally opposed aircraft engine in section. See text for details.

is constructed so that the pistons all move in and out simultaneously. The reaction due to accelerating the piston on one side is then balanced by that from the other side and vibration is reduced. The cylinders are not precisely opposite one another to give room for the crank web.

The large bearing in the connecting rod is called the big end. The aluminium pistons have slots fitted with rings which are springy metal strips. These press outwards against the cylinder wall and ensure a good pressure seal. In a piston engine four distinct stages are needed in the complete cycle, and in the four stroke or Otto cycle engine, the piston traverses the cylinder four times, which requires two revolutions of the crankshaft. The cylinder head is fitted with valves allowing fresh charge to enter the cylinder and burned charge, or exhaust, to leave. The valves are operated by a camshaft fitted above the crankshaft. This carries a series of rounded bumps called lobes. As the camshaft turns, each lobe presses against the flat end of a small piston-like object known as a tappet. In the tappet there is a swivel joint that connects to the push rod. This travels in a tube up the outside of the cylinder to a rocker in the cylinder head that pushes the valve open. When the cam lobe retreats, the valve is closed by the valve springs, which also push back the rocker, the pushrod and the tappet. When the valve is closed there must be a little slack in the pushrod, known as valve clearance, so that the whole spring pressure is keeping the valve shut.

Since the full four strokes require two crankshaft revolutions, the camshaft is driven through a 2 : 1 reduction gear so the sequence of valve openings resulting from one camshaft rotation is spread over the two crankshaft revolutions. On the intake stroke, the inlet valve is opened and charge is drawn into the cylinder. On the compression stroke both valves are closed and the charge is squeezed into the remaining space above the piston. On the power stroke, pairs of spark plugs ignite the charge, and it burns and expands, driving down the piston and turning the crankshaft. On the exhaust stroke, the exhaust valve opens and the piston pushes out the spent charge. The cycle then repeats.

In practice momentum of the gases means that the valves operate at a slightly different time than this simple explanation would indicate. Figure 6.6 shows a typical valve-timing diagram. During the induction stroke the charge entering through the inlet manifold is travelling at high speed and cannot stop easily. The inlet valve is left open until after bottom dead centre (BDC). Momentum will continue to force charge into the cylinder so that extra charge can be admitted allowing more power to be generated.

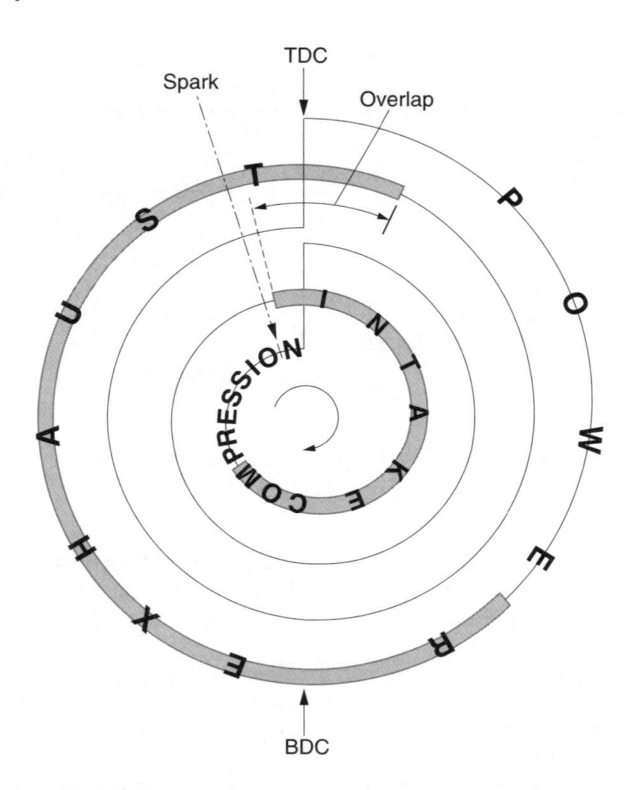

Fig. 6.6 A valve-timing diagram. Owing to the inertia of moving gases, the valve timing may be adapted to obtain better mass flow. Note the overlap at the end of the exhaust stroke where exhaust momentum is used to draw fresh charge into the cylinder.

The exhaust valve opens before BDC so that the exhaust can start to accelerate out of the cylinder in time for the exhaust stroke. The exhaust gases leave at high speed, and at the end of the exhaust stroke the momentum of the gases will carry them onwards even when the cylinder is empty. This causes a reduction in pressure in the cylinder. The inlet valve is opened before the exhaust valve closes so that the momentum of the exhaust leaving will pull fresh charge in. This is known as valve overlap. Racing engines use camshafts having a lot of valve overlap to help charge transfer at high speed. This makes them very rough at lower speeds. Helicopter engines are required to have good throttle response over a limited RPM range rather than ultimate power, and will be fitted with milder camshafts.

The valve mechanism in early engines required an adjustment to take up slack in the pushrod as wear took place. A small clearance was left to allow the parts to expand without the valves being pushed open when they were supposed to be shut. If the valve clearance was excessive the valves would not open for as long, and power would be reduced. The mechanism would make a lot of noise. Tappet adjustment was frequently required to maintain performance.

In later engines a self-adjusting mechanism was devised. The tappet contains a small piston that drives the pushrod. Engine oil pressure is communicated to the tappet through a small drilling so that the piston will move until it has taken up the slack in the valve train. When the camshaft opens the valve, the force is transmitted through the oil pressure. The drilling is too small to allow a significant amount of oil to be forced back out of the tappet. Hydraulic tappets are self-adjusting and the periodic maintenance is eliminated.

6.7 The ignition system

Ignition of the charge is achieved by the generation of a spark between a pair of electrodes in the cylinder head. A very high voltage is needed to break down the insulation of the compressed charge so that a spark can occur. The high voltages present in ignition systems cause stress to the components, and as a result the ignition system is probably the least reliable part of the engine. This is overcome in aircraft engines by having a completely duplicated ignition system. There are even two spark plugs per cylinder, one for each ignition system. This also allows a small increase in power because the charge is ignited in two places at once.

In cars, the high ignition voltage is generated by an induction coil driven by the electrical system. If the electrical supply is cut off the engine stops. This is not reliable enough for aircraft, and these use an alternative system that generates its own electricity directly from engine shaft power using a rotating magnet. This magneto-dynamic ignition unit is invariably known by the abbreviation of magneto. In a magneto system the engine will keep running even if the electrical system of the helicopter has totally failed. A pair of magnetos are fitted on the back of an aircraft engine, driven by the gear cluster that drives the camshaft and the oil and fuel pumps. When the aircraft engine is mounted backwards in a helicopter, the magnetos can be found adjacent to the cockpit firewall.

The principle of operation of the magneto is based on the same physics as electricity generation. When current flows through a coil of wire, it generates a magnetic field. The field can be amplified by winding the coil around a soft iron core. If the magnetic field passing through a coil is constant, there is no effect, but if the field changes, a voltage proportional to the rate of change of flux is induced. The two phenomena described above combine to give a coil the property of inductance. An inductor is a sort of magnetic flywheel. A mechanical flywheel stores kinetic energy proportional to the square of the speed. An inductor stores magnetic energy proportional to the square of the current. In the same way that a flywheel resists changes of speed, an inductor resists changes of current.

Imagine an inductor with a current flowing through it. If some external resistance attempts to reduce the current, the magnetic flux in the coil is reduced. The falling flux induces a voltage in the coil that attempts to increase the current. On the other hand if something attempts to increase the current, the magnetic flux will increase, and the change of flux in the opposite direction will induce a voltage that opposes the current. As a result the inductor tries to keep the current passing through constant.

Figure 6.7 shows the construction of a magneto. The input shaft turns at camshaft speed and carries a powerful permanent magnet. As the magnet turns its magnetic field passes through stationary poles in alternate directions. Around one of the poles is a winding of thick copper wire. The ends of this coil can be shorted out by a pair of contacts, known as points, operated by a cam on the input shaft. A capacitor is fitted across the points that are closed when the magnet flux in the stationary pole is changing from one direction to the other. The changing flux induces a voltage in the coil, but the coil is shorted out so there is very little resistance. The coil builds up a high current at a rate determined by the inductance.

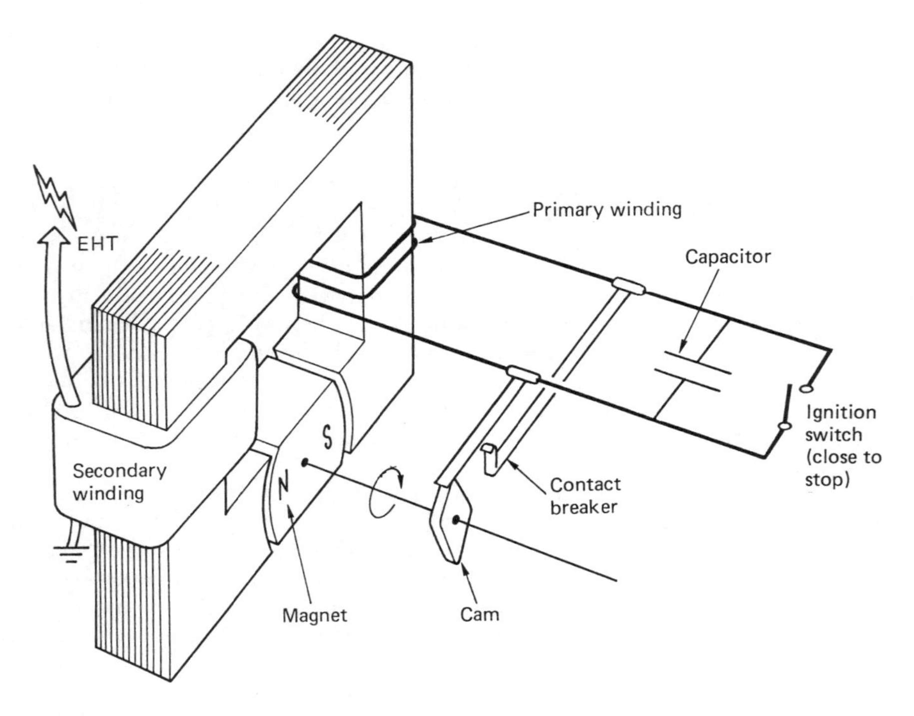

Fig. 6.7 The magneto generates its own electricity and so is potentially more reliable than an ignition system using the aircraft's electrical supply.

When maximum current is reached, the cam opens the contacts. The inductance attempts to keep the current flowing, but it cannot pass the open points. The only place the current can go is through the capacitor. As the capacitor charges up, the current reduces. This reduces the inductor flux and increases the induced voltage. When the current finally stops flowing, the capacitor is charged to a high voltage. The magnetic energy has been exchanged for electrical energy stored in the capacitor. The voltage on the capacitor now reverses the current, which builds up in the inductance. The current keeps flowing after the capacitor voltage has fallen to zero and charges the capacitor up in the reverse direction. The inductor and capacitor form a resonant circuit where the energy repeatedly exchanges between the inductor and the capacitor and an alternating current flows in the windings. The same effect is achieved mechanically in a tuning fork.

As the capacitor is quite small, the frequency of oscillation is quite high, and the rate at which the current changes is also high. Thus a high rate of flux change is achieved in the inductor. A secondary coil consisting of many turns of fine wire is wound on top of the first coil. The rapidly alternating flux induces a very high alternating voltage in the secondary coil, and this results in a rapid succession of sparks at the plug electrodes. The sparks continue until the magnetic energy is dissipated. In a multi-cylinder engine, this high tension (HT) current is directed to the appropriate cylinder by a rotary switch called a distributor on the end of the shaft. There is no mechanical contact in the distributor; the metal parts come close enough for the HT to jump across the gap, so in fact there are two sparks in series, one at the plug and one at the distributor.

The capacitor is a vital part of the system, because without it the current would have nowhere to go when the points open, and most of the magnetic energy would be dissipated in a spark at the points, which would then be rapidly eroded away. The spark in the cylinder would then be very weak. This can happen if the capacitor becomes disconnected or fails open circuit. If the capacitor fails short circuit the points are bypassed, and no spark will be produced at all. This type of failure is rare.

The HT wiring to the spark plugs needs to be well insulated so that the HT does not jump to the nearest piece of metal instead of going to the plug. At high altitude the reduced air pressure reduces the insulating ability of the air, and as a result the ignition leads must have somewhat better insulation than those found in cars. Part of the pre-flight check is an inspection of the condition of the HT leads. Needless to say they should not be touched when the engine is running.

Whilst the magneto has the advantage that it is completely independent of the helicopter's electrical system, it does suffer from one disadvantage. When starting, the engine speed is low, and the rate of change of flux in the magneto will be correspondingly low and a poor spark results. One solution to this problem is a device known as an impulse coupling which is basically a torsion spring in the magneto drive shaft. The presence of the permanent magnet makes the torque needed to rotate the magneto vary with the angle of rotation. This can easily be demonstrated by attempting to turn a bicycle dynamo by hand. At starting speed, when the torque is large, the spring in the impulse coupling is wound up as the magneto lags behind the shaft. As the torque falls, the spring tension is released, and the magneto is temporarily turned faster than the shaft. This allows a larger spark to be generated. A further use of the impulse coupling is that the lag delays the time at which the spark is generated. Under normal conditions, the spark is supplied a little before top dead centre (TDC) in order to allow the charge time to start burning before the power stroke. At low starting RPM, the advanced spark might cause the engine to kick back, or fire during the compression stroke. The delay of the impulse coupling prevents this happening. At normal operating speed the torsion spring does not have time to twist and the magneto runs at constant speed with the correct timing.

Dual ignition systems are provided to increase the reliability of the engine. This reliability increase is only present if both ignition systems are working properly. Part of the pre-flight check is to test the ignition system. This is done using the ignition switch. As the magneto is self-contained, the only way it can be switched off is by shorting out the points. The ignition switch is connected across the points and the magneto is switched off by closing the ignition switch contacts. This is a safer approach since a broken wire from the ignition switch will not stop the engine. The ignition switch is constructed so that either or both magnetos can be switched on. In aircraft the ignition switch may have a further position to operate the starter, but this is not used in helicopters. Instead the starter button is fitted on the end of the collective lever where it may be operated by the pilot without letting go of the controls.

When the engine is warm, the dual ignition may be tested. At a specified RPM, the ignition switch is turned from 'both' to 'left'. This disables the right magneto. The result should be a slight reduction in RPM, known as a magneto drop, because only one spark is being generated per cylinder. The drop is normally that which is measured after 5 seconds running on one magneto. The engine is run on both systems again for a short time to clear any fouling which may have built up on the plugs whilst they were out of use. The switch is then set to 'right', when an identical drop should be obtained. If the drops are grossly different or do not occur, there is a problem which must be rectified before flight.

To take an example, if on switching from 'both' to 'left' there is no drop, either the right magneto was not working or it has not been switched off due to a broken wire. Switching to 'right' will show which. If the engine cuts, the right magneto is inoperative. If a drop is obtained, the right magneto is permanently on due to a wiring fault. This can be proved by switching off the ignition. If the engine continues to run, the wire to the ignition switch is broken. The engine can be stopped by the idle cut-off as normal.

If the drops are not identical, one ignition system is working better than the other for some reason. If only an RPM loss is noted, the different drop could be due to a timing difference between the magnetos. If one magneto gives a larger drop and a misfire, consideration should be given to the possibility of a failed spark plug or capacitor or a fractured HT lead.

The cylinders are numbered from what would be the front in an aircraft, which in a helicopter corresponds to being numbered away from the drive pulley. In a Lycoming engine mounted back to front in a helicopter, the odd numbered cylinders are on the port side. The sequence of power strokes is designed to reduce torsional stress in the crankshaft by firing cylinder pairs alternately. This gives the firing order of 1-3-2-4.

6.8 The starter

The starter is basically a powerful electric motor used to turn the engine. A reduction gearing system is needed so that the starter can develop enough torque to overcome the engine compression. A small pinion on the starter engages with a large ring gear on the engine flywheel. The gears are only engaged during starting. This can be done by a helical thread mechanism that screws the pinion along the motor shaft or by a solenoid that slides it. In the latter type a one-way clutch prevents the engine driving the starter. Recently geared starter motors have become available. These have a high speed electric motor and a reduction gearbox and are lighter than conventional starters and consume less current. The weight saving may be compounded because they allow a smaller battery and thinner wiring to be used.

The motor must be quite powerful, and it will require a current of several hundred amperes to operate it. In order to minimize power loss in the cables, these run direct from the battery to the motor. This current is beyond the capability of a switch, and a relay is used. The starter button on the collective lever switches a small current through a solenoid coil, and this attracts the solenoid armature, sliding the starter gear into engagement with the flywheel and pulling the main motor contacts closed. Normally the starter motor gear is driven through a one-way clutch so that it can run at high speed when the engine fires without damaging the starter motor. A warning light is fitted which should go out if the starter button is released. If it does not the starter may still be engaged.

If the starter fails to operate, this could indicate a faulty solenoid. It is generally possible to hear the solenoid operating. If the solenoid chatters or vibrates this indicates a loose battery connection or a discharged battery.

The starter motor dissipates a lot of heat when starting, and if the engine does not start promptly because it is flooded or vapour locked, prolonged cranking will flatten the battery and overheat the starter motor. It is better to allow a cooling period between starting attempts than to burn out the starter motor. The starter motor should never be operated when the engine is running as engaging the starter pinion with the fast turning gear on the flywheel could remove teeth. The engine must be allowed to come to a complete halt before making a further starting attempt.

6.9 The oil system

The oil system is responsible for lubrication and cooling of the moving parts of the engine. Heavy forces are involved in the connecting rods and crankshaft bearings, and if metal-to-metal contact took place the life of the parts would be very short. Oil possesses viscosity, which is a resistance to flow. If viscous oil is pumped between two surfaces it can carry a load without them touching because the viscosity resists the oil being squeezed out. The relative movement of the surfaces causes shearing of the oil. Shearing a viscous liquid develops heat, and this heat load adds to the heat picked up where the oil is used to cool parts of the engine. An oil cooler is necessary to control the oil temperature.

Figure 6.8 shows a typical engine oil system. Oil is held in the sump below the engine. The sump contains a filler neck whose cap holds a dipstick so the oil level can be checked. This should be done before flight. Oil is drawn from the sump through a strainer to prevent particles of dirt from damaging the pump. The pump may consist of a pair of meshing gears driven from the camshaft/magneto drive cluster at the rear of the engine. The capacity of the pump exceeds the requirements of the engine, and surplus oil is returned to the sump by a spring-loaded relief valve that opens when the correct

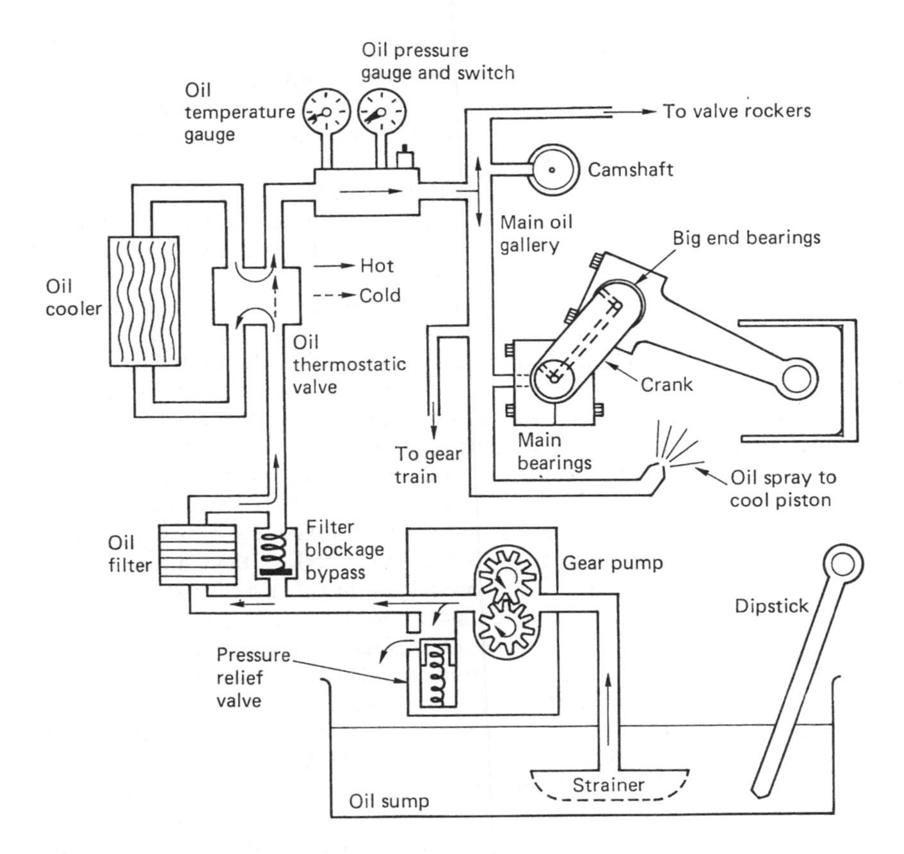

Fig. 6.8 Lubrication system for a typical piston engine. See text for details.

oil pressure is reached. The oil then passes through a filter that removes microscopic particles of metal due to engine wear and combustion products that have found their way past the piston rings into the oil. The filter needs to be replaced periodically. If the filter becomes clogged, a spring-loaded valve opens allowing oil to bypass the filter.

The oil next passes to the oil cooler. This is a small heat exchanger that is often fitted in the engine cooling baffles so that it gets a cold air feed from the engine driven fan. The cooler is bypassed by a thermostatic valve that is open at low temperatures to allow the oil to warm up quickly. The valve can also be forced open by oil pressure if the cooler blocks up. At the outlet of the cooler the oil is ready to be fed to the engine. At this point the oil pressure and temperature are monitored and displayed on gauges in the cockpit. Some of the oil is fed to the main crankshaft bearings. The crankshaft is hollow and oil is pumped in from the main bearings and flows to lubricate the big ends. The connecting rods may also be hollow and oil carries on up the rods to lubricate the little end. An oil jet is drilled in the top of the connecting rod to spray oil on the piston for cooling and to lubricate the cylinder wall. Oil is also supplied to the camshaft bearings and the valve mechanism as well as to the gear train driving the camshaft and the magnetos. The oil will eventually fall back into the sump.

The engine oil has a finite life primarily because it builds up an increasing load of acids, which are by-products of combustion, and water from condensation. Oil is characterized by its viscosity. Viscosity is measured by the time taken in seconds for a standard quantity of oil to flow through an orifice standardized by the Society of Automotive Engineers (SAE) at a standard temperature. Oil may also be classified by a commercial aviation number that is twice the SAE number.

The viscosity of oil falls with increasing temperature. So-called multi-grade oils have a smaller viscosity loss with temperature. The correct viscosity of oil is essential and it must be used at the correct temperature. If the oil is too viscous or used at too low a temperature, it may not flow readily to all the necessary parts of the engine. If the oil is too thin or used at too great a temperature, it may be squeezed from bearings and permit contact. The engine oil must be warmed up before flight so that correct lubrication is achieved when full power is demanded.

The flight manual will specify the viscosity of oil that must be used in temperate conditions. A mandatory placard on the machine adjacent to the filler will state the viscosity needed. If the machine is consistently operated at elevated ambient temperatures, it may not be possible to keep the oil down to the required temperature. In this case thicker oil will be used which will thin down to the right viscosity at the elevated temperature. In a similar way if the machine is to be used in cold conditions the oil may run too cold and thinner oil will be specified to get the right viscosity at a lower temperature. It is important that the oil level is checked regularly and kept topped up with the same type of oil. Oil grades should not be mixed and it is better not to mix brands. Oils contain a host of additives to improve performance and sometimes the additives in one kind of oil react with those in another. Turbine engines need oil with different additives and viscosity. Turbine oil cannot be used in piston engines.

The oil pressure and temperature gauges tell a great deal about the health and working conditions of the engine and should be monitored regularly as part of the pre-flight check and whilst airborne. On starting the engine, it may take some time for oil pressure to be established as the oil is cold and thick and the pump has difficulty dragging it from the sump. Expect the delay to be longer in cold conditions. Generally if oil pressure has not established in 30 seconds the engine should be shut down and the cause investigated.

When cold the oil resists flowing through the engine and once pressure is established the pump has no difficulty supplying enough pressure. Paradoxically, once the pressure has established, it may become excessive in cold conditions or if using straight (non-multi-grade) oil in a new engine. In this case the engine revs must be kept to a minimum after starting until the oil has warmed up. Another cause of excessive oil pressure at idle is if the relief valve has stuck. In any case of excess pressure, engine revs must not be increased as a further increase in pressure could burst the oil filter or the cooler.

In a hot engine the oil flows more freely and worn bearings will increase the oil flow. At idle speed the oil pump may not be able to meet the flow demand and the pressure will fall. Thus low oil pressure with hot idle may indicate bearing wear.

After a cold start, the engine should be run on the ground until the oil temperature has reached a minimum level for flight. If it takes an excessive time for this temperature to be reached the thermostatic valve in the oil cooler could be stuck and making the cooler work continuously.

Sudden and total loss of oil pressure in flight is rare and will be due to a mechanical failure such as a sheared pump drive or a burst pipe. The engine will be destroyed unless it is shut down immediately. If a suitable landing site is available, the engine should be stopped and an autorotating landing will be needed. A conscious decision to autorotate is safer than trying to land using power because without oil the engine could sieze up without warning and this might occur in the avoid curve. If no suitable forced landing site is near, it will be necessary to sacrifice the engine in order to attempt to fly to safety. Flying at minimum power speed will buy time.

A more likely fault situation in flight is where the oil temperature slowly rises and the oil pressure slowly falls. This is an indication of oil loss due to a leak or oil being burned in the cylinders. A landing should be made promptly so that an investigation can be made.

When assessing the oil pressure and temperature gauges, the stress the machine has been under should be considered. Following a lengthy climb at maximum all up weight (MAUW) one might expect hot engine oil. The stress on the machine would be reflected in elevated transmission oil temperature and cylinder head temperature. If all of these readings are high, the machine is being driven hard and should be flown in a way that will reduce the power requirements.

On landing the engine should be run at idle for a short while in order to cool it. If this is not done oil inside the engine will be exposed to heat that it cannot carry away and the oil could be carbonized and lose effectiveness. This is known as heat soak.

6.10 The carburettor

Engines produce power by burning charge. The purpose of the carburettor is to mix air and fuel to create charge. Charge will burn over the following range of fuel/air mixes. A ratio of 1 part fuel to 8 parts air is called a rich mixture, and not all the fuel will be burned. A ratio of 1:20 is called a lean mixture and not all of the oxygen will be burned. A ratio of about 1:12 is called a stoichiometric ratio as there is just enough fuel to combine with all the oxygen.

Clearly a chemically accurate ratio will give the most power and a lean mixture will give the best fuel economy. Unfortunately it is not possible to take full advantage of this because of a number of practical considerations. At high power the engine temperature cannot be kept at a safe level unless a rich mixture is used. The surplus fuel absorbs heat by evaporation.

The four-stroke engine compresses the charge before combustion to increase the power output. The greater the compression ratio, the more power can be released. When charge is ignited it should proceed to burn smoothly as a flame front travelling away from the spark plug. Unfortunately when charge is compressed the temperature will be raised and it can detonate instantly when ignited. This causes a rapid pressure step that creates shock waves travelling through the engine. A metallic ringing noise called 'pinking' or knock can be heard. If sustained, engine damage will result.

The ability of a fuel to resist detonation must be matched to the compression ratio of the engine it will be used in. The detonation resistance of fuel is measured in a special variable compression test engine. The fuel under test is used to drive the engine and the compression is increased until detonation commences. The engine is then run on iso-octane, a relatively detonation resistant hydrocarbon, and this is steadily diluted with heptane, a detonation prone fuel, until detonation commences. The percentage of octane at which this takes place is the octane number of the fuel. Since the test was devised, more detonation resistant fuels have been developed, and the scale has been extended above 100 by extrapolation.

Using fuel of inadequate octane number for the engine causes damage through detonation. The machine carries mandatory placards adjacent to the fuel fillers stating the octane number of the fuel that must be used.

A rich mixture helps to prevent detonation by reducing the temperature increase during compression. Fuels are often specified by two octane numbers; the second is for a rich mixture. For example, many piston helicopters burn 100/130 LL fuel. At normal mixtures this is 100 octane, but at full rich mixture it is 130 octane. LL stands for low lead. In the 1920s it was discovered that the addition of tetra-ethyl lead to fuel increased the octane number. Decades later it was proved that the resultant lead pollution was slowing the rate of brain development of children living near busy roads and less damaging alternatives have had to be found.

Whilst fuel of inadequate octane rating causes damage, occasional use of higher octane fuel is acceptable. Prolonged use of high octane fuel may result in spark plugs being fouled by the octane-boosting additives. No power increase will be observed because the power output is determined by the engine design, particularly the compression ratio.

The carburettor has to mix the air and fuel in the chosen proportion. Figure 6.9 shows that it is a fairly simple device. Engine power is controlled by throttle (3): a disc that pivots in the intake tract. With the throttle wide open, the pressure in the inlet manifold will be nearly atmospheric, whereas with the throttle closed down to the idle position, the manifold pressure will be near zero. Manifold pressure is displayed on a gauge in the cockpit as it indicates the proportion of maximum power the engine is producing. The pilot needs to know how much reserve power is necessary before committing himself to a manoeuvre needing higher power: a procedure called a power check.

The result of throttling the engine is to reduce the flow of air through the carburettor. The fuel flow must remain proportional to the mass flow. This is done by the creation of a small constriction or venturi in the inlet tract. The air must go faster through the smaller cross-section. This increases the dynamic pressure of the air. As the static and dynamic pressures sum to a constant, the static pressure must reduce, as described by Bernoulli's theory. The reduction in static pressure causes a pressure difference across the fuel in the carburettor body making it flow through a metering orifice or jet (2) and mix with the air. Clearly the depth of fuel could affect the pressure and the carburettor has a float valve (1) that admits fuel as it is used to prevent the level falling.

The system is self-balancing, because greater airflow causes greater suction at the venturi and increases the fuel flow. If the throttle is opened suddenly, the fuel inertia prevents a rapid increase in fuel delivery and the mixture will weaken momentarily. Weakening is prevented by the action of a small pump (4) that produces a jet of fuel when the throttle is opened quickly.

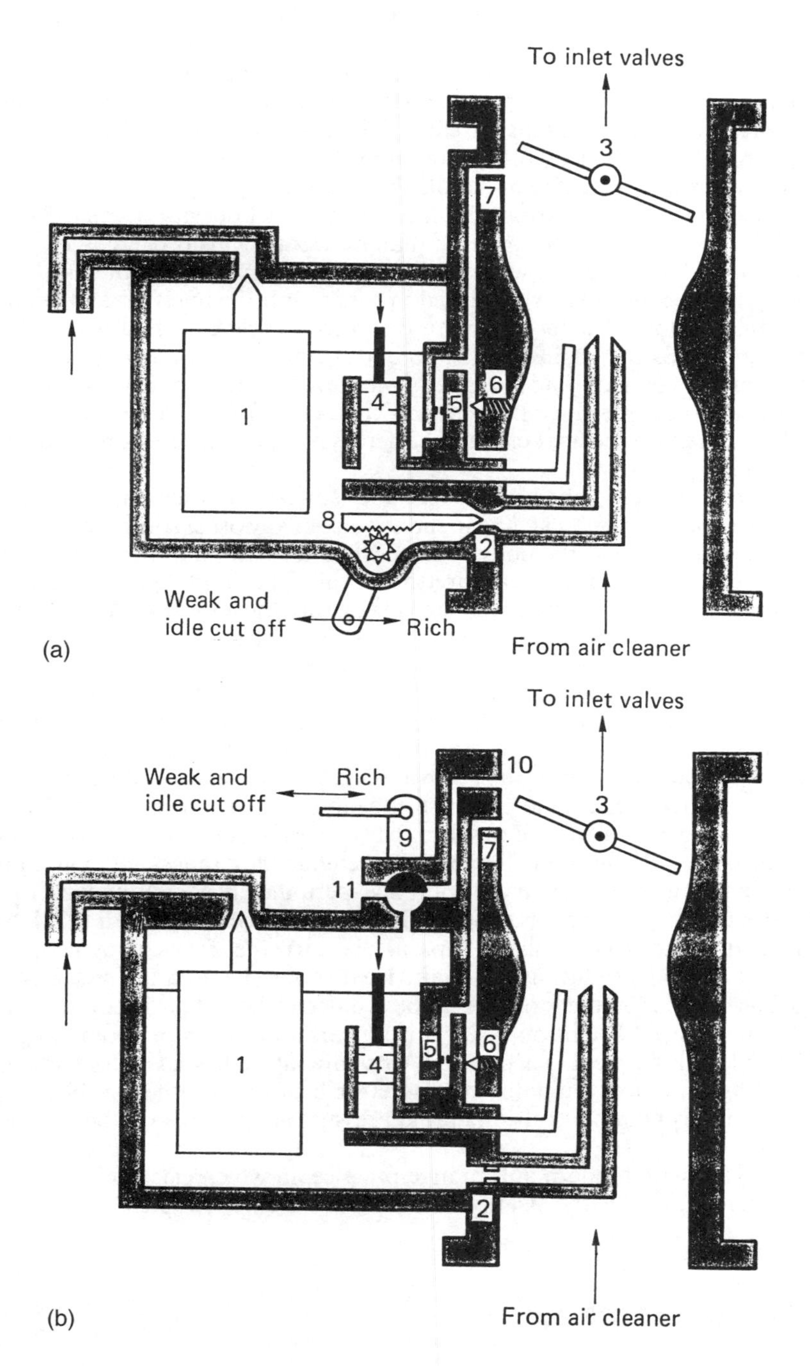

Fig. 6.9 The carburettor is a simple device which uses a venturi to create a pressure drop that will suck fuel from a jet. The jet may be variable (a) for mixture control, or this may be done by air bleed changing the pressure exerted on the fuel (b).

When the engine is at idle, the throttle will nearly be closed. The minimum throttle opening is set by throttle stop screw that prevents the engine inadvertently being stopped. With the throttle closed the airflow through the venturi is very small and insufficient to draw fuel. As the throttle disc reaches the closed position it admits low manifold pressure to a drilling in the side of the tract. This draws fuel from the float chamber through the idle jet (5), and air from the idle mixture screw (6). In effect a miniature carburettor is created to handle the small charge flow at idle.

As altitude increases, the air density falls but the fuel density doesn't. The carburettor tends to produce a richer mixture than necessary. This is overcome by fitting a mixture control. There are two ways of controlling the mixture. In the first, shown in Figure 6.9(a), a needle valve (8) is placed in series with the main jet and changes the resistance to fuel flow. Placing the mixture control to idle cut-off stops the fuel flow altogether and stops the engine. In the second method, shown in Figure 6.9(b), low pressure from the inlet manifold is applied to the float chamber by a valve (9) to oppose the venturi suction and reduce the fuel flow. Idle cut-off is achieved by allowing full manifold pressure into the float chamber. There is no resultant pressure across the fuel and flow ceases.

The carburettor is a simple device but it does suffer from one major problem. The evaporation of fuel requires heat, and in the absence of any other source the heat is taken from the incoming air and the body of the carburettor. In humid conditions this can lead to the formation of ice in the carburettor venturi. The rapid flow around the edge of a nearly closed throttle and subsequent expansion into the low pressure in the manifold duplicates the mechanism of the domestic refrigerator and can cause throttle icing. This is particularly dangerous because at low power on a descent the ice build-up will not be noticed. When full power is required to arrest the descent it will not be available. The engine will artificially be throttled by ice and the mixture will be unpredictable.

The traditional approach to icing is to supply hot air to the carburettor intake. This is conveniently obtained by drawing air over the exhaust system. Unfortunately hot air is less dense and reduces engine power so it cannot be used all the time. The pilot is therefore provided with a control to select the carburettor heat. If there is the slightest suspicion of icing, full heat should be applied. If engine smoothness and power improve, there was icing. It may be necessary to apply heat periodically or continuously to control icing. If power reduces, there was no ice and the engine is running on air that is too hot.

Heat control requires a disciplined approach because the effects of an incorrect decision may not be evident until it is too late. Heat should be used for descents, but the prolonged use of heat on the ground is to be avoided if the hot air is not filtered. Some machines, such as the Robinson R-22, have a carburettor temperature gauge which is useful for deciding if heat is necessary. Air temperature has less effect on the probability of icing than does humidity. Carburettor heat has another problem which is that it affects interpretation of the manifold pressure gauge. This will be considered in section 6.14.

In comparison with modern automotive practice, in which electronic fuel injection is used almost exclusively, the aviation carburettor is a museum piece.

6.11 Fuel injection

In a fuel injected engine there is no carburettor. The fuel is sprayed into the inlet manifold just before the inlet valve by a fuel nozzle. There is one nozzle per cylinder. The use of fuel injection has a number of advantages. Fuel vaporizing takes place in a hot part of the engine where icing cannot occur. A fuel injected helicopter needs no hot air system and full power is always available with a corresponding increase in safety. Throttle response is instant. The flow measuring venturi restricts airflow much less than a carburettor, and maximum power is increased.

There are two types of fuel injection system. In mechanical systems, the fuel delivery is continuous and is determined by the fuel pressure in the fuel manifold (also known as the fuel distributor or flow divider). Fuel manifold pressure is displayed on a gauge in the cockpit, often combined with the inlet manifold pressure gauge. In electronic systems, the fuel delivery is controlled by a solenoid valve, one incorporated in each nozzle. In this case the fuel pressure is held constant relative to manifold pressure and the amount of fuel delivered is determined by the period of the pulses that operate the solenoids.

The engine power is controlled as usual by a throttle disc in the inlet duct. In a mechanical system, a small venturi is created in the duct and the suction created is a function of inlet airflow. The fuel pressure is made proportional to the venturi suction in the fuel control unit shown in Figure 6.10(a). The pilot's mixture control changes the balance in the fuel control unit and the idle cut-off function stops all fuel flow.

The only real drawback of mechanical fuel injected engines is that hot starting can be difficult because residual engine heat vaporizes the fuel in the pipes leading to the nozzles. The solution is to purge the pipes and cool them with fresh fuel by running the boost pump for a few seconds with the throttle 'cracked' (partially open). The surplus fuel will flow into the inlet manifold and must be purged by operating the starter with the ignition off and the throttle wide. The throttle is then returned to idle, ignition is switched on and the starter operated again. Once the engine is running the boost pump is switched on again as a flight backup for the mechanical pump.

In electronic systems, a signal processor computes the correct amount of fuel to inject. It does this based on the mass flow of air into the engine. Figure 6.10(b) shows that engine RPM is measured by sensing ignition pulses or with a sensor adjacent to a toothed wheel. In the inlet tract, the air mass flow may be measured by a hot-wire sensor. Alternatively the manifold pressure may be measured. The air temperature is also measured to compensate the computation. As a result the mixture is always correct and no human intervention is required. In automotive applications, there is only one processor. This represents a single point of failure not acceptable in aviation. However, it is quite possible to provide one processor per cylinder each having its own sensors. It is also necessary to provide electrical power to the fuel injection system even if the main electrical power system fails. This is easily done using power-generation windings in the magnetos as is the practice in motorcycles.

Whilst the fine nozzles of injection systems are more prone to blockage than a carburettor, with proper filter and fuel management this is not an issue.

6.12 The turbocharger

The power a piston engine can develop is a function of the charge burned on each power stroke. Clearly a greater charge can be obtained by using a larger engine, but this will be heavier, negating some of the increase in power. A more efficient way of increasing the power is to get more charge into a given engine.

In a naturally aspirated engine, when the piston goes down on the intake stroke, charge is pushed into the cylinder by atmospheric pressure. The elements of the induction system such as the air cleaner, the venturi and the valve stems all impede the flow.

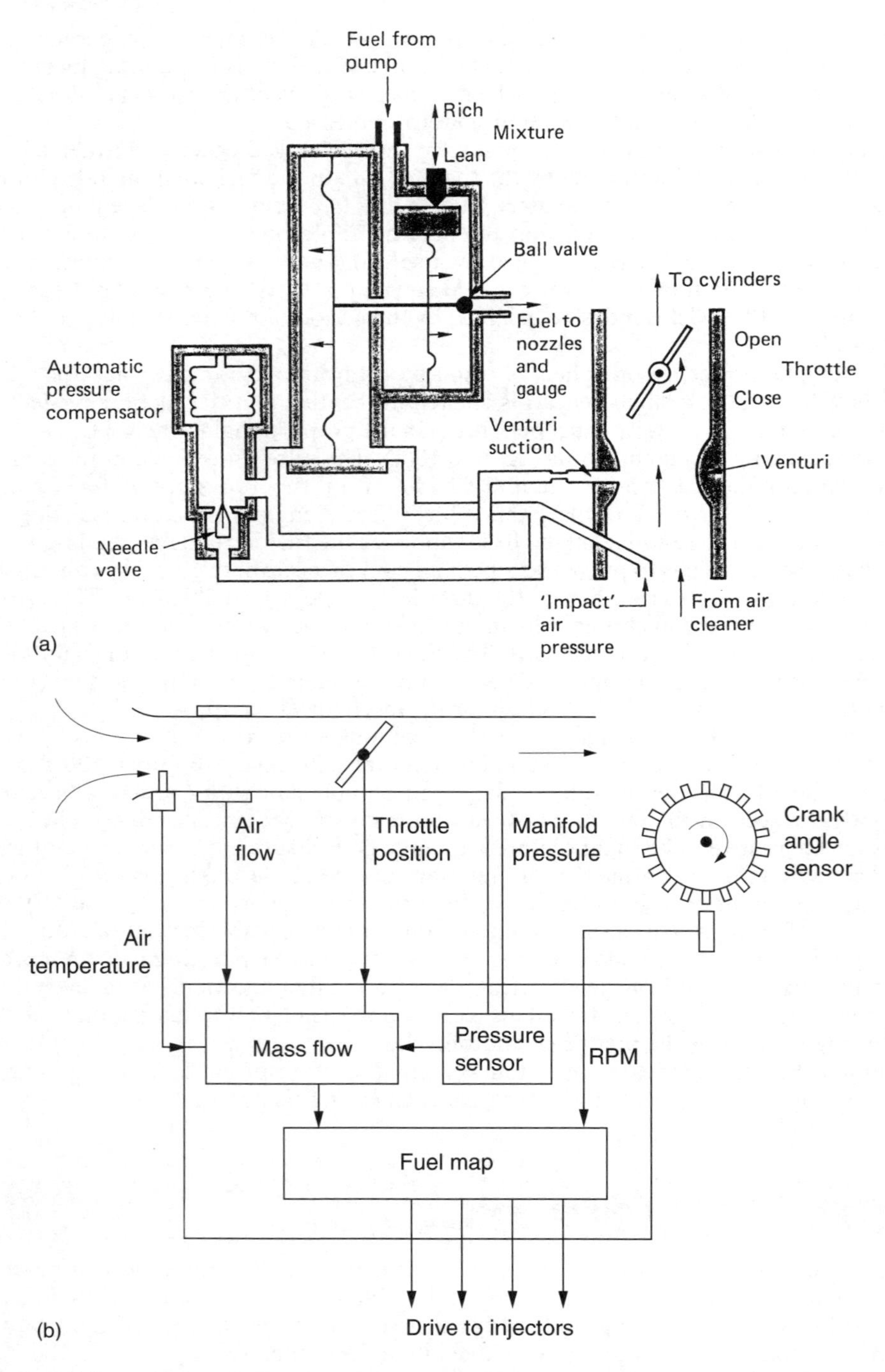

Fig. 6.10 (a) A mechanical fuel injection system virtually eliminates intake icing because most of the induction system contains only air. (b) In an electronic fuel injection system the mass flow is calculated to determine how much fuel to inject. This is controlled by the length of time for which the injector solenoids are energized.

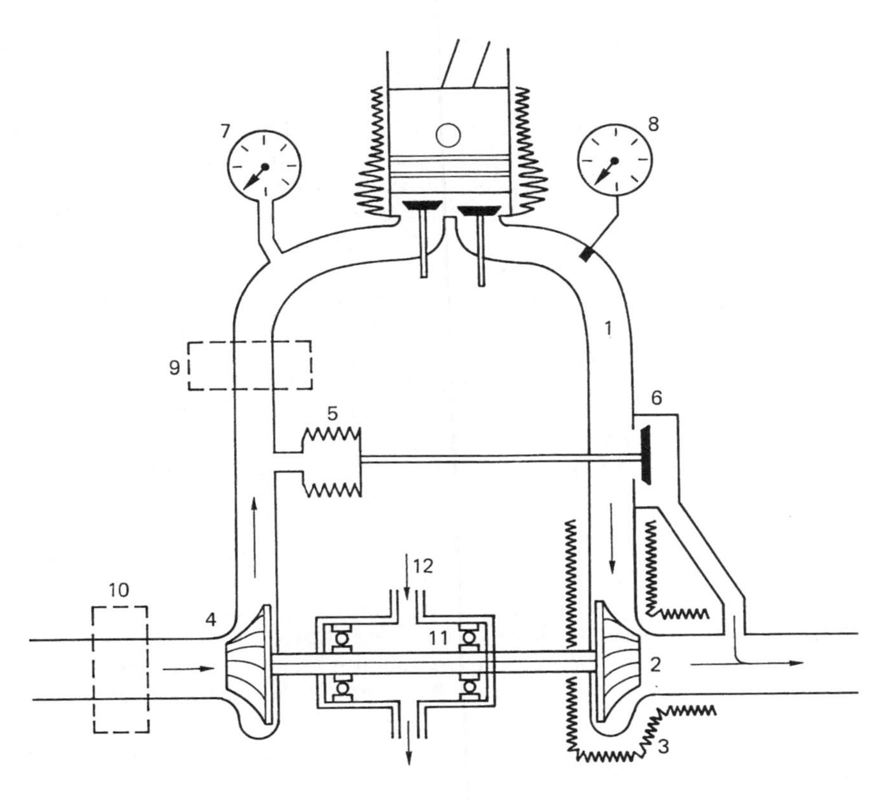

Fig. 6.11 The turbocharger uses energy from the exhaust gases to increase the mass flow of the induction process. See text for details.

As altitude increases, atmospheric pressure reduces, so there is less to push the charge into the cylinder. The air density falls, so the mass of the charge is even further reduced. Without some help, the piston engine simply cannot breathe at altitude. Any device which forces more charge into the cylinder is a supercharger. These were originally mechanical pumps driven by the engine crankshaft. Later it was realized that the residual energy in the exhaust could be harnessed to drive the pump. This is the principle of the turbocharger shown in Figure 6.11.

Exhaust gases 1 pass through a turbine 2 before passing to the atmosphere. The presence of the turbine extracts energy from the exhaust gases, lowering the temperature and the pressure. The pressure drop needed to make the turbine work means that the pressure in the exhaust manifold, known as back-pressure, is higher than normal. Heat insulation 3 is usually needed around the exhaust turbine. The exhaust turbine drives a centrifugal compressor 4 that takes in air through the air cleaner 10 and forces it into the induction system. The boosted induction pressure is displayed on the boost gauge 7 and increases the charge entering the cylinder. Back-pressure in the exhaust prevents the charge blowing straight through during valve overlap.

The turbocharger is a simple device, but it may turn at over 20 000 RPM. The stresses due to this rotational speed are enormous, yet the material of the exhaust turbine also has to withstand the elevated exhaust temperature due to raising the power output.

Turbochargers only became economic when suitable materials were found which could withstand this kind of treatment for a reasonable lifespan.

A turbocharged engine without some form of control system could be dangerous. The more power the engine develops, the more turbocharging can be achieved, which develops even more power and so on. In practice this is prevented by a system controlling the induction pressure. The exhaust turbine may be bypassed by a valve known as a wastegate 6. If the induction pressure exceeds the design limits, the wastegate actuator 5 begins to operate to cut the power received by the exhaust turbine. At sea level the air is sufficiently dense that only a little exhaust energy is needed for charging and the wastegate will open relatively wide. As the machine climbs, air pressure and density fall, and the induction pressure tends to fall with it. The control system senses the fall in pressure and starts to close the wastegate. This diverts more of the exhaust through the turbine and drives the compressor faster until the pressure is brought back to the sea-level value. Clearly engine power is maintained with increasing altitude until the wastegate is completely closed, and after that, power will naturally fall off.

The compression process in the induction system raises the air temperature and reduces its density. A further power increase can be obtained by cooling the induction air between the turbocharger and the engine. This is done by an intercooler 9 that is a form of heat exchanger. These are uncommon on helicopters because of the extra complexity, but provided the weight is moderate the effects are beneficial.

It is sometimes claimed that turbocharged helicopter engines are less reliable because the turbo stresses the engine. Whilst this is true of racing car engines, where brute power for one race is the goal, the helicopter engine does not use the turbo so much to increase power as to maintain the same power under all conditions. This can only be a good thing.

From the pilot's point of view the turbocharger does not cause any difficulty. In the cockpit the only change is that the figures on the manifold pressure gauge can go above sea-level pressure. There may be a boost pressure limit denoted by a red line. An overboost warning light is sometimes found. Turbocharged machines display transient droop. When more manifold pressure is called for, the wastegate has to close down to increase the turbo power, and it takes a finite time for the turbocharger RPM to increase.

Part of the pre-flight check is to inspect the turbocharger for mounting security (when it is cold!). The rotor bearing 11 needs a supply of oil from the engine oil system 12, and the small pipes should be checked for cracks or leaking unions.

Power loss due to high altitude or air temperature has less effect on a turbocharged machine and when in-flight power checks are performed it will be seen that there is more reserve power than in a naturally aspirated machine.

6.13 Gasoline engine instruments

In air-cooled gasoline engines, the cylinder head temperature (CHT) gauge is used after starting to determine if the engine is warm enough to run properly with the rotors engaged. After flight it is used to check that the engine has cooled down sufficiently to prevent heat soak when it is switched off. In flight CHT rises with the power the engine is asked to develop and if it reaches the red line on the gauge the power should be reduced. If the CHT reaches an unreasonable value for the applied load the possibility of an excessively weak mixture or a blockage or other defect in the cooling should be considered.

In water-cooled engines there will be a water temperature gauge which indicates the temperature of the water leaving the engine. Temperature should climb steadily after starting and level off at the designed temperature. This will often be above 100°C since the coolant is not actually water but a mixture of ethylene glycol and the system is also pressurized so that the boiling point is raised.

On turbocharged machines an exhaust gas temperature (EGT) gauge (8 in Figure 6.11) may be found. This will also rise with engine power and the gauge may be red-lined to protect the exhaust valves and the turbocharger.

The oil pressure gauge indicates the performance of the oil pump. If the pump can meet the flow demanded by the engine, the pressure will be normal. If it cannot, the pressure will fall. Low pressure could be due to a worn pump or bearings, but it could also be due to the oil being extremely hot so that it is thinner and runs through the engine faster than the pump can deliver it. The oil pressure and temperature gauges should be read together to form a meaningful picture.

The manifold pressure gauge and the fuel pressure gauge in a fuel injected gasoline engine both indicate the proportion of full power being developed and so the reserve power can be deduced. The fuel pressure indicator is also calibrated in fuel consumption rate so that the flying time can be calculated from the fuel load.

Manifold pressure or boost gauges are calibrated in absolute pressure, i.e. relative to a vacuum. In a naturally aspirated engine the manifold pressure cannot go above atmospheric and it is technically incorrect to refer to it as a boost gauge. In the presence of a turbocharger the manifold pressure can go above atmospheric and then the term boost gauge is correct. For a given induction air temperature, the boost pressure is proportional to the power the engine is developing. However, changing the air temperature changes the air density and so the mass flow, which really controls the power, will be different for a given boost pressure.

Consequently if carburettor heat is in use, the air entering the engine will be less dense and so less power will be developed for a given boost reading. If the engine is derated by setting an induction pressure limit, adhering to that limit with carburettor heat selected will result in unnecessarily limiting the available power. It is the actual power developed which determines the stress on the engine, not the induction pressure. With carburettor heat on, induction pressure can be increased by the correct amount to restore power without any risk of engine damage. It is important to know how much the induction pressure rises with carburettor heat for the same power. This can easily be established by hovering with and without carburettor heat.

With a traditional carburettor heat system, the pilot can be placed in a difficult situation. In the case of a fully loaded machine on a hot, humid day, full power will be required to hover. However, the landing approach will require the use of carburettor heat. At the end of the approach as the machine enters the hover, the textbooks explain that carburettor heat must be deselected. However, the textbooks do not explain how this is to be done. At the end of an approach and upon entering the hover, the pilot must manipulate the collective lever to arrest the descent and use the pedals to prevent yaw with the increased collective. He must also use the cyclic control to arrest forward speed. With both hands and both feet occupied, deselecting carburettor heat is difficult.

In the case of the Robinson R-22 there is a simple solution. In this machine the engine is derated by induction pressure limits to achieve reliability. The degree of derating is such that, even up to a few thousand feet above sea level, maximum rated power can be obtained even with full carburettor heat. Consequently under many real conditions the R-22 can be flown with hot air permanently selected. The boost gauge limit must be readjusted upwards. This is done by hovering under otherwise identical conditions with and without hot air to establish the percentage by which the induction pressure limit must be increased to keep the power the same. Using this technique carburettor icing cannot occur, pilot workload is reduced and fuel economy is improved because feeding the engine with hot air reduces pumping loss. Given that there have been a number of fatalities in this machine owing to icing it is difficult to see why this solution has not been proposed earlier. The only circumstances under which hot air would be deselected would be where ambient air temperature is extremely high or for operation at high altitude. These conditions can easily be deduced from the flight manual.

6.14 The aeroDiesel

The Diesel engine is a type of piston engine in which the induction process admits air only. There is no throttle and the maximum amount of air is admitted at all times. As only air is being compressed, a phenomenal compression ratio can be used without danger of detonation on the compression stroke. The high compression ratio which is maintained at all power levels, is responsible for the high efficiency of the Diesel engine. In the gasoline engine, when the throttle is partially closed, low manifold pressure opposes the motion of the piston on the induction stroke. The compression pressure falls, effectively reducing the compression ratio.

The compression raises the air temperature to such an extent that if fuel is forced into the cylinder it will burn immediately. The amount of power is controlled only by the amount of fuel injected. As combustion takes place with excess air, all of the fuel is burned, further raising efficiency. No ignition system is needed at all. Instead a device called an injection pump is required. This device produces the right amount of fuel at sufficient pressure to overcome the compression pressure and at the correct time. Fuel is fed into the cylinder by an injector. This is effectively a non-return valve that opposes the compression pressure in the cylinder. When the pressure from the injection pump has risen high enough, the injector admits the fuel in a conical spray pattern.

Diesel fuel is less volatile than gasoline and represents a reduced fire hazard. On injection, the fuel begins to burn instantly and this results in a rapid pressure rise. The Diesel engine needs to be robustly constructed to withstand the forces generated by the high compression ratio and the sudden pressure step. However, the Diesel has the advantage that the materials are not subject to such high temperatures. As it is thermally efficient, less heat is created in the engine and the exhaust is cooler, reducing stress on the exhaust valves. Water cooling is particularly advantageous with Diesels because the water jacket reduces noise and the reduced heat output allows the cooling system to be light.

As the Diesel engine induces only air and has no throttle, the danger of intake icing is insignificant and cold air will be supplied at all times for maximum power. The excess induction air may be used to sweep all combustion products out of the cylinder ready for the next cycle. This is called scavenging and in a four-stroke Diesel it is done by valve overlap. Some of the induction air follows the exhaust gases out of the cylinder. The exhaust valve is also cooled due to scavenging. This is especially easy in a turbocharged Diesel where turbo pressure assists the scavenge flow. Clearly scavenging cannot be used in the gasoline engine as it would result in unburned mixture entering the exhaust system.

The turbocharger works well in gasoline engines, but the extent to which power can be boosted is limited by the additional heat stress on the engine parts. As the Diesel produces less heat stress the degree to which it can be boosted by a turbocharger is higher. As the compressor in the turbocharger will increase the temperature of the induction air, a further increase in power can be obtained using an intercooler, which is a heat exchanger designed to cool the air between the compressor outlet and the induction manifold. An alternative on water-cooled Diesel engines is a charge cooler which uses the circulating coolant to cool the air from the compressor.

The boost gauge reading is not proportional to power or engine stress in a Diesel engine since there is no throttle. If a turbocharger is fitted, the boost gauge will indicate that the correct turbo outlet pressure exists. A malfunctioning wastegate could cause excessive boost and an abnormally high gauge reading. In this case it would be necessary to fly at a lower power level to prevent engine damage.

In order to display the power level in a Diesel engine it is necessary to measure fuel flow. However, a torque meter could also be used as in turbine practice.

6.15 The uniflow Diesel

The two-stroke engine has some advantages for aviation, not least the saving in weight and complexity due to the elimination of some moving parts. Although power is produced on every piston downstroke, less of the stroke is used and so the increase in power is not as great as is commonly thought. However, the doubling in firing frequency for a given RPM allows vibration to be halved. This may result in transmission weight saving.

At the end of the power stroke, the exhaust gases have to be replaced by fresh charge at one and the same time. Thus effective scavenging is inherent in two-stroke engines. Some incoming charge goes out of the exhaust no matter what. The gasoline engine mixes fuel and air externally and so the scavenging process passes unburned fuel into the exhaust, doing no good to the fuel economy or the environment. A further problem with the gasoline two-stroke is that most designs use the crankcase as part of the induction system. When the piston comes down, the charge in the crankcase is compressed and the pressure is used to drive it into the cylinder. The presence of charge in the crankcase means that lubricating oil will be mixed with the charge, resulting in a smoky exhaust. The conventional gasoline two-stroke engine will eventually be outlawed because of these environmental concerns.

The two-stroke Diesel engine can overcome these problems because it admits air not charge. One approach is the uniflow two-stroke Diesel shown in Figure 6.12. This is mechanically somewhere between a four stroke and a two stroke as it still has a camshaft and valves. However, all of the valves are exhaust valves and a huge valve area can be used for efficient breathing.

Towards the end of the power stroke the camshaft opens the exhaust valves. Exhaust gases exit the cylinder and their momentum causes the cylinder pressure to fall. Shortly after, the piston uncovers ports near the base of the cylinder to admit induction air. Effectively the cylinder is open at both ends so that induction air can sweep up the cylinder until some of it leaves via the exhaust valves in the scavenge process. Next the exhaust valves close but the momentum of the induction air continues to drive it into the cylinder. This continues until the returning piston closes the induction ports. The cylinder is now sealed and the air is compressed. Near TDC, the fuel is injected and the power stroke begins. In an automotive engine there is one injection pump feeding all of the cylinders. However, in aviation applications there will be one per cylinder so that a failure will cause a power loss rather than a stoppage. The injector can run from the camshaft.

The uniflow two-stroke Diesel has a number of advantages. The crankcase is not involved in the induction process and remains at atmospheric pressure just as it does in a four-stroke engine. Conventional recirculating oil lubrication can be used, without

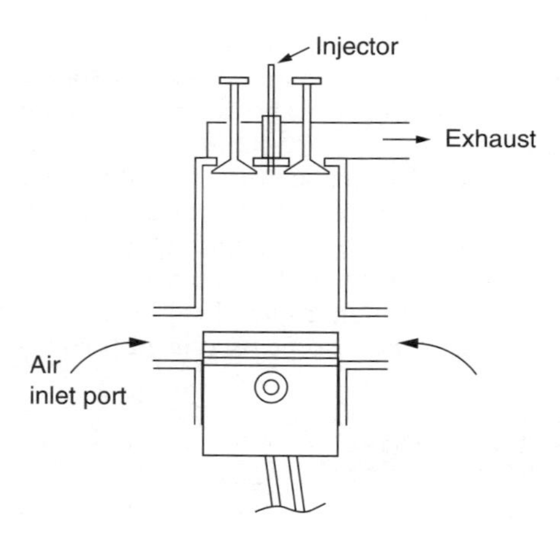

Fig. 6.12 In the uniflow two-stroke Diesel engine, induction is by ports in the cylinder wall, whereas exhaust is via valves in the head. The short scavenging path and potentially large cross-sectional area of the intake and exhaust ports allow high mass flow.

fear of mixing the oil with the charge. Thus the uniflow is a two-stroke technology that can meet environmental regulations.

The uniflow system is more effective at replacing exhaust with charge than either the conventional two stroke or the four stroke. In the conventional two stroke the fresh charge is admitted at the bottom of the cylinder and then has to proceed to the top and turn around to reach the exhaust port. This is called loop scavenging. The four-stroke engine needs two complete strokes to do this. Figure 6.12 shows that the breathing efficiency of the uniflow is high because much of the cylinder head can be occupied by exhaust valves and the intake port can run round the perimeter of the piston save for a few rails to stop the rings popping out. As a result the uniflow Diesel can deliver power over a higher proportion of the cycle than can four-stroke or loop scavenged two-stroke engines.

When a turbocharger is fitted, the efficient gas flow means that the valves can be arranged to open for a smaller angle, lengthening the effective stroke. When this approach is used, the engine produces its power in the form of more torque at a lower RPM. In the helicopter this is useful as the amount of gearing to the rotor may be reduced.

The uniflow engine is not self-starting and it needs some externally produced induction flow during cranking. The turbocharger may be fitted with an electric motor powered during starting.

In the two-stroke uniflow Diesel engine developed by Teledyne Continental Motors, some additional engineering features are incorporated. As the force on the piston is always downwards, a slipper big end can be used. Figure 6.13 shows that the big end bearing does not encircle the crankshaft. This means that both cylinders in a horizontally opposed engine can share the same crankpin and be exactly in line. The thrust of a piston on a power stroke drives straight through the crankpin to push the opposite piston up on the compression stroke. In a four-cylinder engine the two crank pins are at 90° to give four evenly spaced power strokes per revolution. This configuration requires balance weights on each end of the crankshaft to make the mass centroid of the moving parts align with the shaft axis. The result is very low vibration.

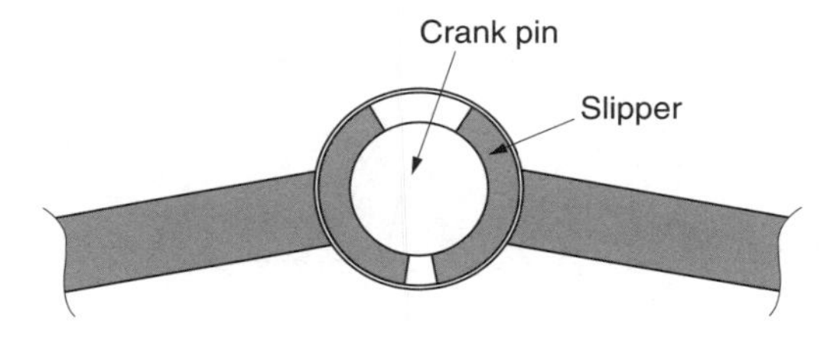

Fig. 6.13 The slipper big end of the Continental aeroDiesel allows both connecting rods to share the same crank pin.

6.16 Cooling systems

Most aviation piston engines are air cooled because in a fixed-wing aircraft a supply of cooling air is easy to come by. This argument is not so powerful in a helicopter in which the relative merits of air and water cooling need to be carefully examined.

Air-cooled engines reach higher peak temperatures and need looser tolerances to prevent seizing. Actually the air-cooled engine obtains a significant amount of cooling from the oil. The oil-cooling system will be heavier in consequence. The cylinders need to be further apart to allow space for the cooling fins needed to increase the surface area of the cylinder and head. This lengthens the engine and makes it heavier.

The in-line configuration is not good for air cooling because the cylinders at the back run hotter and the crankcase is very long. The radial engine is the ultimate air-cooled configuration because all of the cylinders receive equal cooling and the crankcase is short and compact. However, the frontal area and drag are significant. In high performance piston-engine aircraft, such as the Spitfire, the frontal area of the engine had to be reduced as much as possible. This was done using in-line or vee construction that dictated water cooling. In the Spitfire the drag of the radiator and the vee engine together is less than if air cooling was used.

Water has a much higher specific heat than air and so a smaller mass of water can carry away the heat. This means that the cylinders can be closer together because no fins are required. This makes the crankcase shorter and the crankshaft can be lighter. When designed from the outset to be water cooled, an engine with radiator need be no heavier than an air-cooled engine and it will be quieter.

In the helicopter there will be no slipstream in the hover and a cooling fan will be needed whatever the cooling technique. The power loss of an engine-driven fan is considerable, as is the noise it creates and any viable alternative deserves study.

6.17 The fuel system

The fuel system of a light helicopter is reasonably simple and the installation used in an Enstrom F28 is shown in Figure 6.14. The object of the fuel system is to ensure a reliable delivery of uncontaminated fuel at consistent pressure to the carburettor or fuel injection system.

The fuel system begins with the tanks. These are invariably installed as a pair, one each side of the mast. The tanks are as close as possible to the CM so that fore-and-aft trim is unaffected as fuel burns off. The two tanks are cross-connected with fuel and

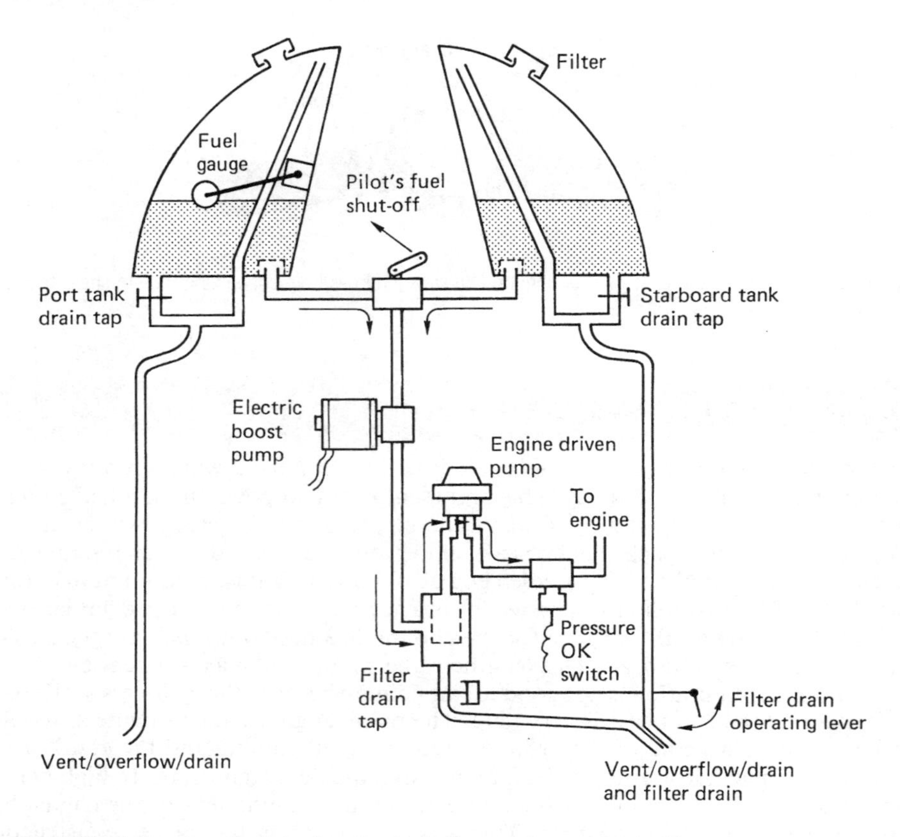

Fig. 6.14 The fuel system of a light piston-engine helicopter. See text for details.

vent pipes so that fuel is used equally from both tanks. This prevents a lateral trim change. As the fuel level is equal in the two tanks, only one fuel level gauge is needed. The transmitter in the tank is connected electrically to the gauge on the instrument panel and so the gauge only works when the master switch is on. It is wise not to place too much trust in float gauges and it is good practice to check the level in the tank itself. Some machines have sight gauges in the walls of the tank. Otherwise a good old-fashioned dipstick can be used.

Mandatory placards are placed near the tank filler caps to indicate the type of fuel to be used. RED for AVGAS, BLACK for AVTUR. It is vital that the correct fuel is loaded. Whilst a turbine will run on AVGAS, a piston engine will stop if AVTUR reaches it. Since AVTUR is denser than AVGAS, it will probably find its way to the engine whilst the machine is still on the ground. A mixture of the two fuels has a very low octane, and the engine could suffer damage from detonation before it stops.

When filling the tanks some consideration should be given to the fact that fuel will expand as the temperature increases. If the machine is to be flown immediately, there is little need for concern, but if the machine is to be left in the sun for any length of time there must be headspace for the fuel to expand into or it may overflow.

The commonest form of fuel contamination is water. Whilst in storage, as a tank is emptied, air fills the space above, bringing moisture with it. Condensation due to low temperatures results in water in the fuel. Condensation can also occur in the helicopter's own tanks, but this can be minimized by filling the tanks before leaving the machine overnight.

Water does not readily mix with fuel and it is denser, so it will be found at the bottom of tanks. The point where the engine feed is taken from the tank is slightly above the lowest point of the tank so that fuel is drawn from above any water that may settle out in the tank. As a result there will always be a quantity of unusable fuel in the tank, typically about a gallon. The lowest point of the tank is fitted with a drain. This is a spring-loaded valve which is normally closed but which can be pushed open momentarily by hand. Part of the pre-flight check procedure is to operate the tank drains and to catch the fluid that comes out in a transparent container. Any water in the tank will come out first and a clear boundary will be seen in the container between the water and the fuel once all the water has come out.

The tanks are also fitted with air vent plumbing to allow air to enter as fuel is used. In the absence of a vent atmospheric pressure would crush the tanks as the fuel was pumped out. The fuel pickup pipes from the two tanks go to the fuel shut-off valve. This is in the engine compartment, but operated by a rod passing through the firewall to a control often mounted near the pilot's shoulder. The main fuel line runs down from the fuel shut-off valve through the electric boost pump to the main filter; also known as a strainer. The filter contains a fine mesh element that prevents debris entering the carburettor or fuel injection system where it could block the fine jets. Debris is either trapped in the mesh until it is removed during maintenance, or falls into the bowl. Any water that gets past the measures designed into the tanks will also collect in the bowl. The engine feed is from the top of the filter, so water is left behind. A further vital pre-flight check is to open the spring-loaded drain valve on the filter bowl again catching the results in the glass container. Draining is continued until pure fuel flows. The bowl is then free of water and can trap the maximum amount.

Fuel from the filter passes next to the engine-driven pump. This is mechanical and only works when the engine is running. The combination of electric and mechanical pump one after the other means that fuel pressure is always available even if one of the pumps fails. In addition the electric pump can run without the engine, and this can be useful in fuel-injected machines to purge the injection pipes of fuel vapour prior to a hot start.

The fuel pump contains a pressure regulating mechanism so that constant fuel pressure is available. A pressure switch operates a 'pressure OK' light on the instrument panel once fuel pressure exceeds a set threshold. Fuel is then delivered to the carburettor or the fuel injection system.

Figure 6.15 shows the fuel system of the JetRanger. Twin electric boost pumps (1) lift fuel from the tank and check valves (2) allow one to carry on if the other fails. The main fuel valve (7) is remote from the cockpit and is motorized. The airframe filter (8) has a pressure drop monitor (9) that operates a warning light (10) if blockage occurs. The engine-driven pump (15) also has a filter (11) and blockage switch (13). The delivery pressure is regulated in the fuel control unit by returning surplus fuel to the inlet side of the gear pump (16).

In larger helicopters the fuel system will inevitably become more complex. In order to limit the consequences of damage, for structural reasons and to control the CM position, there may be a large number of smaller tanks, each needing provision for refilling, supply and a quantity gauge. A full fuel tank is heavy and may break free in a crash. A deliberately weak point may be introduced in the pipes along with self-sealing valves so that the tank will not leak if it breaks free. Wiring to pumps and gauges will also be designed to detach. In military machines, self-sealing tanks may be fitted. These are essentially double skinned and between the skins is a compound that expands on

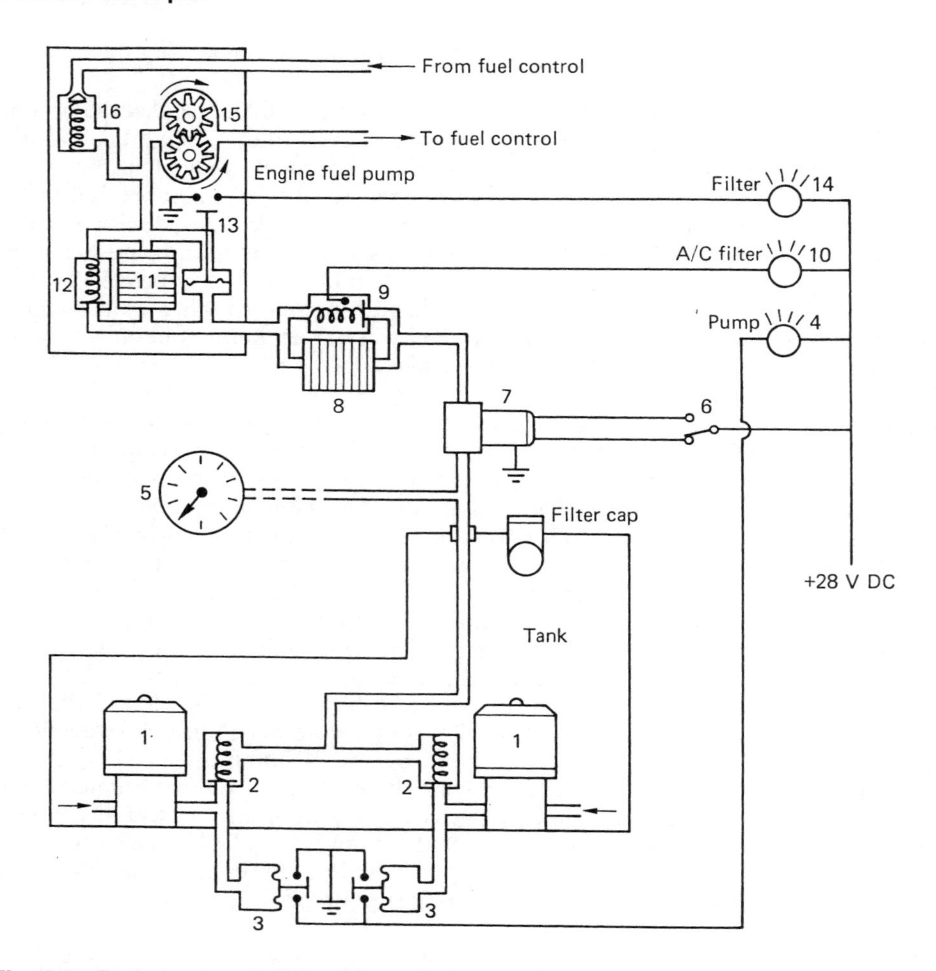

Fig. 6.15 The fuel system of a light turbine-engine helicopter. See text for details.

contact with fuel. If a tank is pierced, leaking fuel contacts the compound and this will expand, sealing the hole.

Figure 6.16 shows the fuel system of the Chinook. This is complex because there are six tanks, two main engines, an APU and a fuel-burning cabin heater. In addition there is provision for pressure refuelling from a central point as well as connections in the cabin to allow long-range tanks to be carried internally. The fuel is carried in the prominent side sponsons. These cannot contain a single tank for a number of reasons. First, a long single tank would be dangerous because when partially filled all of the fuel could flow to one end and cause a huge trim shift. Second, a single tank would be vulnerable to battle damage. Finally, the front undercarriage leg and its supporting structure occupies some space in the sponson. As a result the Chinook has three tanks along each side. These are not simply interconnected as if one were to be holed this would result in loss of all the fuel on one side. Instead the forward and aft tanks each have a pump which discharges into the main, or centre, tanks. The centre tanks have two pumps in parallel and valves ensure that one can fail without affecting the other.

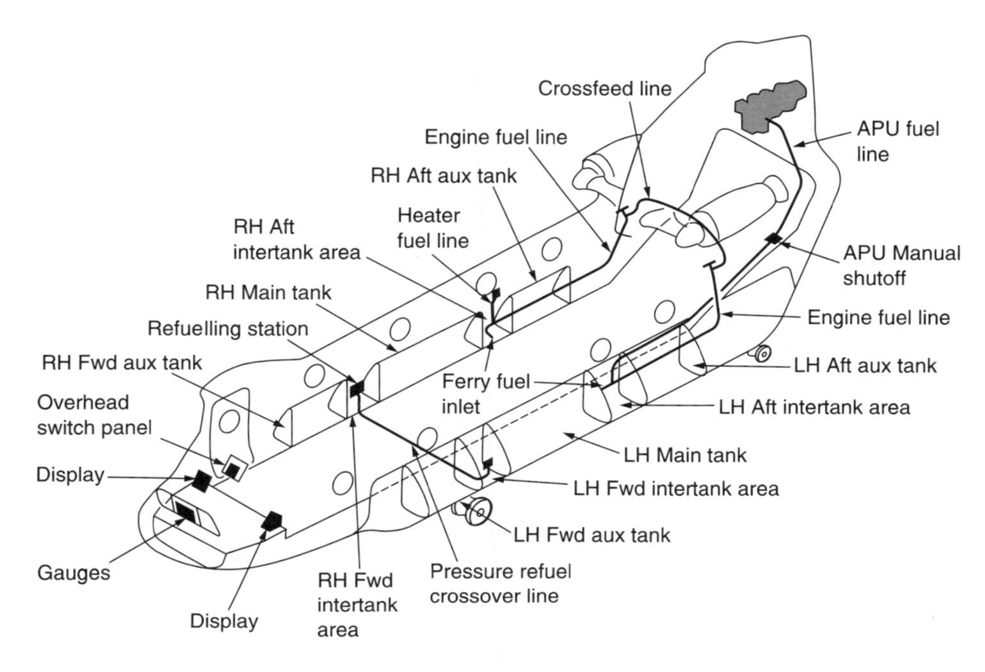

Fig. 6.16 The complex fuel system of the CH-47. See text.

Each main tank supplies the engine on the same side, but there is also a cross feed line between the engines so that both can run from one main tank. This line can be closed by the pilot to prevent fuel loss in case of major damage at one side of the machine.

6.18 The turbine engine

The turbine engine is a more recent development than the piston engine. Although the idea is older than the piston engine, the first aircraft turbines were not perfected until much later when the necessary materials were available. Whittle in England and von Ohain in Germany independently developed practical turbine engines during World War II.

The Otto cycle or four-stroke piston engine draws charge in through a throttle, compresses it, burns it to release power and then exhausts. In the Diesel engine only air is drawn in and compressed, and the fuel is injected directly into the cylinder. There is no throttle, and the power is controlled by the amount of fuel injected. Detonation cannot occur, and low octane fuel can be used. In a piston engine the four phases are conducted sequentially on a fixed quantity of charge in the same cylinder. The turbine is like a continuous version of a Diesel engine. Instead of operating on charge one phase at a time in the same place, charge passes through the machine continuously and the four phases take place at different points in the machine. Like the Diesel, the turbine can use low octane fuel and generally runs on AVTUR which is basically kerosene. Also like the Diesel engine, the power developed by a turbine engine is controlled by the amount of fuel fed to the burners.

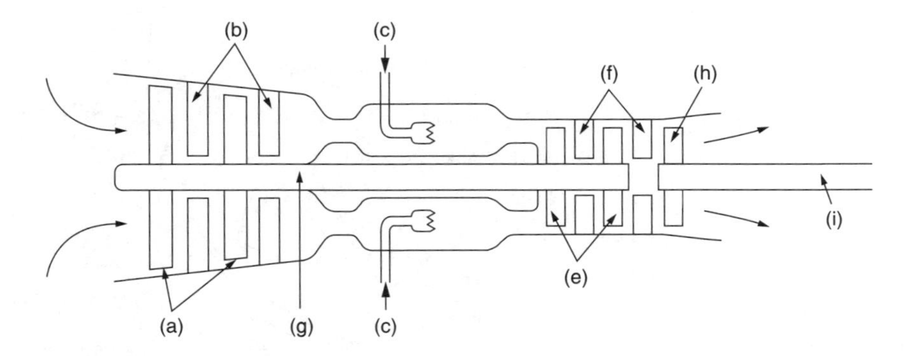

Fig. 6.17 The turbine engine carries out the processes of the four-stroke engine continuously. Air is drawn in and compressed, fuel is burned in it and power is extracted before exhausting the gases. The continuous operation and purely rotating parts result in freedom from vibration. Air is drawn in by compressor blades (a) and stationary blades (b). Compressed air is delivered to combustion chamber along with fuel (c) which burns at (d). Hot gases drive the turbine (e) which has stationary blades (f). Turbine (e) drives the compressor by means of shaft (g). Energy in hot gases exceeds power needed to run the compressor by a considerable amount, and also turns power turbine (h) to produce shaft power at drive shaft (i). Concept is very simple being little more than a turbocharged blowlamp. Practical engines require special materials which can withstand enormous rotational forces and high temperatures.

Figure 6.17 shows that the turbine engine is effectively a turbocharged blowlamp; in fact simple turbojet engines have been made using turbocharger components. Air is drawn into the compressor: basically a very powerful fan. This compresses the air continuously so that a steady pressure is maintained at the compressor outlet. Compressed air passes to the flame cans or burners where fuel is sprayed in and burned continuously. The charge temperature increases and the gases expand but they cannot overcome the pressure of the compressor and must flow onwards. The expansion results in the gases flowing much faster than the flow rate through the compressor. The hot gas then encounters the power turbine. This converts the energy in the hot gases into shaft power. Some of this is wasted driving the compressor, but the remainder is available to drive an external load.

In the free turbine engine there are two power turbines. One drives the compressor and the other drives the load. Effectively the first turbine and compressor together form a gas generator that powers the free turbine. The gas generator and the free turbine do not generally turn at the same speed. The free turbine should run at constant speed because it is geared to the rotor, whereas the gas generator will turn faster if more drive torque is required. As a result the free turbine engine has two RPMs and these are called N_1, the gas generator RPM, and N_2, actually the rotor RPM but proportional to the free turbine RPM. In practice, rotor RPM is controlled by adjustment of N_1 in order to stabilize N_2.

The turbine engine works with steady pressures at each stage and develops continuous power with only low octane fuel, unlike the gasoline engine which needs high octane fuel to prevent detonation and which encounters a serious pressure increase during the power stroke and needs to be strongly constructed to withstand it. The turbine can be much lighter than the piston engine. In turbines, the equivalent of the compression ratio is the pressure ratio: the ratio of compressor outlet pressure to inlet pressure.

Figure 6.18 shows a section through an Allison free turbine engine. Air enters through a multi-stage axial compressor followed by a single-stage centrifugal compressor.

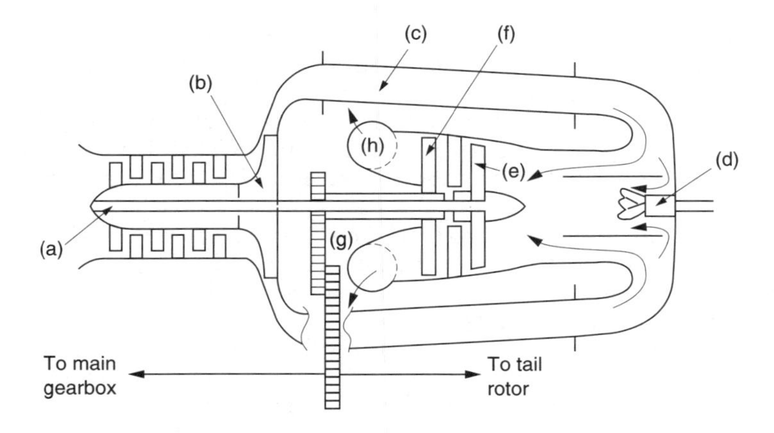

Fig. 6.18 The Allison C-250 light turbine. The compressor has a centrifugal stage and the flow through the power turbine is reversed to shorten the rotating assembly. Compressor is two stage, first stage (a) is axial, second stage (b) is centrifugal. Compressor air is ducted to the rear of the engine by two pipes (c) and led to the burner (d) where fuel is injected. Hot gases drive turbine (e) which powers the compressor, and turbine (f) which is the free power turbine. Power turbine runs on the outside of the mainshaft and transfers power through geartrain (g) to output shaft. Exhausts (h) are at the centre of the engine and lead upwards. Mainshaft has its own geartrain (not shown) to power accessories.

The outward flow of compressed air enters two large tubes leading to the burner. The flow is turned around to lead forward prior to entering the burner where fuel is injected. The hot gases then lead forward through the gas generator turbine that drives the compressor and then through the free turbine that powers the helicopter. The exhaust gases are then turned upwards and exit through the top of the engine compartment.

In the centre of the engine are two independent gear trains. One of these transfers power from the power turbine to the rotors and drives the power turbine governor. The other transfers power from the gas generator spool to the fuel pump, the oil pump, the gas producer governor and the hydraulic pump for the powered flying controls. This gear train also allows the electric starter motor to spin the gas producer spool.

A significant advantage of the turbine is that there are no reciprocating parts so that it can turn a lot faster than a piston engine. More charge can be passed at high speed, and more power developed. The output torque is unvarying and allows the transmission to be more lightly built than a piston engine transmission. The power to weight ratio of a turbine engine is quite incredible. In fact a turbine engine has no natural power limit because it has no induction throttle and if not externally controlled by restricting fuel flow could produce so much power that it would reach excessive internal temperatures or RPM and destroy itself. Practical engines are controlled or governed so that they can only produce a fraction of this power so that they will have a reasonable life. An indication of the stress in a turbine engine can be obtained from the turbine outlet temperature (TOT).

When an engine is controlled in this way the power output is said to be flat rated. As power is artificially limited, it can remain constant as the density falls with altitude since this initially limits only the maximum power. Governed power will fall only when the maximum power of which the engine is safely capable falls below the governing threshold. The altitude performance of turbine engines is consequently good. As altitude increases, the reduction in density is beneficially offset by the reduction in temperature so that mass flow is not reduced as seriously. At high altitude the governor will attempt to maintain power by increasing fuelling. The reduced mass flow reduces the dilution of the combustion products by excess air with the result that the gas temperature entering the power turbine tends to rise. Consequently although the engine could produce power at very high altitudes, in practice power will have to be limited to protect the power turbine. It should be appreciated that the power needed by the airframe may also fall with altitude so this effect is hardly a problem, especially in helicopters.

Power is controlled through limiting fuel admission to the burners. As kerosene in not very volatile it is either atomized at the burner nozzle or heated by passing it through pipes exposed to the burning gases on the way to the burner nozzle. It then vaporizes on leaving the nozzle and burns readily.

Starting a turbine requires an electric motor which will spin the gas generator spool fast enough to make the compressor operate. Once a suitable compressor speed is established (typically 15% of flight idle) the fuel is sprayed through the nozzle and the igniter is operated. This results in hot gas generation that will increase the turbine speed until engine power can take over from the starter motor. The starter motor and igniter are typically turned off at 50% of flight idle. If insufficient compressor speed is achieved before the ignition attempt, combustion pressure will overcome the compressor pressure and the turbine equivalent of a backfire takes place. In some cases the starter motor is permanently connected and becomes a generator when the engine is running.

6.19 Compressors

The job of the compressor is to provide a steady flow of air under pressure to the burners. Compressor design is fraught with compromise primarily because of the need to deliver different flow rates depending on the power required. It is relatively easy to design a compressor that is very efficient under one specific set of conditions, but it may become very inefficient under other conditions. In practice it may be better to design a compressor which is a little less efficient, but which maintains that efficiency over a wide range of flow rates. The compressor can be centrifugal, axial or a combination of the two.

The centrifugal compressor was used in early helicopter turbine engines as it allows a shorter assembly and could be designed using experience from turbochargers. Figure 6.19(a) shows a single-entry centrifugal compressor. The air enters axially near the eye or centre of the impeller at a speed approaching the speed of sound. The blades may be twisted at the eye to allow a smoother entry thus avoiding compressibility effects. As the air moves away from the axis the impeller blades impart higher tangential velocity. In practice the compressor impeller delivers air in a direction having both radial and tangential components. This high velocity is then converted into pressure using an assembly known as a diffuser: basically a divergent duct operating according to Bernouilli's theorem. The diffuser may at least double the pressure at the impeller outlet and more in some designs.

It is possible to construct a double-entry centrifugal compressor as shown in Figure 6.19(b). This allows twice the mass flow with the same diameter, but is less efficient because the blades run hotter than in the single-entry design which obtains some cooling at the rear of the impeller. A single centrifugal stage may produce a pressure ratio of up to 4.5 : 1. Stages may be cascaded to produce higher pressure ratios, but then the advantage of shortness is lost. The centrifugal compressor is relatively easy to

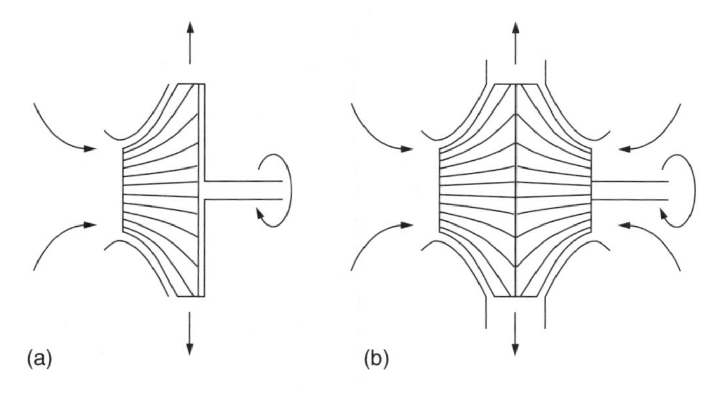

Fig. 6.19 (a) The single-entry centrifugal compressor. (b) The double-entry centrifugal compressor. Centrifugal compressors are less efficient than axial compressors and today are only used in APUs.

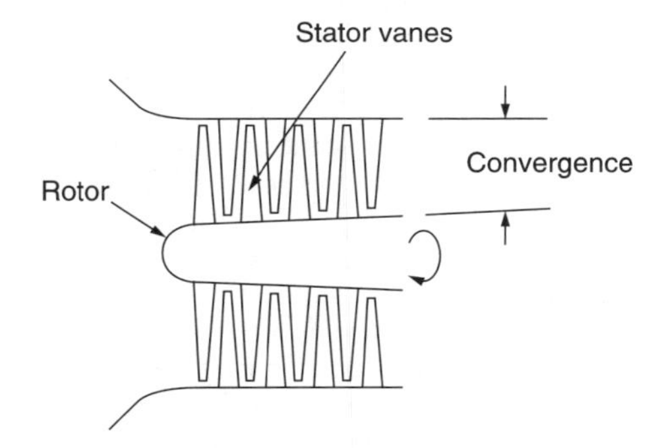

Fig. 6.20 An axial compressor is built with the stator in two halves so it can be fitted around the rotor.

make, and is less prone to icing, but is less efficient than the axial compressor which requires more components and may require anti-icing measures. The small area of the inlet eye of the impeller restricts the mass flow unless the overall diameter is made very large. Whilst overall engine diameter is less of a problem in helicopters, the weight of a large centrifugal compressor is unwelcome. Centrifugal compressors are obsolete in main engines, but may be found in APUs.

In the case of an axial compressor, shown in Figure 6.20, the rotor consists of several sets of blades or vanes attached at their inner ends to discs set on a common shaft. Fixed vanes are set between the rotating vanes. The fixed vanes are attached at their outer ends to a stator assembly that is made in two halves so that it can be assembled around the rotor. The fixed vanes act as diffusers for each stage of the rotor to prevent excessive velocity being reached. The fixed vanes also perform a swirl recovery function. It will readily be seen that the axial design allows a large cross-sectional area to be used at the inlet. The mass flow must be the same in each stage, so as the pressure builds up at each stage, the cross-section of the compressor can be reduced. This is done, for example, by increasing the diameter of the mounting discs and by using shorter blades. In order to maintain efficiency over a range of mass flows, the pitch of the first stage of stator vanes may be adjusted by an actuator.

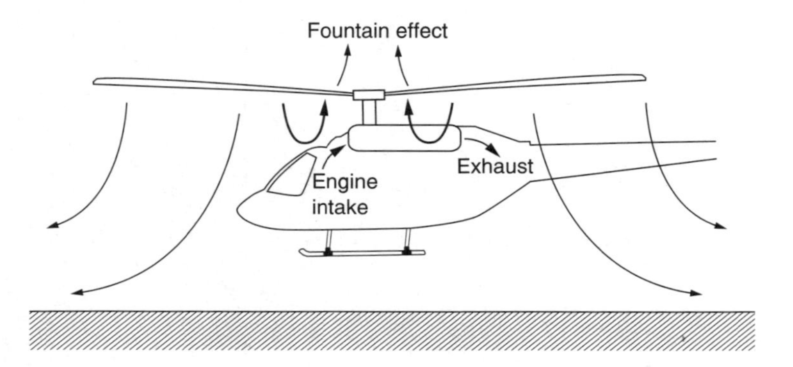

Fig. 6.21 In the hover, exhaust gases can recirculate back to the air intake and reduce power.

Turbine compressors may be affected by a phenomenon known as surge. This is an unstable or oscillating condition that replaces the usual steady-state conditions in the compressor. Surge may be triggered if the airflow into the compressor becomes disturbed. In helicopters the main rotor is very effective at disturbing the air and it is necessary to take some care over the siting and design of the engine intakes. In ground effect, airflow under the main rotor has some toroidal characteristics, including the fountain effect, in which pressure below the rotor leaks through the hub and blade shanks. Figure 6.21 shows that this can draw exhaust gases back into the engine air intakes. Surge may also be initiated during starting if, for some reason, ignition is delayed and a substantial quantity of fuel suddenly ignites.

Surge is characterized by periodic stalling of compressor blades. A stalled blade will reduce the compressor output pressure and combustion pressure may cause reverse flow into the compressor output. This reduction in mass flow through the compressor then causes the compressor pressure to increase, stemming the reverse flow until flow breaks down again. The oscillations due to surge are felt as vibration along with power loss and an increase in TOT as the governor adds fuel in an attempt to maintain power. Surge may be arrested by opening bleed valves in the wall of the compressor stator. These vent pressure to atmosphere and allow the mass flow to increase thereby unstalling the blades.

Compressors work at their highest efficiency when the blades are clean. Over time, the blades get dirty owing to various contaminants, including smog, salt and insects, and efficiency falls. The solution is periodically to wash the compressor. The turbine is motored with the starter motor and water is sprayed into the air intake. The compressor pressure-sensing pipe to the governor must be detached or sealed off during washing to prevent water entering the governor mechanism. The washing procedure may also require the bleed valves to be wedged open.

6.20 Combustion

The turbine is a continuous flow engine and relies on a delicate equilibrium being maintained. There are two related problems to be overcome. The first is that the velocity of flame propagation in kerosene is very slow compared to the air velocity from the compressor, which can be at several hundred feet per second. If directly exposed to compressor flow, the flame would literally be blown out. The second problem is that

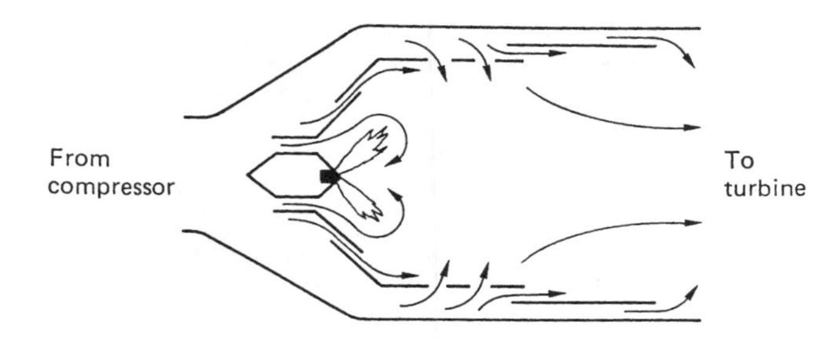

Fig. 6.22 The combustion chamber diverts a small proportion of the mass flow so that combustion can take place at low airspeed. The burning gases are then recombined with the main flow.

kerosine will only burn efficiently at an air/fuel ratio of about 15:1, when temperatures of as high as 2000°C are reached. In practice a range of fuel ratios between 45:1 and 130:1 will be required, and lower temperatures must be arranged at the power turbine to avoid blade damage.

Figure 6.22 shows how the combustion chamber is designed to overcome these problems. Only a small part of the compressor delivery enters the area of the burner, and this restricted flow is slowed down so that combustion can take place. The remainder of the airflow passes around the outside of the burner thereby cooling it. Within the burner, fuel leaves the nozzle in a conical spray and begins to burn with a roughly optimal mixture. The burning gases are joined by the annular airflow from the compressor and the result is that the flame turns inwards and produces a toroidal vortex. Burning fuel can recirculate in the vortex to ensure complete combustion. The high temperatures reached during this process are insulated from the walls of the flame tube by further air from the compressor. As combustion is completed, the hot gases are diluted by further compressor air. As a result the temperature of gases reaching the power turbine is controlled, and a much weaker overall mixture is achieved.

The burner nozzle does a similar job to the carburettor in a piston engine. In order to achieve atomization, the burner nozzle contains a small chamber into which fuel is admitted under pressure by a series of tangential ports. The chamber diameter reduces to the orifice. As the rotating fuel moves to the orifice, the radius of rotation must reduce and conservation of momentum suggests that the rate of rotation must increase. As it emerges from the orifice the fuel is spinning rapidly and is thrown into a cone-shaped spray. The burner has a similar problem to the carburettor when handling the reduced fuel flow needed to sustain idle power. Passing a small fuel flow through a burner intended for full power results in slow rotation, poor atomization and combustion. Figure 6.23 shows that the solution is to build a slow running jet into the nozzle. A nozzle of this kind is called a duplex burner and is controlled by a spring-loaded valve that prevents fuel flowing from the main nozzle until sufficient pressure has been applied to make it work properly.

Several small annular combustion chambers may be arranged around the engine, each fed from the compressor. The combustion chamber outputs are then merged together to drive the power turbine. As an alternative, one large annular combustion chamber may be used. This has several advantages. Flow from the compressor is annular, as is the flow into the power turbine so an annular combustion chamber gives better pressure and flow distribution. The surface area of an annular combustion chamber is also less than that of a multi-chamber system and this reduces losses.

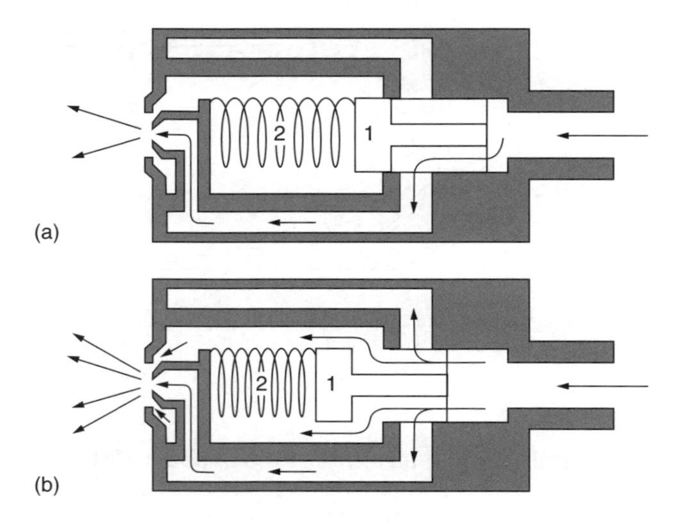

Fig. 6.23 The duplex nozzle. (a) At idle the fuel flow is small and spring 2 forces the shuttle 1 to the right allowing fuel only through the idle nozzle. (b) As flow increases the shuttle admits fuel to the larger nozzle also.

6.21 Power turbines

The power turbine is the most highly stressed part of the engine as, like the compressor, it has to operate at high RPM but with the additional problems of working with high temperature gases and at higher power. Turbine inlet temperature may be between 700 and 1200°C. As the temperature is so high, the speed of sound is correspondingly high and gas velocities in the turbine may reach 700 metres per second. The power turbines are almost always axial to provide sufficient flow. Like compressors, power turbines may have cascaded stages with stator vanes between. The power turbine extracts power from the gas flow by reducing its velocity, pressure and temperature. As the gases give up energy, they expand and so the cross-sectional area of the turbine must increase towards the exhaust. As was seen in Chapter 3, a rotor is at its most efficient with uniform inflow. One of the functions of the stator is to modify the radial distribution of gas velocity and pressure so that the gas emerging from the adjacent rotor disc has uniform pressure and velocity from root to tip. The stator vanes also induce swirl in the direction of rotation. The rotor blades are not shaped like conventional airfoils, but have much deeper curvature or camber characteristic of reaction wheels.

One of the greatest challenges in power turbine design is to control the temperature of parts exposed to the hot gas flow. In general oil cooling cannot be applied to the rotating blades because of sealing difficulties and the extremely high oil pressure that would be built up due to the high rotational speed. Thus in practice the parts must be air cooled. Air from the compressor can be used for this purpose. Figure 6.24 shows that the stator vanes are hollow and can be fed with cooling air from both ends. The air can be arranged to exhaust into the gas flow via small drillings in the face of the vane. In this way an insulating boundary layer of air is provided which protects the thin trailing edge of the vane.

The rotor blades are subject to extreme temperatures and extend in length considerably due to expansion. If sufficient tip clearance is provided for expansion, losses due

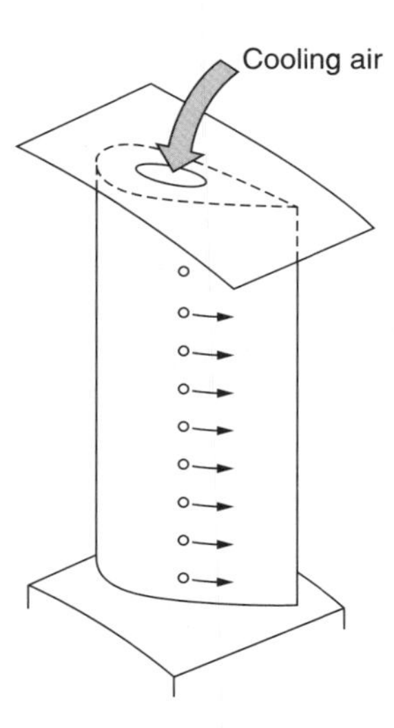

Fig. 6.24 The components of the power turbine are subject to extreme temperatures. The stator vanes are cooled by passing air along them through internal passages. The air may be exhausted to the surface of the vane where it creates a cooler air film over the trailing edge.

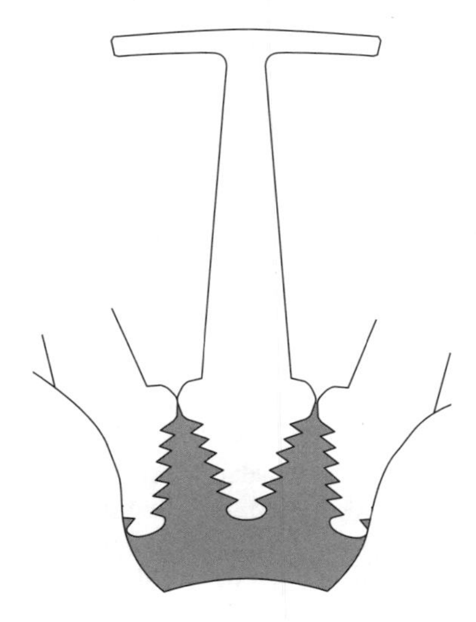

Fig. 6.25 The fir-tree method of blade attachment shares the load over a large surface area which also assists cooling.

to gas flow around the tip will be excessive. Rotor blades are generally constructed with an outer ring or shroud to combat tip loss. The blades are cooled by conduction to the hub and shroud and by radial airflow. The hub discs are air cooled to dissipate the conducted heat from the blades.

The blades are rotating at such speed that the root attachment force is of the order of 1000 kilograms. Various means for reliable blade attachment have been developed. Figure 6.25 shows the common method known as a fir-tree attachment. The dimensions

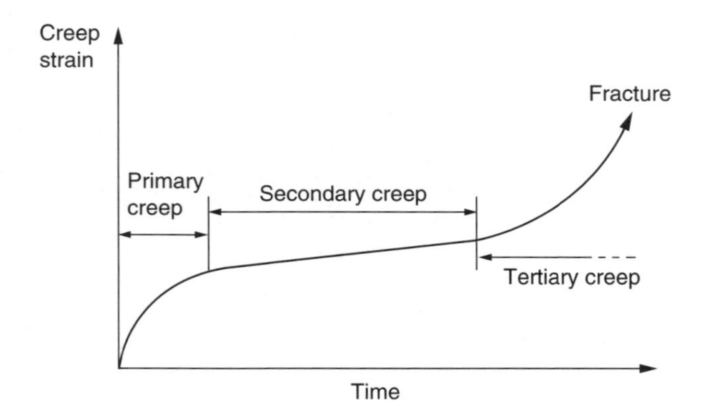

Fig. 6.26 Blade creep takes place in three distinct stages as shown here. Blades thus have a finite life.

are accurately determined such that when at service speed the strain in the root distributes the thrust over all of the teeth. At rest the blade will be loose in the hub disc.

The rotational stress and temperature experienced by the blade cannot be sustained indefinitely. Under these conditions, all known materials slowly extend or *creep*. Figure 6.26 shows the creep characteristic of a typical blade. The initial extension or *primary creep* is rapid, followed by a more stable regime where the *secondary creep* is relatively slow. At the end of this stage, *tertiary creep* sets in, leading to failure. There are several consequences of this characteristic. The first is that all turbine engines have a finite service life before the blades need replacing. The second is that considerate piloting that avoids excessive TOT will improve the reliability of the engine. Finally in an emergency it is possible to extract substantially more power from a turbine for a short time than it can reliably sustain. The result will be accelerated creep that will diminish or even use up the service life of the engine, but the helicopter can still fly to safety. This topic will be considered further later in this chapter.

6.22 The turbine oil system

The turbine is mechanically straightforward as there are few moving parts. The rotating parts must be carried on bearings designed to withstand the very high speed. Bearings adjacent to the power turbines must be protected from the enormous temperature of the gases flowing past. This is achieved by oil cooling. Heat flows towards the bearings because of the temperature difference, but the oil carries the heat away at the rate it arrives and so the temperature can be controlled. As there are no heavy reciprocating masses, the major function of the oil is cooling and turbine oil is only about one-tenth the viscosity of piston engine oil so that large volumes can be pumped to transfer heat. The low viscosity oil is also an advantage when starting in extremely cold conditions. An oil cooler is used to transfer the heat from the oil to the surrounding air. In helicopters the oil cooler may be provided with airflow from a transmission-driven fan. The same fan may also cool the transmission oil. Hot oil from the engine may also be used for anti-icing. In some engines, the oil tank forms part of the air intake.

It is vital that the correct shutdown procedure is followed for a turbine engine. The engine must run off load for a time after flight in order to cool down before it is stopped. If this is not done the heat from the power turbines will soak through to the bearings and carbonize the oil. Synthetic oils have been developed which resist carbonization better than conventional oil. Heat soak may be prevented with an electric pump to keep the oil flowing after shutdown until the hottest engine parts have cooled.

The temperature at the bearings is such that conventional elastomeric oil seals would have a very short life. Instead oil sealing is performed by the use of threaded sections on the shafts that have the effect of screwing the oil back to where it should be. The designer also uses the fact that much of the interior of the engine is at high pressure to keep the oil in place.

6.23 Turbine fuel control

Turbine engine power is ultimately controlled by the amount of fuel injected. If the fuel flow is increased gradually, the burner temperature rises and exhaust gases have an increased velocity. This drives the turbine faster and in turn raises P_C, the compressor pressure. The greater the pressure between compressor delivery and atmospheric, the more power the turbine can produce.

The air/fuel ratio cannot go outside the limits shown in Figure 6.27 or combustion could cease. If the mixture is too weak the airflow blows the flame away, if it is too rich the cool fuel quenches the combustion, resulting in a flameout. Neither of these extremes is desirable, but can be brought about if the fuel flow fails to match the airflow. This can happen if an attempt is made to change the engine power too rapidly. The accurate fuelling need is provided by a constant-pressure fuel pump feeding a fuel control unit. The fuel pumps are variable displacement devices. Figure 6.28 shows that the pump pistons are driven by a swashplate. When the swashplate is square to the shaft the pistons do not oscillate and there is no flow. As the swashplate is tilted, the pistons create a flow proportional to the tilt. A spring tries to tilt the swashplate to the maximum flow position, but the fuel delivery pressure is applied to a piston opposing the spring. When the fuel flow is low the delivery pressure rises by the small amount necessary to compress the spring and reduce the pump displacement.

The fuel flow control of a turbine engine has to ensure that the fuel flow is always within the limits required for combustion. Power is controlled by slightly disturbing the equilibrium toward one or other of the limits. If it is required to increase power, a slight increase in fuel flow will begin to accelerate the spool, and as it turns faster the

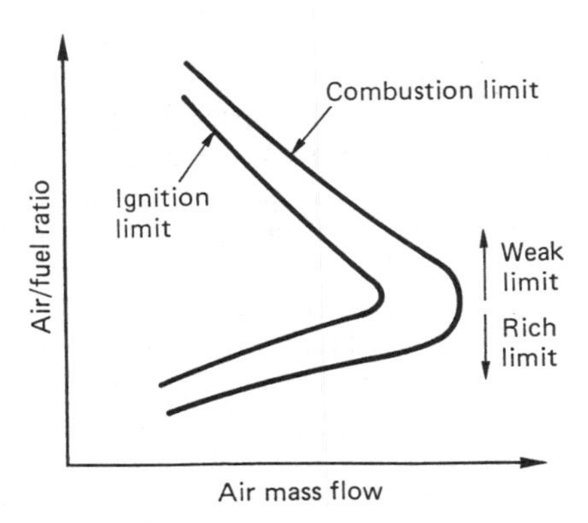

Fig. 6.27 The range of air-to-fuel ratio that allows combustion is quite small for kerosene or AVTUR.

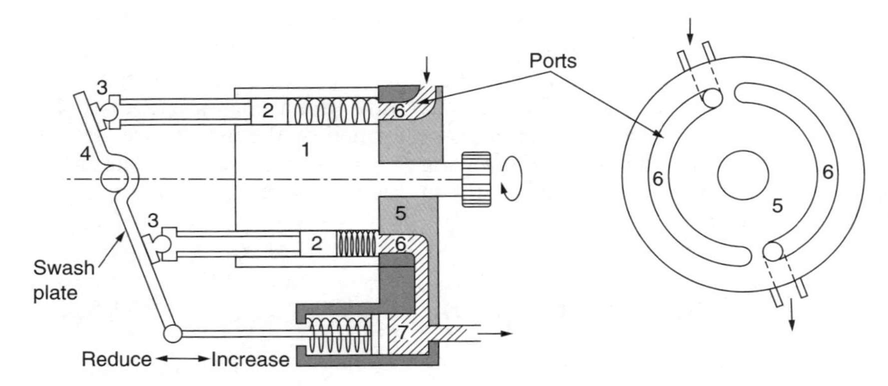

Fig. 6.28 A swashplate pump has variable displacement controlled by the angle of the plate. Rotor 1 is turned by the drive shaft. The rotor contains several pistons 2 which have ball-jointed slippers 3 which slide over the stationary swashplate. This causes the pistons to move in and out of the rotor. The backplate 5 is fitted with ports 6 which allow fuel to enter the cylinders when the piston moves to the left and direct fuel to the pump outlet when the piston moves to the right. The angle of the swashplate determines the stroke of the pistons and the amount of fuel delivered in one rotation. As delivery pressure builds up the piston 7 overcomes spring pressure to reduce the swashplate angle and the pump delivery.

fuel flow must increase more rapidly to match the increasing airflow. When the desired power output is reached, the fuel flow is reduced to the new equilibrium level. In order to reduce power, the fuel flow is slightly reduced, and the spool slows down. As it does so, the fuel flow rate must reduce to match the airflow. When the new power level is reached, the fuel flow must be raised slightly to arrest the deceleration.

Figure 6.29 shows how the fuel flow is controlled. Compressor outlet pressure, P_C, is delivered to a pair of chambers through small flow restrictor orifices. A large diameter bellows is fitted between the chambers, and a small diameter bellows is fitted between the second chamber and atmosphere. The bellows are joined with a rod that opens the fuel valve as it moves downwards. Air is allowed to spill from the chambers by two valves controlled by the governors. These spill valves are in series with the flow restrictors in the compressor line. An increase in P_C collapses the bellows and opens the fuel valve further, admitting more fuel to match the increased airflow. A reduction P_C has the opposite effect. The fuel flow is maintained approximately correct by this mechanism. Power output is increased or reduced by disturbing the equilibrium of the bellows system using the spill valves.

If the governor wishes to increase power, it closes the spill valves slightly. This restricts the flow of compressor air and raises the pressure in the bellows chamber, opening the fuel valve. If this is done too rapidly, the mixture will be too rich and the flame will be quenched. This is prevented with the accumulator: a tank preventing the air pressure in the upper bellows chamber changing rapidly. The air has to flow in and out of the accumulator through small orifices and this takes time. As a result when the spill valves are first closed, pressure in the upper bellows chamber rises slowly, and lags behind the pressure rise in the lower chamber.

The pressure difference acts on the upper bellows to oppose the rate of increase of the fuel valve, preventing quenching. As the gas generator accelerates, the accumulator pressure rises and ceases to oppose the increased fuel delivery. Increased compressor pressure due to the higher spool power causes the fuel flow to be further increased. When the governor decides the power has increased enough, it opens the spill valves

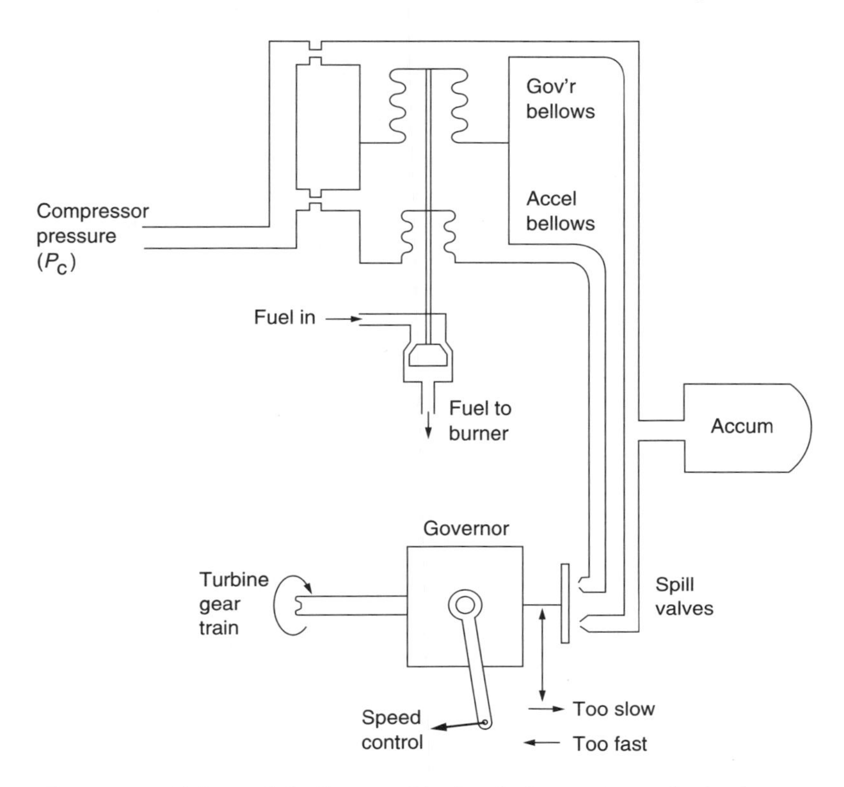

Fig. 6.29 To prevent rapid changes in fuelling, a combination of a flow restrictor and a chamber or accumulator is used. Pressure in the accumulator can only change slowly.

once more. The pressure in the bellows chamber falls, but rapid closure of the fuel valve must be prevented or the flame will blow out. This is achieved once again by the pressure lag due to the accumulator.

In some free turbine engines, transient droop is reduced using compressor bleed. If the governor develops a large underspeed error it may operate valves to allow some of the compressor output to leak to atmosphere. This unloads the compressor and allows the spool to accelerate. This technique is used, for example, in the T-55 engines of the Chinook.

The rate of acceleration of a turbine engine is relatively slow and this explains the aircraft carrier pilot's technique of advancing the throttle to maximum just before touchdown. If the arrester hook fails to catch, the engine will have spooled up to full power in time to go around. If this technique is not used the plane may fall off the front of the ship.

6.24 FADEC

Full Authority Digital Engine Control (FADEC) is a system in which the operation of a turbine engine is completely controlled by a processor. The goal is reduced pilot workload along with greater reliability. With FADEC the engine starting procedure is automated and the pilot only has to turn it on. The stabilization of rotor RPM is

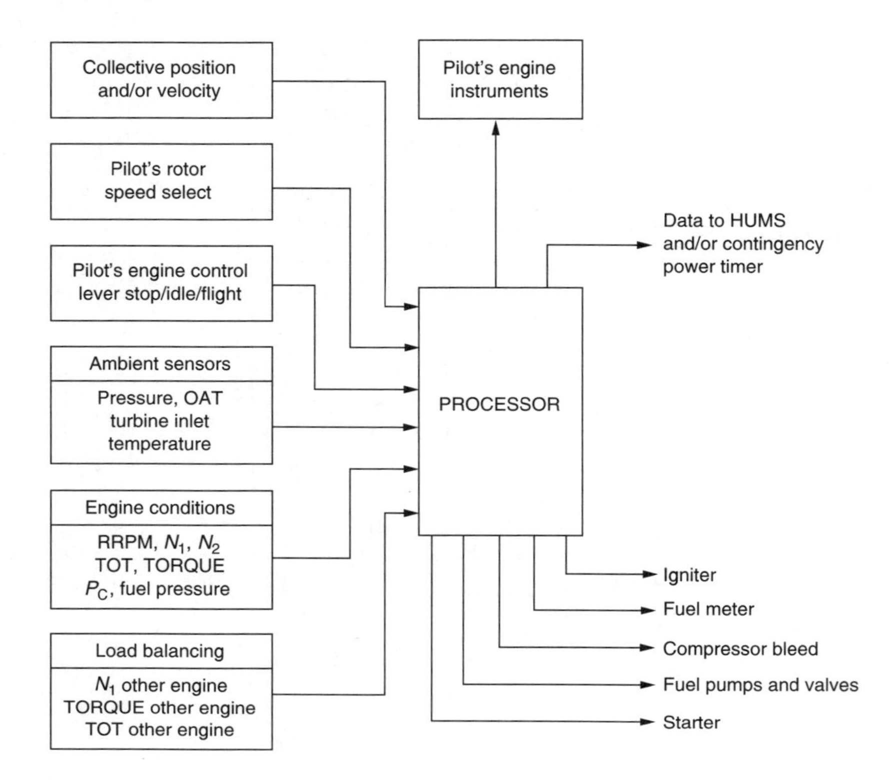

Fig. 6.30 In a FADEC system all relevant parameters are sensed and converted to binary codes to form inputs to the processor. The processor then controls the fuelling.

also a FADEC function, along with automatic power balancing in the case of a twin or multiple engine installation. FADEC systems may also be able to handle engine failures by automatically increasing the power of the remaining engine.

Figure 6.30 shows that all of the parameters measured by the instruments and the fuel control, including N_1, N_2, P_C and TOT are converted to digital codes and fed to the processor. This will determine what the fuel flow should be as well as controlling the starter motor, fuel valves and igniters. Needless to say the power supply to the processor becomes critical. The power controller will be wired to multiple sources so that it can run from any one.

Additional feedforward signals may come, for example, from a sensor on the collective lever used by the computer to anticipate a power demand to reduce transient droop. In a conventional governor if the rotor speed goes above normal the engine power will be reduced to contain it. If autorotation is entered, the engine power may be reduced to nearly zero and this will mean that the gas generator spool will be running very slowly. The turbine engine takes considerable time to return to full power from this condition. At the end of the autorotation a conventional governor would not respond until the RRPM had fallen below normal and the result would be droop until the turbine responded. However, in a FADEC machine the computer could anticipate the end of the autorotation by detecting the upward movement of the collective lever and it could start spooling up the gas generator before the RRPM fell below normal. This would reduce the amount of droop.

Sensing the collective lever velocity is also useful in the case of a rejected landing. This is where the machine is landed but the ground is found to be unsuitable. Perhaps the machine has started to break through a crust of ice on top of snow, or the landing area turns out to be waterlogged. Ordinarily the reduction of collective at touchdown would cause a slight overspeed causing the gas generator to spool down. Pulling collective to reject the landing would cause droop. Sensing the collective velocity would help the fuel control anticipate the power requirement.

In twin-engine machines, it is necessary to adjust the fuelling in such a way that the engines share the load. If the FADEC system is sensing TOT and N_1, either of these can be used to achieve a load share. In practice, engine tolerances will mean that a balance of TOT may result in a small difference between N_1 and vice versa.

Mechanical fuel controls and correlators are complex and expensive and prone to wear. The substitution of stable digital electronics will result in the elimination of wear and allows arbitrarily complex transfer functions to be implemented easily. However, the downside of FADEC is that it requires the control software to be utterly reliable. The difficulty with computers is that the number of states and combinations that can be achieved is staggering and it is therefore very difficult to *prove* that the computer can correctly recover from every possible state. In the development of FADEC systems there have been some spectacular failures, including an in-flight rotor speed runaway on a Chinook that required the pilot to climb at record speed to prevent the rotors being thrown off until the engines could be shut down.

In well-engineered systems a FADEC failure should leave the pilot, or co-pilot, with some form of manual power control, even if RRPM is not as precise.

6.25 Turbine instruments

As there is no throttle an inlet pressure gauge would be pointless. The actual power being generated is a fairly stable function of the reading displayed on the fuel flow meter so this or the torque meter may be used for power checks. An intake air temperature indicator will be provided so that the pilot will have an indication of the maximum power available. N_2 (power turbine RPM) is proportional to RRPM and needs no separate indication. However, N_1 (gas generator RPM) will have its own tachometer. The power turbine inlet temperature (PTIT) is measured by thermocouple in the engine and displayed on the instrument panel and will be red-lined at a safe temperature above which the turbine blades may be damaged. In some cases the turbine outlet temperature may be measured instead. This is simply because the temperature is lower at the outlet and the sensors may be less expensive. In the case of an engine failure, the remaining engine may run at higher power and this will cause PTIT/PTOT to exceed the red line and enter the contingency power range. This will result in a caution light illuminating. Usually a timer also runs to assess the stress caused to the engine. The oil pressure and temperature are displayed as for the piston engine. Oil quantity may also be displayed. There may also be chip detectors in the oil system.

6.26 Fuel management

When refuelling, whoever actually undertakes the task, the responsibility is taken by the commander of the machine. It is essential to specify the quantity and type of fuel required and to check that this has actually been supplied. AVTUR and AVGAS smell quite different and the nose will quickly establish that the correct fuel has been loaded. Gasoline engines will not run on AVTUR, whereas turbine engines will burn almost anything. For low temperature operation turbines may be run on a mixture of one part lead-free AVGAS and two parts AVTUR.

In addition to AVTUR and AVGAS there are MOGAS and ordinary car fuel. Car fuel is made to fairly slack specifications and no warranty of freedom from contamination is given with it. The composition of car fuel may vary with the season. In winter it may have higher volatility than AVGAS and can cause vapour locks in the fuel pipes. It should never be used for any aviation purpose. MOGAS is made to tighter specifications than car fuel and may be suitable for some aircraft piston engines but should not be used without a specific check.

When refuelling from drums, it must be borne in mind that the drums could have been in storage for some time. There will be an expiry date on the label. There could be rust or other dirt in the fuel along with water. If this is suspected, fuel should be taken on through a filter to remove debris, and through a chamois or felt pad to soak up water. After fuelling, the tank and filter drains must be checked for water before flying. In the event of heavy water contamination, the tanks may have to be drained and refilled.

If it is proposed to load only sufficient fuel to make a given flight, perhaps in order to increase payload, an allowance must be made for headwinds along with half an hour for contingencies. The flight time plus allowances must be multiplied by the machine's hourly consumption figure for the planned load. It is important to be quite sure of the units in use. Unfortunately fuel can be measured in pounds, kilograms, imperial gallons, US gallons and litres and there have been forced landings where the refueller assumed a different unit from the captain. In such chaos, the old-fashioned dipstick makes an excellent safety check. However, tanks are often an irregular shape and the quantity is not always proportional to the depth of fuel.

6.27 The transmission

In the pure helicopter, the rotor provides all lift and propulsion. The rotor is an actuator converting shaft power into thrust. The primary functions of the transmission are to deliver the shaft power and to transmit the thrust to the hull. Secondary functions of the transmission include driving the tail rotor and vital services such as hydraulic pumps and generators. There must be an autorotation clutch that allows a failed engine to stop whilst the rotors carry on turning. In some machines the transmission also drives cooling fans. There may also be a rotor brake to stop the rotor quickly after landing. In piston-engine machines a clutch will generally be needed to allow engine starting. Free turbine engines have no need for such clutches, but in twin-engine machines having no APU, an extra clutch may be fitted to allow one engine to be run without the rotors turning.

Figure 6.31 shows the construction of an autorotation clutch. There are two coaxial races that are cylindrical. The outside member is typically engine driven whereas the inner is attached to the transmission. Dog bone-shaped wedges known as *sprags* are positioned between the inner and outer races. The sprags are located with a pressed steel cage. The sprags are slightly asymmetrical. As the figure shows, if a sprag turns one way, its effective thickness increases, whereas if it turns the other way the thickness reduces. The sprags are biased in the former direction with light springs. If the outer race turns one way, the sprags are twisted against the springs and slip so that no drive is transmitted. If the outer race turns the other way, the sprags are twisted in such a way

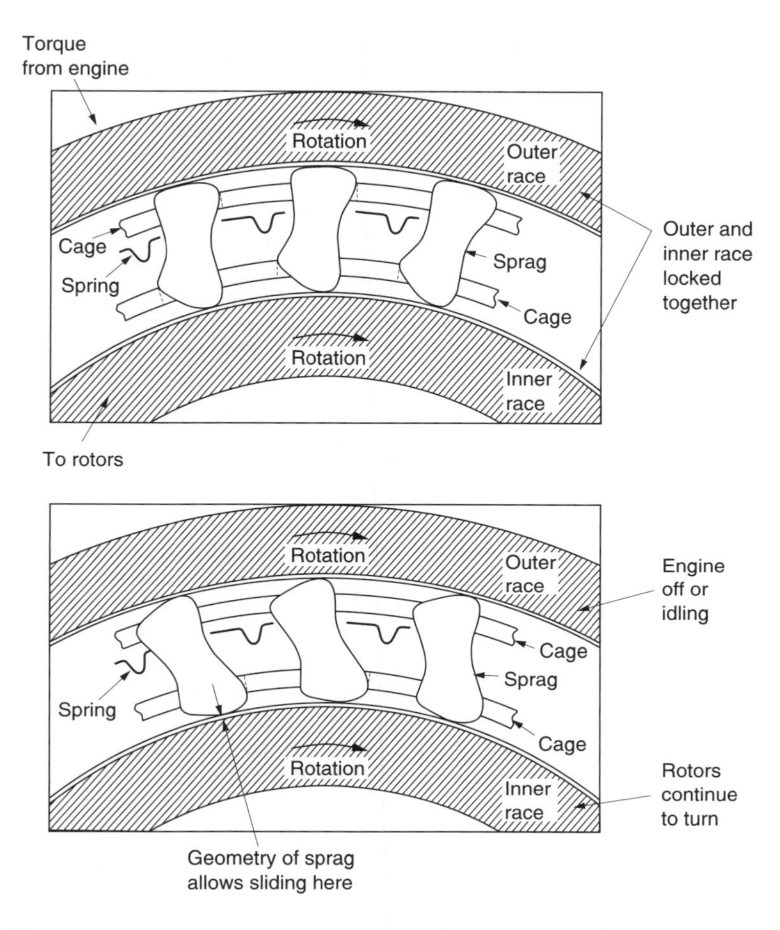

Fig. 6.31 The autorotation clutch or sprag clutch only transmits drive one way. The dog bone-shaped elements twist and lock between the inner and outer races to transmit the drive. If the relative drive direction reverses, the dog bones turn the other way and slip.

that they expand to fill the space between the races that are thus locked together. The more torque that is applied the harder the sprags grip. If the engine stops, the inner member runs on, releasing the wedging force so that it is free to turn. The action of the autorotation clutch is fully automatic and needs no control. It should, however, be checked for correct operation before flight. With the rotor at flight RPM, the throttle is briefly closed to prove that the engine can slow down whilst the rotor runs on.

The main rotor of a light helicopter needs to turn at something like 300 RPM which is rather slower than the piston engine which runs at more like 3000 RPM, so the transmission performs a speed reduction. The turbine engine runs much faster than the piston engine and the reduction ratio to the main rotor shaft will be of the order of 100 : 1. In a large helicopter the disparity between rotor and engine RPM will be higher. The continuous rotary power generated by the turbine is free of the torque impulses of the piston engine and the transmission can be lighter. The tail rotor usually has a similar tip speed to the main rotor and so needs to run at an RPM between those of the engine and the main rotor. In a light helicopter, the transmission may incorporate two reduction stages as was shown in Figure 6.1. The first will reduce engine RPM to approximately tail rotor RPM, and the second will reduce the tail rotor RPM to main rotor RPM. The tail rotor gearbox contains a pair of bevel gears to turn the drive through a right angle. This has usually a near 1:1 ratio, but some variation is possible. In development the tail rotor RPM may need to be changed to obtain a different compromise between noise and available thrust, or if the solidity is changed. This can be achieved by changing the tail gearbox ratio. In machines with cranked tails a further gearbox may be needed to turn the drive at the base of the fin.

Although simple at one level, gearboxes incorporate some subtle detailing. Consider a gearbox with a ratio of 2 : 1. The gears have, for example, 40 and 20 teeth. As these numbers factorize, each tooth on the small gear only touches two teeth on the large gear and any irregularities would form a wear pattern. However, if the number of teeth were to be relatively prime, every tooth would touch every other tooth and this would result in the irregularities averaging out. This could be achieved by using 20 and 41 teeth. For this reason gearboxes seldom have simple ratios.

If the gear teeth were parallel to the shafts, the gearbox would make a lot of noise and vibration as the drive jumps from one pair of teeth to the next. In practice the teeth are twisted so that the point of contact slides along the shaft as the gears rotate. Contact between the next pair of teeth begins just before the current pair part company and the result is a more even power transmission and less noise and wear. Helical gears produce end thrust on the shafts because of the angular contact. The angled contact faces of the gear teeth also cause the shafts to be forced apart. The gear case, the shafts and the bearings must be stiff enough to contain the forces generated without distorting. Distortion will reduce the accuracy of tooth mesh and destroy the uniformity of the tooth loading. Tapered roller bearings resist combinations of radial and axial forces. When used in opposed pairs, the outer race of one of the bearings can be adjusted axially to remove all play in the shaft and apply a slight *pre-load* to the rollers to stop them skidding.

Figure 6.32 shows that in most gear systems both gearwheels have convex teeth making the contact patch between the teeth very small. In conformal gearing, one of the gearwheels has concave teeth designed to rotate around the convex teeth of the other gear. This allows a much larger contact area so that a given size of gear can transmit more torque. The Westland Lynx has conformal gearing.

The epicyclic or planetary reduction gearbox is popular in helicopters because it can be built with low weight for the torque handled. Figure 6.2(a) showed that there is a fixed internally toothed gear which is coaxial with the input or *sun* gear. Between these gears are three *planet* gears that are fitted to pins on the output shaft. As the sun gear rotates, the planets orbit the ring gear and revolve at the same time. The part of the planet gear in contact with the sun gear is travelling twice as fast as the axis of the planet gear. As a result the planetary mechanism itself gives a 2:1 speed reduction which may further be increased by reducing the size of the sun gear. The torque is shared between the three planets and applied at three different places on the sun gear and the ring gear. This makes the epicyclic gearbox ideal for the final stage of rotor drive where the torque is greatest. Figure 6.33 shows a Chinook gearbox, in which the final stage epicyclic gear ring can be seen.

The gear teeth slide over each other, and metal-to-metal contact is prevented by special gear oil known as extreme pressure (EP) oil. The viscosity of the oil means that it is not squeezed out from between the teeth by the contact pressure. The sliding then takes place in the shearing of the oil film. Shearing viscous oil produces heat and the gearbox casing may be made with fins to dissipate the heat. In large machines an oil cooler may be fitted, perhaps with a fan and oil pump. In some machines the oil

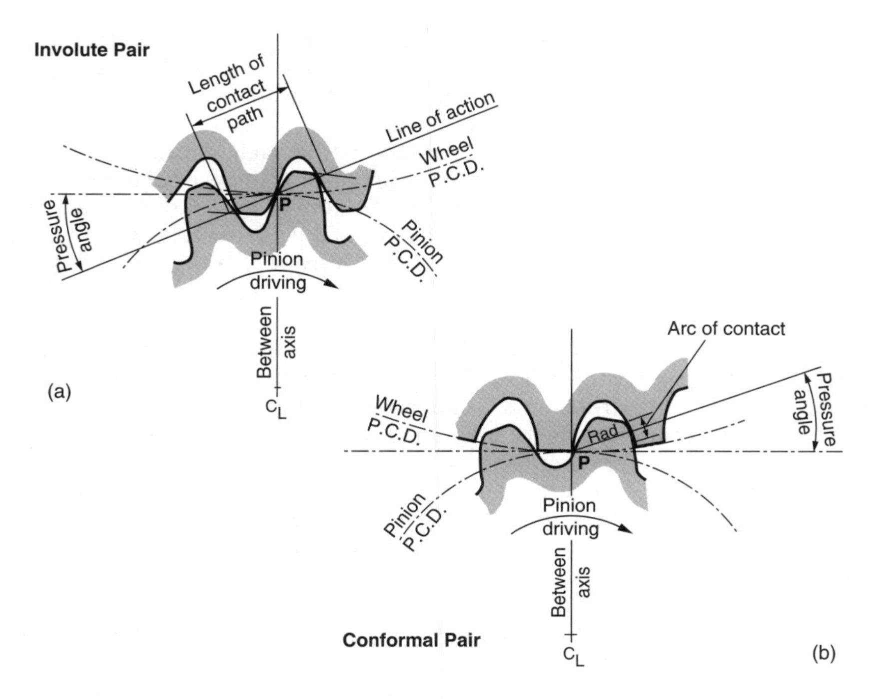

Fig. 6.32 At (a) the conventional gear tooth is convex and has a small contact patch. At (b) the conformal gear has one wheel with concave teeth and one with convex teeth. This increases contact area and reduces the distortion of the tooth allowing higher loading.

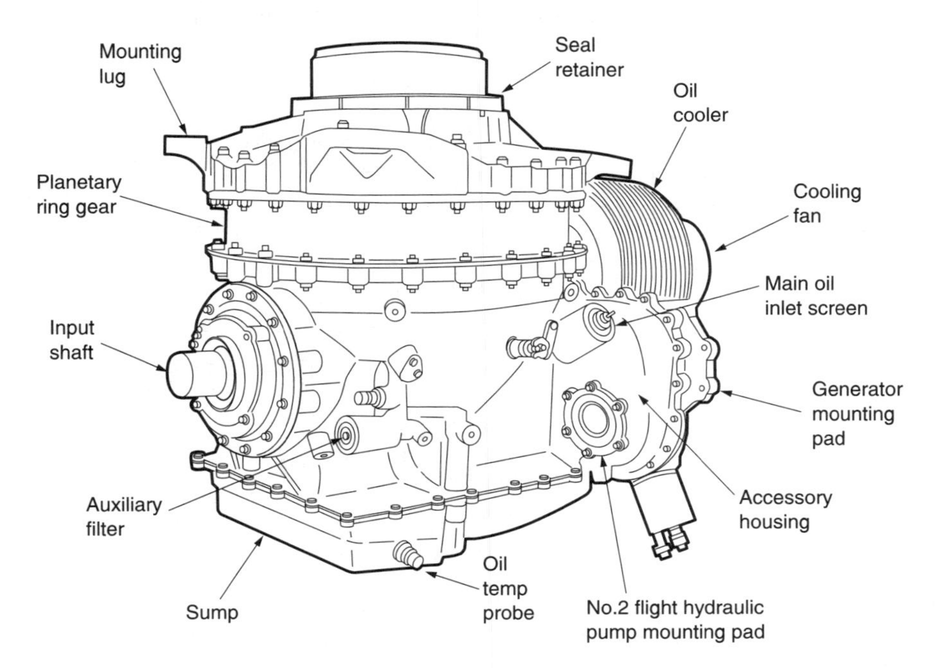

Fig. 6.33 The epicyclic or planetary gearbox of a Chinook, showing the outer ring gear.

tank forms part of the engine air intake so that de-icing is available with no additional power required. The gearbox usually has a temperature probe which drives a gauge in the cockpit. It is absolutely vital that the main gearbox is filled with the right kind of oil. EP oil generally has a sickly smell of almonds.

The gearbox will be fitted with a sight glass so that the oil level can be seen from the outside. Some machines have a chip detector. This consists of a pair of electric contacts that are also magnetic. These are situated at the bottom of the gearbox. If a gear loses part of a tooth, or a bearing starts to break up, the debris will be attracted by the magnet and will bridge the contacts, turning on a warning light in the cockpit. Some machines may also be designed so that the transmission can continue to function for a short period even if all of the oil is lost.

As a helicopter rotor gets larger, it will turn at lower RPM as the tip speed tends to remain fixed. The larger rotor will need more power but at a lower RPM. Thus the torque delivered by the gearbox increases disproportionately with size, and may limit the maximum size of conventional helicopters. The twin-rotor helicopter has an advantage which is that for a given disc area the rotor radii will be smaller and the RRPM higher. This reduces the torque needed and the numerical gear ratio, both of which lighten the gearboxes to the extent that the two gearboxes may be lighter than the single gearbox in a conventional machine of the same weight.

Rotor forces must be carried into the hull by a suitably reliable thrust bearing. In many helicopters, the main rotor is fitted directly to the transmission output shaft so that the rotor thrust is carried by the transmission bearings and the gear case. As an alternative, the rotor is carried on a bearing attached to the hull structure. In this case the transmission only provides drive torque and the drive shaft is freed from flight loads. It is advantageous to transfer rotor forces into the hull as directly as possible. An offset CM causes alternating bending loads in the shaft but constant loads in the hull. A further advantage of the use of a separate thrust bearing is that the transmission can be replaced without removing the rotor. Figure 6.34(a) shows the arrangement of

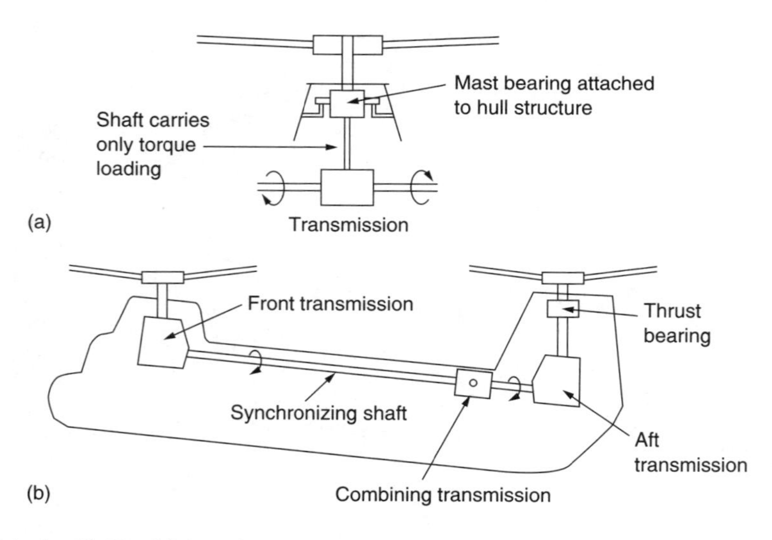

Fig. 6.34 In the AH-64 at (a) the main rotor bearing is attached to the hull and the rotor shaft does not carry the weight of the machine. In the rear rotor of the Chinook, (b), a thrust bearing at the top of the fin carries flight loads.

the AH-64 Apache. In the Chinook, shown in Figure 6.34(b), the front rotor thrust is carried by the transmission, whereas the rear rotor has a thrust bearing at the top of the fin. The drive shaft is fitted with a splined coupling so that hull flexing does not stress the transmission.

6.28 Multi-engine transmissions

There are a number of reasons for installing more than one engine in a helicopter. One obvious result is increased power, but clearly this is only obtained if all the engines are working. Another reason may be to increase safety. If one engine fails, the other(s) will continue to provide some power. Another possibility is to improve range or economy by shutting down one engine in cruise.

In a multi-engine helicopter, each engine will have its own one-way clutch so that the loss or seizure of one engine does not prevent the transmission turning. In most cases twin engines are fitted for safety reasons. Single engine helicopters are not permitted to fly over built-up areas. Ideally in the case of an engine failure the machine would be able to continue normal flight. However, this would mean that each engine would have to deliver the same power as a single engine. In the case of turbine engines this is very inefficient because in normal flight both engines would be delivering only half their rated power. However, the power needed to drive the compressors would be twice the case for a single engine and this would impair the fuel consumption.

In practice a helicopter only needs full power for a short time, typically at take-off with a full fuel load and landing at a hot/high destination. For the rest of the flight less power would be acceptable after a failure by using lower speeds and reduced rates of climb. Turbine engines are very reliable and failures are relatively uncommon. Consequently instead of overengining a helicopter, it is more sensible to design engines that can be overrated for short periods of time. Thus in addition to the continuous power rating an engine would have a higher 'contingency' rating which it could only tolerate for a few minutes. There may also be an even higher emergency rating that might only be sustainable for half a minute. If an engine enters one of these conditions, an indicator operates which can only be reset on the ground and a timer runs to measure the degree of overload.

The use of these contingency power ratings in some cases may put the engine under such stress that it will need immediate overhaul, but the frequency with which this happens is so low that the saving in fuel when the engines are working normally is of more consequence.

Figure 6.35 shows how a twin-engine helicopter might take-off from a rooftop helipad in a built-up area. The take-off is conducted upwards and backwards at first so that the helipad remains in the pilot's view. If, during this initial climb, an engine fails the pilot has the option of returning to the pad and the time for which contingency power is needed would be quite short. The height reached in the initial climb has to be such that if an engine failed just as the machine moved into forward flight it could use the power of the remaining engine and the power obtained by losing height to reach minimum power speed without falling below the height of the pad or nearby buildings.

If another engine of the same type is fitted to a single engine helicopter, the safety aspect will be improved, but the fuel economy will suffer because the losses of two compressors are being borne. At a sufficient height, one engine could be shut down so that the other runs more efficiently. In the event that the running engine fails, provided the second engine can be started promptly, flight could continue to a convenient point.

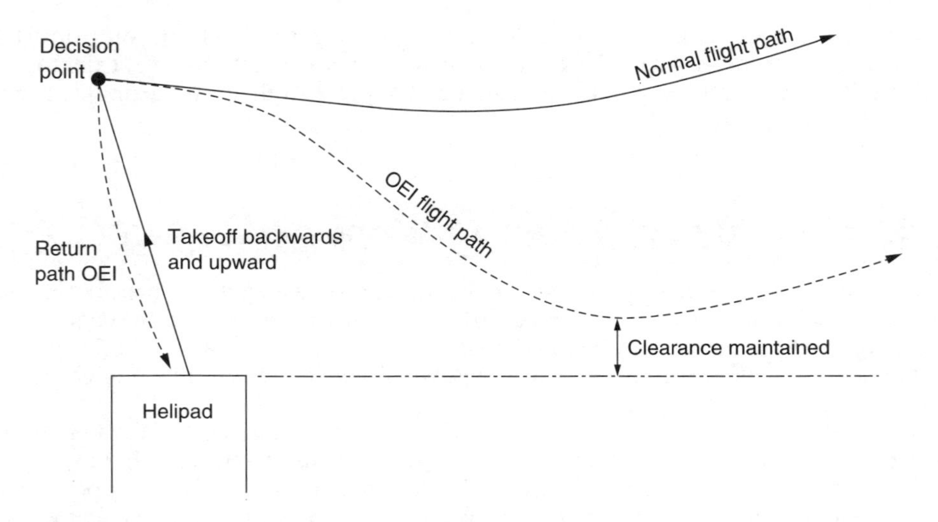

Fig. 6.35 Take-off profile from an urban helipad. By taking off upwards and to the rear the pilot can see the helipad and return to it in the case of an engine failure below decision height.

Clearly before descending below a certain height the second engine would have to be started. The certification process would have to include determination of the critical height and the demonstration of reliable restarts.

One important point is that the redundancy of twin engines is only achieved if they are independent in every respect so that no single failure could stop both. Whilst an engine might fail because of a non-violent problem such as a broken compressor pressure-sensing pipe, it should be borne in mind that turbine engines also fail violently through blade disintegration on occasions. If debris from a failed engine can damage the other engine then it too might fail. In a civil helicopter it would be adequate to have a barrier between the engines or engines designed to contain shed blades. In a military machine, the engines may also expect damage by enemy fire and in this case a significant physical separation is required so that an explosive loss of one engine does not affect the other. The pod-mounted engines of the Apache and the Chinook are good examples of this philosophy.

Figure 6.36 shows the transmission of the Westland Sea King. This twin-engine machine has hydraulically folded blades but no APU. There is one overrun clutch for each engine, but one engine has an extra clutch allowing it to be started and run with the blades folded so that accessory power is available. The accessories are driven through one-way clutches from one engine and from the main transmission. With the first engine running, hydraulic power is available to unfold the blades. The second engine is then started and used to bring the rotors up to speed. When this has been done, the clutch of the first engine can be engaged so that both engines can drive the transmission.

6.29 Transmission instruments

A light helicopter may have no transmission instruments at all and ensuring the correct oil level is all that can be done. In most piston-engine machines, the transmission can withstand more torque than the engine can deliver and so it is difficult to overstress the

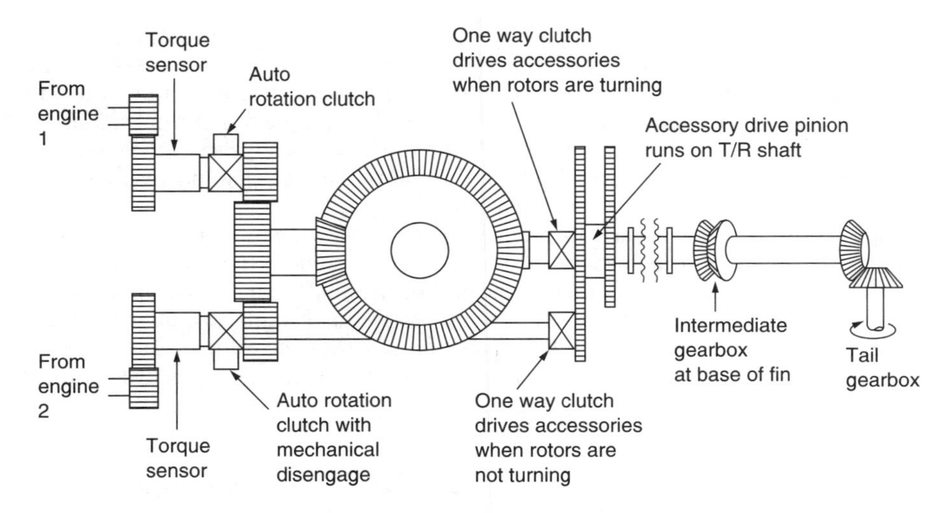

Fig. 6.36 The Sea King has an extra clutch so that one engine can be started without the transmission turning so that power is available to unfold the blades.

transmission. As helicopters get larger, the transmission becomes more complex and under more stress and instruments become necessary.

The turbine has vast power reserves and a speed governor. If the pilot applies a large amount of collective pitch during a high inflow condition, the induced drag on the rotor will be at its greatest. This would tend to slow down the rotor and thus the free turbine, but the governor responds by increasing fuel flow so in fact the result is an increase in engine torque not a reduction in RRPM. This is fine for the rotors but it is less so for the gearbox where a torque overload could distort the casing and misalign the gears or break through the oil film separating the gear teeth. To help prevent this, turbine helicopters are fitted with torque meters to display the gearbox torque on the instrument panel. The gauge will be red-lined at the safe limit for the gearbox and the pilot then should limit the severity of manoeuvres to stay within the torque limit. Helicopters with hingeless rotors may also be fitted with instruments to display the bending stress in the mast due to the application of cyclic control.

The heat dissipated in the gearbox will be roughly proportional to the torque and so a transmission oil temperature gauge would tend to reflect the torque meter history. An excessive transmission oil temperature is the cause for concern because this may permit metal-to-metal contact with serious consequences. If the high temperature is due to known heavy use, the machine must be flown at reduced power to contain the temperature. If the high temperature cannot be accounted for in this way, then there is a possibility of a problem such as oil loss, a blocked cooler or loss of airflow through the cooler and investigation will be necessary.

Gearboxes are commonly fitted with a chip detector. This is a plug fitted in the oil sump carrying insulated magnetic electrodes. In the event of any ferrous debris finding its way into the oil, the magnetic electrodes will attract it and the debris will bridge the electrodes, completing a circuit and lighting an indicator.

There are two basic types of torque meter. In the first type, engine torque twists a slim shaft in the drive train, and the amount of twist is measured with a set of strain gauges. These are fine metal wires embedded in an insulating material. Four such strain gauges are mounted in a diamond-shaped pattern on the shaft and connected electrically as a bridge. Power is applied to the ends of the bridge and the outputs are taken from the centre to a differential amplifier. As the shaft temperature increases, all of the strain gauges will increase in length but this will have no effect on the differential signal as all four gauges are equally affected. However, if the shaft is twisted, two of the gauges will contract and two will extend, producing a differential signal. The power and signals are communicated using rotary transformers.

In the second, the end thrust on one of a pair of skew gears is measured. One of the shafts is allowed to move axially. As it does so it reduces the area of a port that allows oil to escape from a chamber at the end of the shaft. The oil is supplied under pressure from the engine oil pump, through a restrictor. Engine torque will slide the shaft into the chamber until the restriction of flow causes the oil pressure to rise and oppose any further motion. The oil pressure is now proportional to the torque and can be displayed on a gauge.

6.30 The helicopter rev counter

The rotor RPM must be maintained at the correct setting during all modes of flight since the response to the controls and the available lift are both affected by it. Low rotor RPM makes the controls sluggish and reduces tail rotor authority. Vibration control techniques employing tuning only operate at one RRPM.

The rotor rev counter is one of the most vital instruments in the helicopter. It is generally concentric with the engine rev counter. Each instrument is independent and has its own scale and needle, but the scales are graduated so that when the engine is driving the rotors the two needles overlap and move together. In a piston engine machine, the engine rev counter is crankshaft driven, whereas in a turbine engine the RPM of the power turbine (N_2) is measured. The rotor rev counter is driven by the transmission after the autorotation clutch.

Figure 6.37 shows some typical displays. At (a) both engine and rotors are stopped. At (b) the clutch is disengaged, and the engine has been started and is warming up. At (c) the engine is warm and the pilot is starting to engage the clutch. At (d) the clutch is fully engaged. The needles are said to be *married*. At (e) the pilot has advanced the throttle to flight RPM with collective fully down. At (f) the throttle is closed from flight RPM and the rotors continue turning whilst the engine slows down. This action tests the autorotation clutch. At (g) the machine is in normal flight. At (h) the pilot has let the revs drop and the bleeper has sounded. At (i) the pilot has initiated a practice autorotation. The engine has dropped to idle, but the rotors maintain flight RPM. This is called '*splitting the needles*'. At (j) the machine flares in autorotation. Rotor revs have gone up to maximum permissible. If the pilot wishes to abort the auto and resume powered flight, he must not advance the throttle at this stage, because the high rotor revs will allow the engine to overspeed. The correct procedure is to reduce the rotor revs first, and then to bring the engine revs up slowly so that the needles meet at flight RPM as shown at (k). At (l) is shown the result of an engine failure. The engine has stopped and the rotors are slowing down. It is necessary to enter autorotation. Since the rotors are turning at normal speed, an in-flight restart does not require the clutch to be disengaged. Not shown is a condition resulting from transmission failure between the engine and gearbox. The rotor slows down and the engine may overspeed as the correlator attempts to maintain RRPM. It will be necessary to enter autorotation and shut off the engine.

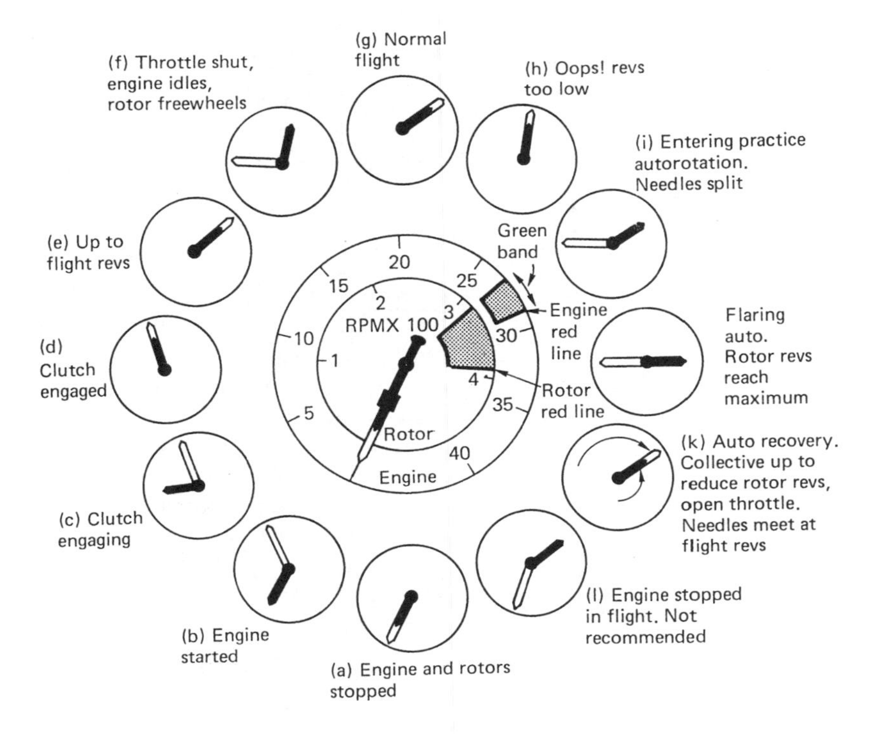

Fig. 6.37 Displays which may be seen on a coaxial helicopter tachometer. See text for details.

6.31 Tip jets

The use of thrust directly applied at the blade tips has an obvious appeal to designers because the conventional transmission is eliminated along with its weight and torque reaction. The tips of a helicopter blade move at a high subsonic speed and jet thrust matches this kind of application well. Despite those advantages, the use of tip jets remains uncommon because they introduce other problems instead.

Figure 6.38 introduces the various options available for tip jet drive. At (a) is the tip-mounted rocket in which fuel and oxidant are supplied along the blade. These react in the rocket motor to produce a vast increase in volume. At (b) is the ramjet and its cousin the pulse jet which take in air locally so that only fuel needs to be supplied along the blade. The tip-mounted turbojet is in the same category. At (c) is the pressure jet system in which compressed air produced in the hull is ejected at the blade tips. This is known as a cold-cycle system. At (d) in the tip-burning system fuel is piped down the blade and burned in the compressed air supply to produce greater thrust. At (e) the exhaust of a turbine engine is piped down the blades. This is a hot-cycle system.

Figure 6.39(a) shows that the ramjet or athodyd (aerothermodynamic duct) is a simple device in which air entering at the front is compressed by virtue of the forward motion through the air. This dynamic pressure is enough to overcome the pressure due to combustion so that exhaust gases are ejected from the rear to produce thrust.

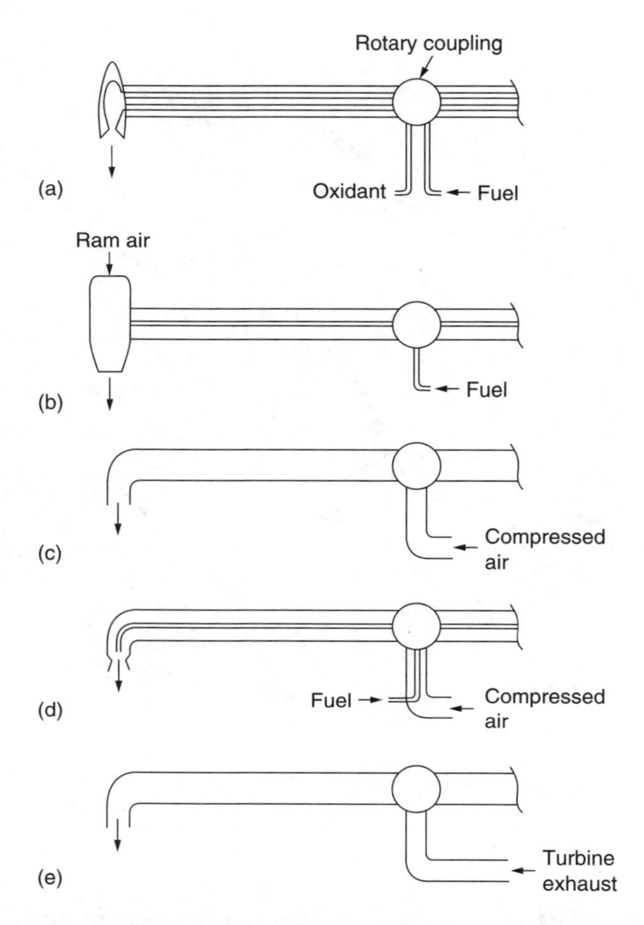

Fig. 6.38 Various options for tip jet propulsion. All of these eliminate torque reaction. At (a) the tip rocket is fed with fuel and oxidant down the blade. At (b) the ramjet, pulse jet and turbojet all take in air and only need a fuel supply. At (c) the cold-cycle pressure jet system pipes compressed air to the blade tip. At (d) the tip-burning system. At (e) the hot-cycle system feeds turbine exhaust down the blade.

The ramjet is not self-starting and in a helicopter application the rotors must be brought to an appreciable tip speed before the ramjets can be started. This might require an electric motor or an auxiliary piston engine. The McDonnell Little Henry experimental helicopter used ramjets, as did Stanley Hiller's Hornet.

Figure 6.39(b) shows that the pulse jet is an intermittent combustion device. Forward motion or a supply of compressed air is needed to start. Air enters the combustion chamber through a flap or reed valve and is mixed with fuel and ignited. The pressure increase shuts the reed valve which causes the combustion products to be ejected rearwards, producing thrust. However, when the thrust cycle is over, the momentum of the exhaust gases in the tailpipe causes a pressure drop in the combustion chamber. This causes the reed valve to open, admitting more air for the next cycle. An engine of this type was used in the Fieseler Fi-103 flying bomb of World War II; an early cruise missile aka the V-1.

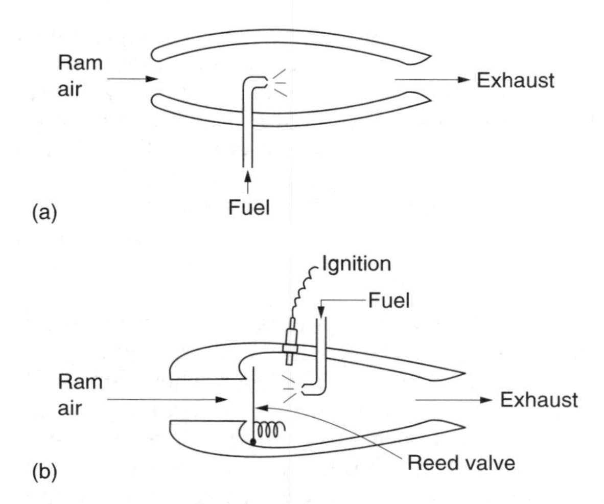

Fig. 6.39 (a) The ramjet operates using dynamic pressure from travelling through the air at high speed to achieve compression. It thus has no moving parts. (b) The pulse jet uses a non-return valve at the inlet to contain combustion pressure.

Hiller buried the mechanism of a pulse jet inside the length of a rotor blade with the fuel inlet and valves at the hub and an angled nozzle at the tip. This was known as a powerblade. Hiller also developed valveless pulse jets in which the gases in the engines resonated like organ pipes. This overcame the tendency of the valves to malfunction because of the high-*g* environment. Needless to say pulse jets and sonic engines also produce phenomenally high sound levels that preclude their use today.

In the pressure jet system the hull contains an air compressor. This may be driven by piston engine, or it may be the compressor stage of a modified turbine engine. The compressed air is fed through a rotating seal into the rotor and ejected rearwards at the tips. This makes much less noise than tip-burning or tip-mounted motors. The French SNCASO Djinn used this system and reached production.

An alternative approach to pressure jet drive is to obtain the jet pressure chemically. Machines have been built which are propelled by the catalytic decomposition of hydrogen peroxide. If liquid hydrogen peroxide is passed over a catalyst such as calcium permanganate or potassium chromate, it spontaneously decomposes into steam and free oxygen. The noise level of a rotor driven in this way is very low, but the running cost is fantastic and hydrogen peroxide is corrosive and prone to explode if impurities are present.

The Doblhoff tip-burning helicopter used a piston engine in the hull driving a supercharger to provide compressed air for the tip motors. Rotor power was only used for hovering. Once forward flight was established, the machine was driven by a conventional propeller and the rotor would windmill like an autogyro. Doblhoff's machine mixed the fuel with the air before it entered the blades. Whilst this eliminated the need for fuel piping in the blades, it meant that each blade was full of potentially explosive mixture. In the Hughes XH-17 flying crane the safer route was taken and fuel was piped to the tip burners.

The tip jet helicopter has not achieved wide use for a number of reasons. One of the most serious of these is noise. The pulse and ramjet are never going to be silenced, especially as allowable noise levels are falling. Pulse and ramjets are not fuel efficient at the speeds of helicopter blades, as they cannot reach a high enough compression ratio with only dynamic pressure at the intake. Their fuel consumption is about ten times that of conventional engines, which outweighs the fact that they will burn almost anything.

Hiller's powerblade was an unsolvable compromise because low drag required a slim blade section too small to contain an efficient engine. The additional constraints of the hot-cycle approach make blade design harder because the blades have to withstand high temperatures as well as flight loads. Tip-mounted motors such as pulse and ram-jets create enormous drag when the motor stops. This makes the rate of descent in autorotation rather too fast for comfort and as a result no machine of this type was ever certified.

Pressure jet rotors don't have a serious noise problem and their autorotation performance isn't impaired. However, the frictional losses of ducting the air to the blade tip are serious, with the result that the overall gain when compared to a conventional transmission is small. Finally, although there is no torque reaction, tip jet helicopters may need a tail rotor in order to meet crosswind hover requirements.

Hiller has correctly pointed out that in order to construct a very large helicopter, with a rotor diameter measured in hundreds of feet, the transmission weight would be intolerable with a conventional design. Instead Hiller proposed the installation of turbojet engines at the blade tips. With large, very slow turning rotors, the *g* force at the blade tips would not be serious and a turbojet engine would work well there.

6.32 The electrical system

The electrical system is needed for engine starting, lighting, instrument power and to operate the avionics. When the engine is running, electrical power comes from the alternator. When the engine is stopped, the battery will be used. Most light helicopters use a nominal 12 volt battery which is constructed from six 2 volt lead–acid cells in series. Some machines have nickel–cadmium batteries and these will use ten 1.2 volt cells in series. Late Enstroms and JetRangers have 24 volt systems.

A lead–acid battery only has a terminal voltage of 12 volts when it is supplying a light load. In order to charge the battery, the terminal voltage must be raised to about 14.4 volts. Current will flow into the battery instead of out of it, and it will recharge. Conveniently the battery voltage rises slightly as recharging is completed, reducing the charging current automatically. If the applied voltage is maintained constant, the battery will take current until it is recharged, when the current will fall to a trickle. A lead–acid battery evolves hydrogen and oxygen when it is recharging, and provision must be made to vent these gases to avoid explosion risk. In the Enstrom the battery is beneath the starboard seat in the cockpit. Charging a flat battery in the machine is forbidden since it could fill the cockpit with an explosive mixture. The level of the electrolyte must be checked periodically.

A nickel–cadmium battery has different charging characteristics. When it is fully charged, the voltage does not rise very much, and charging current continues to flow. This current is converted to heat instead of charge. The battery temperature is sensed and displayed on the instrument panel. If the battery temperature reaches an excessive value, the alternator is switched off until some cooling takes place. An advantage of Ni–Cd batteries is that they do not outgas or need topping up.

The regulator controls the system voltage when the engine is running and the alternator is switched on. This may be a separate unit or integral with the alternator. An alternator (Figure 6.40) consists of a rotor (3), driven by the engine, which can be magnetized by direct current flowing through a coil. This field current is fed in through sliprings (2) and carbon brushes. The rotation of the resultant magnetic field causes flux reversals in the stationary magnetic circuit or stator surrounding the rotor. A winding (4) on the stator will produce an alternating voltage. This cannot be used directly in a system using a DC battery, but must be rectified using diodes (5). Clearly rectified alternating current is very irregular, and in practice a more constant current is obtained by mounting three sets of coils on the stator at 120°. This results in a three-phase alternator and if the output is rectified with six diodes as shown in Figure 6.40 the voltage is much more constant. The diodes are usually incorporated into the alternator body so that they may be cooled by fan.

The output voltage is proportional to the field current. The voltage regulator (1) senses the system voltage, and if it is excessive, the field current is reduced. If it is too small, the field current is increased. As a result the system voltage will stabilize at a fixed value even if different loads are switched on and off. A voltage regulator failure could result in full field current being applied. In this case the alternator output current would be limited by the inductance of the windings. The battery would absorb much of this current and eventually boil dry. Some machines have an overvoltage sensor to cut out the alternator if this happens.

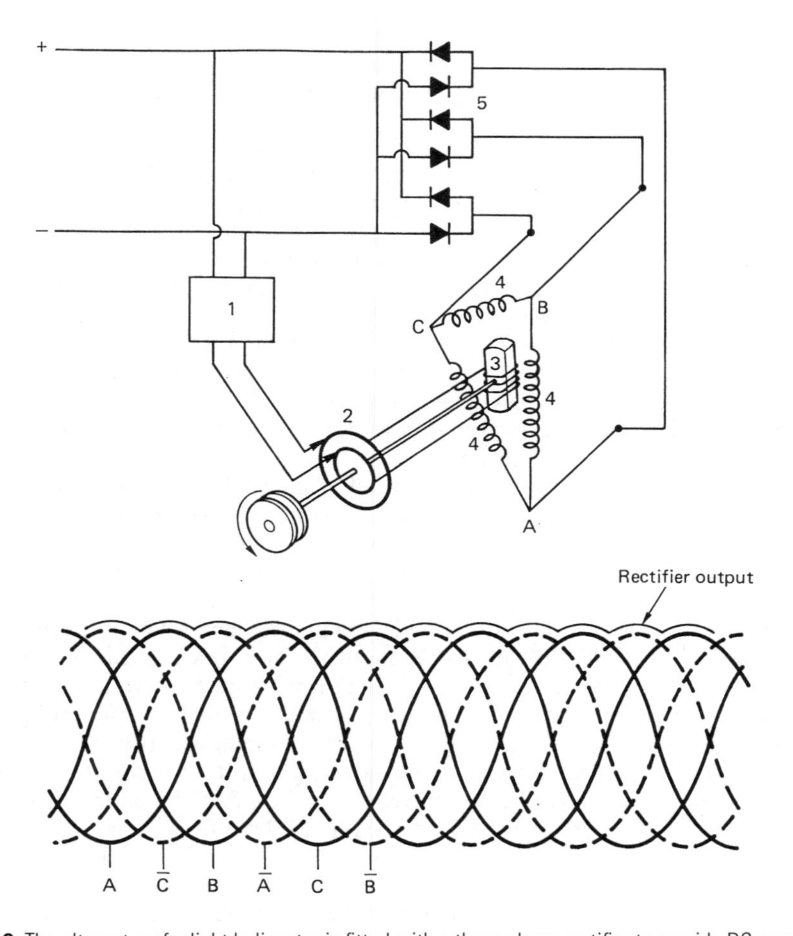

Fig. 6.40 The alternator of a light helicopter is fitted with a three-phase rectifier to provide DC power.

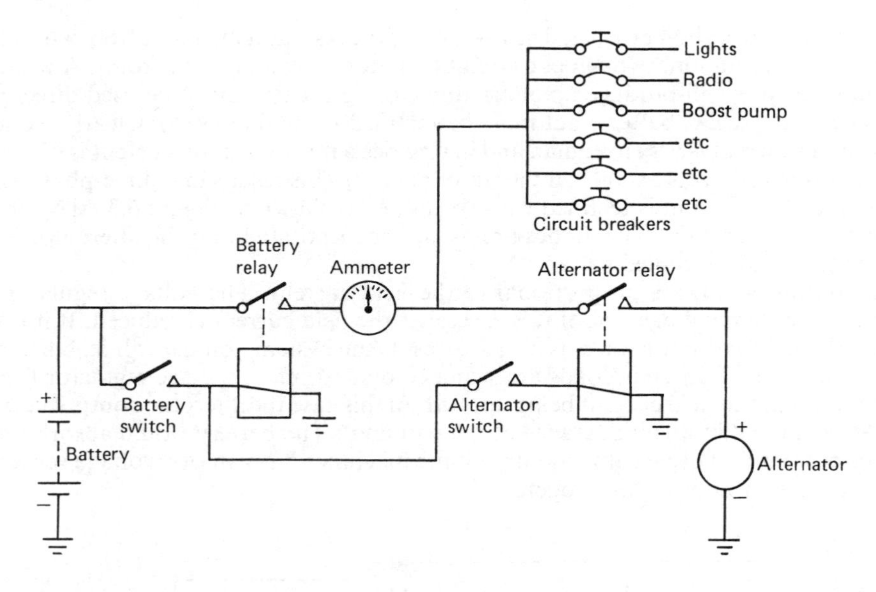

Fig. 6.41 Electrical system of a light helicopter showing the position of the ammeter and the master switch.

The state of the battery and alternator can be assessed using an Ampere meter, always abbreviated to ammeter, which is connected as shown in Figure 6.41. When the engine is stopped, electrical loads are supplied by battery, and current flows out through the ammeter which will register discharge (−). When the engine is started, the battery will have been partially discharged by the starter motor, and current will flow back from the alternator to charge the battery, registering charge (+). As the battery becomes recharged the magnitude of the charging current will fall and the ammeter will return to the null position where it will remain. If the ammeter displays a discharge with flight RPM, the alternator is not supplying enough current, it is switched off or some load has been applied beyond its capacity. A common explanation is that the belt driving the alternator has broken or is slipping.

Figure 6.41 also shows that the master switch isolates all electrical devices from the battery (except the clock). In some cases the total electrical load is too large for a switch and a relay or contactor is used. This is a small version of the starter motor solenoid and is controlled by a small current from the switch itself. Generally only one wire feeds current to each device and current returns through the metal frame of the machine.

If a failure of the insulation occurs somewhere in the electrical wiring, current could flow direct to the frame with little resistance to stop it. The high fault current could cause a fire in the wiring. This is prevented by splitting the wiring into sections and protecting each with a fuse or circuit breaker. In the case of a fault, the section containing the fault will carry an excessive current that melts the fuse or magnetically trips the breaker. The current is cut off and the fire is prevented. Equally importantly, the supply is maintained to all other sections; only devices using current from the affected section will lose power.

A fuse is cheap and effective, but it has to be replaced if it fails. A fuse should only be replaced with one of the same rating. If a fuse is replaced with one of a higher rating, the protection it gives is lost and the wiring could burn in the case of a fault. A circuit breaker pops out when tripped and can be reset by finger pressure. In addition breakers can be used as switches by pulling them out. A tripped breaker does not always indicate a fault. Some loads, particularly electric motors and electronic equipment, draw a surge current when first switched on which is higher than their running current. If the surge current is close to the trip current, once in a while the breaker will trip when a particular load is applied. The breaker can simply be reset, and if all is well it will not trip. If there is a genuine fault, the breaker will refuse to reset and it should be left tripped and the cause investigated. If a breaker trips in flight, the appropriate action depends on how necessary that circuit is. Loss of the engine instrument power calls for a landing at the next airfield for repairs, whereas loss of the landing light on a summer morning is less urgent. Nevertheless all electrical problems should be treated with the utmost suspicion in case the apparent fault is a symptom of something else. The ammeter is a good friend at times like these. If the ammeter is showing a null reading, a heavy fault current cannot be flowing and there is unlikely to be a fire risk.

Some machines have provision for a ground power source to be used. This would ease the load on the battery when starting in extreme cold, for example. It is important that the ground power unit is compatible. Some aircraft use 24 volt DC electrical systems, and if a 24 volt ground supply is plugged into a 12 volt machine extensive damage will result.

In larger machines, AC power systems are used. These will operate at 400 Hz rather than the 50 or 60 Hz land-based power because the mass of components can be reduced as the frequency rises. Alternators may be driven from the engines or the transmission. As the alternator output frequency is proportional to shaft RPM, system frequency will vary with RRPM. Alternators will usually disconnect themselves if the shaft speed falls below about 85% of nominal.

6.33 Hydraulic systems

As helicopters become larger, heavier and faster, the forces necessary to control the rotor head become too great for the pilot to manage reasonably, and some form of power operation is necessary. If stability augmentation or autopilot functions are required, powered controls will also be needed so that the control information can fly the machine.

Control forces may come from a variety of sources. The advancing blade may bend back putting a significant area at some distance from the feathering axis. This will feed feathering loads back into the swashplate. The advancing blade tip can enter the region of compressibility and this may also result in a pitching load. It was shown in Chapter 3 that rotor blades tend to return to flat pitch and a significant thrust is needed in the control system to obtain positive collective pitch. The enormous tensile forces in the blade root require adequately strong feathering hinges. These inevitably must be stiff to move. In larger helicopters, power operation will be required beyond the flying controls. Winches, cargo doors and ramps, underslung load release, wheel brakes and steering may all be powered.

There are two basic types of hydraulic systems: fully powered and power assisted. In large helicopters the controls are fully powered. The consequences of control loss are serious and real systems have to be designed so that failure of any one part (and often more) still leaves at least some measure of control. Many systems have a completely duplicated hydraulic system powered by two pumps driven independently from the transmission. Whatever happens to one hydraulic system, the other should remain functional. In addition hydraulic systems may be interconnected by motor/pump combinations known as power transfer units. In the case of a main pump failure, the power transfer unit driven by the remaining system can pressurize the failed system.

In smaller helicopters the controls are power assisted and in the event of hydraulic failure the pilot can still operate the controls, albeit with greater effort. The JetRanger, for example, uses a single power assisted control system and can be flown without it, whereas the CH-47 has duplicated fully powered controls. This section first considers hydraulic power principles and uses the machines mentioned as examples of actual practice.

There are many ways of obtaining power operation, but the high pressure hydraulic system has the advantage that large forces can be developed in compact actuators. The linear action of hydraulic rams is easy to integrate into real mechanisms. In hydraulic systems, losses are dominated by viscosity. The work done by an hydraulic ram is the product of the pressure, the piston area and the travel, whereas the volume of fluid used is the product of the area and the travel. As the flow is inversely proportional to the pressure, it follows that the smallest flow losses will be experienced when the highest practicable pressure is used. This is limited by the availability of sealing materials. A pressure of 3000 pounds per square inch is not unusual.

Simple hydraulic pumps can be made with meshing gears, but higher flow hydraulic pumps use pistons. Figure 6.28 showed a swashplate fuel pump. The same approach can be used for hydraulic systems. The pistons are fitted with ball-jointed slippers that contact the angled swashplate. As the pump body turns, the pistons oscillate with amplitude controlled by the swashplate angle. The swashplate may be tilted with a pressure-sensitive actuator. As target system pressure is approached, the eccentricity of the swashplate is reduced. This is more efficient than a fixed delivery pump with a relief valve used in small systems.

If hydraulic pressure is supplied to a pump of this kind it will act as a motor. The swashplate can then be fixed. As an alternative, hydraulic motors may have radial pistons acting on an eccentric. Hydraulic motors are used for winches and for engine starting.

Fully powered systems operate with larger forces and the losses in the system result in heating of the fluid. Excessive fluid temperature may cause boiling and loss of control in addition to damaging the seals. The fluid reservoir may have cooling fins. Alternatively the fluid circuit incorporates an oil cooler which will typically have a fan driven from the transmission. In most systems, fluid returning to the reservoir passes through a back-pressure valve which maintains the return flow at a significant pressure. This prevents cavitation in the pump, reduces the probability of boiling and prevents dissolved air from causing frothing and airlocks in the system. Any dissolved air will come out of solution in the low pressure region following the back-pressure valve and can escape into the reservoir.

Filters are used to remove any foreign bodies from the oil as these could cause seal damage and premature wear as well as malfunctioning of delicate parts such as valves.

Many systems incorporate an accumulator or hydraulic energy reservoir. An accumulator consists of a reinforced cylinder or sphere in which a flexible membrane separates the hydraulic oil from pressurized gas which is typically nitrogen. The accumulator absorbs rapid fluctuations in oil pressure, allows peak flow in excess of the pump capacity and also provides continued operation for a time after a pump failure. In the Chinook an hydraulic accumulator is used to start the APU. The starter motor acts as a pump when the APU is running.

In almost all cases each hydraulic pump will feed a number of actuators. A single leaking pipe or actuator would allow all of the oil to be lost and cause all of the actuators to fail. A number of safety measures can be incorporated. First, the oil reservoir is fitted with a float switch to warn the pilot if the oil level is falling. This would happen if a pipe fitting was 'weeping' oil slowly. An oil temperature gauge may also be fitted. Oil loss will result in the remaining oil circulating more often and this will result in a temperature rise. If a seal fails inside an actuator there may be no net loss of oil, but there will be a short circuit from the high pressure feed to the return and this will also result in heating.

Sometimes hydraulic systems fail catastrophically because of a burst hose or battle damage. The solution here is to install the hydraulic equivalent of fuses. Each actuator is fed by an independent pipe run from the pump manifold. This manifold is fitted with flow-sensitive valves. These contain a spring-loaded ball that normally allows fluid flow. In the case of a burst, the violent flow due to escaping fluid will push the ball onto its seat causing it to seal off the leaking pipe. The pressure in the leaking pipe will then be lost and the pressure differential will hold the ball valve shut indefinitely. The back-pressure valves will prevent oil loss via the return pipes. In this way a burst pipe or hose only affects one actuator and the others remain operational because total fluid loss is prevented.

In the same way that electrical faults can be isolated by pulling circuit breakers, in some cases the pilot can isolate parts of the hydraulic circuit with valves so that faulty or damaged units are prevented from impairing the remainder of the system. These valves may be remotely driven by electric actuators, but a manual operating lever will be provided in case of electrical failure.

Figure 6.42 shows the basic components of a JetRanger hydraulic system having power-assisted actuators. The machine can be flown manually in the event of hydraulic failure. The hydraulic fluid is held in a finned reservoir that acts as a cooler. A sight

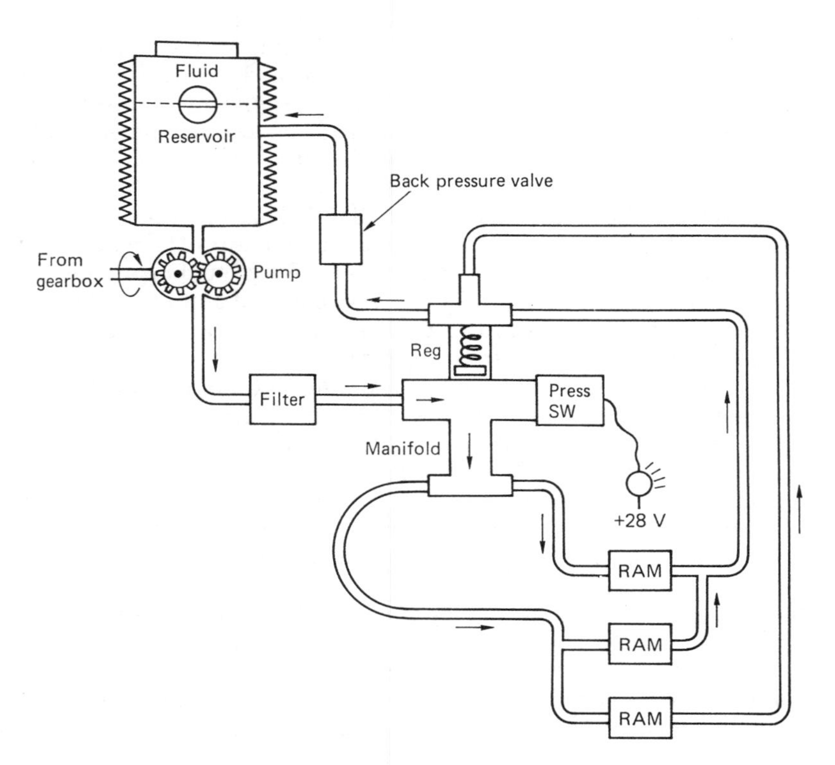

Fig. 6.42 Hydraulic system of the Bell 206 is not duplicated as the machine can be flown manually in the case of failure.

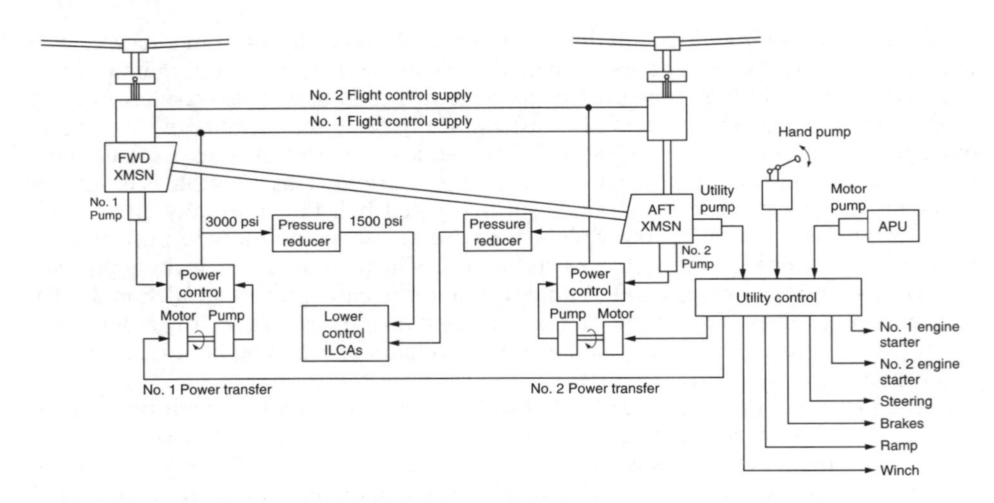

Fig. 6.43 Hydraulic system of the CH-47 is fully powered and extensively duplicated. An APU is also provided to allow hydraulic pressure to be available without the rotors turning.

glass allows the level to be checked. The reservoir feeds a constant flow pump driven from the main gearbox so that power is still available during autorotation. The pump output passes through a filter to remove contamination and is fed to the manifold. This unit contains a pressure relief valve to allow fluid back to the reservoir when the pump delivery exceeds demand. The manifold also acts as a connection point for the flow and return pipes to the rams. A pressure switch on the manifold operates a warning light should the pressure fall.

Figure 6.43 shows the hydraulic system of the CH-47. This large transport has tandem rotors of such size that fully powered controls are mandatory. The machine has twin engines and an APU (auxiliary power unit) intended for ground operation. There are three hydraulic systems, No. 1 and No. 2 flight systems and a utility system. All three systems are permanently pressurized to prevent pump cavitation and a small hand pump is provided to allow refilling from a central location. No. 1 system is driven by a pump on the front transmission; No. 2 by a pump on the rear transmission. The APU has a motor/pump unit and there is a further pump on the aft transmission. Utility pressure is then available if the rotors are turning or if the APU is running. All three systems have variable displacement pumps driven by pressure regulators and have filters, accumulators and oil coolers. The utility system has more accumulator capacity because it is used to start the APU using the motor/pump. The APU will recharge the starting accumulator when it is running, but a hand pump is provided in case pressure has been lost.

The utility system provides all non-flight critical hydraulic power and drives the undercarriage steering, swivel centring and locking and the wheel brakes, the winch, the cargo door and ramp actuators and the cargo hook release as well as driving hydraulic motors for main engine starting. In addition, power transfer units allow the utility system to pressurize the flight systems so that these can then be operated without the main engines running or the rotors turning. A power transfer unit consists of an hydraulic motor driving an hydraulic pump. This allows the hydraulic fluid circuits to be kept separate.

The pilot can isolate the ramp actuators and the brake/steering circuits in case of damage or failure so that utility power remains available. The brake system contains a small accumulator so that limited brake operation is still possible after a utility failure.

The hydraulic display panel shows for each system the level in the reservoir, the oil pressure and temperature. There are warning lights for each of the four pumps and warning lights operated by excessive pressure drop across the filters.

6.34 HUMS

Health and Usage Monitoring Systems are an important development that promises improved safety and running costs. The use to which helicopters are put varies tremendously. This means that the wear and fatigue suffered are difficult to assess simply from a flight hour timer. As a result components have to be conservatively lifed so that no failures occur whatever the circumstances. This means that a machine used in an undemanding role has to replace parts that are still serviceable at the same rate as a machine in arduous service that has worn the parts out. The solution is to assess the actual conditions of service of each machine so that the degree of wear on major components can be predicted. In this aspect of HUMS, sensors are located at strategic points and the stresses on the machine are recorded.

The unacceptability of failure leads to the regular replacement of parts. However, not all parts fail instantly. Many components deteriorate gradually and do not actually fail. In others failure is preceded by symptoms that are often present for some time before the failure itself. For example, a failure may result from a crack. Cracks tend to start slowly and propagate at increasing speed as the remaining material lessens. If it is possible reliably to detect those symptoms, an early warning of a failure can be given. For example, gears and bearings make slight characteristic vibrations when they are working normally, but these will change in the case of a defect. The presence of a crack will alter the way a device vibrates. By analysing the vibrations from a gearbox, it is possible to determine if a part is failing and even to suggest which part. Given that most helicopter flights are quite short, if the vibration analysis is sufficiently sensitive the failure will be anticipated sufficiently in advance that the flight can be completed. A more sophisticated analysis could determine the rate of deterioration and suggest a precautionary landing.

7

Control

7.1 Introduction

The physical control requirement in a helicopter is no more than the application of appropriate cyclic, collective and tail rotor pitch settings. Implicit in this is the need for some form of rotor speed governing. However, the physical control requirement is only a small part of the overall process that is inevitably more complex. In order to perform useful missions in safety, the pitch settings of the rotors must be such that at all times the machine has, for example, the desired attitude, airspeed, altitude and geographical location.

In order to control the flight, information must be gathered about the present conditions, and these must be compared with the desired conditions. Any discrepancies must result in changes to the physical controls that would tend to reduce discrepancies to zero or at least as near as makes no difference. This is, of course, classical negative feedback as was introduced in Chapter 2. The essential components of this process must be information gathering, or sensing, information transmission, decision-making, or information processing, all of which takes place with small amounts of power, followed by a power amplification stage that allows the control information actually to operate the controls.

Figure 7.1(a) shows the case of a simple light helicopter. It will have instruments such as airspeed indicator, altimeter and compass whose output is purely visual. The pilot looks at the instruments and out of the canopy in order to establish the actual conditions, but has to perform all of the actions necessary to stabilize the machine and hold it on a desired course. The machine itself may actually be unstable, in that left to itself it would diverge from its original course, but the combination of the pilot and the machine can be a stable system. In other words the pilot is working out the difference between what is wanted and what is actually happening. As a result of a difference of this kind the pilot then has to operate one or more of the controls in a sense that reduces the difference. An unskilled pilot may overcontrol so that the machine oscillates about the desired course, whereas the experienced pilot will have learned the dynamics of the machine and applies just the right control inputs smoothly to bring the difference to zero.

In Figure 7.1(a) the pilot is part of a feedback loop and is not only computing the servo error, but is also mentally modelling the machine's response to avoid instability. The pilot may also be providing all of the control power with his own muscles. This task must be performed for the duration of the flight, as the machine will diverge if attention lapses for more than a few seconds.

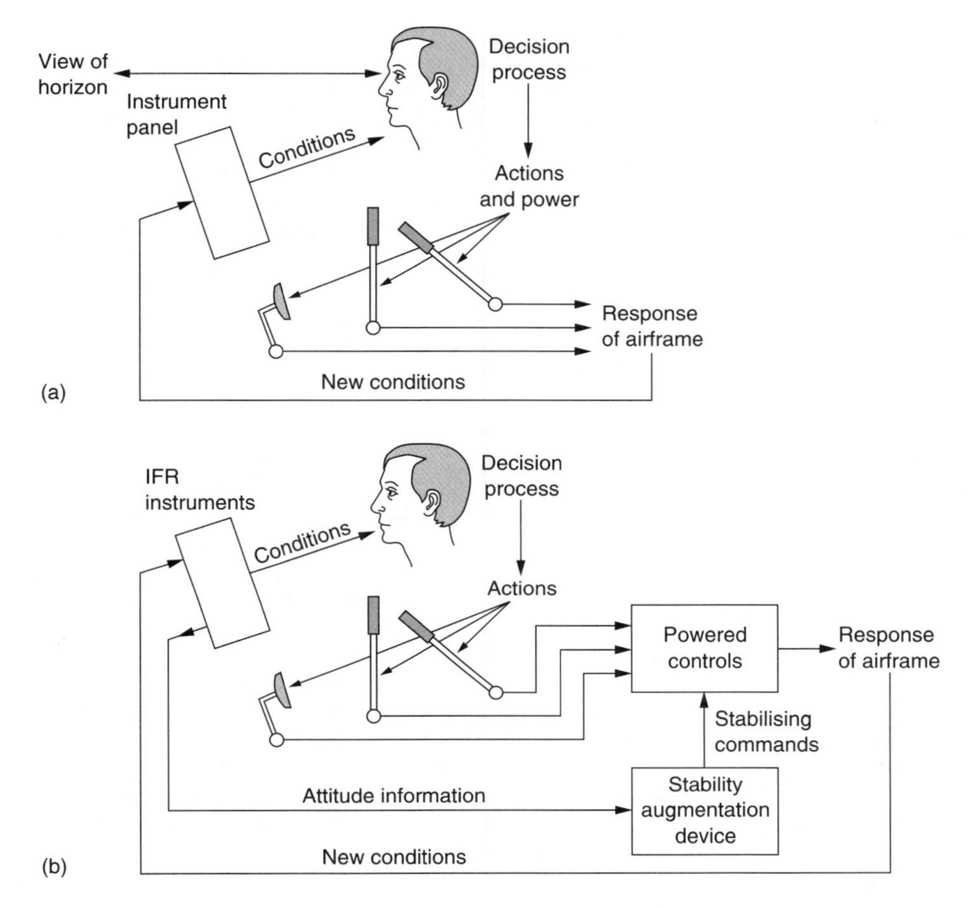

Fig. 7.1 (a) A minimal control loop in which the conditions are sensed from simple instruments and the view from the canopy. (b) Power assistance is provided for the controls, making possible the use of a stability augmentation device.

Put this way, the pilot's task sounds onerous, but it need not be so. In a simple light helicopter in good conditions this process is extremely enjoyable. However, many helicopters have to operate in poor conditions, and in addition to flying the machine the pilot may be required to perform other tasks associated with the mission. The goal of the designer may then be to make flight possible in all weathers, and/or to keep the pilot's stress or workload down to an acceptable level for the expected mission.

The pilot's assessment of the machine's attitude is assisted by the provision of instruments such as an artificial horizon and a direction indicator. These will allow the pilot to maintain the machine's attitude despite poor visibility and are essential if the machine is to be flown under instrument flight rules (IFR). The addition of power-assisted controls to Figure 7.1(a) will beneficially reduce the pilot's physical workload on long flights. IFR instruments improve the quality of information presented, but the stability of the machine is still down to the continuous concentration of the pilot. Flying a simple helicopter on instruments is possible, but difficult. Two pilots might be required, one to fly, the other to handle navigation and communications.

Once powered controls are available, it is then a small step to make them accept signals from an automatic stabilizing system that can reduce the pilot's mental workload. The stabilizing action of the pilot is augmented or even replaced by a signal processor programmed to respond to disturbances in the way a skilled pilot would. However, the signal processor has no eyes and needs information about the flight conditions in the form of input signals. Figure 7.1(b) shows that the aircraft instruments no longer just have a visual indication, but also output signals representing the current condition. It will also be seen in Figure 7.1(b) that the outputs of the signal processor can act upon the powered controls.

In such a machine there are two parallel control paths. Figure 7.2(a) shows that the control of the machine can be entirely by the pilot, entirely by the signal processor, or by a combination of both, with the possibility that the amount of contribution from each control path may also change throughout the flight. This must be the source of further complexity. It is fundamental to negative feedback that there can be only one overall feedback loop in a system. Two negative feedback loops around the same process will fight each other. For example, if the outputs of the pilot's stick and the processor are simply added, by fitting what is known as a series actuator in the pushrod, the pilot's controls cease to function. Figure 7.2(b) shows why. It is the goal of the processor to maintain the attitude of the machine. If the pilot applies, say, left cyclic, the machine will begin to bank left, but the processor will sense this as an error and apply right bank. The machine will continue to fly straight, with the processor precisely opposing everything the pilot does. If the pilot releases the stick, the series actuator may succeed in moving the stick instead of the controls.

In early systems using series actuators, engaging the autopilot must lock the stick in the neutral position to give the series actuator a mechanical reference against which to react. The pilot has to disengage the processor if he wants to resume control.

In the parallel actuator system shown in Figure 7.2(c) the actuator moves the mechanical reference of the stick centring springs. With the signal processor off, the actuator may perform the trim function. With the signal processor on, the pilot releases the stick and the actuator moves both stick and controls. This gives visual confirmation that the system is working, but more importantly the pilot can override just by moving the stick against the centring springs.

In modern systems there is a more convenient way. Figure 7.2(d) shows that the processor is permanently on, but the pilot's stick is connected differently. If the stick is released, it has no effect and the processor controls the course. However, in order to override the processor, the pilot simply moves the stick as normal. This generates a false error that is fed into the processor. For example, if the pilot wishes to bank left, he moves the cyclic stick left, and this generates a signal that is added to the processor's bank angle error to indicate falsely that there is a right bank condition. The processor acts to cancels that condition by performing a left bank.

Thus it will be seen that the way for the pilot to override a feedback system is not to attempt to oppose the output, which cannot work, but to modify the reference value or the parameter the system is trying to hold constant. In this way the system still stabilizes the override manoeuvre. For example, in the example of the left bank above, the system now acts to hold the bank angle constant. In the event of gusts disturbing the bank angle, the system would correct for them. In this way the combination of control system and real pilot is close to the ideal.

Figure 7.1(b) showed a simple *stability augmentation* system in which the goal is to reduce the naturally divergent behaviour of the helicopter as well as external disturbances from gusts using a parallel actuator. The reduced pilot workload may make single pilot instrument flying possible. Clearly the stability augmentation system must

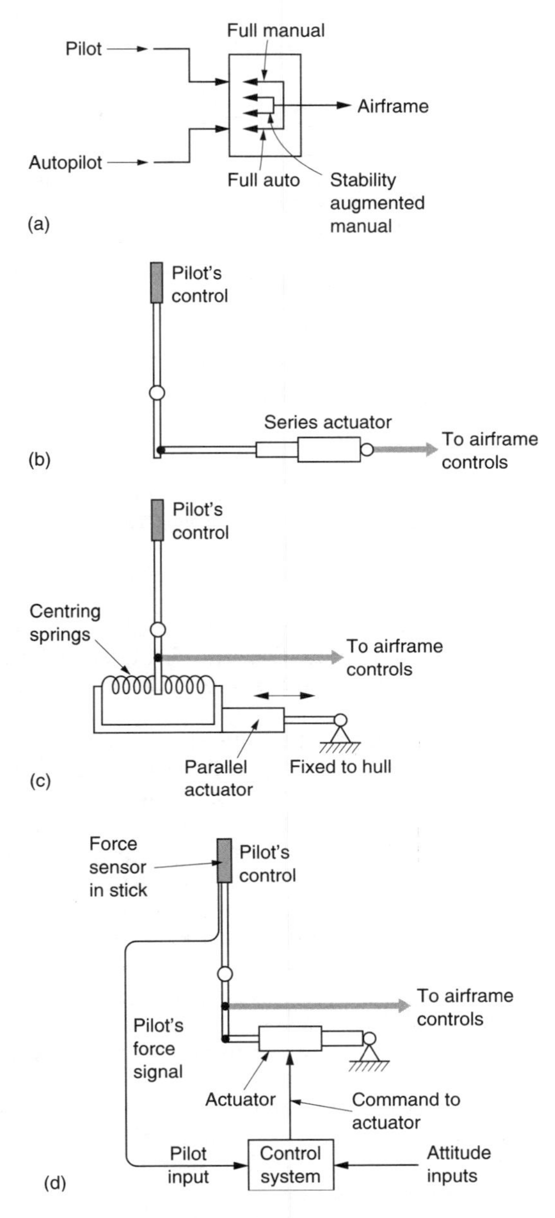

Fig. 7.2 (a) When attitude conditions from the artificial horizon can act directly upon the powered controls, the result is an autopilot. Under different circumstances, the pilot may wish to have complete control or no control or even a partial control as he wishes. The control system must be able to react to those different requirements. (b) In a series actuator system, a servo extends or shortens a pushrod from the control column. If the autopilot is on, it will simply cancel everything the pilot does. If the pilot releases the stick, the actuator may move the stick instead of the controls. Series actuators must lock the control column when the autopilot is engaged. (c) A parallel actuator system in which the actuator moves the centre reference of the cyclic stick springs. (d) Modern system in which the pilot's input produces a false error to the autopilot. By cancelling the false error, the autopilot carries out the pilot's wishes.

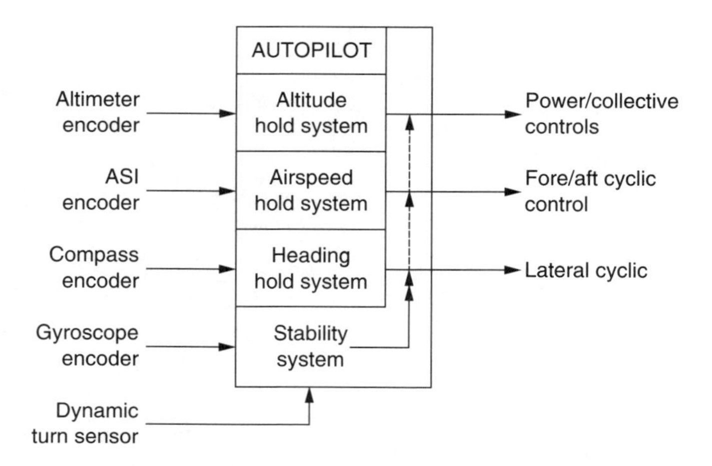

Fig. 7.3 With an autopilot, attitude, altitude, airspeed and heading can be controlled automatically, but the pilot must still select the course. With an autopilot coupled to a flight director, navigational inputs can control the autopilot so that a course can be held.

understand the dynamic response of the helicopter in order to avoid any overshooting or oscillation when correcting the attitude. The attitude of the machine is still controlled by the pilot, but the stability of the machine is improved by using attitude signals from the artificial horizon.

In Figure 7.3 the process has been taken one stage further. Here the augmented stability achieved in Figure 7.1(b) makes it possible for the machine to be controlled automatically. Inputs to the signal processor from the altimeter, the airspeed indicator and from a compass allow the machine to hold a heading without any action from the pilot. This system is generally called an automatic pilot or just an *autopilot*.

The system of Figure 7.3 relieves the pilot of the need to fly the machine continuously, but can only hold a heading and would not compensate for a change in wind strength or direction. However, with further inputs from a navigation system or from a *flight director* the machine could hold a course determined by a GPS receiver, VOR beacon or an instrument landing system (ILS). In this case the flight director creates a further feedback loop. The actual course is computed by navigational equipment and compared with the desired course. Any error is then fed to the autopilot as a modified heading reference.

In early helicopters, the provision of stability augmentation and autopilot systems was frequently optional, whereas in modern machines such systems are often designed in from the outset. One modern trend is away from instruments with a direct visual readout. Instead instruments may be used which have only an electrical output and are best called sensors or transducers. These outputs are supplied not only to the autopilot, but also to a display processor that produces a virtual instrument panel on a computer graphics-type display. This is known colloquially as ‘glass cockpit’ technology. One advantage is that the instruments themselves can be located anywhere convenient and do not have to be in the instrument panel.

In helicopters, the stability augmentation and autopilot capabilities may be extended into the hover. In this case altitude and airspeed information from atmospheric pressure is not good enough and instead RADAR will be used to measure height and ground-speed. In flying cranes and in search-and-rescue helicopters the ability to remain at a fixed point in the hover despite external disturbances is extremely valuable. In advanced search-and-rescue helicopters the pilot simply presses a button as he flies directly over the victim and the helicopter will automatically perform a 360° turn and come to the hover at exactly the same place.

A helicopter with a suitable degree of automation does not need a pilot on board if the commands he would have given the autopilot can be transmitted by radio. This makes possible a range of devices from the simple radio-controlled helicopter, which must remain within the pilot's view, to the autonomous drone that can undertake an entire mission without human intervention.

The provision of various control, stabilization or autopilot systems depends upon a number of fundamental technologies. These include the signalling of control positions from one place to another, power operation of controls, attitude sensing, altitude and airspeed sensing, parameter signalling and feedback control. The subject of safety must also arise. What happens if any of these mechanisms go wrong? These concepts will be considered in this chapter.

7.2 Flight sensors

There are four main categories of flight instrument into which virtually any device can be placed. These will be contrasted prior to a detailed discussion of each one.

(a) Heading sensing devices to display information relating to the direction the helicopter is pointing in. The most important of these is the magnetic compass that displays the magnetic heading or direction with respect to the earth's magnetic field. The direction indicator (DI) is a gyroscopic device arranged to display the same heading as the compass. It is less affected by manoeuvres than the compass and is easier to read. The rate of turn indicator, usually just called the turn indicator, is a gyroscopic device that displays the rate at which the heading is changing and the direction of the change.

(b) Height sensing indicators display the vertical distance between the helicopter and some reference. The altimeter is a pressure-sensing device whose display is calibrated in feet or metres. It is important to use the appropriate reference or the display will be misleading. The RADAR altimeter is an electronic device which times the reflections of radio waves emitted downwards from the helicopter and so displays the height above ground. The vertical speed indicator measures the rate of change of air pressure and displays rate of altitude change and direction, usually in hundreds of feet per minute. Height information is also available from GPS receivers.

(c) The airspeed indicator (ASI) is a device that measures the dynamic pressure of the air caused by forward motion of the helicopter. The scale is calibrated in knots or mph. It is also possible to measure ground velocity using Doppler RADAR and this will allow the helicopter's track to be established. In this way the heading can be adjusted to cancel the effects of wind.

(d) Attitude-sensing instruments display the attitude of the helicopter in pitch and roll with respect to the earth's gravitational field. These are commonly gyroscopic and include the artificial horizon, which displays pitch and roll on one instrument, and the simpler rate of turn display.

The above classification is of most use to the pilot since in flight one is more concerned with the readings themselves than how the instruments work internally. As well as classification by the quantity displayed, instruments can also be categorized by the internal operating principles, and this is more appropriate to a technical description. The internal operating principle can be magnetic, pressure, gyroscopic, optical or radio and these principles will be examined in turn.

7.3 The magnetic compass

The simple compass is a small freely suspended magnet that attempts to align itself with the earth's magnetic field. Figure 7.4(a) shows that the earth acts as if a large bar magnet were buried inside it. The north-seeking pole of the compass magnet (generally just called the north pole) is so called because it points to the north. As opposite poles attract, clearly the magnetic pole at the north end of the earth is actually a south pole. It is called the north magnetic pole because of where it is. Figure 7.4 also shows that as the lines of magnetic force return to the poles, they angle sharply into the ground. This is called magnetic dip and in the UK lines of force enter the ground at about 65° to the horizontal. A simple bar magnet suspended at its CM would adopt that angle of dip. Only the horizontal component of the earth's field is useful for navigation so dip is clearly a nuisance. There is no dip at the equator since lines of force are parallel to the ground there. Near the magnetic poles compasses are useless.

The axis of the earth's magnetic field is not aligned with the rotational axis, and so the magnetic poles are some distance from the true poles. The north magnetic pole has been wandering about in northern Canada in recent history. Maps are made with latitude/longitude grids that relate to the true poles, and so a line of longitude by definition points true north. As the magnetic pole is not at the true pole, a magnetic compass does not point to true north. Variation is measured in degrees and direction. For example, if the compass points 8° to the west of true north, the variation is 8°W.

In order to use the magnetic heading for navigation, maps are drawn with additional lines which show the variation. A line joining all points having the same variation is called an isogonal line. On a map covering a large area, isogonals are drawn every 2°. There is one line joining places of zero variation called the agonal line and which currently passes through Sweden. On charts covering a small area in great detail the variation is shown once and can be considered to be the same over the whole chart.

Direction is expressed in degrees clockwise from North or by the cardinal points North, South, East and West and can be true (T) or magnetic (M). The degree symbol is often omitted and heading will simply be a three-digit number followed by the magnetic/true descriptor, e.g. 295M, 310T.

Runways are numbered according to their magnetic direction but rounded to two-digit accuracy. The runway name is painted at the landing end. Thus a runway running east–west would have 09 painted at the western end since this would be visible to an aircraft flying on a course of 90°. The number 27 would be painted at the opposite end where it would be seen by an aircraft flying on a course of 270°.

A simple bar magnet will also oscillate about true north if disturbed because the earth's field acts like a weak spring and the magnet has rotational inertia. The balance wheel of a watch oscillates in the same way. In order to damp oscillations, aircraft compasses incorporate a liquid. Figure 7.4(b) shows the construction of a typical aircraft compass. The magnet is actually a number of magnets, typically four, embedded in a disc made of some non-magnetic material such as plastic. The outside of the disc is engraved with the magnetic heading which will be read off against a mark on the case

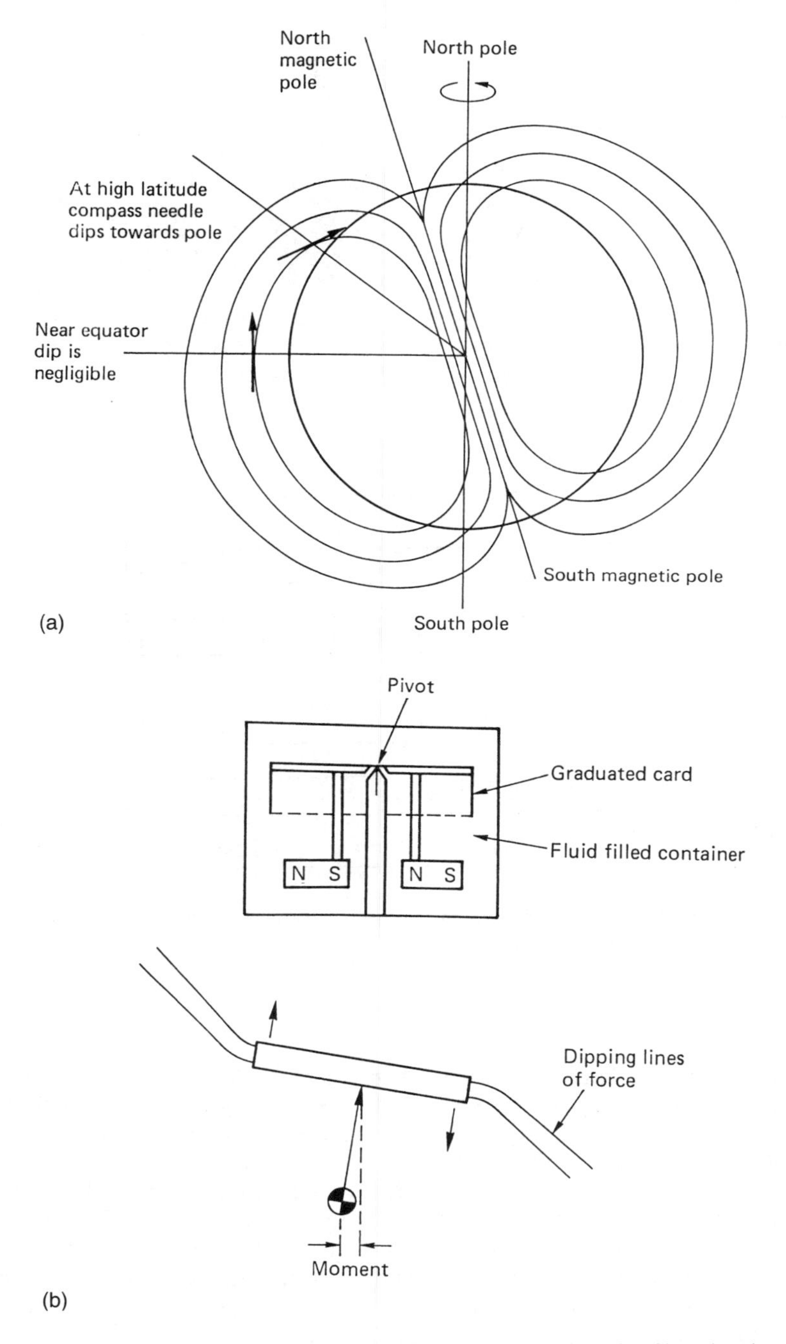

Fig. 7.4 (a) The earth behaves as if it contained a bar magnet between the poles. Note that the geographic north pole is a magnetic south pole. Note the dip angle which is a function of latitude. (b) An aircraft compass. Dip is opposed by supporting the needle some distance above its CM.

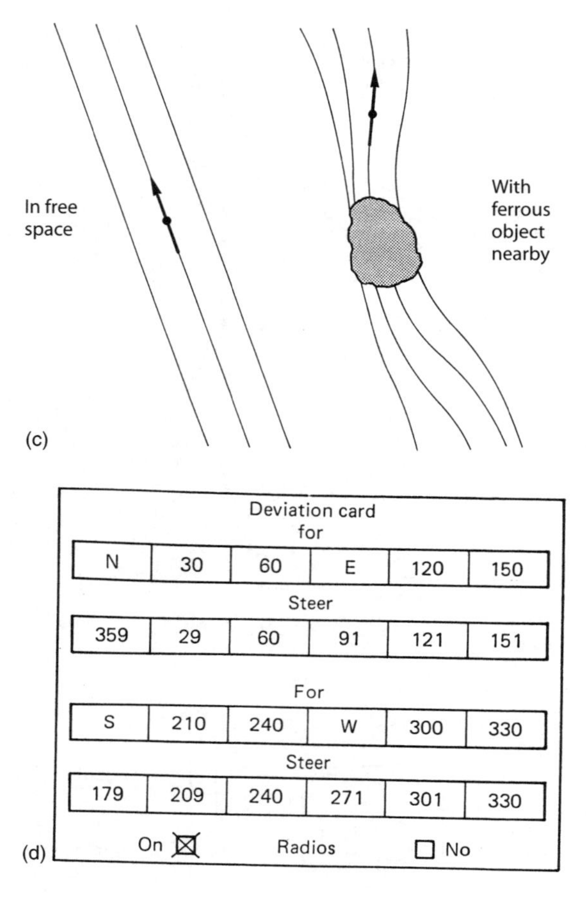

Deviation card
for

N	30	60	E	120	150

Steer

359	29	60	91	121	151

For

S	210	240	W	300	330

Steer

179	209	240	271	301	330

On ☒ Radios ☐ No

Fig. 7.4 (Continued) (c) Deviation is an error in a magnetic compass due to a nearby ferrous object. (d) A deviation card shows what compass errors exist.

called the lubber line. Since the compass is invariably mounted ahead of the pilot, the lubber line will be at the back of the unit, and so the engraving is rotated 180° with respect to the magnet. Thus if the helicopter and pilot and magnet are all facing due north, the part of the disc with 'N' on it will be facing the pilot.

Compass dip is minimized by pivoting the disc well above the CM. If the dip exerts a tilting couple, the CM moves from below the pivot to balance it. The disc remains within 2 or 3° of horizontal as the vertical component of the earth's field is thus opposed. The disc rotates until the magnet is parallel to the horizontal component of the earth's magnetic field. Oscillation is prevented by the viscosity of the liquid acting on the large periphery of the disc.

The contact pressure between the pivot and the disc is reduced by buoyancy of the liquid and this reduces friction. When visualizing how a compass works, it is important to remember that the compass disc tries to keep pointing the same way as the helicopter turns around it.

7.4 Compass errors

A magnetic compass is simple and reliable and needs no power, but it has a number of characteristics that need to be understood if it is to be used correctly. The compass can only align itself to the field it experiences. If something external to the compass disturbs the direction of that field, the compass cannot know and will point to a false north. The result is known as deviation and should not be confused with variation. This is shown in Figure 7.4(c). A helicopter contains a wealth of components that are necessarily magnetized such as the magnetos, the alternator, the electric fuel pump, the motors in the trim system and gyroscopic instruments and the moving coil drive units in the headsets.

Additionally many of the parts of a helicopter whilst not magnetic themselves are ferrous and can distort the earth's field by their presence. Any steel tubing in the hull, the engine and gearbox and the crankshaft, con rods and gears are all ferrous. Aluminium, brass, glass fibre and plastics are non-ferrous and have no effect.

The sum of all of the effects of the helicopter's structure on the compass determines the deviation. Before a helicopter can be released to service the deviation must be measured and displayed on a deviation card (Figure 7.4(d)) mounted next to the compass. The deviation card is completed at the end of a procedure called swinging the compass. The helicopter is taken away from buildings and aligned with each cardinal point in turn. The deviation due to the machine's own magnetism can be detected by this process as at some points it will increase the compass reading and at other points it will reduce it. Sometimes the deviation can be reduced by the installation of ferrous compensators adjacent to the compass, but in any case the remaining deviation must be measured every 30° and recorded on the deviation card. This should be done with the engine running as a rotating permanent magnet has less effect than when it is stationary. Since radios can also generate magnetic fields, these must be tested for deviation effects.

It was stated above that the dip of the compass is overcome provided the disc stays horizontal. Unfortunately there are occasions in the normal flight of a helicopter when this is not the case and the compass dip is not overcome but acts to give an erroneous reading. The problem is caused whenever the machine accelerates in a horizontal plane. The acceleration can be due to a speed change or due to flying at steady speed in a turn.

Figure 7.5(a) shows a helicopter which has just taken off and is accelerating forwards. The rotor thrust is inclined well forward and is accelerating the machine in a horizontal plane. The helicopter hangs from the rotor head like a pendulum, and the compass disc CM hangs from the pivot in the same way. As a result the compass disc in a pure helicopter usually stays very nearly parallel to the cockpit floor. The effect of the tilted disc during acceleration depends on the direction of flight and the hemisphere in which the machine is flying. If the machine in the example were to be heading 000°M the disc tilt will be in the same direction as the dip and there is no effect. However, in the example here the machine accelerates along 270M and the disc is inclined west down/east up. The vertical component of the earth's field will rotate the north pole of the disc downwards and increase the heading shown against the lubber line. In the southern hemisphere, or if accelerating east, the opposite effect is obtained and the heading shown will reduce. On the equator the effect is absent.

Acceleration in the horizontal plane will also occur in a banked turn, which is the only kind of turn a helicopter can make at speed. Figure 7.5(b) shows a helicopter in a sharp bank to port. As the nose of the machine passes through magnetic north the acceleration is to the west and the disc tilts west down/east up. The vertical component of the earth's field once more rotates the disc down at the north pole and increases the heading displayed.

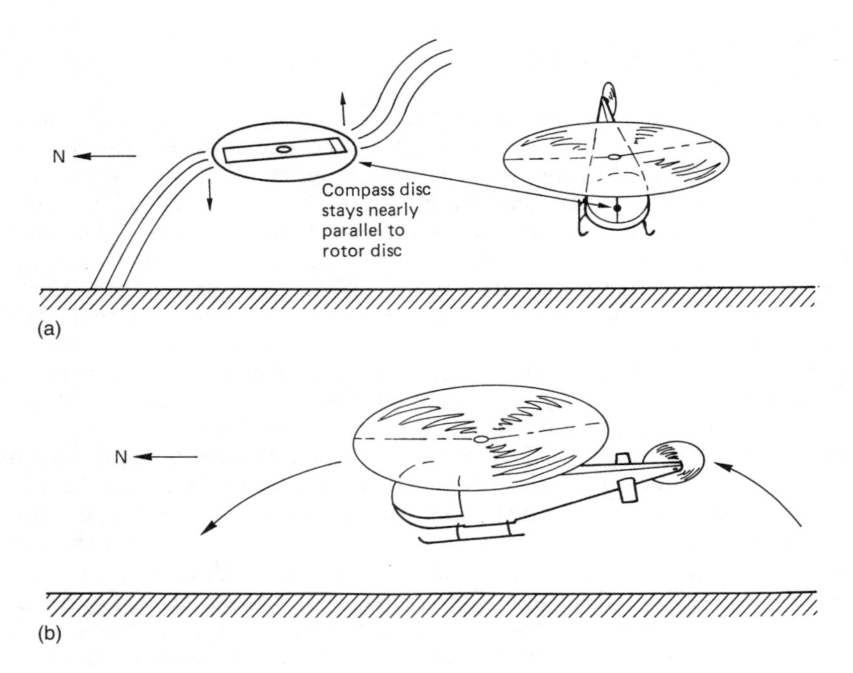

Fig. 7.5 (a) When a helicopter accelerates at 270M in the northern hemisphere, dip will pull the north pole of the compass down and increases the apparent heading. (b) In a sharp bank there will also be acceleration towards the centre of the turn and a further acceleration error will occur.

Erroneous explanations will be encountered attributing some of the rotation of the compass disc to the acceleration alone, but a moment's thought will confirm that a couple cannot be conveyed through a single pivot. It is almost certain that there will be a question on compass acceleration errors in one of the pilot's examinations, and the correct answer will be obtained following the process below:

(a) Visualize the manoeuvre and the attitude of the rotor disc. The compass disc will be very nearly parallel to the rotor disc.
(b) If the disc is tilted north down/south up or vice versa there is no effect. If the tilt is east down/west up or vice versa there will be an error.
(c) Visualize the tilted compass disc in the earth's dipping magnetic field. In the northern hemisphere the north-seeking pole dips; in the southern hemisphere the south-seeking pole dips.
(d) The dip will turn the tilted disc. Visualize whether this increases or reduces the heading read off against the lubber line.

Here is an example: a machine turns to port through north in the southern hemisphere. The rotor disc and compass disc will be tilted down on the west side, so the dip in the southern hemisphere will pull the south-seeking pole down. The compass disc is turned clockwise when seen from above. This causes the heading in degrees to be less than it should be. The pilot should turn to a greater reading or 'overshoot the turn' to compensate. In practice the error is countered by turning to a greater or a lesser indicated heading that will become the correct heading when the turn ceases and the compass settles.

In most cases the machine will have a gyroscopic direction indicator (see section 7.12) set to the same heading as the compass during steady flight. The direction indicator is used for turns because it does not suffer from acceleration errors so it is only necessary to invoke the theory if the DI fails.

In everyday use there is very little that can go wrong with a compass. It should be inspected for leaks of the liquid, for bubbles or for some obvious broken or cracked component, and if the transparent housing is plastic, this may darken as the material degrades with age and indicates that it is becoming brittle as well as difficult to see through. The deviation card must be present and the machine must have been swung on the appropriate occasions.

7.5 The flux gate compass

Figure 7.6(a) shows how a flux gate compass works. There are three radial pole assemblies and the side view shows that these are split so that a single coil with a vertical axis can be wound at the centre. There are a further three coils, one on each limb. The pole pieces are made of a highly permeable material such as permalloy. The entire assembly is suspended below a Hooke joint so that in steady flight it will hang level. An alternating excitation signal, typically 400 Hz, is applied to the central coil. In the absence of any external magnetic field, there would be no signal induced in the three sense coils because they are wound around both poles and the flux from the excitation coil would cancel out. However, in the presence of the earth's magnetic field, the highly permeable material is more prone to saturation in one direction than the other. Figure 7.6(b) shows that the earth's field adds to the excitation and shifts the operating point along the characteristic curve of the magnetic material. It will be seen in Figure 7.6(b) that the curvature of the magnetization characteristic has the effect of asymmetrically compressing the flux waveform so that there is a resultant flux in the sensor coil. The relative amplitudes of the three alternating signals induced in the sensor coils are a function of the direction of the earth's field.

The three signals are sent to any remote instrument needing magnetic heading information. This could be a simple display or a tied gyroscopic DI. Figure 7.7 shows how a remote display works. The three signals from the flux gate are supplied to three coils arranged at 120°. There is also a second single coil that can rotate with respect to the three input coils. The resultant of the flux produced by these three coils will have a direction determined by the direction of the earth's field at the sensor. As this is an alternating flux it will induce a voltage in the secondary coil. However, if this coil is exactly at right angles to the resultant flux, no voltage will be induced.

Figure 7.7 shows that the secondary coil signal is connected to an amplifier that drives a motor to turn the secondary coil and the display card. The amplifier will drive the motor until the secondary coil is transverse to the resultant field of the three input coils. At this point the induced signal will disappear and the motor will stop. Thus as the magnetic heading of the machine changes, the signals from the flux gate change the resultant flux direction in the display. The secondary coil is driven until it once more is in a null. In this way the motor and secondary coil follow the angle of the earth's field with respect to the flux gate.

The flux gate compass has no controls and can be mounted anywhere convenient. In a fixed-wing aircraft the wing tip is a favoured location as it is well away from heavy ferrous objects. In a helicopter the tail boom could be used. Some flux gate compasses contain additional coils fed with carefully calibrated currents from a control box in

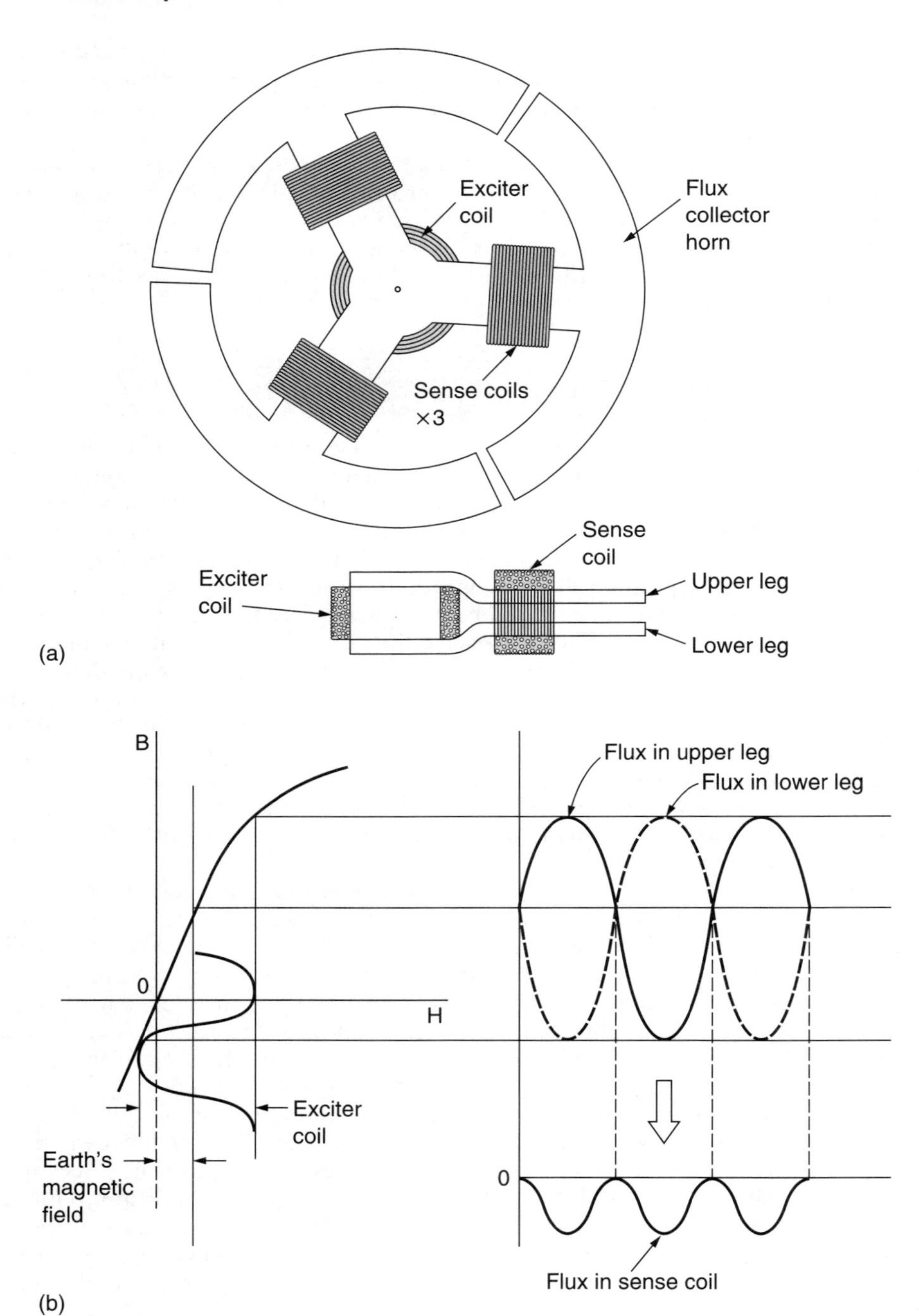

Fig. 7.6 (a) In the flux gate compass, the earth's magnetic field disturbs the conditions (b) in each of the three limbs of the device differently and its direction can be established from the relationship between the three output signals.

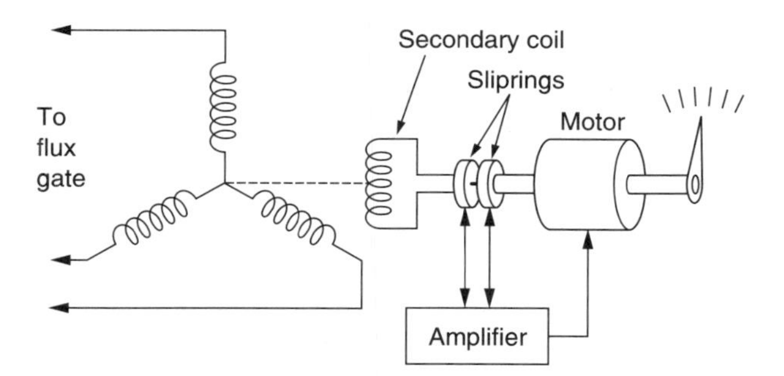

Fig. 7.7 The remote indicator for a flux gate compass works by aligning its rotatable coil at right angles to the resultant flux generate by the three signals from the compass. At all other angles a voltage will be induced in the coil and this will be amplified to drive the motor. The coil also carries the direction indicator.

order to cancel out deviation. In this case during swinging of the compass the goal is to find the correct values for these currents.

It is also possible to remove the variation in the earth's magnetic field and obtain a true heading rather than a magnetic heading. This requires a device called a *differential transmitter*. This contains two sets of windings at 120°, but one set can be turned by an operator control. If the two sets of windings are aligned, the relative amplitudes of the signals coming out will be the same as those entering, but if the secondary winding is turned, the output signals are effectively rotated. Thus if the differential transmitter is set to the amount of local magnetic variation, the remote display will read true heading.

7.6 Pressure instruments

The instruments operated by pressure are the altimeter, the vertical speed indicator (VSI) and the airspeed indicator (ASI). Of these, the first two measure static pressure piped to the instruments from the static vents, whereas the ASI measures the dynamic pressure due to motion sensed as the pressure difference between the pitot head and the static vents. The necessary pipework is shown in Figure 7.8. The static vents are usually installed one each side of the fuselage in order to cancel any effect due to asymmetrical airflow. Each carries a mandatory placard stating that the vent is to be kept clear. The static pipes initially run upwards from the vents to prevent the entry of water.

The pitot head is installed facing forward at some point where the airflow is reasonably undisturbed. It may be heated to prevent icing. The pitot should be checked for blockage as part of the pre-flight checks. It is good practice to cap the pitot with a suitable cover when the machine is on the ground, in which case the checks will include the removal and stowage of the cover.

7.7 The altimeter

The altimeter is to all intents and purposes an aneroid barometer. The mechanism is shown in Figure 7.9. It contains a sealed capsule that is corrugated to allow it to flex. The capsule is evacuated, hence the term aneroid, and so ambient pressure attempts to crush it. The stiffness of the corrugations and a separate spring oppose the pressure. The capsule will expand as ambient pressure falls with rising altitude, and a system of

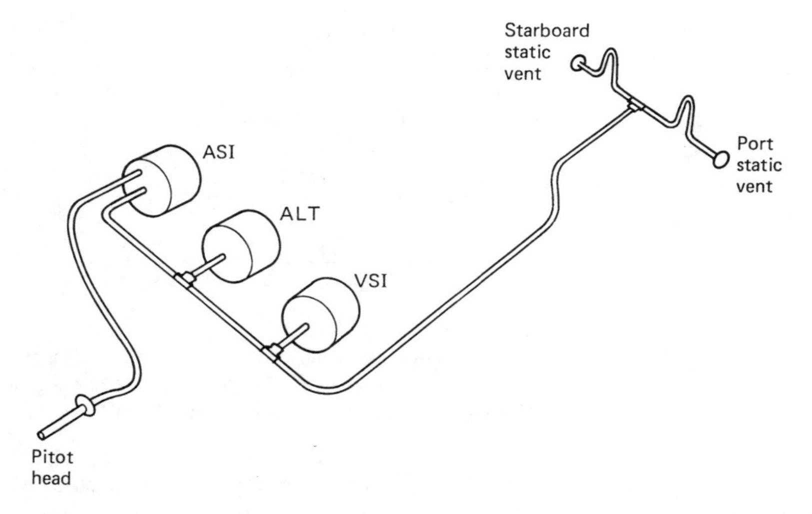

Fig. 7.8 The pipework involved with pressure-sensing instruments. These require a dynamic pressure feed from a forward facing port or pitot head and a static pressure feed typically from ports on the side of the hull.

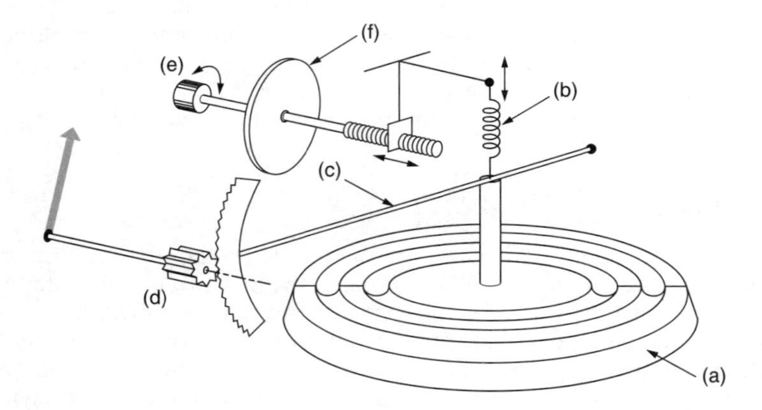

Fig. 7.9 The altimeter is essentially a barometer fitted with a control that can adjust the reference. This allows it to display altitude with respect to a pressure that has been input on the setting scale. Evacuated bellows (a) is sensing element in conjunction with spring (b). Rising air pressure collapses bellows and stretches spring. Falling air pressure allows the spring to expand the capsule. Capsule movement is amplified by lever (c) which operates the display pointers via a system of gears (d). Altimeter is compensated for ambient pressure by turning knob (e) according to subscale setting required on the card (f). Subscale knob is arranged to stretch or relax the spring slightly in order to give zero reading at a range of barometric pressures. The case of the instrument is connected to the static vent pipe.

links and gears amplifies the small capsule expansion to move the pointer over a scale. The aneroid altimeter needs no power except to illuminate the scale.

A conventional barometer measures absolute pressure for meteorological purposes, whereas the altimeter is required to measure the pressure difference between its present location and some reference pressure in order to compensate for barometric changes. The pressure difference is, however, displayed in feet (or metres). The reference pressure is set by the use of a control knob. Turning the knob will simultaneously change the reference pressure displayed on the subscale and the aneroid capsule spring tension.

If, for example, the reference pressure is to be raised, the act of setting the higher pressure on the subscale slightly extends the spring and the capsule will reach equilibrium showing a higher altitude. Conversely if a lower reference pressure is to be used, the spring is allowed to contract slightly and the capsule reaches equilibrium with a lower altitude reading. Throughout this process the absolute pressure at the altimeter did not change, but the pressure difference and the altitude did because the subscale setting was changed. Clearly without an appropriate subscale setting an altimeter reading is meaningless. It should also be clear that, in general, height and altitude are different.

There are three main subscale settings used with an altimeter and these are shown in Figure 7.10(a). If the subscale is set to the ground pressure at the elevation of the airfield, the altimeter will read zero prior to take-off and in flight will measure height above the airfield. The barometric pressure at an airfield is called QFE. The 'Q' dates from the days of carrier wave radio and Morse code and indicates a question. An aircraft arriving at an airfield may ask for QFE in order to set the altimeter subscale so that the altimeter reads height above the field. QFE is easily remembered as 'field elevation'. In some parts of the world this technique is not used. For example, the QFE of a mountain airfield may be so low that it is beyond the end of the altimeter subscale.

During a flight at moderate altitude the overriding concern is clearance of terrain and structures. These are quoted on maps as height above mean sea level (AMSL). In order to establish the clear height above such obstacles the altimeter reference is set to QNH; the barometric pressure in the area 'reduced to mean sea level'. This means that the actual barometric pressure is measured at a known elevation and the value the pressure would have had at sea level is computed from the height and temperature. With QNH on the subscale, the altimeter reads feet AMSL and subtracting the height of obstacles in feet AMSL gives the clearance. QNH is easily remembered as 'nautical height'.

Figure 7.10(b) shows a typical low-level flight. The machine takes off with QFE in the subscale to show height above the airfield. Once away from the field the altimeter is set to regional QNH and the machine is trimmed to cruise at 2000 feet AMSL. On the chart is a hill rising to 1200 feet AMSL and the machine will clear it by 800 feet. On approaching the destination airfield, the pilot asks for QFE and sets this on the altimeter. This should read zero as he touches down.

Each country is divided up into altimeter setting regions which each have their own QNH value as a function of the weather system across the country. The QNH quoted takes into account the rate at which barometric pressure is changing. If pressure is falling, a machine flying at constant pressure altitude based on QNH would actually descend and not have the expected clearance. QNH is adjusted upwards by the authorities to ensure that clearance is still achieved at the end of the period for which a setting is valid even if barometric pressure falls. On a long flight an altimeter setting region boundary may be crossed. The pilot must obtain the new QNH for the area to ensure his altimeter still reads feet AMSL. The subscale is generally calibrated in millibars (mb) or, more recently, hectoPascals (hP) that are numerically identical. In America, scales calibrated in inches of mercury (inHg) will be found.

For flight in airways, the overriding concern is not terrain clearance, but vertical separation between aircraft. For this reason the same reference pressure is always used in airways. This is the pressure of the International Standard Atmosphere (ISA) at Mean Sea Level (MSL): 1013.25 mb. Older texts may use the term QNE to describe this pressure. Aircraft flying with ISA on the subscale are flying at pressure altitude. The flight level is the pressure altitude with the last two digits removed. For example,

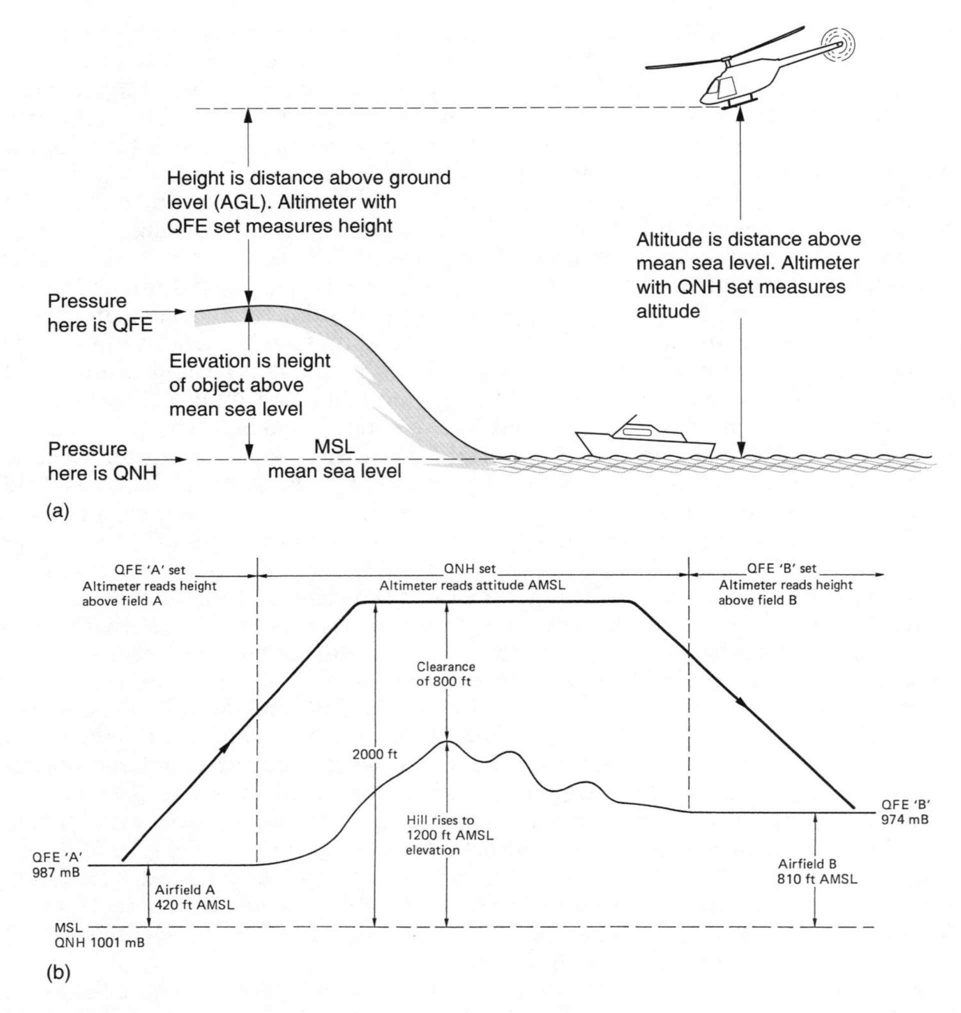

Fig. 7.10 (a) QFE is the local barometric pressure at ground level at an airfield. When the subscale is set to QFE, the altimeter reads height AGL (above ground level). (b) Once away from an airfield, the altimeter may be set to QNH. This is local pressure reduced to MSL (mean sea level) and the altimeter reads height above MSL. As obstructions on maps are described by their height AMSL, the QNH setting allows the pilot to establish that terrain clearance exists. QFE may be used for landing. This may be different from the QFE that was used at take-off.

an aircraft at FL 350 is 35 000 feet above ISA MSL. Another aircraft at FL 340 would always be 1000 feet lower whatever the atmospheric conditions. Very few helicopter operations take place in airways, although as tilt-rotor and tilt-wing machines enter service this will change.

It should be noted that helicopter altimeters only sense local atmospheric pressure when the rotors are not turning. In the hover, especially in ground effect, the pressure under the rotor disc will be appreciably above ambient. Going from flat pitch on the ground to a low hover will result in a noticeable reduction in the reading on the altimeter, of the order of 30 feet. The effect is negligible in translational flight.

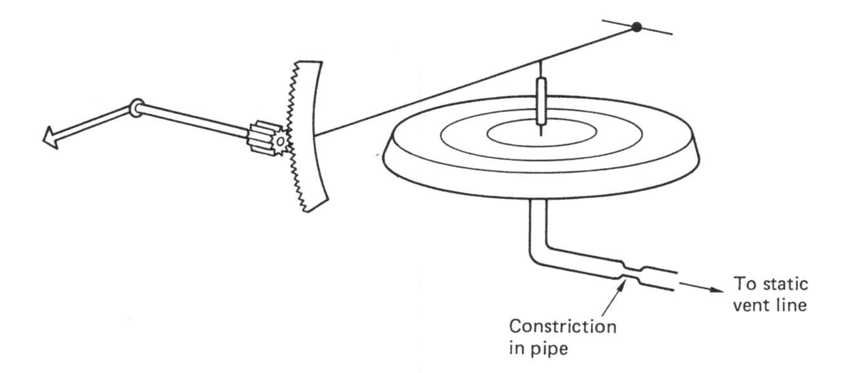

Fig. 7.11 The VSI (vertical speed indicator) is a leaky barometer that responds to a rate of pressure change. At constant altitude the pressure in the instrument will be the same as ambient owing to the leak. In a climb, air will try to flow out of the leak and the pressure across the orifice will be sensed to operate the display.

7.8 The vertical speed indicator

As its name suggests, the vertical speed indicator (VSI) displays the rate at which altitude is changing. Whilst this can be established by looking at the speed with which the hand of the altimeter turns, this takes some time and it is quicker to look at a separate display. Figure 7.11 shows how the VSI is constructed. The case of the instrument is connected to the static port of the aircraft. The mechanism includes a thin metal capsule sealed except for a very small orifice of carefully determined size. Should the aircraft climb, ambient pressure will fall and the capsule will expand, moving a pointer across a scale. The reduced ambient pressure will cause the air inside the capsule to flow out, but the flow is slowed down by the restriction. At constant altitude, the capsule will eventually have the same pressure inside and out, and the pointer will read zero. The faster the ambient pressure changes the more the restriction opposes the balancing airflow and the greater the deflection of the pointer. As it is self-balancing the VSI needs no pilot adjustment and has no controls. The VSI is particularly useful in cruise where it should read zero. Any discrepancy indicates that the helicopter needs retrimming to fly at constant height.

7.9 The airspeed indicator

The airspeed indicator (ASI) is connected between the forward facing pitot head and the static vents. This arrangement cancels ambient air pressure so that it measures only the dynamic pressure due to forward flight. Dynamic pressure is proportional to air density and to the square of the velocity. As a result the expansion of the pressure-sensing capsule is proportional to the square of the speed, and it is necessary to convert the square law response to drive a practical scale. Figure 7.12 shows that this can be done using a straight spring (a) whose effective stiffness increases with deflection. At low speeds the whole length of the spring is active (b) and the restoring force is relatively low. When the spring contacts a stop (c) the active length is reduced and the spring becomes stiffer.

The system does not achieve complete linearity. Practical ASIs have decidedly non-linear scales and the designer positions the stretched part of the scale at the most useful

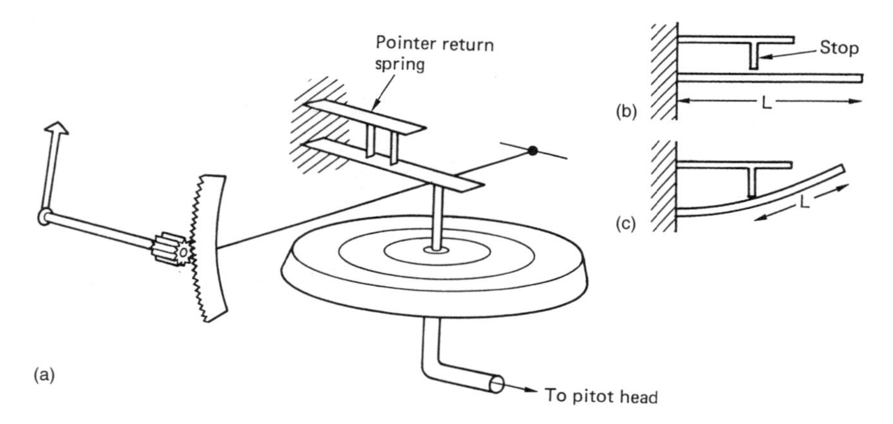

Fig. 7.12 Dynamic pressure rises as the square of the airspeed and instrument manufacturers use various techniques to linearize the scale to some extent. This may include the provision of stops to make the return spring progressively stiffer as speed increases.

range of flying speeds. In a light helicopter the stiffer spring may come into operation at about 100 mph. Pointer movement from 70 to 80 mph may be twice that from 100 to 110 mph. A further consequence of the square law is that the dynamic pressure at low speed is very small indeed and comparable to the friction in the linkages. The ASI is unreliable at low speeds because of the low dynamic pressures involved and because of the effect of downwash. Thus an ASI generally has no calibrations below 20 mph. According to the country of origin ASIs will be found calibrated in mph, knots and kph, some have dual scales like car speedometers.

The reading of the ASI is also proportional to air density, so it will underread at high altitude. The scale on the ASI displays indicated air speed (IAS) proportional to density and the square of the speed. Since the rotors behave proportionally to these parameters, IAS is the best way of displaying speed from the point of view of the airframe. Clearly IAS is not the actual speed through the air, and for navigational purposes it needs compensating for ambient density to give true air speed or TAS. Density is affected by altitude and temperature. At 10 000 feet the IAS must be increased by around 17% to compensate for density change. An increase of 10°C raises TAS about 2%. If TAS, windspeed and direction are known, the groundspeed (GS) can be calculated.

In real ASIs the scale markings are not perfect, and frequently the position of the pitot head with respect to the hull causes errors at certain airspeeds. A calibration table is supplied so that IAS can be corrected by looking it up on the table. The result is called rectified or calibrated airspeed (RAS or CAS) and is the value the IAS would have read if it were free of error. In a helicopter airspeed is relatively slow and uncertainty about windspeed causes more error than the ASI. In this case it is probably not worth correcting the ASI for navigational purposes.

7.10 Airspeed and altitude sensing

Airspeed and altitude are both sensed by measuring pressure, one dynamic and one static. Both of these processes are often combined in a single unit. As was seen above, pressure is measured using the deflection of a corrugated capsule. The amount of

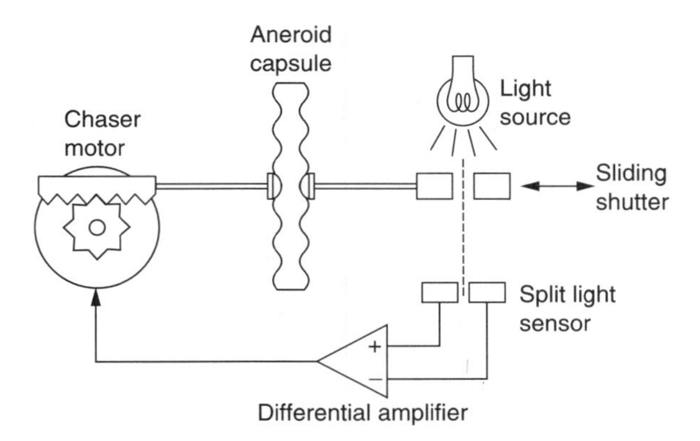

Fig. 7.13 A chaser system which allows the position of an altimeter mechanism to be measured without adding friction. See text.

movement and the power available is small and it is difficult to drive a conventional encoder directly. Instead a form of electromechanical amplification is used. Figure 7.13 shows one example. The sensing capsule is mounted on a platform that can be moved by an electric motor. The moving diaphragm of the capsule carries only a low mass vane or plate placed in a light beam. The light falls on differential photocells so that very small off-centre movements of the diaphragm produce an electrical signal because more light falls on one cell and less on the other. This signal is amplified and supplied to the electric motor such that the motor is driven in a direction that recentres the optical sensor. Alternate types of sensing may be used, for example a differential transformer as will presently be described.

A motor used in this way is known as a *chaser*. Effectively the chasing action means that the diaphragm of the capsule remains in the same place. Pressure changes must then be reflected in the position of the motor drive, which has enough power to drive the pressure encoder. In addition to the encoder output, the speed of the motor may be measured to produce a rate output. When used for the altitude hold function in an autopilot, the chaser motor may be disconnected and the signal from the optical sensor will be used to make the aircraft climb or descend as needed to maintain altitude.

Altitude is sensed using an evacuated capsule, and means are required to allow the appropriate reference pressure to be set. Airspeed is sensed using a capsule exposed on one side to dynamic pressure at the pitot head and to static pressure on the other.

For automatic approach or landing systems a RADAR altimeter is a necessity. In fact it is not an altimeter as it does not consider air pressure; it is a height meter because it works by timing the reflection of a radio signal to give height AGL which is the requirement for landing.

7.11 Gyroscopic instruments

Gyroscopic instruments are those depending on a rapidly rotating flywheel for their operation and include the direction indicator (DI), the turn and slip indicator (T + S), also known as the turn and bank indicator, and the artificial horizon. These display, in various ways, the attitude and heading of the helicopter. It is possible to fly a helicopter in good conditions without them, relying only on the magnetic compass for heading information and assessing the attitude of the machine from the view of the world outside. In poor conditions or at night the horizon may be indistinct and without attitude clues even the finest pilot could lose control owing to disorientation. The gyroscopic instruments provide sufficient clues that in conjunction with the other flight instruments the machine can be flown without looking outside at all. Whilst the instruments are optional in a VFR light helicopter, most have them fitted and it is worth knowing how they operate as they could make a significant difference if an unanticipated deterioration in the weather takes place or if cloud is inadvertently entered.

The basic principle of the gyroscope was considered in Chapter 2. In practical instruments the gyro rotor is spun by an air jet or by an electric motor. Many light planes have a venturi in the slipstream that generates suction. This is applied to the case of the gyro instrument. Filtered cabin air is drawn into the instrument through a nozzle by the suction and blows the rotor round. Clearly this arrangement is not suitable for a helicopter as there is not necessarily a slipstream. A further alternative is to have a vacuum pump driven by the engine. The solution adopted in most helicopters is to drive the gyroscopes electrically. Power is fed through the gimbals by miniature sliprings.

As the operation of the gyro is dependent on the rotational speed, many gyros, but not all, have a flag coloured red or orange which swings into view if the rotor is not running fast enough. No flight action should be based on the reading of an instrument showing a flag.

If a gyroscope is mounted in gimbals and set running with the rotor axis parallel to the earth's surface, it will not remain in this condition due to *drift*. Drift has two unrelated components. First, the gyroscope is built to finite accuracy and minute imbalances in the gimbals and bearing friction will result in the spin axis changing slowly in an unpredictable manner; a genuine drift. Second, as the earth is turning at 15° per hour the gyro axis fixed in space will display *apparent drift*.

Figure 7.14 shows that the effect of the earth's rotation on apparent drift is a function of latitude. At the equator the gyro axis and the earth's axis are parallel and no drift is observed. However, at the geographic pole, the full 15°/hr will be observed. The apparent drift due to earth rotation is a sinusoidal function of latitude. The drift may

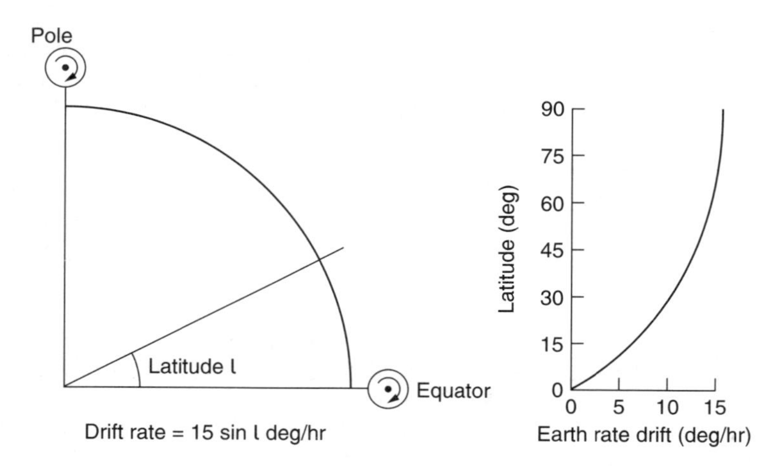

Fig. 7.14 Apparent drift in a gyroscope is not a random effect but is due to the earth's rotation. At the poles the full rotational rate of the earth (15°/hr) will be apparent. This falls to zero at the equator as a sinusoidal function of latitude.

be modified if the gyro is in a moving aircraft. The track of the aircraft will have an east/west component and movement in this direction will have to be added to the rate at which the earth turns. This is known as *transport error*. The term transport wander will also be found, but this is misleading as the effect is entirely deterministic.

A completely free gyro will suffer from three forms of error: an unpredictable error due to non-ideal construction, and predictable errors due to the rotation of the earth and to being transported with respect to the earth. In a helicopter the low airspeed and relatively short flights mean that transport error is generally insignificant.

For earth-related activities such as flying, something will have to be done about drift. The solution depends upon the application, but essentially the rigidity of the gyro is used to offer short-term stability, whereas the long-term stability comes from elsewhere. A gyro externally stabilized in this way is described as *tied*. Where the long-term stability comes from the earth's gravitational field, the result is called an *earth gyro*.

7.12 The direction indicator

The direction indicator is a gyroscopic device that is designed to overcome acceleration errors in magnetic compasses and to be easy to read. Figure 7.15 shows the general arrangement of a DI. The rotor axis is maintained horizontal and the shaft is mounted in a pair of gimbals where the outer one pivots about a vertical axis. A gearing system couples the rotation of the outer gimbal to the disc or card visible to the pilot. If the

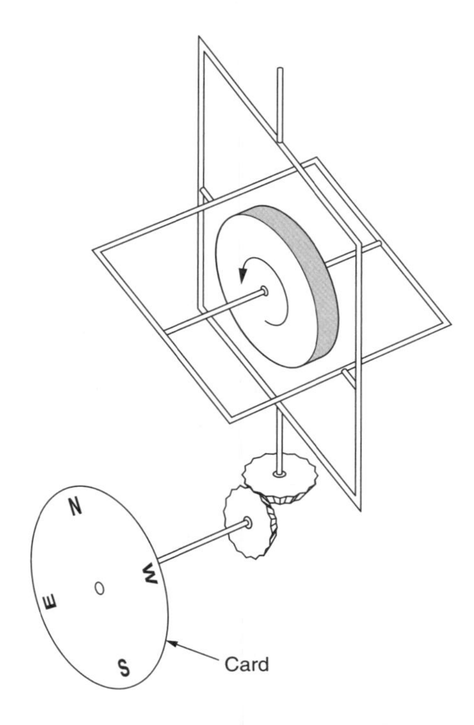

Fig. 7.15 The DI (direction indicator) is a gyroscopic instrument in which the gyro axis is maintained horizontal. As the aircraft changes heading, the gyro axis will remain the same and can drive a pointer to indicate the new heading.

instrument is turned bodily about a vertical axis, the rigidity of the gyro will keep the gimbals fixed, and so the action of moving the instrument case around them will rotate the scale and indicate the angle through which the unit has turned.

The gyro axis is maintained horizontal by tying the gyro to the vertical gimbal ring. In an air-driven gyro there is already a stream of exhaust air leaving the rotor shroud. In an electrically operated gyro the rotor can be designed to act as a blower to produce an air jet. The air stream from the shroud impinges upon a wedge-shaped plate on the vertical gimbal ring. If the gyro axis (or the horizontal gimbal ring) is not at right angles to the vertical ring, the airflow will produce an imbalanced force on the vertical ring, tending to turn it. The direction of this force is such that it will cause the gyroscope slowly to precess in a direction that will bring the inner ring to rest at 90° to the outer or vertical ring.

As a result the gyro axis is aligned with the long-term average attitude of the helicopter about the roll axis. Compared to the slow rate of precession in the tying process, manoeuvres of the helicopter are very rapid and average out. In flight a turn would be accompanied by banking, but the gyro would maintain its axis because the inner gimbal can pivot in the outer.

Although more accurate in a manoeuvre than a magnetic compass, the DI does suffer from minor geometric errors. The outer gimbal has two sets of bearings at 90°, and essentially is acting as a Hooke joint between the gyro and the display card. As has been seen elsewhere in this volume, the Hooke joint is not a constant velocity joint, but suffers from cyclic angular errors when the input and output axes are not parallel. The result is that the DI reading is slightly out due to what is called *gimballing error*. It should be pointed out that gimballing error is very small and disappears when straight and level flight resumes.

The occurrence of gimballing error depends on the orientation of the gyro axis with respect to the helicopter's yaw axis. In the general case where the gyro axis is not transverse to the flight path, a roll will cause gimballing error. In the special case where the gyro axis is transverse, a combined roll and pitch change will be needed to cause a gimballing error.

Figure 7.14 showed that with the DI gyro axis horizontal, the rotation of the earth causes apparent drift at a rate given by multiplying the earth rate, 15°/hr, by the sine of the latitude. For example, on the south coast of England at Lat. 51° N there will be a DI earth drift rate of just under 10°/hr.

In some units it is possible to compensate the DI for earth rate mechanically. The inner gimbal ring is deliberately imbalanced by a nut on an arm so that gravity produces a slight rolling torque on the gyro. The gyro will precess this into a yaw and when correctly adjusted the rate of yaw will be equal to the earth rate for the latitude at which the machine is normally operated so that the drift is cancelled. In more sophisticated DIs, there may be an adjustable earth rate compensator having a control which is set to the current latitude and a switch to select the sense of the compensation for the appropriate hemisphere. A latitude-dependent current is generated and sent to a torque motor in the DI fitted between the inner and outer rings. This applies torque to roll the inner ring, but the gyro precesses this into yaw. If the correct torque is applied for the latitude, the resultant precession rate will equal the earth rate and the DI drift will be minimal. In principle if the groundspeed, latitude and heading are known transport error can also be cancelled. This may be useful for flying near the magnetic poles.

However, most DIs rely on periodically being reset to the same reading as the magnetic compass and are fitted with a control knob which is pushed in and turned to force the gyro round to the correct heading. Pushing in the knob also forces the gyro axis to be at right angles to the vertical ring, a process known as *caging*. Clearly the machine must be in steady flight when the DI is aligned to the compass or the gyro axis will be incorrect and it will be set to repeat the compass turning error.

7.13 The gyromagnetic compass

More sophisticated DIs can tie themselves to a magnetic flux gate compass to produce a gyromagnetic compass which combines the best features of both devices. The flux gate compass was described in section 7.5. The tying process is heavily filtered so that compass errors during manoeuvres do not affect the reading. Gyromagnetic compasses have the advantage that they do not need to be reset.

Figure 7.16 shows how the DI follows the magnetic compass. The signals from the flux gate compass are applied to a set of three coils encircling the shaft driving the display card in the DI. On the shaft is a single secondary coil. When the DI is accurately aligned with the compass, this coil will be transverse to the resultant flux of the three input coils and so no voltage will be induced in it. However, if any angular difference develops, a signal will be induced in the secondary coil and this can be amplified and used to drive torque motors that apply a rolling torque to the inner gimbal of the gyro. The gyro will precess this torque into yaw so that the display card will turn, and with it the secondary

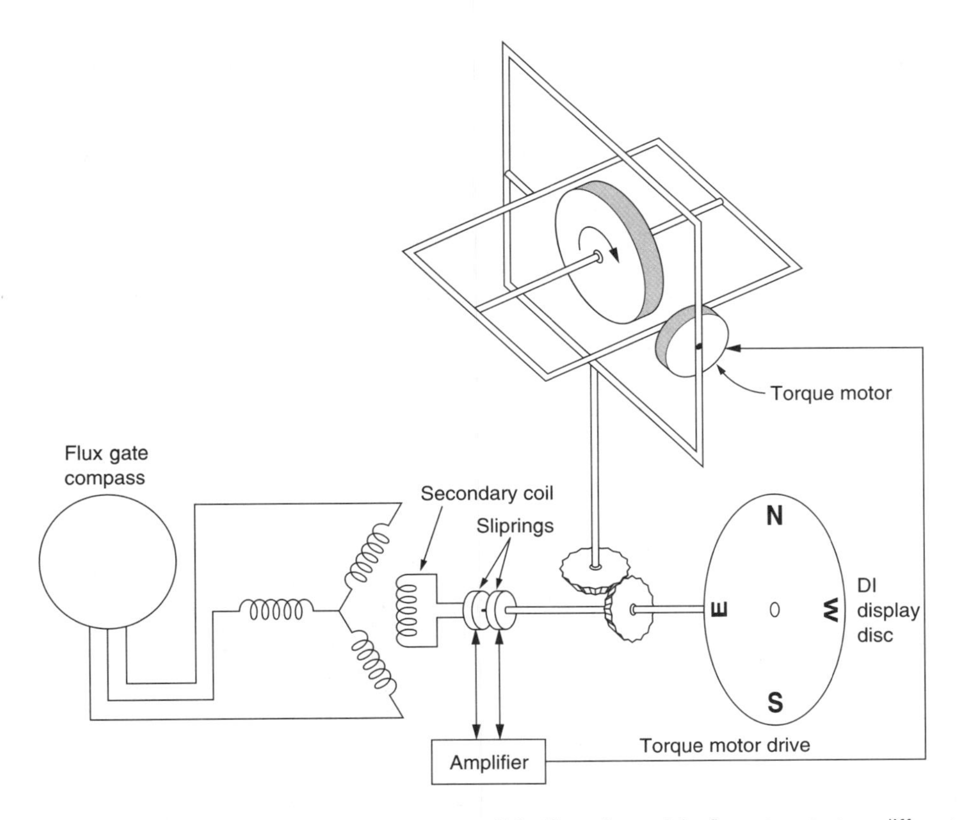

Fig. 7.16 A slave DI that follows a flux gate compass. If the DI reading and the flux gate output are different, a torque motor precesses the DI until the two have the same indication.

coil. In this way the gyro is precessed until the signal in the secondary coil nulls and the precession stops. At this point the display card will be showing the heading coming from the compass. The gain in the system will be quite low so that the gyro follows very slowly.

In some units it is possible to rotate the three coils using a front panel variation control so that the DI card reads true heading rather than magnetic. In most conditions tying the DI to magnetic north is useful, but when flying very close to the magnetic poles it may be better to turn the slaving off and to rely on the earth rate and transport error compensation only. A switch will be provided for this purpose.

A gyromagnetic compass may also form the heading reference for an autopilot. In this case the display unit has additional controls allowing the pilot to enter the desired heading. This is displayed with respect to the DI card markings by an indicator or *bug* on the periphery of the instrument. If the autopilot is engaged and a new heading is selected, the machine will carry out a two-minute rate turn to the new heading at which point the bug will align with the lubber line of the DI.

7.14 The artificial horizon

The artificial horizon is a gyroscopic instrument displaying a realistic copy of what the real horizon does for use in poor visibility. When the helicopter banks, the horizon in the display remains parallel to the real one. When the machine dives, the displayed horizon rises up the face of the instrument. A small symbolic aircraft on the face of the instrument then has the same attitude to the horizon as the real one, and if the pilot flies the symbolic aircraft the real one will follow. In a machine equipped with an autopilot, in addition to the visible display, the artificial horizon may provide attitude output signals that are fed into the control system to stabilize the flight attitude.

The gyroscope in an artificial horizon must be compensated for drift. This is done by an *erection* mechanism to maintain the axis of the gyro vertical with respect to gravity. For this reason, this type of instrument is sometimes called a vertical gyro. The addition of gravity sensing creates what is known as an *earth gyro*. Figure 7.17 shows how gravity sensing works. The rotating wheel is fitted inside a case. Two small pendula are fitted to the case in such a way that they can swing in orthogonal directions. The electrically driven rotor is fitted with vanes at one end so that it acts as a centrifugal blower. Air from the blower is arranged to leave the casing at four equally spaced points. Each pendulum has two vanes that block or reveal the blower orifices. The vanes are

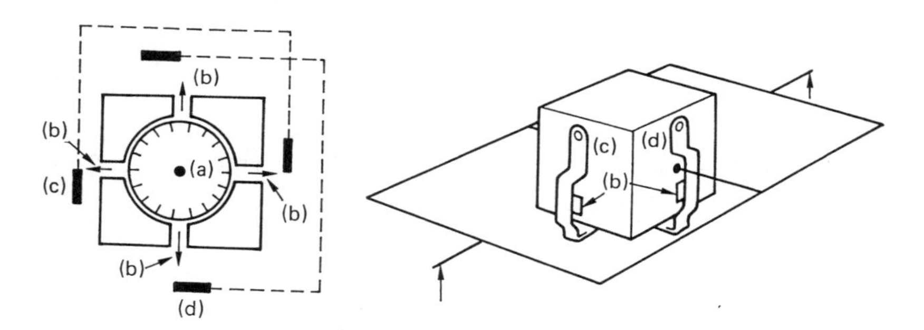

Fig. 7.17 Gravity sensing is used in earth-tied instruments such as the artificial horizon. Small pendula (c) and (d) hang in the efflux (b) from a blower (a) built into the gyro rotor. If the rotor axis is not vertical the pendula will tilt and the reaction from the blower will be imbalanced so a precessive torque is created.

arranged on opposite sides of the pendulum so that if the pendulum swings one way the area of one orifice is reduced whilst that of the other is increased; the situation is reversed if it swings the other way.

If the pendulum is central, the two orifices have equal areas revealed, and the thrust from the air jets will cancel. Should the gyro tilt with respect to gravity the pendulum will swing away from the centre position and the jet reaction will no longer balance. The net reaction is at right angles to the angular error, but this is exactly what is necessary as it causes the gyro to precess and right itself. As there are two orthogonal systems working simultaneously, the gyroscope maintains itself along the earth's gravitational field. The erection system needs only to be able to work somewhat faster than the earth rate, and so the air jets do not need to be very powerful. In fact it is better if the erection process is slow because then momentary disturbances of the pendula caused by manoeuvres and turbulence are simply too rapid for the gravity-sensing system to respond to and they are filtered out. In some gyros the earth tying is done by tilt switches that operate torque motors acting on the gimbals. In more sophisticated instruments the erection process is switched off if the tilt switches tilt by more than a few degrees as this must be due to a manoeuvre.

As the gravity sensing is so subtle, a gyro takes about ten minutes to settle if started up with random orientation. In order to speed the process, the instrument is fitted with a caging knob that is operated when the aircraft is level. This operates a system of cams to force the gimbals back to the approximately correct attitude. When the cams retract, the gravity sensing finishes the job. The caging knob should not be operated when the machine is not level as a false attitude will be indicated. Gyros with tilt switch erection systems may have a fast erection mode for use on the ground or in level flight in which the precession rate is increased.

Figure 7.18 shows the general arrangement of the artificial horizon. The outer gimbal is pivoted on the roll axis of the aircraft and the inner gimbal is pivoted on the pitch axis. The gyro axis is vertical. Movement of the inner gimbal relative to the outer operates

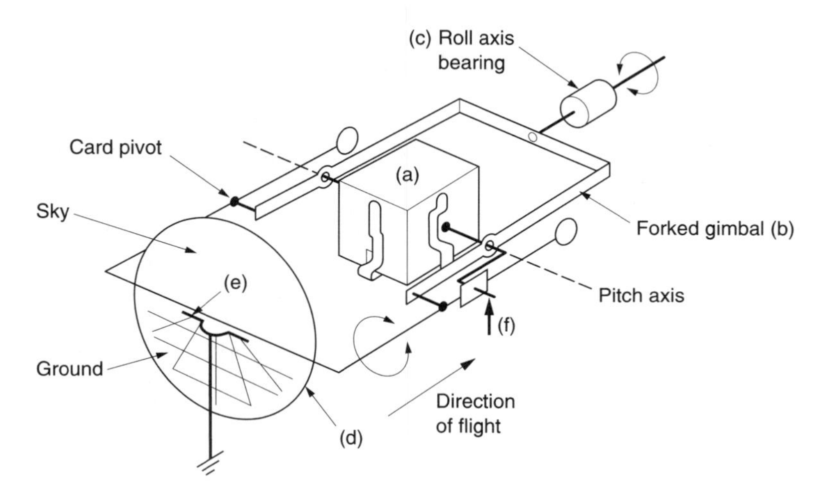

Fig. 7.18 In an artificial horizon, a gravity-sensing gyroscope (a) is mounted on a forked gimbal (b), which is free to turn in the roll axis bearing (c). If the machine rolls, the horizon on card (d) will remain horizontal. If the machine pitches, the card and gyro will remain vertical, but the card will move in the viewing window with respect to the symbolic 'wings' (e). The crank (f) reverses the sense of the card movement so that when the machine dives the 'ground' on the card moves up.

levers to move a large card painted with the earth at the bottom and the sky at the top. In a dive the instrument is tipped nose down and this causes the horizon card to rise with respect to the symbolic aircraft. In a climb the card falls. During a roll the outer gimbal remains horizontal, as does the horizon card. The instrument effectively banks around the card.

Clearly if a gyroscopic instrument is to work properly it must be fitted in the helicopter with the correct orientation, which is usually with the instrument face vertical. The artificial horizon is particularly sensitive to mounting attitude and if fitted to a sloping instrument panel a wedge plate is used to keep the instrument level. Certain artificial horizons can be adapted to work on a sloping panel. The card linkage is offset and the caging cam is turned around the inner gimbal by the angle of tilt so that the gyro remains vertical with the instrument tilted and the horizon in the centre of the display. In a helicopter the attitude of the machine about the pitch axis depends on forward speed. At high speed the nose down attitude would fool the instrument into thinking the machine was diving. To overcome this, the symbolic aircraft is fitted with an adjusting knob that allows it to be moved up or down slightly so that at the chosen airspeed it can be set to the displayed horizon.

The gyroscope can only maintain its rotational axis if the gimbals allow it sufficient freedom. This is true for moderate manoeuvres, but not for aerobatics. If a machine equipped with an earth gyro performs a quarter loop so it is going straight up, the two gimbals will become parallel and there will only be one degree of freedom. If, as this condition of *gimbal lock* is approached, the machine also rolls, there will be violent precession known as *toppling* as the gyro attempts to conserve momentum without the necessary freedom. After toppling, the gyro will be useless until it has erected again. If a gyroscopic instrument is expected to operate under aerobatic conditions, it will need additional outer gimbals. These are servo driven from sensors on the inner gimbals so that the latter are maintained at right angles at any attitude so that gimbal lock can never occur.

7.15 The turn and slip indicator

The turn and slip indicator is actually two completely independent instruments but combined in one housing as the two would be used together to make a turn. Figure 7.19 shows the mechanism of a turn indicator; actually a *rate* gyroscope. There is only one gimbal pivoted on the fore-and-aft axis of the helicopter and the gyro shaft is transverse. The gimbal is held central by a light spring and rocking of the gimbal moves a pointer over a scale. If the helicopter turns, the gyroscope precesses and rocks the gimbal. The faster the turn, the further the precessing gyro will be able to extend the centring spring. A small dashpot is provided to damp the motion of the rocking gimbal. This may consist of a piston sliding inside a cylinder having a small air bleed hole at the end. This reduces pointer movement due to vibration or turbulence and displays the average turn rate only. Helicopter turn indicators are generally designed to reach a scale mark at a rate of 180° per minute, corresponding to the two-minute turn rate commonly used in instrument flying. As the gyro is controlled by the centring spring, no action is needed to compensate for earth rate and the instrument has no controls.

The slip indicator is no more than a weight in a fluid filled curved glass tube. In a correctly banked turn, the apparent gravity should remain perpendicular to the cockpit floor and the weight stays in the centre of the curved tube at the lowest point. If the amount of yaw does not match the amount of bank, there is sideslip and this causes the weight to slide away from the centre of the tube. Figure 7.20 shows that, in a fixed-wing

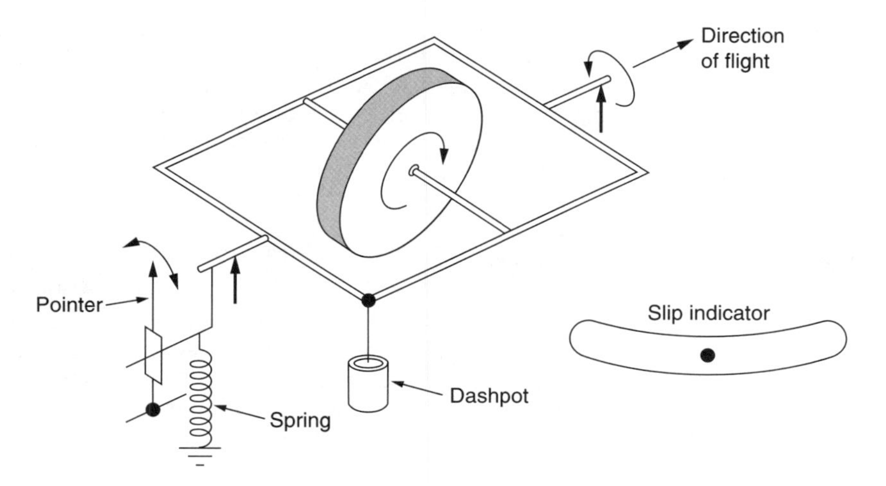

Fig. 7.19 The turn indicator is spring centred. In a turn, the rigidity of the gyroscope attempts to oppose the turn but it is overcome by a force from the spring. The tighter the turn, the greater the spring extends and the higher the reading. A conventional slip indicator is a curved tube containing a weight and damping liquid.

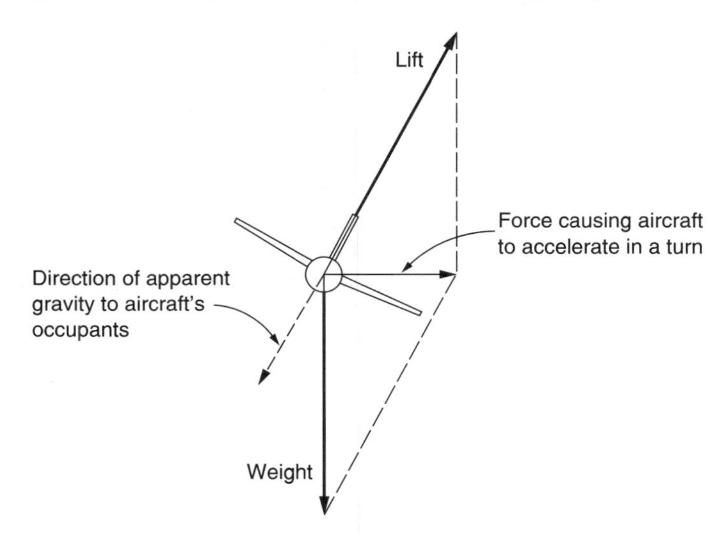

Fig. 7.20 In an aeroplane making a properly co-ordinated turn, the weight in the slip indicator remains central.

aircraft in a properly co-ordinated turn, the acceleration into the centre of the turn and the acceleration due to gravity combine to give an apparent gravity in the cockpit that is straight down through the floor. Thus it is possible to use a damped pendulum as a slip indicator.

As was seen in section 5.2, the conventional helicopter cannot balance main rotor torque with the tail rotor alone. There has to be another force in conjunction with the tail rotor in order to create a couple. In forward flight this may come from a slight rotor tilt or from side slipping the hull or from some combination of both. The least drag is obtained with zero slip, but in many helicopters the zero slip condition results in a hull that is not level in straight flight. If the slip indicator is installed vertically with respect to the hull, and if the machine is flown with the slip ball centred, the result may be sideslip, increased drag and a navigational problem because a side slipping machine is not going the way it is pointing. In addition, certain conditions of fuel level or loading may cause the CM of the machine to be offset to port or starboard of the centreline. This will further affect the slip gauge.

7.16 Attitude sensing

An autopilot needs to sense the attitude of the helicopter in each axis being stabilized, and the rigidity of the gyroscope makes an excellent attitude reference. In some cases it is possible to use the existing vertical axis gyroscope in an artificial horizon to provide pitch and roll data for an autopilot. When this is done, in addition to driving the visual display the gyro contains encoders to provide electrical signals proportional to the amount of pitch and roll.

Where high accuracy is required, a separate gyro having no display but providing attitude signals only can be used. Clearly the location of such a device in the airframe is subject to a good deal of freedom. If an inertial navigator is carried, the inertial platform must be tied to the earth and so can act as a reference to provide attitude signals in all three axes. The attitude data may be sensed by synchro generator or by digital encoder, according to the signalling technology used.

Automatic turn co-ordination in forward flight requires essentially a slip ball with an electrical output. This is the function of a device known as a dynamic vertical sensor. Figure 7.21 shows that it consists of a pendulum whose axis is fore and aft and which carries a transducer such as a synchro or other encoder. The pendulum will be damped against turbulence and vibration. In a correctly co-ordinated turn the pendulum will not move *with respect to the sensor* because the direction of the acceleration will be along the pendulum axis.

An alternative to the dynamic vertical sensor is to use a vane-type sensor to measure the direction of motion of the hull through the air.

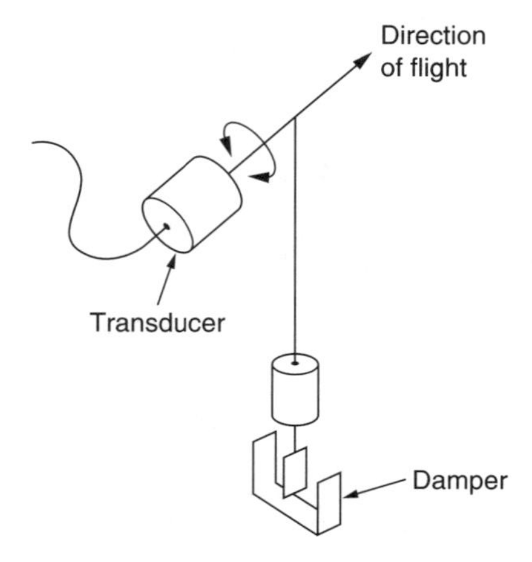

Fig. 7.21 The dynamic vertical sensor is essentially a slip indicator with an output signal. It consists of a damped pendulum having a transducer.

7.17 Airflow-sensing devices

The helicopter differs from the fixed-wing aircraft in that it can move in virtually any direction in three dimensions and at very low speeds. This makes airspeed measurement difficult. The conventional ASI with its pitot head is only accurate when the relative airflow approaches within 10–15° of the straight-ahead position. At greater angles, the pressure caused by flow of the rotor downwash around the hull may result in major errors, especially at low speeds where the dynamic pressures due to airspeed are very low. Helicopter pilots learn to treat ASI readings with suspicion or even amusement at low speeds especially when hovering in cross- or tail-winds.

With a conventional ASI, it is impossible to establish if a zero-airspeed condition exists, yet this is important to helicopter operations because of the possibility of entering a vortex-ring condition. Another requirement for accurate low airspeed information is in systems that control automatic stabilators. Special sensors have been developed for helicopters to overcome some of these difficulties.

Figure 7.22 shows a system that can measure horizontal airspeed in any direction. This consists of a rotating assembly with pitot heads on each end of an arm. The assembly can be mounted above the rotor to turn with it, or on the hull and turned by a motor. The dynamic pressure will only be constant if the horizontal airspeed is zero. In all other cases there will be a sinusoidally varying difference in dynamic pressure between the two pitots. The amplitude will represent the airspeed and the phase with respect to the rotational phase of the arm will represent the direction. A two-dimensional indicator may be used to display the fore-and-aft and lateral components of airspeed.

Figure 7.23 shows an alternative system consisting of a small finned swivelling body that will align itself with the downwash. This contains a pitot to measure the downwash velocity. From the downwash velocity and the angle of the body it is possible to compute the horizontal component of the downwash and this is the airspeed.

The slip string is a simple piece of string attached to the centre of the windscreen. If the machine is being flown without slip, the string will align with the airflow and take on a vertical attitude. If there is any slip the string will turn. Despite its apparent crudity, the slip string is a useful device because it does not suffer from the problems of the pendulum-based slip indicator. A machine flown with the slip string centred will have lower drag than one flown with the slip ball centred. This may be significant if power is limited. The machine will also fly along its heading, making navigation easier.

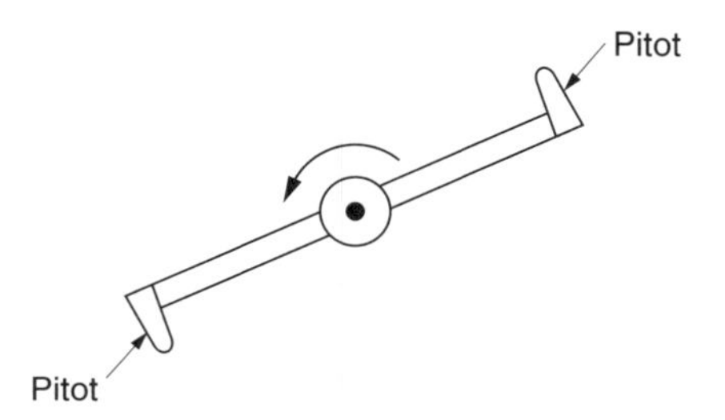

Fig. 7.22 A rotating airspeed indicator that can measure airspeed in any direction down to zero. The difference between the dynamic pressures at the two pitot heads determines the airspeed. The phase of the dynamic pressure signals with respect to the rotation reveals the direction.

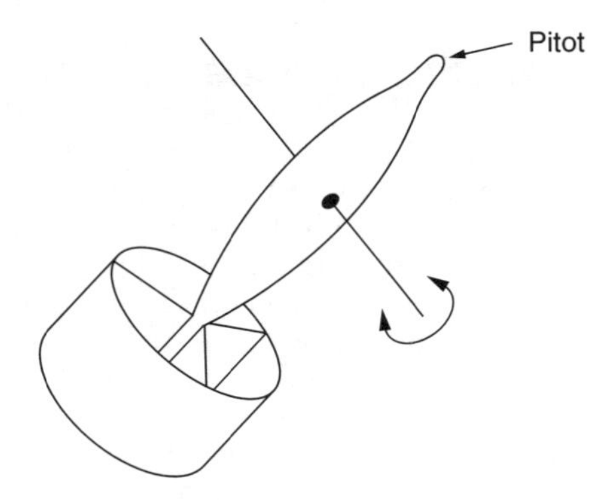

Fig. 7.23 A swivelling pitot head aligns itself with the rotor downwash. From the angle it makes to the vertical the airspeed can be calculated.

In a fancy version of the slip string, a small weathervane on the nose of the machine having an angular sensor drives a slip meter on the instrument panel.

7.18 RADAR sensors

The principle of radio direction and ranging (RADAR) is simply the analysis of reflected radio waves. In active RADAR, the radio signals are generated at the analysis equipment, whereas in passive RADAR signals may come from other sources. Radio signals are electromagnetic waves and these have the characteristic the way they interact with objects is a function of the relative size of the object and the wavelength. Very long wavelengths simply diffract around objects returning very little energy to the transmitter, whereas short wavelengths are reflected more efficiently, hence the use of short wavelengths in RADAR.

Figure 7.24 shows that two variables can be extracted from returned radio signals. At (a) a pulse is transmitted and the time taken for the reflection to arrive will allow the distance to the target to be computed. At (b) a continuous signal is transmitted and any relative velocity between the target and the transmitter will cause the frequency of the return to be shifted by the Doppler effect.

The RADAR altimeter is based on the first principle. Despite its name, it is not an altimeter. The transmitter and receiver are directed downwards and the signal reflected from the earth is used to compute height above ground, not altitude. This is extremely useful for terrain avoidance and for landing in poor conditions and may be used as the actual height input to an altitude hold system.

RADAR altimeters have some limitations. When flying over a tree canopy, the reading may be anywhere between ground level and treetop height. When flying with an underslung load, the RADAR altimeter may measure the length of the load cable.

Figure 7.25 shows how a Doppler RADAR works. The received signal is amplified and multiplied by the transmitted signal in a mixer. Any difference in frequency between the transmitted and received signals will be output by the mixer. This frequency is proportional to the axial component of relative velocity between the transmitter and the target. Figure 7.26(a) shows that as a practical matter, an airborne Doppler RADAR must use a transmitted beam which is angled downwards. The forward motion of the helicopter is not in the same direction as the beam. The system will measure the actual

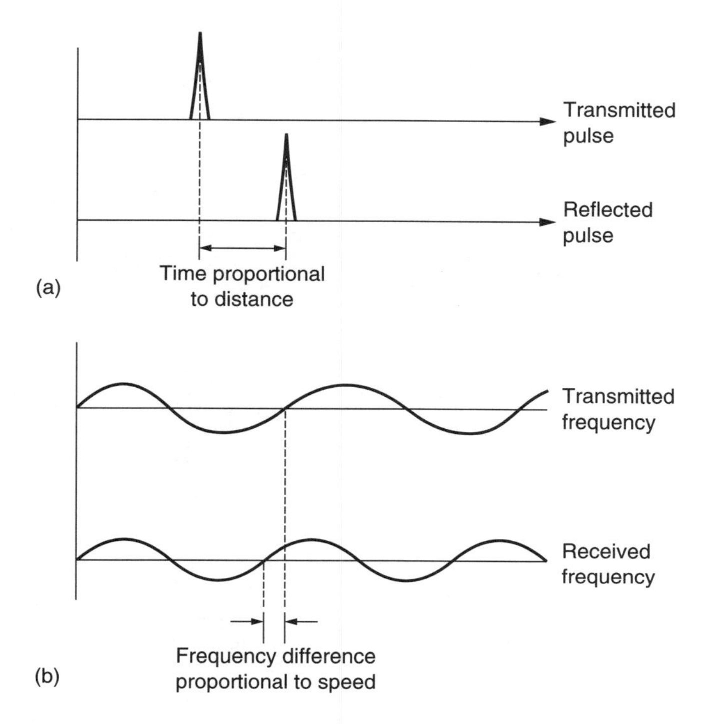

Fig. 7.24 Using RADAR signals. At (a) a transmitted pulse is timed to measure distance. At (b) the amount of Doppler shift measures velocity.

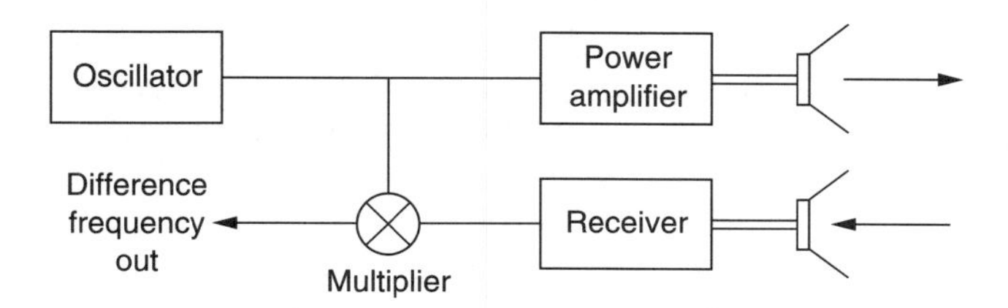

Fig. 7.25 Doppler RADAR mixes the transmitted and received signals to create a difference frequency proportional to target velocity.

ground velocity multiplied by the cosine of the beam angle. A signal processor in the Doppler receiver can make the necessary correction.

If two orthogonal (mutually at right angles) beams are transmitted, it will be possible to resolve the angular difference between heading (direction of travel through the air mass) and the track (direction of travel over the ground) from the relationship between the two Doppler velocities. If this information is compared with the true airspeed and the heading it will be possible to compute the wind velocity.

Manoeuvres will result in the angle of the beams changing with respect to the earth and this must be compensated using attitude signals from a vertical gyro or with a third beam.

Another useful characteristic of Doppler RADAR is that the number of cycles of the signal emerging from the mixer is proportional to the distance flown in the

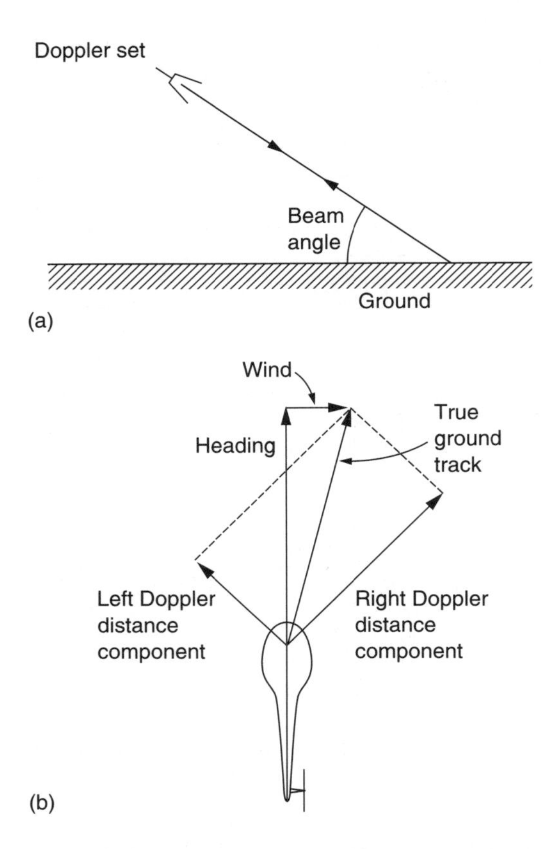

Fig. 7.26 (a) As the beam is angled downwards, the measured velocity must be geometrically compensated. (b) If two orthogonal Doppler beams are used, it is possible to extract distance and direction from the relationship between the outputs from the two beams.

direction of the beam. If the heading output of a gyromagnetic compass is combined with the Doppler-derived track-to-heading angle, the true track can be computed. Figure 7.26(b) shows that during a flight in wind, the orthogonal Doppler beams will both output distance components. The resultant of the two will give the ground distance flown. This is the principle of the Doppler navigator that has the advantage of needing no earth-based signals whatsoever.

7.19 Control signalling

Control signalling is the process of communicating the position of a mechanism from one place to another. There are many applications of such signalling in a modern aircraft: the position of one of the pilot's controls may be signalled to an actuator, the actual position of an actuator may be signalled to a feedback control system and the attitude of an artificial horizon may be signalled to an autopilot and to a RADAR set so that the attitude of the RADAR antenna remains stable during manoeuvres.

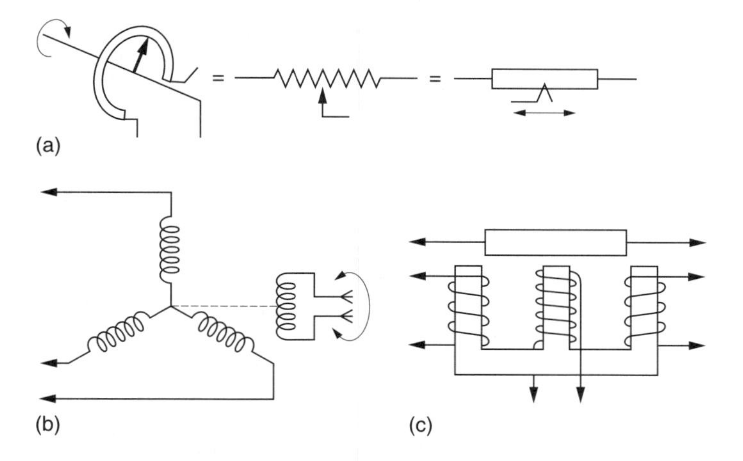

Fig. 7.27 Transducers used in signalling. At (a) the potentiometer or variable resistance. At (b) the synchro outputs three AC signals as a function of the rotor angle. The LVDT (linear variable displacement transducer) shown at (c) is a form of linear synchro. It has the advantage over the potentiometer that there is no sliding contact and no wear mechanism.

Figure 7.27 shows some methods used for control signalling. At (a) is the potentiometer. This is a resistance that outputs a DC voltage proportional to the position of the moving contact or wiper. The potentiometer may be rotary or linear and is simple, but the sliding wiper introduces a wear mechanism. The remaining devices are more complex but wear free. At (b) is the AC operated synchro generator that signals rotational angle to a synchro motor using three wires. The synchro is ideal for measuring the angle of a joystick or side-arm controller or the reading of a compass. In the synchro there is a rotor carrying a single coil which is fed by sliprings and brushes. Surrounding the rotor is a set of three coils mounted at 120° in a star arrangement. The synchro works like a transformer in that it must have an alternating supply. Traditionally this will be the 400 Hz AC used in aircraft and the supply is fed to the rotor via the sliprings. The current flowing in the rotor produces a magnetic field at right angles to the rotor shaft. The voltage induced in each of the secondary windings depends on the angle of the rotor. If the field of the rotor were parallel to a given secondary, the induced voltage would be a maximum, but if the rotor were at 90° to that secondary the induced voltage would be zero. The induced voltage in a given winding is proportional to the cosine of the angle between the rotor and that winding. With three windings there must be a unique combination of voltages for any position of the rotor.

At (c) is the linear equivalent of the synchro which is the LVDT (linear variable displacement transducer). This is also fed with an AC excitation signal. The moveable core alters the magnetic coupling between a pair of input coils and the output coil. The input coils are connected in anti-phase and with the core in the central position the flux from these two is cancelled out. If the core moves off-centre the amplitude of the output increases in proportion to the displacement. The phase of the output represents the direction. A phase-sensitive amplifier can extract a voltage proportional to the position of the core. As the LVDT signals linear position, it is ideal for measuring the extension of an hydraulic ram or of a ball screw actuator.

The synchro motor is almost identical to the synchro generator. The main difference is that it contains a rotary damper to prevent it spinning like an electric motor. The rotor of the synchro motor is fed with the same AC drive as the synchro generator by connecting the two in parallel. When the three windings of the synchro motor are fed with signals from a synchro generator, they produce a resultant flux direction with which the rotor aligns.

The synchro system is capable of high resolution but the synchro motor cannot develop much torque. If a significant load has to be positioned remotely, the synchro system will require power amplification. In this case the receiving device may be a control transformer. This is virtually identical to the synchro motor, the main difference being that the rotor winding is not connected to the AC supply but is instead connected to a phase-sensitive amplifier. Figure 7.7 showed the principle. A motor driving the load also turns the rotor of the control transformer. An amplifier senses the signal from the control transformer rotor and drives the motor.

The three signals from the synchro generator reproduce the flux direction of the generator's rotor in the control transformer. If the rotor of the latter is at 90° to that flux direction it will produce no output. However, if there is a misalignment, say clockwise, the rotor will produce an output which is a sine wave in-phase with the signal exciting the synchro generator. However, if the misalignment is anticlockwise, the rotor output will be in anti-phase with the synchro generator input. The amplitude of the rotor output will increase with the angular error. The phase-sensitive amplifier drives an electric motor to move the load. Using the phase information from the stator, the amplifier can drive the load in the correct direction to make the angular error in the control transformer smaller. When the rotor of the control transformer is at 90° to the induced field from the three windings the rotor output disappears and the motor stops. In this way a significant load driven by a motor can be controlled remotely by the low power signal available from the synchro generator.

7.20 Digital signalling

The potentiometer, the synchro and the LVDT are analog devices in that the signals they produce are infinitely variable. Instead of analog signalling, it is also possible to transmit control position numerically. In digital systems, a binary number proportional to the position of the control is encoded. All of these signals may be carried by electrical cable, leading to the marketing term 'fly-by-wire'. Digital signals can also be carried along optical fibres, leading to the term 'fly-by-light'.

Digital signalling is a technology that arrived in aviation quite suddenly and so there is no traditional knowledge or experience base. As a result many people regard digital signalling as little short of black magic when in fact it is based on relatively simple principles. One of the key concepts to grasp is that digital signalling is simply an alternative means of carrying the same information. An ideal digital signalling system has the same characteristics as an ideal analog system such as a synchro: both of them are totally transparent and remotely reproduce the original applied control movement with negligible error.

Although there are a number of ways in which the position of a control can be represented digitally, there is one system, known as pulse code modulation (PCM) that is in virtually universal use. Figure 7.28 shows how PCM works. Instead of being continuous, the time axis is represented in a discrete or stepwise manner. The control waveform is not carried by continuous representation, but by measurement at regular intervals. This process is called *sampling* and the frequency with which samples are taken is called the sampling rate or sampling frequency F_s. The sampling rate is generally fixed and is thus independent of any frequency in the signal.

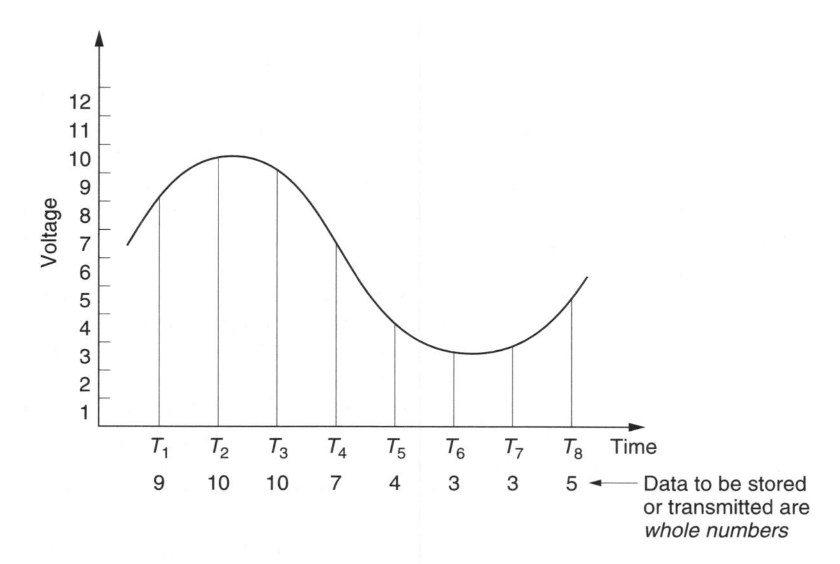

Fig. 7.28 In PCM or digital signalling, the position of a control is sampled periodically and the sample is expressed as a discrete number.

Those who are not familiar with digital signalling may worry that sampling takes away something from a signal because it is not taking notice of what happened between the samples. This would be true in a system having infinite response rate, but no real control system can be like this. All control signals from joysticks, actuators, gyros etc., have a distinct limit on how fast they can change. If the sampling rate is high enough, a control waveform can only change between samples in one way. It is then only necessary to carry the samples and the original waveform can be reconstructed from them.

Figure 7.28 also shows that each sample is also discrete, or represented in a stepwise manner. The value of the sample, which will be proportional to the control parameter, is represented by a *whole number*. This process is known as *quantizing* and results in an approximation, but the size of the error can be controlled until it is negligible. If, for example, we were to measure the height of humans only to the nearest metre, virtually all adults would register 2 metres high and obvious difficulties would result. These are generally overcome by measuring height to the nearest centimetre. Clearly there is no advantage in going further and expressing our height in a whole number of millimetres or even micrometres.

The point is that an appropriate resolution can also be found for a control system, and a higher figure is not beneficial. If a conventional pushrod control system is considered, each bearing and joint in the system has some backlash and the cyclic stick on a typical helicopter can always move some small distance before any motion is detected at the swashplate. If the movement step size of a digital system is of the same order as the backlash of a mechanical system, both systems have the same accuracy.

The advantage of using whole numbers is that they are not prone to drift. If a whole number can be carried from one place to another without numerical error, it has not changed at all. By describing control parameters numerically, the original information has been expressed in a way that is better able to resist unwanted changes. The amount of backlash in a mechanical system increases with the number of ball joints and bell cranks in a control run. The accuracy of a mechanical signalling system would decline as wear occurs. A digital system can transfer whole numbers from one end of the airframe to the other without any loss of accuracy at all and will not be affected by wear.

Essentially, digital signalling carries the original control parameter numerically. The number of the sample is an analog of time, and the magnitude of the sample is an analog of the value of the parameter. As both axes of the digitally represented waveform are discrete, the variations in the parameter can accurately be restored from numbers as if they were being drawn on graph paper. If greater accuracy is required, paper with smaller squares will be needed. Clearly more numbers are then required and each one could change over a larger range.

Humans expect numbers expressed to the base of ten, having evolved with that number of digits. Other number bases exist; most people are familiar with the duodecimal system using the dozen and the gross. The most minimal system is binary, which has only two digits, 0 and 1. BInary digiTS are universally contracted to bits. These are readily conveyed in switching circuits by an *on* state and an *off* state or in optical fibres by the two states of a light source. With only two states, there is little chance of error.

In decimal systems, the digits in a number (counting from the right, or least significant end) represent ones, tens, hundreds and thousands etc. Figure 7.29 shows that in binary, the bits represent one, two, four, eight, 16 etc. A multi-digit binary number is commonly called a word, and the number of bits in the word is called the word length. The right-hand bit is called the least significant bit (LSB) whereas the bit on the left-hand end of the word is called the most significant bit (MSB). Clearly more digits are required in binary than in decimal, but they are more easily handled.

The word length limits the range of a binary number. The range is found by raising two to the power of the word length. Thus a four-bit word has 16 combinations, and could set a control to only 16 positions. Clearly this would not be good enough for a flight control. However, a ten-bit word has 1024 combinations, which is close to 1000. A mechanical flight control that could be positioned to one part in a thousand would be considered remarkable.

In a digital signalling system, the whole number representing the value of the sample is expressed in binary. The signals sent have two states, and change at predetermined times according to some stable clock. Figure 7.30 shows the consequences of this form of transmission. If the binary signal is degraded by noise, this will be rejected by the

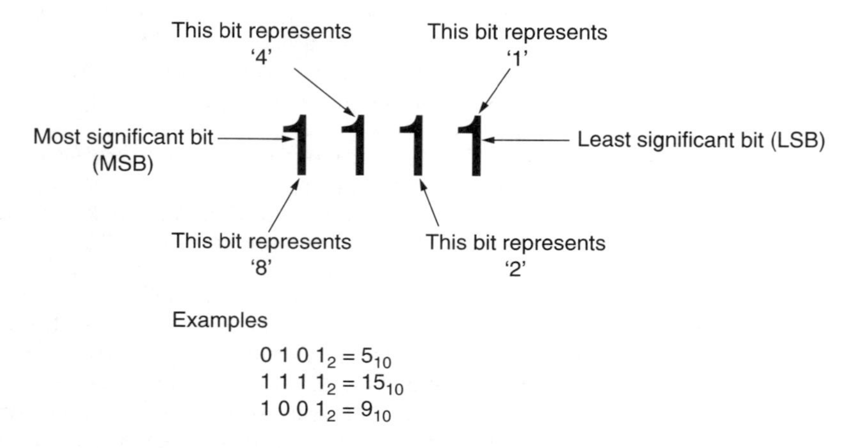

Fig. 7.29 Binary coding principles. See text for details.

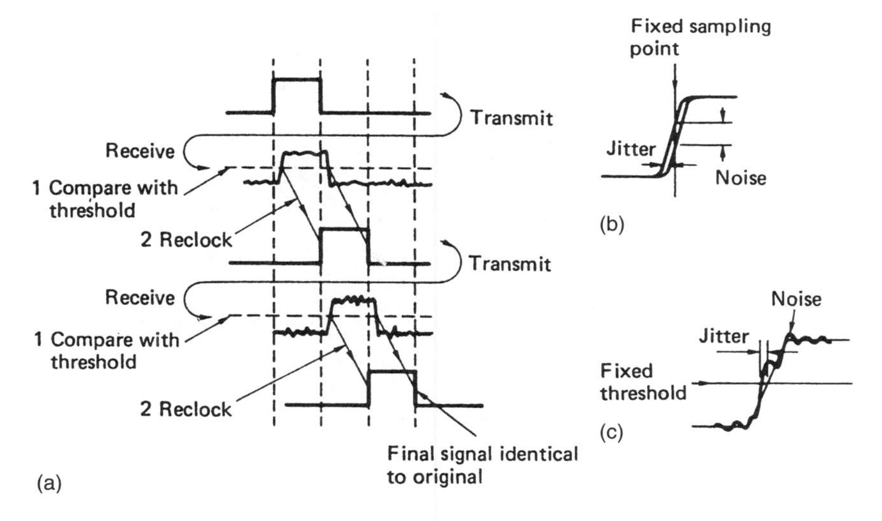

Fig. 7.30 Using slicing and reclocking, a binary signal can be recovered from a real noisy signal path without loss of information. See text for details.

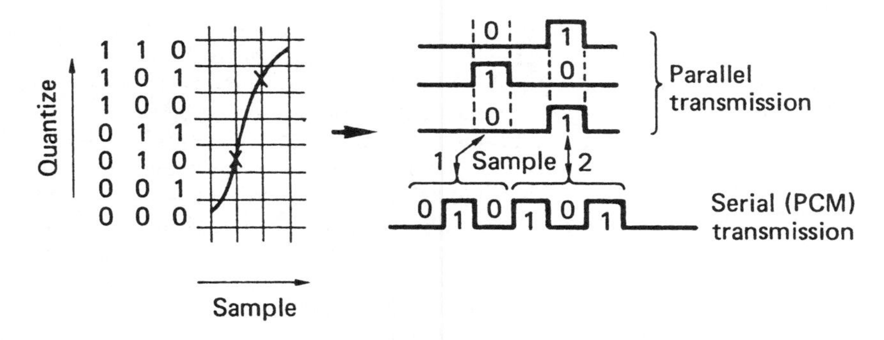

Fig. 7.31 Parallel and serial signalling. It is simpler and lighter to send the bits of a binary number one at a time down a single wire rather than use complicated multi-core cables.

receiver, which judges the signal solely by whether it is above or below the half-way threshold, a process known as *slicing*. The signal will be carried in a channel with finite bandwidth, and this limits the slew rate of the signal; an ideally upright edge is made to slope. Noise added to a sloping signal can change the time at which the slicer judges that the level passed through the threshold. This effect is also eliminated when the output of the slicer is reclocked. However many stages the binary signal passes through, it still comes out the same, only later. Control samples represented by whole numbers can reliably be signalled from one place to another by such a scheme, and if the number is correctly received, there has been no loss of information en route.

There are two ways in which binary signals can be used to carry control samples and these are shown in Figure 7.31. When each digit of the binary number is carried on a separate wire this is called parallel transmission. The state of signal on the wires changes at the sampling rate. Using multiple wires is cumbersome and heavy, and a

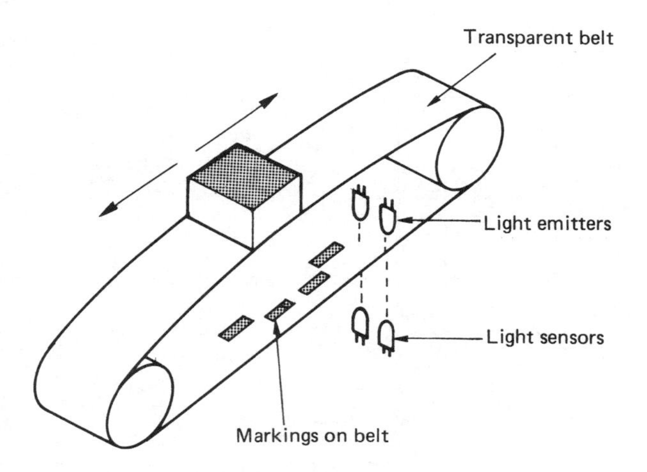

Fig. 7.32 A linear encoder in which the position of the moving part determines the binary code created.

single wire can be used where successive digits from each sample are sent serially. This is the definition of pulse code modulation. Clearly the clock frequency must now be higher than the sampling rate.

Whilst control samples can be obtained by digitizing the output of an analog control such as a potentiometer, it is preferable to obtain samples directly in a digital encoder. Digital encoders are a form of displacement transducer in which the mechanical position of the shaft or pushrod is converted directly to a digital code. Figure 7.32 shows an absolute linear encoder. A grating is moved with respect to several light beams, one for each bit of the control word required. The interruption of the beams by the grating determines which photocells are illuminated. It is not possible to use a pure binary pattern on the grating because this results in transient false codes due to mechanical tolerances. Figure 7.33 shows some examples of these false codes. For example, on moving the fader from 3 to 4, the MSB goes true slightly before the middle bit goes false. This results in a momentary value of $4 + 2 = 6$ between 3 and 4.

The solution is to use a code in which only one bit ever changes in going from one value to the next. One such code is the Gray code first devised to overcome timing hazards in relay logic but is now used extensively in position encoders. Gray code can be converted to binary in a suitable look-up table or gate array or by software in a processor.

For digital signalling, the prime purpose of binary numbers is to express the values of the samples representing the original control position. Figure 7.34 shows some binary numbers and their equivalent in decimal. The radix point has the same significance in binary: symbols to the right of it represent one-half, one-quarter and so on. Binary is convenient for electronic circuits, which do not get tired, but numbers expressed in binary become very long, and writing them is tedious and error-prone. The octal and hexadecimal notations are both used for writing binary since conversion is so simple. Figure 7.34 also shows that a binary number is split into groups of three or four digits starting at the least significant end, and the groups are individually converted to octal or hexadecimal digits. Since 16 different symbols are required in hex, the letters A–F are used for the numbers above nine.

There will be a fixed number of bits in a control sample, which determines the size of the quantizing range. In a ten-bit sample there may be 1024 different numbers. Each

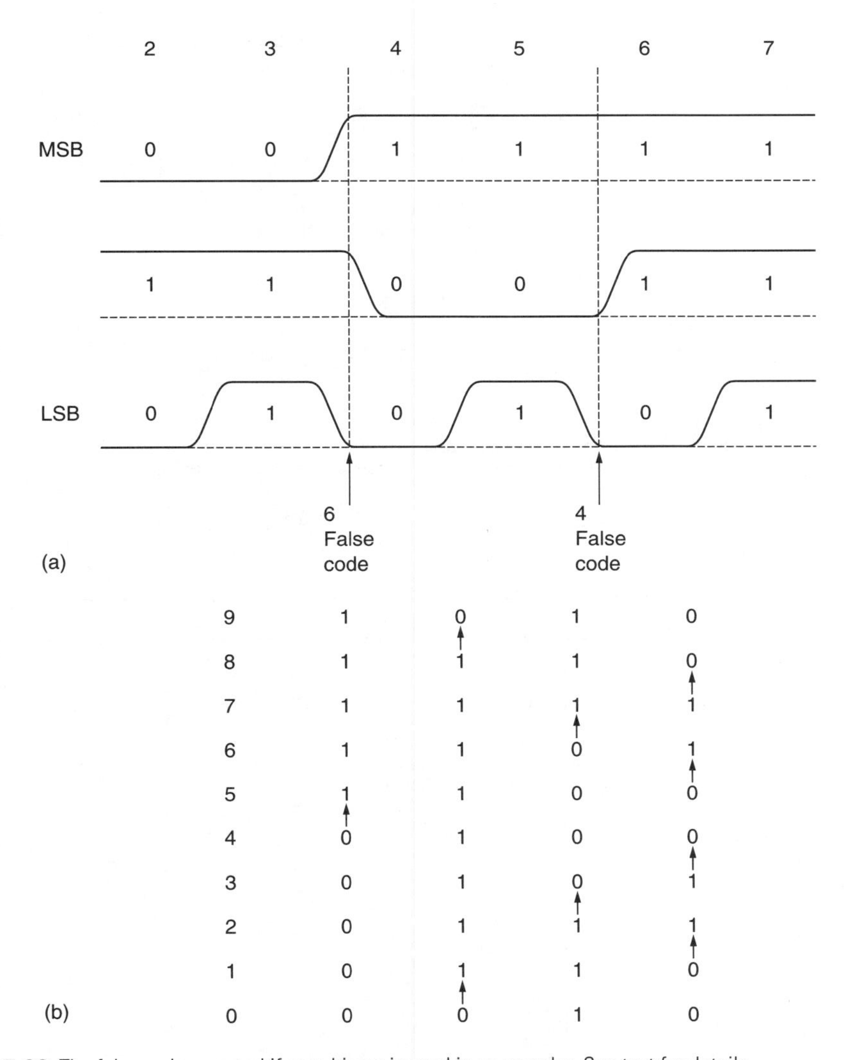

Fig. 7.33 The false codes created if pure binary is used in an encoder. See text for details.

number represents a different control position, and care must be taken during encoding to ensure that the control does not go outside the quantizing range, or it will be limited or clipped.

In aviation many controls are bi-directional and have a centre neutral position. In Figure 7.35 it will be seen that in a ten-bit pure binary system, the number range goes from 000 hex, which represents the largest negative input, through $1FF_{16}$, which represents the smallest negative input, through 200_{16}, which represents the smallest positive input, to $3FF_{16}$, which represents the largest positive input. Effectively, the neutral position of the control has been shifted so that both positive and negative inputs may be expressed as positive numbers only. This approach is called offset binary, and is perfectly acceptable where the control has been encoded solely for signalling from one place to another.

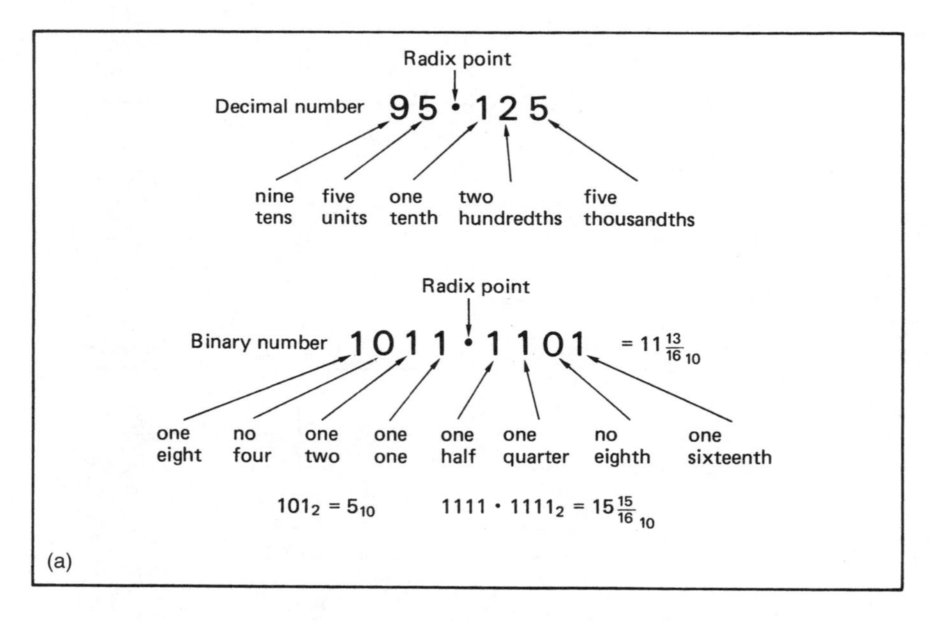

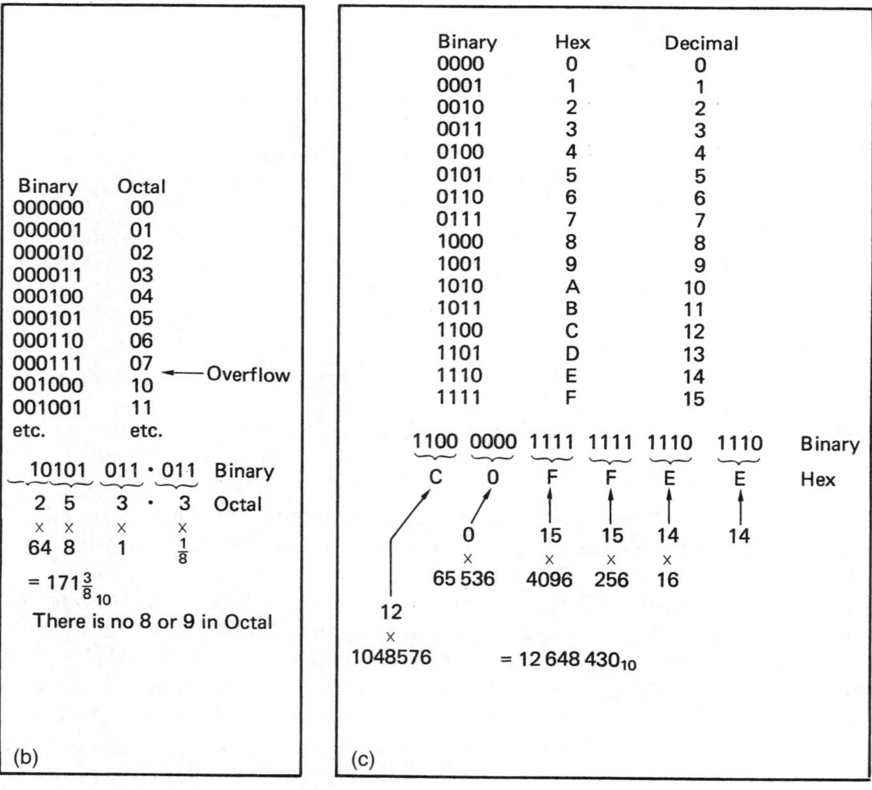

Fig. 7.34 Binary (a), octal (b) and hexadecimal (c) principles.

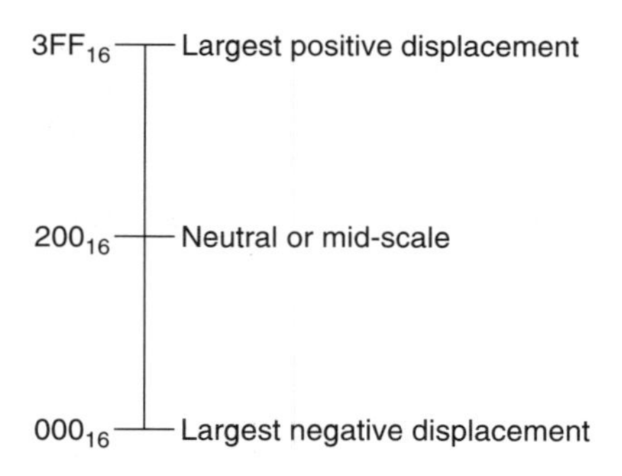

Fig. 7.35 In a pure binary system, only positive values are possible. A bi-directional system can be obtained by adding an offset so that neutral is half-way up the scale.

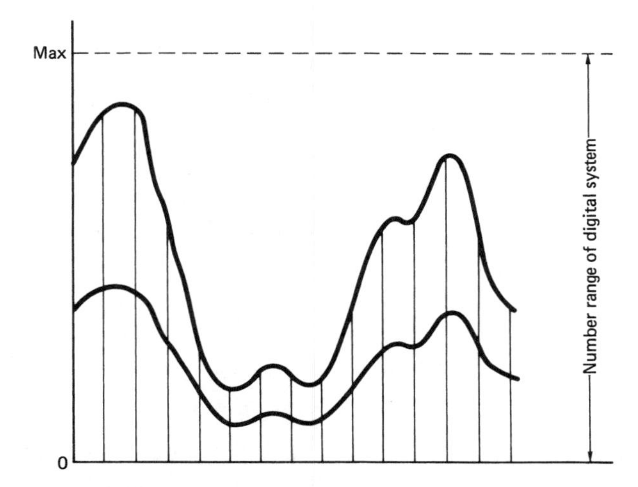

Fig. 7.36 Trying to reduce control travel by dividing all of the sample values by a constant produces an unwanted offset.

In practice it may be required to modify the control samples between, for example, a joystick and the associated actuator. The control may prove too sensitive and by processing the samples a given movement of the joystick could be arranged to produce a smaller movement of the actuator, perhaps as a function of airspeed. A small number might be added to the samples to perform a trim function. Two control signals might be added together to mix cyclic and collective controls into the swashplate or to mix the output of a stability augmentation system or autopilot into the flight controls.

If two offset binary sample streams are added together in an attempt to perform digital mixing, the result will be that the offsets are also added and this may lead to an overflow. Similarly, if an attempt is made to reduce the sensitivity of a control by, say, half by dividing all of the sample values by two, Figure 7.36 shows that the offset is also divided and the waveform suffers a shifted baseline. The problem with offset binary is that it works with reference to one end of the range. What is needed for aircraft controls is a numbering system operating symmetrically with reference to the centre of the range.

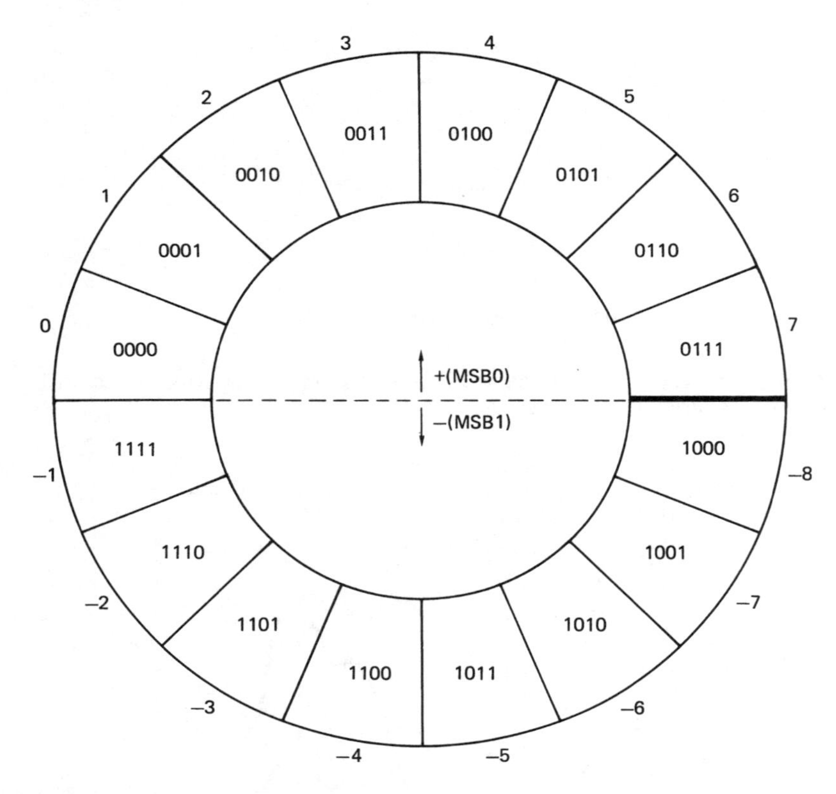

Fig. 7.37 Two's complement coding. See text for details.

In the two's complement system, the upper half of the pure binary number range has been redefined to represent negative quantities. If a pure binary counter is constantly incremented and allowed to overflow, it will produce all the numbers in the range permitted by the number of available bits, and these are shown for a four-bit example drawn around the circle in Figure 7.37. As a circle has no real beginning, it is possible to consider it to start wherever it is convenient. In two's complement, the quantizing range represented by the circle of numbers does not start at zero, but starts on the diametrically opposite side of the circle. Zero is midrange, and all numbers with the MSB set are considered negative. The MSB is thus the equivalent of a sign bit where 1 = minus. Two's complement notation differs from pure binary in that the most significant bit is inverted in order to achieve the half circle rotation.

Controls having limited travel, such as potentiometers and LDVTs, can also have their outputs converted to two's complement. Figure 7.38 shows how a real ADC (analog to digital convertor) is configured to produce two's complement output from, for example, a potentiometer. At (a) an analog offset voltage equal to one-half the quantizing range is added to the bipolar analog signal in order to make it unipolar as at (b). The ADC produces positive only numbers at (c) which are proportional to the input voltage. The MSB is then inverted at (d) so that the all-zeros code moves to the centre of the quantizing range. The analog offset is often incorporated in the ADC as is the MSB inversion. Some encoders are designed to be used in either pure binary or two's complement mode where the MSB inversion may be applied as required.

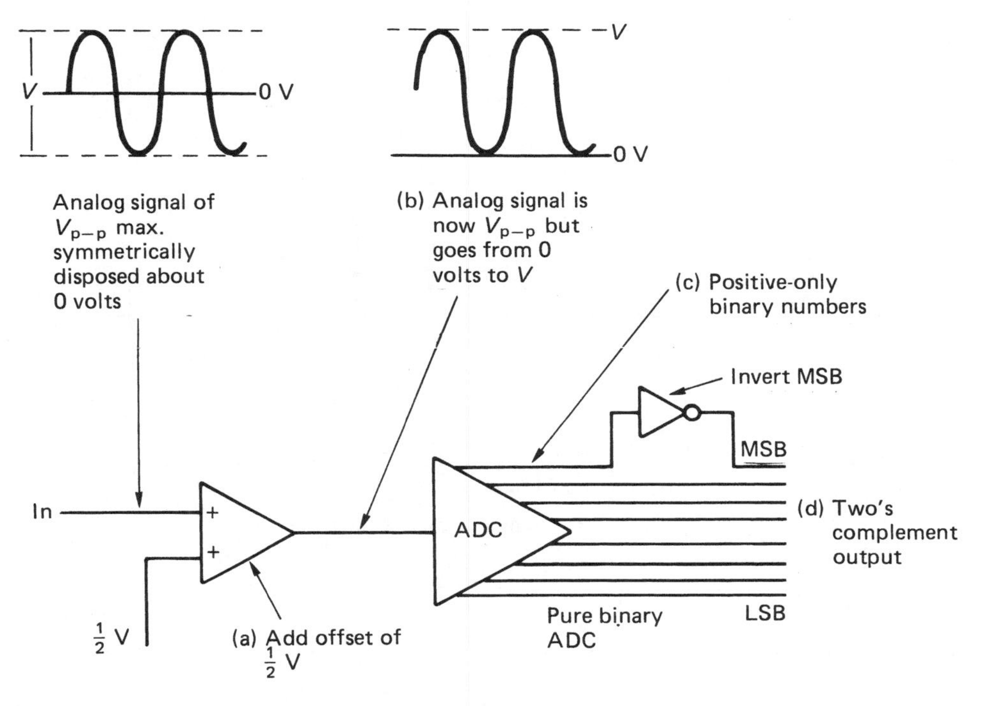

Fig. 7.38 Producing two's complement from real bipolar signals. At (a) an offset of half scale is added to the input. The offset input is converted (b) to pure binary. The MSB is inverted at (c) to produce two's complement.

The two's complement system allows two sample values to be added, or mixed, and the result will be referred to the system midrange; this is analogous to adding analog signals in a synchro differential or adding pushrod movements in a mechanical mixer.

Figure 7.39 illustrates how adding two's complement samples simulates the control mixing process. The waveform of input A is depicted by solid black samples, and that of B by samples with a solid outline. The result of mixing is the linear sum of the two waveforms obtained by adding pairs of sample values. The dashed lines depict the output values. Beneath each set of samples is the calculation that will be seen to give the correct result. Note that the calculations are in pure binary. No special arithmetic is needed to handle two's complement numbers.

It is often necessary to phase reverse or invert a control signal, for example in a cyclic/collective mixer the lateral cyclic signal needs to be added to the collective signal on one side of the swashplate but subtracted on the other. Using inversion, signal subtraction can be performed using only adding logic. The inverted input is added to perform a subtraction, just as in the analog domain. This permits a significant saving in hardware complexity, since only carry logic is necessary and no borrow mechanism need be supported.

The process of inversion in two's complement is simple. All bits of the sample value are inverted to form the one's complement, and one is added. This can be checked by mentally inverting some of the values in Figure 7.37. The inversion is transparent and performing a second inversion gives the original sample values. When a binary counter is incremented it will eventually reach the all-ones condition. A further count will result in an overflow condition. For example, if a four-bit counter is at 1111 (15 decimal) an

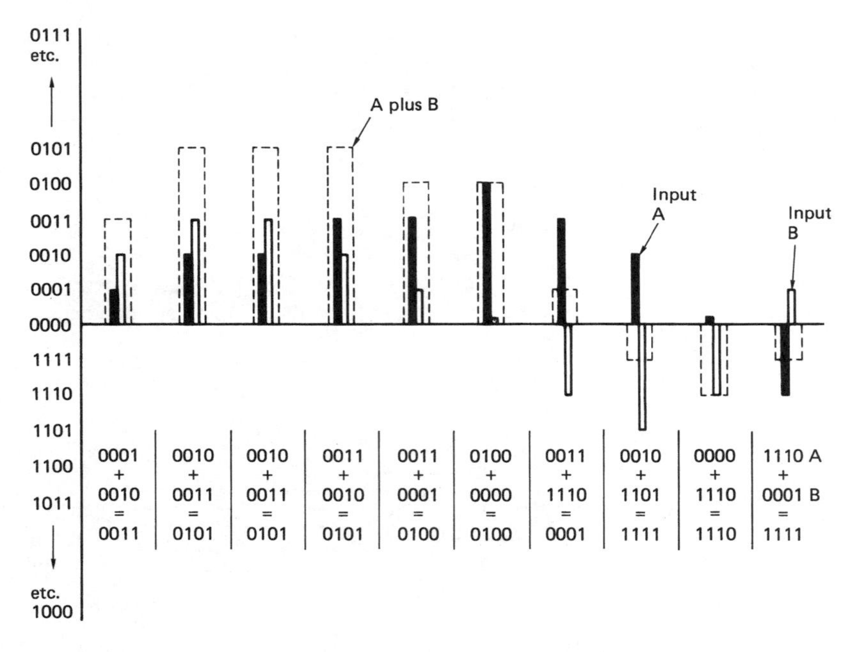

Fig. 7.39 Two's complement addition of bipolar signals for control mixing.

extra count will result in 10 000 (16 decimal) requiring five bits. Clearly there is no facility to handle the leading 1 in a four-bit system and so the remaining four bits will all be zero. Consequently the count starts again. This gives a finite word length counter a circular structure allowing it to represent the operation of an angular sensor. Using finite word length binary, angular parameters such as magnetic heading, variation, and track angle can be incorporated in navigational computations.

7.21 Power-assisted controls

Figure 7.40 shows the interior of an hydraulic power-assisted ram. The output rod carries a swinging arm operating the error valve. The input is connected to the arm by a ball joint. The error valve contains a pair of discs and is called a spool because it resembles a film spool. The spool controls the flow of fluid to the cylinder down passages in the pushrod. If the pilot moves a control such that the input moves to the left, the spool will slide left and the left disc will admit fluid under pressure to the right-hand side of the piston. The right disc will allow fluid from the left-hand side of the piston to return to the reservoir. The piston will move to the left under hydraulic pressure, but in doing so it will move the upper end of the input lever to the left, and will close the spool valve. In servo terminology, the linkage driving the spool valve is comparing the actual load position with the desired position to produce an error.

As a result of the error the output of the ram is made to follow the input. If the input reverses, the spool valve moves the other way and admits fluid to the other side of the piston. As the pressure is so high, only tiny discrepancies need exist between the input and output position for fluid flow to occur and so there is considerable loop

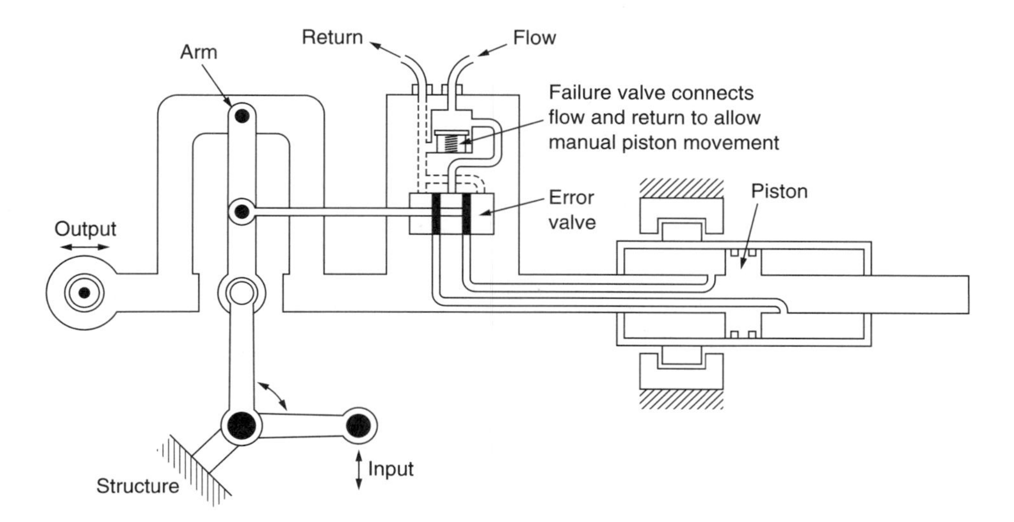

Fig. 7.40 A power-assisted hydraulic actuator. In the case of loss of hydraulic pressure the pilot's input still operates the control but without assistance.

gain. Clearly the actuator body moves and the hydraulic supply needs to be fed through flexible hoses.

In hydraulic rams as the error becomes smaller the orifices through which fluid flows become very small and the viscosity of the fluid provides damping. When the error is zero the flow is negligible and the main ram is effectively locked because there is no path for fluid to pass around the piston. As a result hydraulic actuators tend to be self-damping and in simple systems no tachometer feedback is needed.

It will be seen from the figure that the flight load is shared between the ram and pilot effort by the proportions of the swinging arm. In the case of loss of pressure, the input lever moves until it comes into contact with stops on the piston rod. The input force is thereby transmitted to the output. The controls will be much heavier and will display a slight backlash, but they are still effective.

7.22 Fully powered systems

In a fully powered system a direct mechanical connection between the pilot's controls and the load to be moved is pointless as without power the load could not be moved by the pilot alone. Instead the control signals from the pilot are compared with the actual position of the load, and any difference causes the power system to operate in such a way that the difference is minimized.

Figure 7.41(a) shows an example of a fully powered system in which a spool valve is operated by a balance lever which subtracts the load position from the commanded position.

Figure 7.41(b) shows a duplex actuator which consists of two hydraulic rams in tandem on a common pushrod. Each ram is supplied by a different hydraulic system. If pressure in one system is lost, the other ram can continue to provide control provided that the failed ram does not obstruct the working ram with an hydraulic lock. This can be arranged by providing valves to allow fluid to flow from one side of the piston to

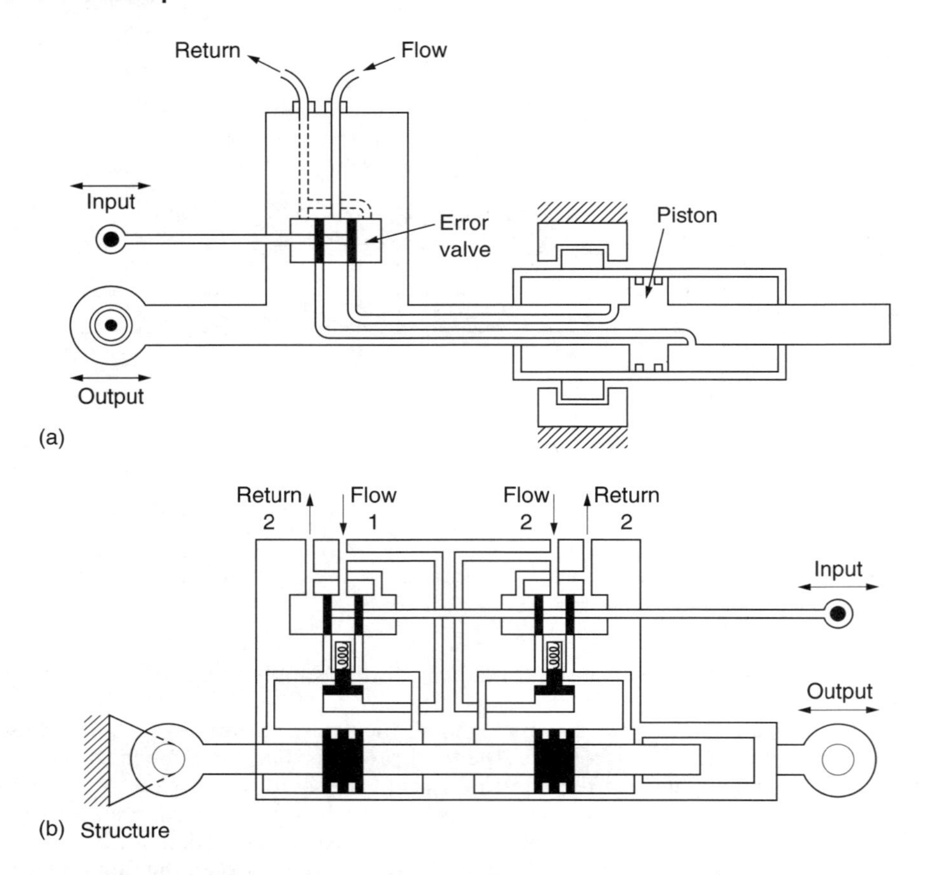

Fig. 7.41 (a) A fully powered actuator which is needed for large machines where the pilot could not operate the controls unaided. (b) For reliability, fully powered actuators are duplicated with separate sources of hydraulic power. Note ram bypass valves which open if hydraulic pressure is lost, allowing the other system to move the piston.

the other. This valve can be seen in the figure. When hydraulic pressure is available, the valve is forced shut, but if pressure is lost the valve opens so that the functioning actuator is not impeded.

In Figure 7.41 it would be possible to replace the pushrod from the pilot with some other type of signalling. If electrical signalling is used, the control ram will contain an electro-hydraulic valve (EHV) so that electrical signals can control the hydraulic power source. Figure 7.42 shows how an electro-hydraulic valve works. The valve is pulled or pushed by magnetic fields from the drive coils, and admits hydraulic fluid to the spool valve controlling the main ram. As fluid is allowed to flow, the resulting fluid pressure on the valve output is arranged to oppose the magnetic force. This tends to close the valve. Thus the fluid flow rate is made roughly proportional to the electrical current in the coil so the ram extends at a controlled speed.

Another possibility is for the mechanical pushrod input to an hydraulic actuator to be driven by a low powered electrically driven servo. This approach may be used in machines where stability augmentation or autopilots are optional. The electric servo makes it possible to input control signals to a conventional hydraulic actuator.

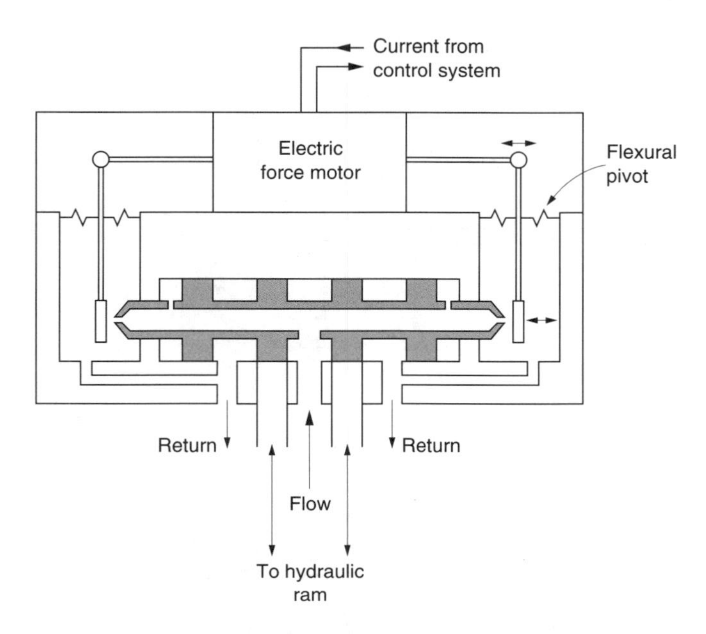

Fig. 7.42 An electro-hydraulic valve. Small electrical signals can control very powerful actuators with this mechanism.

7.23 Stability augmentation

Early helicopters were beset by weight and stability problems where solving one often made the other worse. A further issue is that by definition there was not initially a pool of experienced pilots to fly the first helicopters and any extra stability would be very welcome.

The flybar was an appropriate and elegant solution at the time and played a pivotal role in the development of the helicopter, but subsequent developments have made it obsolete in the full size, although it is still found in most model helicopters.

Early helicopter blades were relatively lightly built because power was limited. This resulted in a rotor with low inertia having a high following rate in response to a cyclic control input or an external disturbance. Making the blades heavier would reduce the following rate, but the weight could not be accepted. The flybar essentially stabilizes the machine by reducing the rate at which the rotor can change its attitude. Although the flybar works, it raises the drag and complexity of the rotor head and causes a performance penalty. Later systems were developed which sensed the machine's attitude using miniature gyroscopes. Correcting signals were then fed into the powered control system. In a sense the flybar was miniaturized and put in a black box.

7.24 The Bell bar

Figure 7.43(a) shows how the Bell flybar system works. This was developed by Arthur Young, using initially electrically driven models. It was subsequently used on a number of full-sized machines including the Bell 47 and the Huey. Use is restricted to two-bladed rotors. The flybar is mounted across the rotor on a bearing parallel to the feathering

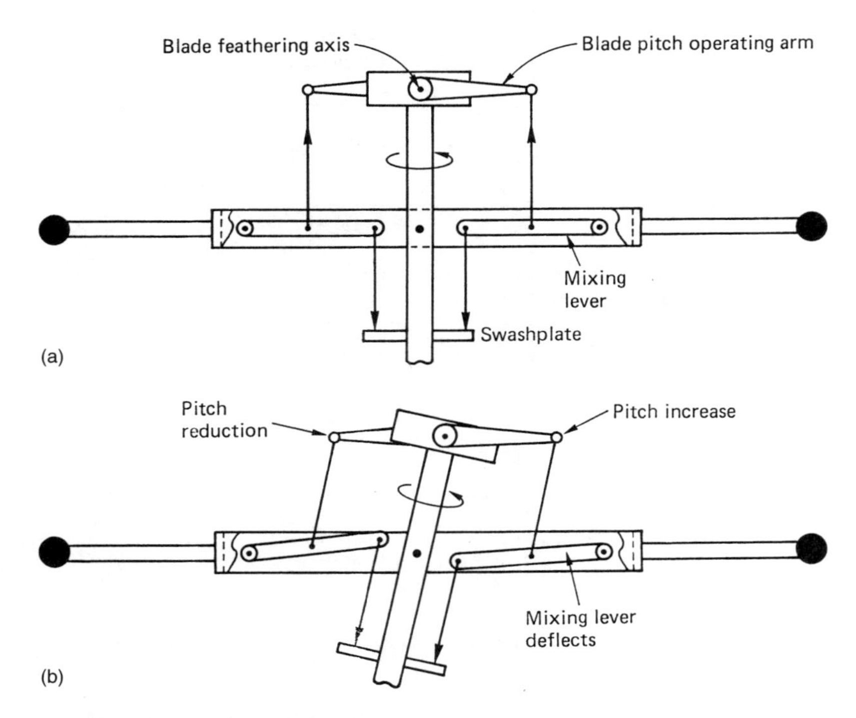

Fig. 7.43 (a) The Bell/Young flybar system. The flybar is essentially a gyroscope. (b) In this example the main rotor will stay parallel to the plane of the flybar because there is a unity relationship between the flybar attitude and the amount of cyclic pitch applied.

axis of the blades. The control rods from the swashplate are attached to a pair of mixer arms pivoting in the flybar. Further rods from the mixer arms then operate the pitch control arms of the blades.

It will be seen from Figure 7.43(b) that if the swashplate remains fixed, with a suitable linkage the rotational axis of the flybar effectively can become the control axis of the rotor, because the blades tend to stay parallel to the flybar. This is called a unity mixing system. A smaller stabilizing effect is still obtained if the feathering of the blades is some proportion of the flybar motion.

The flybar is influenced by dampers on a bar fixed to the rotor shaft. If the flybar rotational axis and the shaft axis are different the dampers will oscillate because the flybar is effectively flapping with respect to the shaft. This action tries to bring the flybar axis parallel to the shaft axis at a rate proportional to the divergence between the axes.

If the flybar is tilted in any axis, the maximum flapping velocity is reached at 90° to that axis. As the dampers produce a couple proportional to velocity, this couple will be about an axis at 90° to the flapping axis. This couple will cause the flybar to precess into alignment with the rotor shaft. Thus the flybar is a gyro loosely tied to the rotor shaft. The flybar axis will align with the shaft axis when the machine is in equilibrium, but in any manoeuvre the divergence will be proportional to the rate of change of the attitude of the shaft axis, i.e. the roll rate or the pitch rate of the hull. Thus the angle of the flybar relative to the mast is a measure of the roll rate of the helicopter.

The flybar is gyroscopic so it will tend to hold a constant rotational axis. If, in steady flight or in the hover, the tip path plane of the rotor is disturbed by a gust, the tip path axis and the flybar axis diverge, which must result in cyclic feathering. This will counter the disturbance and restore the attitude of the disc. The resultant helicopter is very stable because the action of the flybar resists disturbances to the disc attitude.

The action of the flybar when the cyclic control is applied is also beneficial. The mixing levers also allow the flybar to oppose the cyclic input from the swashplate. If the swashplate is tilted to initiate a roll, the blades will cyclically be feathered through the mixing levers and the roll will commence. However, the flybar is measuring the roll rate, and the flybar axis will diverge from the shaft axis. In fact the flybar axis lags behind the shaft axis. This divergence applies an opposite feathering action to the swashplate.

For a given application of cyclic control, the blade cyclic feathering will be nulled for a given angle of divergence between the flybar axis and the mast axis. As this divergence is proportional to the roll rate of the helicopter, it follows that the roll rate actually achieved is proportional to the swashplate tilt. Thus the rotor system is turned into a rate control whose following rate is determined by the inertia of the flybar and the viscosity of the dampers. As the flybar axis can only roll slowly, the following rate of the rotor to the cyclic controls is reduced, giving the effect of a heavier rotor but without the actual mass.

The response of the rotor can be adjusted by changing the degree of damping. If the dampers are made more viscous, the following rate increases because the flybar is more powerfully tied to the shaft axis.

The action of the flybar can be analysed by treating it as a servo. Figure 7.44 shows that the swashplate supplies the desired roll rate, and the flybar measures the actual

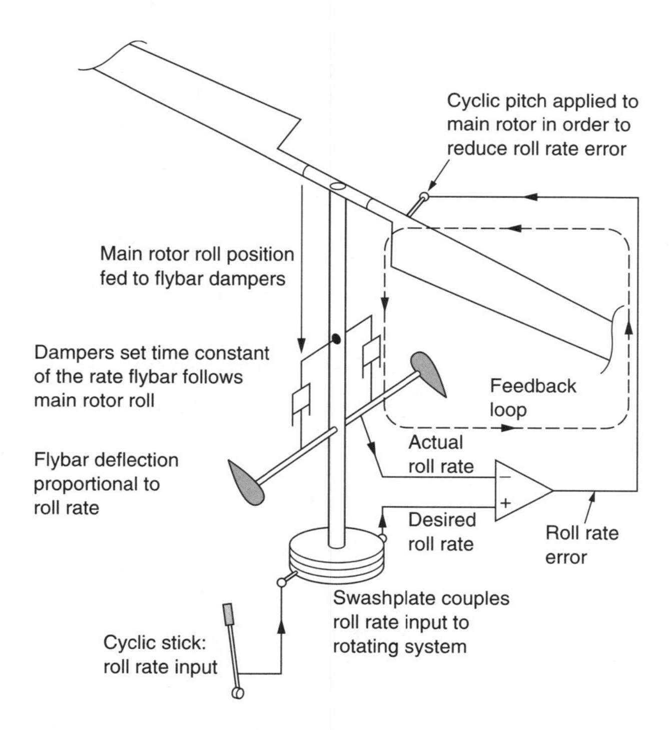

Fig. 7.44 The Bell flybar acts like a servo, adjusting cyclic pitch to obtain the desired roll rate. If the flybar is tilted with respect to the mast, the dampers will gradually bring it back into alignment. The rate of response to the controls is proportional to the degree of damping.

roll rate. The mixing levers subtract the actual roll rate from the desired roll rate to produce the roll rate error. This applies cyclic feathering in the sense that reduces the error. Thus the error is kept small such that the actual roll rate is equal to the desired roll rate.

7.25 The Hiller system

The second major flybar development was due to Stanley Hiller and this was used in many of the early Hiller machines. The main difference between the Young and the Hiller systems is that in the Hiller the flybar has aerodynamic properties as well as gyroscopic properties and so is properly called a control rotor. Figure 7.45 shows that the control rotor is mounted transversely on a bearing as before, but there are two major contrasts with the Young system. First, the Hiller control rotor has full authority over the cyclic pitch of the main rotor, such that the control rotor axis completely determines the control axis of the main rotor. Second, the Hiller control rotor has small blades or paddles and it has feathering bearings. Pitch control arms on the control rotor lead to the swashplate so that cyclic pitch control of the control rotor is obtained. The pilot's cyclic control actually flies the control rotor, whose attitude will then be followed by the main rotor.

The following rate of the control rotor can be made as slow as required by changing its Lock number, i.e. the relationship between the mass of the paddles and their lifting area. Thus as before, relatively light blades can be made to respond relatively slowly. The control rotor will not respond rapidly to gusts and so the attitude of the machine is stabilized as it is in the Young system.

The Hiller system has the advantage that the pilot only has to produce enough force to feather the control rotor. The control forces to feather the main rotor are created by lift on the paddles. Thus not only does the Hiller system provide following rate control

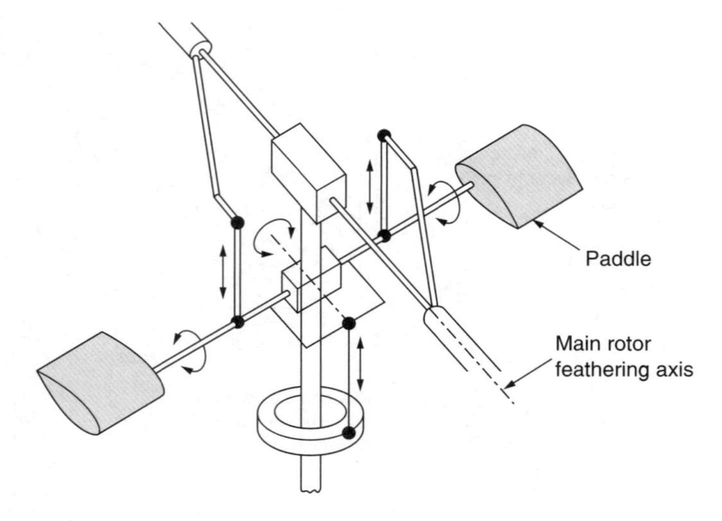

Fig. 7.45 In the Hiller system, the flybar is gyroscopic as in the Bell system, but damping is aerodynamic. The flybar also provides power assistance through the application of cyclic pitch to the paddles.

and autostabilization, it also provides power-assisted cyclic control, with the penalty of an increase in drag due to the paddles.

The Hiller control system incorporates another unusual feature in that the swashplate works in reverse to that of a conventional helicopter. Figure 7.46(a) shows that in order to simplify the control runs, Hiller took an arm from the non-rotating part of the swashplate straight down to the pilot. The pilot would move the end of this arm in the same way as he would a conventional floor-mounted cyclic stick. In order, for example, to apply aft cyclic, the pilot would pull the end of the arm towards himself. This, however, had the effect of tipping the swashplate *down* at the front. Hiller solved this problem by connecting the rotating part of the swashplate to the control rotor with a scissors linkage that reverses the action.

Figure 7.46(b) shows that when the swashplate tilts forwards, the control axis of the control rotor tilts backwards as required. When the overhead control was replaced by a conventional floor-mounted cyclic in later machines, the reversing swashplate was retained and the cyclic pushrods were simply rigged to make the swashplate move in the correct sense.

It is interesting that many examples of the Bell 47 and the Hiller Raven are still flying, although the Bell 47 first flew in 1943 and the Hiller in 1948. The longevity of these designs is testimony to the original concept. The flybar was also adopted in the UH-1 Iroquois, better known as the 'Huey' which itself achieved legendary status.

7.26 The Lockheed systems

The Bell and Hiller systems are ideal for two-bladed teetering rotors, but it is not easy to apply them to rotors with a larger number of blades. Irven Culver at Lockheed developed stabilizing systems to work with rotors having any number of blades. Culver had rightly concluded that flapping and dragging hinges were not necessary and set out to build hingeless rotors. Without flapping hinges, the cyclic response of a rotor can be extremely rapid and exert huge forces on the hull. Gyroscopic stabilization was essential to contain control response.

The Lockheed CL-475 was a flying test bed that sported a variety of rotors and stabilization systems. Initially a two-bladed rotor was used having no flapping hinges and no collective pitch control. Lift was controlled by rotor speed. Cyclic pitch was provided by a single span-wise hinge. A flybar was mounted transversely to the rotor. This superficially resembled a Bell flybar, but only because of the streamlining of the tip weights. The attitude of the flybar with respect to the mast controlled the cyclic pitch of the rotor and the pilot's cyclic stick applied forces to the flybar via springs and the swashplate to make it precess. As the main rotor changed its attitude, pitch link loads would make the gyro bar follow, serving the same function as the dampers in the Bell system. A second gyro stabilized the yaw axis by controlling the tail rotor pitch.

In the Bell system some of the travel of the flybar is lost in the mechanical mixer needed to add in the cyclic control from the swashplate. The direct control of the main rotor by the gyro in the Lockheed system allowed greater loop gain and was thus more stable. However, the two-bladed hingeless rotor suffered from excessive 2P hop in forward flight and work concentrated on hingeless rotors with more than two blades.

In 1960 a new three-bladed rotor was designed, again having no flapping hinges. Initially, each blade had its own single-ended flybar known as a gyrobar. This is shown in Figure 7.47. In the gyrobar system, the feathering axis is not parallel to the blade, but instead the feathering axis is swept back with respect to the blade. As was shown in section 4.7, this causes coupling between the flapping and feathering axes. If the rotor

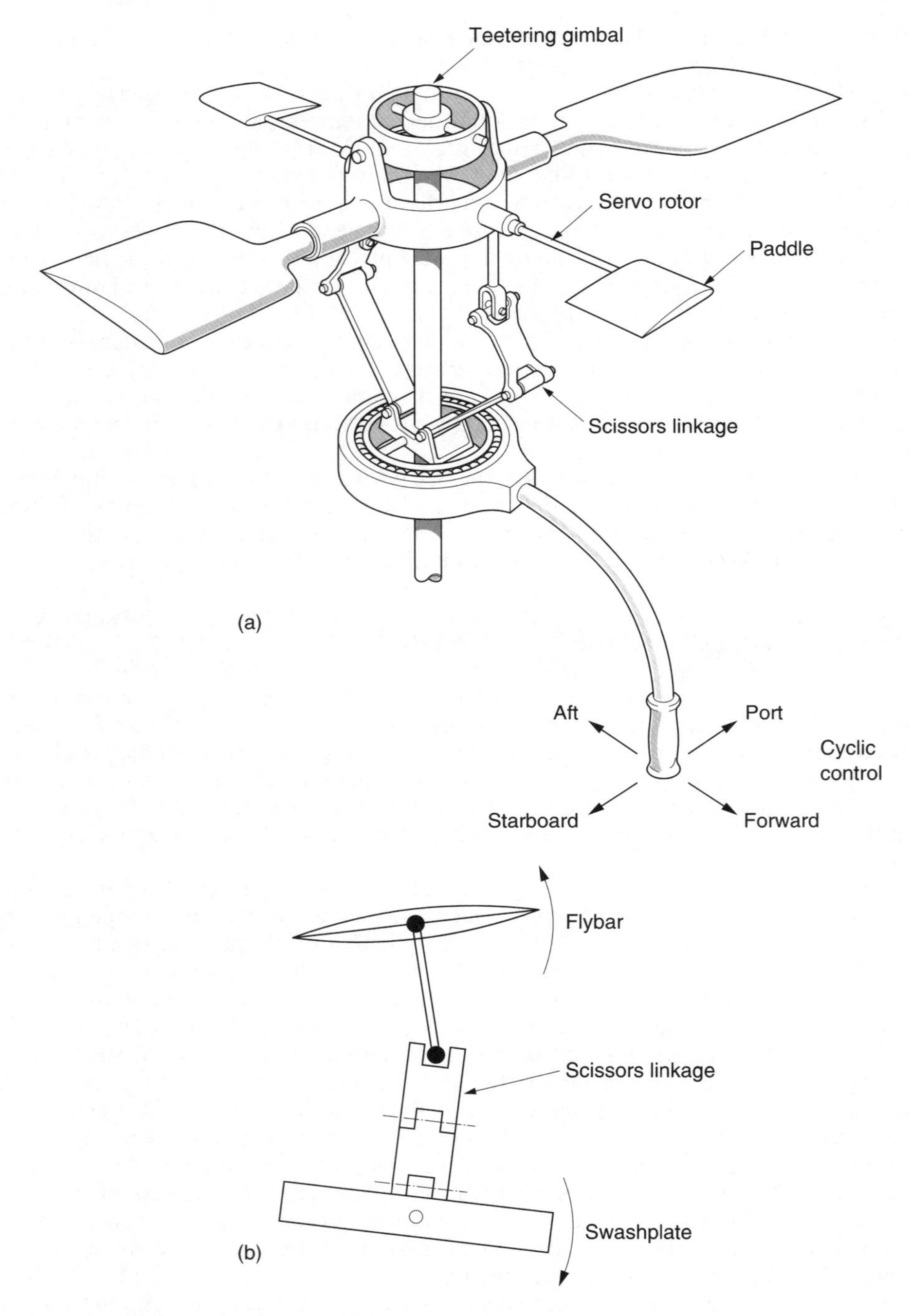

Fig. 7.46 (a) Early Hiller helicopters had an arm leading directly from the swashplate to the pilot. (b) A reversing mechanism is needed between the swashplate and the flybar to allow the control system of (a) to work in the correct sense.

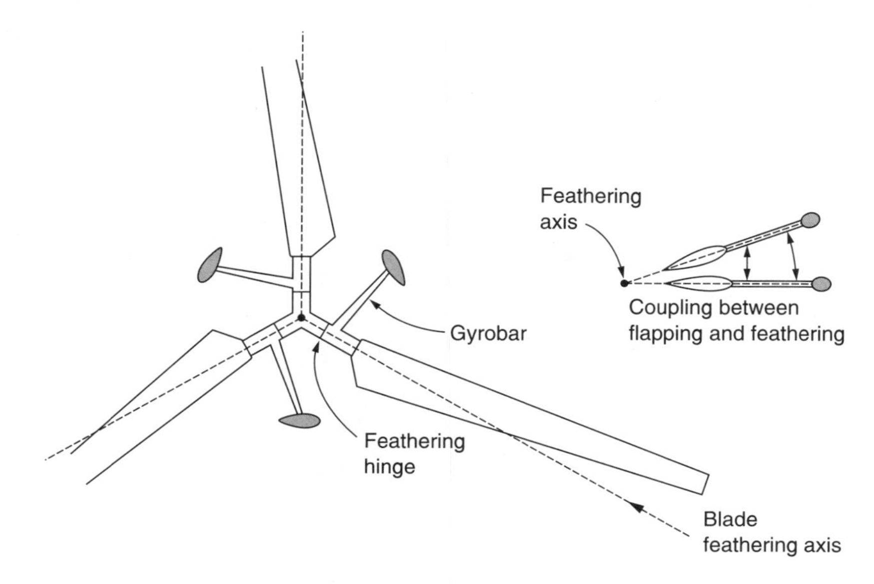

Fig. 7.47 The Lockheed gyrobar system in which each blade has its own flybar. See text for details.

disc should tilt for any reason, the inter-axis coupling causes the feathering bearing to turn slightly, and a couple is applied to the gyrobar. If a given blade is imagined to rise, an upward couple will be applied to its gyrobar. The gyrobar will then precess by 90° causing the blade pitch to increase 90° after the stimulus. As the blade is also gyroscopic, the blade would respond by rising a further 90° after the stimulus. As a result the blade responds at 180° to the stimulus and is thus stable.

Cyclic control is obtained by applying couples to the gyrobar using the swashplate. The inertia of the gyrobar means that it will control the following rate of the main rotor.

In 1961 a ring gyro having direct control of blade cyclic pitch was fitted below the rotor. A collective pitch mechanism was also fitted so that constant RPM could be used. In this form the CL-475 was widely demonstrated and achieved a reputation for stability that represented the state of the art at the time. The machine still exists with the ring gyro and is at Fort Rucker at the time of writing.

Subsequently the Model 286 (XH-51) was developed. This hingeless machine initially had three blades, but changed to four blades again for vibration reasons. The ring gyro is replaced by a rigid four-legged flybar, again directly coupled to the cyclic pitch arms of the rotor blades such that the flybar axis forms the control axis of the main rotor. Gyroscopic rigidity of the flybar means that any unwanted disturbance of the main rotor attitude due to gusts is automatically opposed.

The Lockheed system achieves cyclic control by applying control couples to the flybar via the swashplate to make it precess in the required direction. The cyclic control system is unlike that of any other helicopter because it provides forces, not displacements. This is no more than a scaled-up version of the use of torque motors to erect gyroscopic instruments as was seen earlier in this chapter.

Figure 7.48 shows that couples are applied to the flybar by a system of springs. Each of the control runs, pitch and roll, contains a pair of springs, one positive and one negative. With the cyclic stick at neutral, the geometry is such that if the swashplate moves because the flybar has tilted, one spring is compressed and the other extends

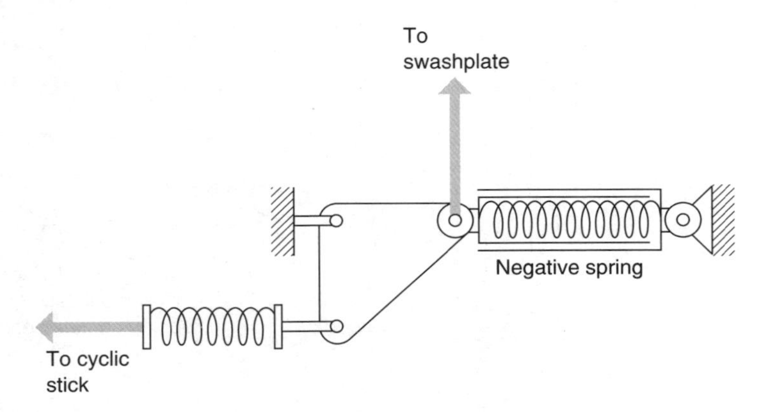

Fig. 7.48 In the Lockheed system the flybar is caused to precess by the application of couples. A double-spring arrangement allows the precession without interference from the other cyclic axis. See text.

by a similar amount. The springs will balance each other over a significant range of movement of the flybar. Thus if the flybar tilts it will not experience a restoring force from the controls. However, if the controls are displaced from neutral, the spring imbalance is upset and the result is a couple applied to the flybar.

Control is achieved as follows. If the pilot wishes to roll the machine, this can only be done if the flybar is made to roll. As the flybar is gyroscopic, it can only be made to roll by applying a couple in the pitch axis. Consequently the controls are rigged such that the roll cyclic pushrod applies a pitch couple to the flybar through the spring system. The flybar will sense this couple and correctly precess with a 90° lag to perform the commanded roll. The flybar is allowed to precess in this way by the balanced springs in the pitch pushrod.

The hingeless model 286 was aerobatic and performed loops and rolls with ease, whilst remaining easy to fly. Essentially the same control approach was adopted in the first models of the Lockheed Cheyenne compound helicopter (see Chapter 9). The high speed of this machine increased the loads that the rotors fed into the gyro through the pitch links, undermining gyro authority. The solution was the system shown in Figure 7.49. Lockheed called this the AMCS (advanced mechanical control system). Here the gyro has been moved inboard, below the transmission. The absence of an external flybar reduces drag. The gyro is fitted with a swashplate through which it controls the mechanical inputs to hydraulic actuators. These provide an irreversible control over the cyclic pitch so that pitch link forces from the blades cannot reach the gyro. The hydraulic actuators operate a spider that passes through the centre of the gyro and the transmission and emerges above the rotor head where it operates pitch links.

The gyro is supplied with couples by the pilot's cyclic control using the positive and negative balanced spring system as before. These are not shown in the diagram. Having eliminated pitch link loads with the hydraulic actuators, the gyro is now made to follow the hull attitude using springs attached to the mast.

Lockheed's AMCS represented the ultimate mechanical helicopter gyrostabilizer, but even as it was being tested, systems were being developed using miniature gyros and electronic amplification and the future of helicopter stabilization would lie in that direction.

Flybars are still found in flying model helicopters. Models are harder to fly than the full size and the flybar allows a welcome degree of stability for the pilot who won't have

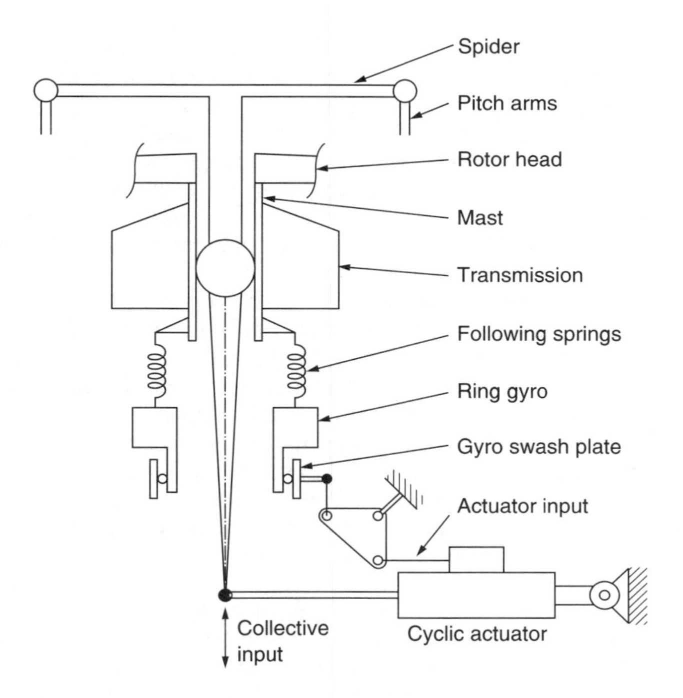

Fig. 7.49 In the Lockheed AMCS, the gyro is caused to precess by springs as before, but the gyro motion controls actuators that drive the swashplate. This prevents control forces disturbing the gyro.

the same level of training as his full-size counterpart. Beginners often fit extra weights to the flybar to reduce the following rate even further. The flybar also allows the use of inexpensive and lightweight wooden blades whose following rate alone would be too fast.

The Bell system was difficult to replicate in miniature because of the hydraulic dampers, and the pure Hiller system was found to produce a very low following rate in models. As a result a hybrid system combining both and aptly called the Bell-Hiller system evolved. This uses the cyclic mixing of the Bell bar, as well as the Hiller paddles and control rotor feathering. There are thus three rods coming from the swashplate: one to each of the flybar mixer arms, and one to feather the control rotor. This arrangement gives a good balance between stability and response, although the number of links involved makes it look like a dog's dinner. The falling cost of the piezo-electric gyroscope and its minute size mean that the flybar can now be eliminated in the model helicopter.

7.27 Autopilots and AFCS

Autopilots vary in complexity, but the general goal is that the pilot is relieved of the need to maintain some parameter of the flight. In a helicopter it is possible to begin with stabilization of the roll axis only, and then add extra features. Automatic control of the tail rotor allows balanced turns and flight at minimum drag. Control of the pitch axis allows the airspeed to be held, and altitude can be held with collective control. Some systems can achieve a stable hover in a fixed location without pilot intervention.

Achieving any or all of this requires an intelligent combination of flight parameter sensing, error measurement, power-operated controls stabilization and ergonomics. In addition to these basics there are some additional criteria. In the same way that two pilots cannot simultaneously have control, it is not possible for a human pilot and an automatic system to have full authority over the helicopter at the same time.

In stability augmentation systems, also called AFCS (automatic flight control systems), pilot and AFCS *share* control so that the pilot actually flies the helicopter but the added inputs from the AFCS make the helicopter follow the pilot's wishes more readily.

In simple autopilots there may be no control sharing at all and some mechanism needs to be provided so that the pilot can *engage* or *disengage* the autopilot so that one or other has full authority. These control reconfigurations must take place smoothly so that there is no sudden disturbance of the machine when the autopilot engages or disengages. Clearly safety is paramount and autopilots must use extensive interlocking so that they will not engage if correct conditions are not present and so that they will disengage and hand back control to the pilot in the case of a failure. In a simple system the servo output is physically disconnected from the controls when the pilot has control, but is connected by a clutch when the servo has control. Clearly the clutch is either engaged or disengaged and so this is not a control sharing arrangement and would not be appropriate for an AFCS.

An autopilot that maintains the machine on an absolutely stable course is in one sense ideal, but in practice the course or altitude needs to be changed from time to time. If the autopilot had to be disengaged and re-engaged every time a course change was required its value in reducing pilot workload would be diminished. Thus it is necessary to provide a means whereby the pilot can direct the flight *through* the autopilot. In fixed-wing practice this happens relatively infrequently and it is adequate to give the pilot an auxiliary set of control wheels or knobs which are used to direct the autopilot to new flight conditions without touching the main controls.

Figure 7.2 showed ways of dividing control between a servo and a pilot. Either a series or parallel arrangement can be used to share control between a small electric servo and the pilot on the mechanical input to a conventional hydraulic actuator. This approach is appropriate if the AFCS or autopilot are optional equipment because the same hydraulic actuator can be used with or without the servo. Where the autopilot is standard equipment, the electric servo can be dispensed with. Figure 7.50 shows a parallel-connected hydraulic actuator designed to allow full control authority by the pilot with the autopilot disengaged and full autopilot authority when engaged. Fluid flow to the main ram is controlled by one of two spool valves. One of these is mechanically operated by pushrod from, for example, the cyclic stick to give pilot authority. The other is operated by electro-hydraulic valve (EHV) to give autopilot authority. Note that the ram is grounded to the airframe and the body of the actuator moves.

Figure 7.50 shows that with the autopilot disengaged the EHV and its spool valve are isolated from the main ram by the autopilot select valve which is closed by spring pressure. If the pilot moves the input pushrod this will displace the manual spool valve and admit fluid to the ram so that the actuator moves bodily ('follows up') until the spool valve is in the neutral position again. In this mode the system is working as a conventional power-assisted actuator as was described in section 7.21.

Provided all of the interlock conditions are met, when the pilot engages the autopilot, the engage interlock output signal operates a solenoid valve admitting hydraulic pressure to a cylinder. This drives a tapering lock pin that moves the manual spool valve to the neutral position, locking it to the actuator body so that no relative movement is

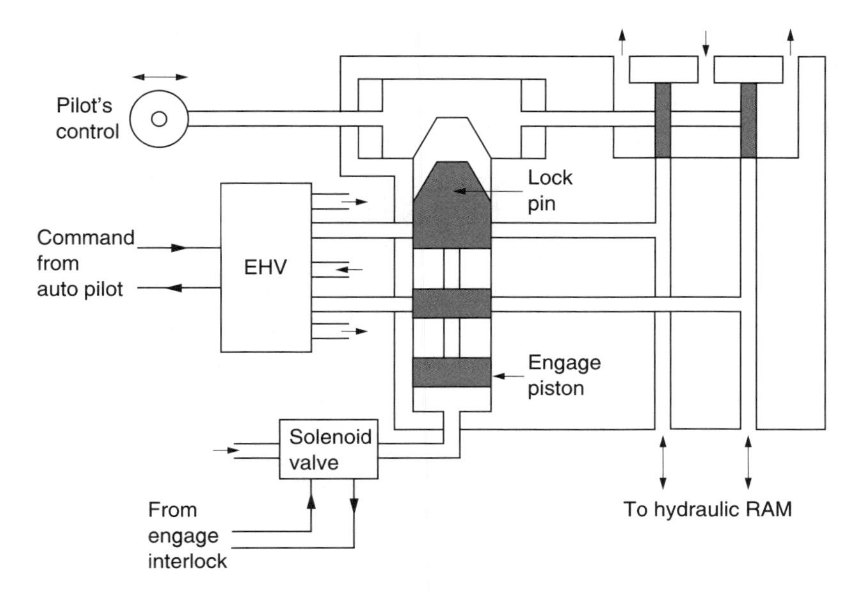

Fig. 7.50 With the autopilot disengaged, the pushrod from the pilot's controls operates the spool valve upper right. With the autopilot engaged, the solenoid valve admits pressure to the engage piston which moves up, locking the pilot's pushrod to the actuator and connecting the EHV outputs to the hydraulic ram.

possible. The second opens the autopilot select valve connecting the EHV spool to the main ram.

Now the autopilot can move the control by supplying a signal to the EHV and can measure the control position using the feedback signal from the LVDT in the ram. Note that as the pilot's controls are locked to the actuator body by the engage cylinder, they will move as the autopilot operates, giving the pilot confirmation that the machine is literally flying itself.

Two nested feedback loops are involved here. The outer loop consists of the attitude gyro to produce an error if the machine turns around the stabilized axis for any reason. The attitude error will compensate for factors such as the moment of inertia of the helicopter about the controlled axis and will then be fed to the actuator loop. The actuator loop converts the attitude error into a control movement in such a sense that the error is cancelled. The actuator feedback loop stabilizes the controls and gives the servo stiffness against flight loads. The machine will maintain the attitude dictated by the gyro indefinitely.

When the autopilot is engaged it has full authority. If the pilot tries to move the controls of a parallel system, the stiffness of the hydraulic actuator will oppose him. The LVDT will sense an unwanted movement and the autopilot will operate the EHV to oppose it. Instead the autopilot control panel carries knobs the pilot can turn to give the autopilot a new reference. Effectively the pilot's control knob adds an offset to the output of the reference sensor. For example, if the pilot wished to make a constant rate turn, he would set the turn control knob to an appropriate angle. This would add an offset to the roll axis gyro output that the autopilot would sense as an attitude error. The autopilot would roll the helicopter until the attitude error was apparently cancelled. Thus only a 5° bank angle signal from the gyro would cancel a −5° bank input from the control knob. The helicopter would indefinitely remain at a bank angle of 5° and

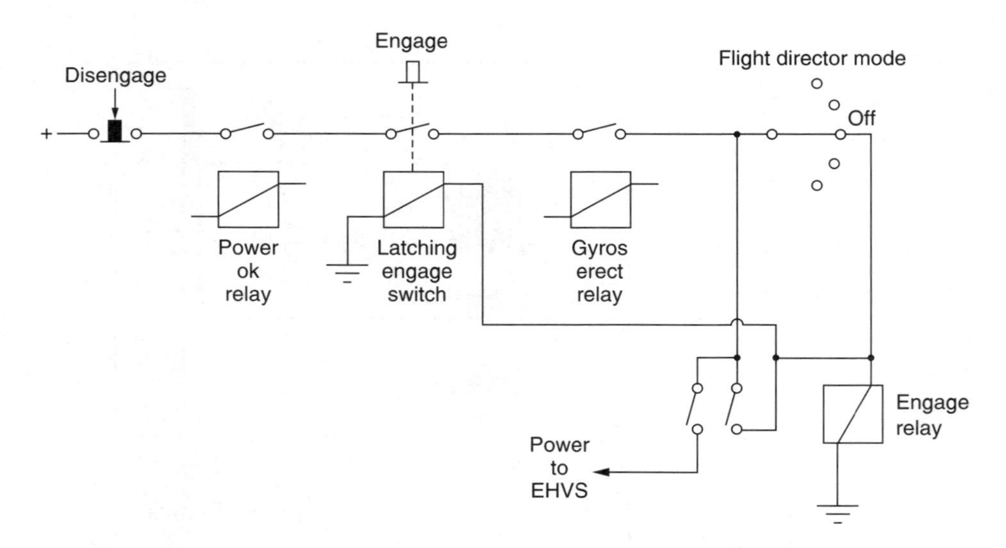

Fig. 7.51 Autopilot interlock. The autopilot cannot be engaged unless all of the necessary conditions are present.

make a constant rate turn. Note that a hazard could exist if the autopilot were to be engaged with the control knobs inadvertently set away from neutral. Accordingly the control knobs have centre detents that operate interlock switches so that engagement is impossible unless the control knobs are in a safe position.

Figure 7.51 shows an illustrative autopilot interlock. A number of conditions must be met before power can reach the engage relay. All necessary power must be present, flight directors must be off, the gyros must be erected and the control knobs must be neutral before the engage switch receives power. The engage switch is spring biased to return to the off position, but if all the interlock conditions are met it will be held in the engage position by a solenoid. When the engage switch is operated this feeds power to a relay which applies power to the solenoid, holding the switch in the engaged position. If the interlock conditions are not met the engage switch will return to the off position as soon as it is released.

The relay has further contacts which bypass the control knob centre switches so that these can subsequently be moved. Power is also fed to the autopilot solenoid in the actuator to enable the EHV. If while the autopilot is engaged any autopilot interlock condition fails, the holding current to the engage switch will be interrupted and the switch will spring back to the off position. The pilot may also disengage by pressing a button on the control column to break the holding current.

In the modern AFCS, the control knobs have been dispensed with. Instead the autopilot reference can be changed intuitively using the normal controls. When the autopilot is engaged, in a traditional system the manual controls are locked and any attempt to move them would be met by the rigidity of the actuator. However, if a force sensor is fitted in the control stick or in a pushrod it is possible to sense that the pilot is trying to move the stick. Figure 7.2 showed that the stick force signal can be used to modify the AFCS reference attitude so that the AFCS acts to implement the new attitude. In doing so, the AFCS moves the pilot's stick. When all of the parameters are correctly set, the pilot's stick moves by virtually the same amount for a given manoeuvre with or without the AFCS. Systems of this type are intuitive and much more appropriate for helicopters which are generally flown at lower altitude and in closer proximity to potential hazards. Such autopilots can be left engaged at all times. If the pilot wishes to perform an emergency manoeuvre, it is not necessary to disengage the autopilot; an instinctive movement of the controls is all that is needed. The autopilot acts as an AFCS and makes the machine execute the manoeuvre more accurately. However, if a small dead-band is incorporated in the stick transducer, the autopilot can sense when the pilot has released the stick and will then maintain the last attitude indefinitely.

An autopilot that can only maintain attitude is not enough. For prolonged use the ability to maintain airspeed and altitude is also required. For instrument landing the ability to maintain a specified rate of descent is essential, as is the ability to move smoothly from altitude control to descent control as the glide slope is entered.

An encoding altimeter as described in section 7.10 produces an actual altitude signal and this is subtracted from the desired altitude set on a dial by the pilot. The resulting altitude error will be cancelled by climbing or descending using the collective control. When cruising at a fixed altitude the necessary rates of climb and descent will be very small, but if the pilot directs the system to a new altitude, a significant rate of climb may result. In the absence of a stabilator, this will increase the downwash on the tail plane and give an aft trim effect. However the vertical axis gyro will sense this as an attitude error and try to cancel it with forward cyclic. This will produce undesirable mast bending and vibration. A solution is to feed a small proportion of the collective climb command across to the cyclic system to modify the pitch axis reference allowing a slightly tail down attitude for the duration of the climb.

The climb will also require more torque and although the yaw control of the autopilot would handle this a better result may be obtained if the measured torque is fed into the yaw system.

Airspeed is easily obtained from the dynamic pressure at the pitot head. In a sophisticated system, the static pressure may also be used to compensate the indicated airspeed (IAS) to true airspeed (TAS). The pilot enters the desired airspeed on a dial and a subtraction produces the airspeed error. An increase in airspeed requires the application of forward cyclic to tilt the thrust vector, an increase in collective pitch to maintain altitude, a change of lateral cyclic trim because of inflow and a change of tail rotor pitch because of the increased torque.

In a simple system the airspeed control might simply apply a forward pitch command to the vertical attitude reference. The resulting loss of height would then be sensed by the altitude hold system that would increase collective to maintain height. The increased torque would cause the yaw stabilizer to operate and the roll attitude would be held against inflow changes by the roll axis stabilization. However, all of these changes could take some time to stabilize. Instead of letting an error occur and then allowing the automatic system to cancel it, it is more accurate to anticipate the error by issuing an appropriate command. This is known as feedforward. For example, if the amount of tail rotor pitch needed to cancel torque is known for all combinations of power and airspeed, then if a power change is commanded, the appropriate pitch change can be added directly to the tail rotor. In this way the tail would not yaw at all during a power change unless the feedforward had not been correctly assessed. Part of the development phase of an autopilot system is to map the feedforward parameters over the entire flight envelope.

Feedforward of this kind may be used in manually controlled helicopters. For example, the collective lever position can usefully be linked to the tail rotor pitch. An increase in collective pitch will require more rotor drive torque and a corresponding increase in tail rotor pitch. By passing the tail rotor cables over a system of pulleys operated by the collective lever, the amount of pedal input needed, for example when entering autorotation, is reduced.

Fast forward flight requires a combination of forward cyclic to oppose the advancing/retreating lift difference and raised collective to handle increased inflow. In some helicopters, as the collective lever is raised a degree of forward cyclic may automatically be applied to reduce the amount of retrimming needed.

7.28 Coupled systems

Once an autopilot is available which is sensitive to external direction, automatic navigation becomes possible. Using inertial sensing, radio navigation, Doppler or other means the course can be corrected automatically to bring the machine to a desired destination. This is known as *coupling* and a unit known as a *flight director* allows the pilot to select which control source is coupled to the autopilot.

In the *heading* (HDG) position of the flight director, the yaw reference for the autopilot becomes the magnetic-compass-locked DI such that the helicopter will maintain a fixed magnetic heading. In the VOR position of the flight director, the yaw reference becomes the bearing of a VOR (VHF omni-range) beacon so that the machine can fly directly towards (or away from) the beacon. Figure 7.52 shows that a VOR beacon radiates in the VHF aircraft band and is modulated with two low frequency sinusoidal signals. One of these is radiated in the same way in all directions, whereas the relative phase of the other signal is a function of the direction of radiation. The in-phase condition is aligned with magnetic north and the phase lag is equal to the magnetic bearing on which the radiation leaves the beacon. The beacon also transmits its identifying signal in Morse code.

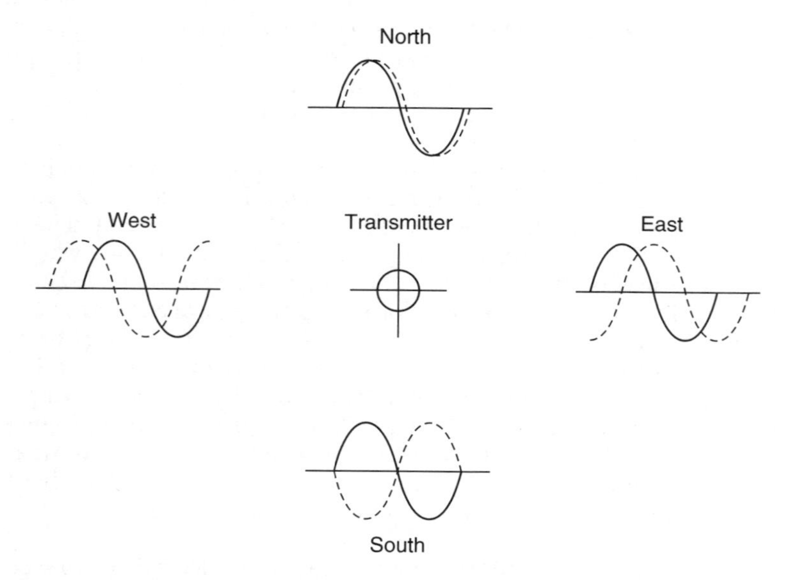

Fig. 7.52 In the VOR (VHF omni-range) system there are two transmitted signals. One of these is the same in all directions whereas the phase of the other is a function of direction. The receiver can determine the direction along which the transmitter was sending and the bearing of the transmitter must be the reciprocal of this.

Thus if a VOR receiver detected a 90° lag in the signals, this would indicate that the receiver lay on a bearing of 90° *from* the beacon. The beacon would then be at 270° with respect to the receiver. A cockpit indicator makes this reciprocal indication automatically. In order to fly along a selected radial, the pilot turns the omni-bearing selector knob on the display until the desired radial heading is at the top. If the autopilot is now coupled to the VOR receiver, should the track deviate from the required radial, a phase change will be picked up which will modify the roll reference of the autopilot until the correct radial is being flown. This will be indicated when the deviation pointer on the display points straight up to the selected radial heading. The display will also indicate whether flight is to or from the beacon.

The signals become erratic as the beacon gets very close. Consequently the VOR coupling is disabled automatically when this condition is sensed and the flight director uses magnetic heading to continue flight until the beacon is behind the machine when VOR coupling will resume with the 'from' indicator showing.

Figure 7.53 shows how an ILS (instrument landing system) works. There are four transmitters associated with each landing site. The *localizer* is sited at the far end of

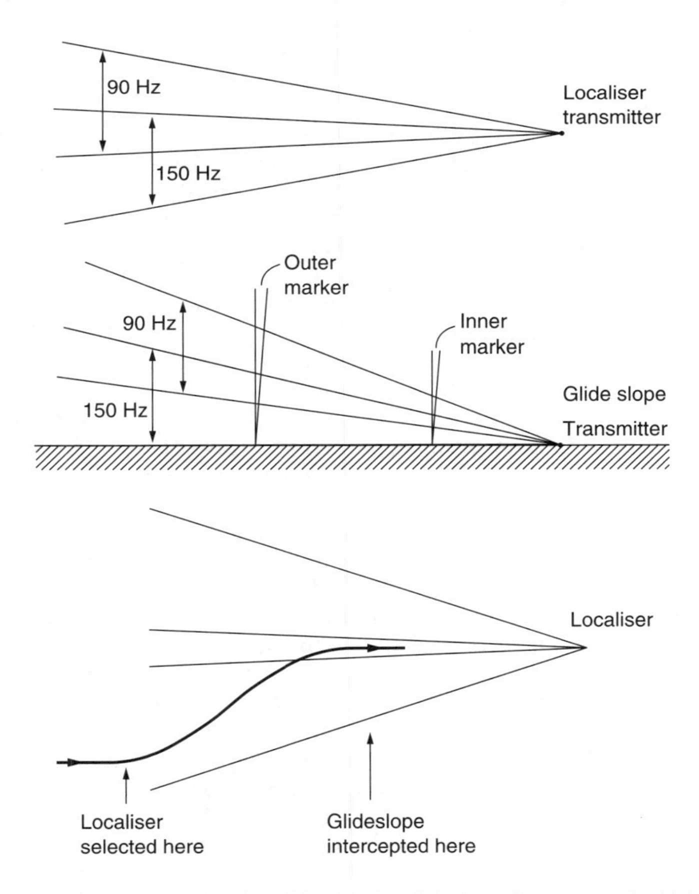

Fig. 7.53 An ILS installation. The localizer guides the aircraft horizontally onto a path which aligns with the runway. The glide slope beams make the altitude a suitable function of distance so the aircraft descends at the correct angle to the threshold. The markers tell the aircraft how far it is to touchdown.

the runway and transmits a pair of beams that are equally off the axis of the correct approach path. These are sensed with equal strength when a correct approach is being made. The *glide slope* transmitter is sited at the approach end of the runway. This also radiates two beams, this time equally off axis with respect to the ideal glide slope. These two beams are sensed with equal strength on the correct glide slope. There are also two highly directional antennae radiating beams directly upwards into the approach path. These are the outer and inner markers.

In order to make an instrument approach the helicopter will initially be on autopilot flying with altitude hold. Once in the vicinity of the approach path, the ILS receiver is tuned to the correct frequency and once identified the autopilot will be coupled to the locator using the LOC position of the selector. The locator alone is not enough to ensure a correct approach and the desired heading of the autopilot needs to be set with the heading of the runway. In this mode the autopilot is influenced by both the localizer signal and the heading error.

As Figure 7.53 shows, the heading could be correct but the helicopter could be approaching parallel to the runway. In this case the heading error would be zero but the localizer error would be issuing a steer left command. The machine will steer left until the localizer error balances the heading error. As the approach path is acquired, the localizer error becomes smaller and so the balancing heading error also gets smaller. The result is that as the machine flies onto the localizer beam centreline it will adopt the correct heading with both errors zero.

The machine will then fly along the beam centre with altitude hold still on. At some point the receiver will intercept the glide slope beams. The point at which this occurs will depend on the altitude. When the glide slope beams are detected, the autopilot coupling must be set to GS (glide slope) mode. The horizontal coupling is unchanged, but the altitude hold is disabled and replaced with coupling from the glide slope receiver. The glide slope information will act on the autopilot by reducing the desired altitude if the machine is above the beam and increasing it if it is below. The result is that the helicopter descends down the glide slope. As the flight path is already aligned with the localizer beam, the helicopter must be on a direct course for the glide slope transmitter.

The markers indicate the progress of the approach. The outer marker is about four NM from the threshold and transmits continuous Morse dashes. The inner marker is about 3500 feet from threshold and transmits alternating dots and dashes.

Using the ILS transmitters the helicopter can make an automatic approach. A machine without an autopilot could make a manual instrument approach with the pilot using left/right and up/down indicators to fly down the glide slope. In the case of an autopilot failure the pilot could elect to continue the approach manually in this way.

At some point the so-called decision altitude is reached. In the case of a manual instrument approach or an automatic approach with a single channel autopilot, if the landing site is not visible by the time the decision altitude is reached, the landing must be aborted. Only a suitably equipped machine may carry on to make a *blind landing*. Such a machine must carry a triply redundant voting autopilot, a RADAR altimeter and Doppler RADAR to measure groundspeed. A pressure altimeter is simply not accurate enough. At a RADAR height of a few hundred feet, the constant glide slope and airspeed are both modified to produce a flare. The disc attitude is pitched back to slow the machine down and collective is lowered to prevent zooming.

As RADAR height falls, forward speed control is switched from dynamic pressure to Doppler RADAR so the groundspeed can be brought to zero a little before RADAR altitude reaches zero. This gives time for the hull to level before touchdown which will be programmed to be within the landing gear limit.

7.29 Fault tolerance

In feedback systems an error is amplified to operate a control. In normal operation this makes the error smaller. However, if there is a failure the error does not get smaller. In the case of an attitude stabilizer controlled by vertical gyro, if the roll attitude signal from the gyro failed and went to an arbitrary value, this would result in a massive roll error which would apply full roll control in an attempt to correct it. This is known as a *hardover* and is an unfortunate failure characteristic of all feedback systems. In practice equipment has to be designed to prevent hardover failures compromising safety. In a simple approach the authority of the automatic pilot is limited so that it cannot apply full control. Slipping clutches may be present in the controls so that the pilot can overcome the incorrect forces and regain control. In more sophisticated systems extensive monitoring is carried out. If a signal deviates from the range of levels in which it normally operates, the signal may be clamped to a null value to prevent a hardover and the autopilot will be disengaged.

In military helicopters natural parts failure is not the only problem. The designer has to consider that the helicopter may come under fire and will take steps to allow control to be retained despite a reasonable amount of damage. In practice this means retaining control after hits from rounds of up to half-inch calibre.

The first line of defence is redundancy. Duplicated hydraulic systems are essential, but it is important to run the pipes far apart so that a single round cannot rupture both systems. The location of twin engines spaced widely apart in the Apache is another example of this philosophy. Mechanical control signalling is vulnerable to small arms fire as a single round can sever a pushrod. It is much harder to disable a multiply redundant digital signalling system which takes different routes through the airframe and where the processing power is distributed in small units all over the airframe. In fact some attack helicopters use the mechanical signalling only as a failsafe in case the fly-by-wire fails. Where control paths cannot be separated, as for example near the pilot's controls, the armour plating which protects the pilot will also be employed to protect critical parts of the control system.

Redundant hydraulic actuators have the additional problem that the failed actuator may be jammed due to impact distortion or a round lodged in the mechanism. In this case the remaining actuator will be unable to move the jammed control. The solution is to construct actuators with frangible pistons. In the event of an actuator seizure, the working actuator can develop enough thrust to break the jammed piston free of the actuator rod.

Where hydraulics are concerned, it is clear when a failure has occurred, and so it is equally clear which actuator should retain authority. However, in electrical signalling systems a failure in a signal processor could result in an entirely spurious voltage or digital code being output. This could have any value over the whole control range. If there are only two control systems it is impossible to know which is giving the right answer. The solution here is triplication where there are three separate systems comparing the actual position with the desired position. In this case a single failure will result in one actuator drive signal being different from the other two. A comparison or *majority voting* system can determine which signal is out of step and disable it.

As with the hydraulics, electrical power sources must be duplicated. Critical electrically powered devices may be fed by multiple sources using blocking diodes. A voting signal processor can take power from the battery wiring or from both generators. If any power source fails its voltage will fall and the diode will reverse bias and block any power drain so that power is still available with any one source still functioning.

There is a good parallel here with the use of one-way clutches and multiple engines. The one-way clutch is a mechanical diode.

There can be more than one voting unit in a real system. The pilot's joystick may be fitted with multiple encoders and a voting system to determine if one has failed. The LVDTs on the hydraulic actuator may have the same system so that reliable feedback signals are available. The outputs of the multiple signal processors will also be subject to voting. In this way the probability of control loss due to natural parts failure is rendered infinitesimal and real protection against battle damage is provided.

When redundancy, electronic processing, automatic stability, flight direction and powered controls are combined, remarkable systems can be realized. It is not inconceivable that future attack helicopters would have the ability to fly themselves home in the event that the crew are disabled.

8

Helicopter performance

8.1 Introduction

Helicopter performance may be assessed in a large number of areas and as these interact to a considerable extent it is difficult to categorize. As helicopters are expensive to operate in comparison with aeroplanes, economic performance is important. Factors such as the fuel consumed, parts worn out and maintenance needed determine the operating cost that is generally calculated on an hourly basis. Operating cost will be assessed by commercial users in proportion to payload. Clearly such economy will only be realized if the machine is fully loaded. Operators will consider machines that offer a payload close to their typical requirement in order to avoid the poor economics of flying partly loaded. In the case of privately owned machines, the payload may be little more than the pilot and passengers, but here different economics apply as the cost of flying may be offset against time saved. In the case of military operators, economics may not be the overriding concern.

Operational performance may be taken to include factors such as the range, airspeed and altitude that can be obtained with various payloads. As the advantage of the helicopter is its ability to hover, the greatest altitude that can be reached in forward flight is somewhat academic and less important that the altitudes at which the machine can hover in and out of ground effect at various weights. Hover out of ground effect (HOGE) altitude is the highest altitude at which the machine can take off vertically, and will be lower than the hover in ground effect (HIGE) altitude. This is the highest altitude at which the machine can get into the hover, which would allow a running take-off if a sufficiently unobstructed area is available. These altitudes, and many other performance factors, are functions of air temperature. If valid comparisons are to be made, test results must be presented in a way that eliminates atmospheric variables. The necessary techniques will be discussed shortly.

Tactical performance may include the manoeuvrability of the machine. Many aspects of this are determined by the load factor, although the power of the tail rotor controls the yawing ability and the ability to hover in crosswinds. Associated factors are the range of positions the CM may take and the weight of underslung load that can be lifted. Military and coastguard users find these aspects of performance more important than commercial users. Achieving high load factor and high tail rotor power may raise weight and operating costs that a commercial operator cannot tolerate. In some cases there are variants of a basic machine each of which is optimized for a particular market. There will be exceptions in the case of specialist operators. Helicopters designed for underslung delivery of stores by the military may be well suited to logging or construction of ski lifts.

Performance and safety are closely related. Operating an aircraft outside its designed performance envelope may result in excessive stresses that can cause damage or even failure. Consequently during the development of a machine, extensive tests are needed to establish just what these limits are. These limits are published in the flight manual in order that pilots are aware of them. A helicopter's performance can be measured in terms of factors that contribute to safety. The degree of stability may affect pilot workload and compromise safety on long flights in poor conditions. If it is intended to operate a helicopter in IFR, the stability performance may have to be improved by passive or active means. Performance is important not just under normal conditions, but also under abnormal conditions. Autorotation performance becomes important in the case of power loss. In twin-engine machines the performance with one engine inoperative (OEI) is crucial if the safety benefit of two engines is to be more than an illusion. The ability of a helicopter to continue flight under icing conditions is also a performance factor.

8.2 The atmosphere

The atmosphere is the medium in which helicopters fly but it is also one of the fuels for the engine and the occupants breathe it. It is a highly variable medium that is constantly being forced out of equilibrium by heat from the sun and in which the pressure, temperature, and humidity can vary with height and with time and in which winds blow in complex time- and height-variant patterns. The effect of atmospheric conditions on flight is so significant that no pilot can obtain qualifications without demonstrating a working knowledge of these effects.

The atmosphere is a mixture of gases. About 78% is nitrogen: a relatively unreactive element whereas about 21% is oxygen which is highly reactive. The remainder is a mixture of water in the gaseous state and various other traces. The reactive nature of oxygen is both good and bad. The good part is that it provides a source of energy for life and helicopters alike because hydrocarbons can react with oxygen to release energy. The bad part is that many materials will react with oxygen when we would rather they didn't. Chemically, combustion and corrosion are one and the same thing. The difference is based on the human reaction to the chemical reaction.

Gases form the highest energy state of matter in which the molecules are no longer bound together strongly as in solids or weakly as in liquids but instead are free to rush around at a high speed that is a function of absolute temperature. The countless collisions between gas molecules and any non-gaseous object result in pressure at the interface. Pressure is measured by physicists and by engineers in units of force per unit of area using imperial units of pounds per square inch or SI units of Newtons per square metre. At sea level, the atmosphere exerts a pressure of about 15 pounds per square inch and has a density of about 0.075 pounds/0.002378 slugs per cubic foot, or in metric units about 100 000 Newtons per square metre with a density of 1.225 kg per cubic metre. Over the years, many other units of pressure have evolved, some from meteorology. One of these is the bar (after barometry) where one bar is the average atmospheric pressure at the place where the bar was defined. In practical use, the bar is divided into 1000 millibars. The bar is slowly being replaced by a numerically identical unit known as the hectoPascal (hPa). The bar and hPa are commonly used in aviation altimetry. The principle of the mercury barometer is that atmospheric pressure supports a column of mercury exposed to a vacuum at the top. Consequently the length of the column is proportional to pressure and can be expressed in inches or cm of mercury. At sea level a reading of about 26 inches of mercury is obtained. This unit may be found in use in altimeters originating in the United States.

The gas law states that the product of pressure and volume is proportional to temperature. Reducing the volume means that external work has to be done to oppose the pressure. This work increases the temperature of the gas. The Diesel engine obtains ignition in this way. Conversely if the volume is increased, work is done by the gas and the temperature must fall. This is why carburettors are prone to icing on part throttle because the air expands on entering the manifold. Air conditioners work in the same way.

If the volume is fixed, as temperature rises, the velocity of the molecules increases and so the impact at each collision with the walls of any container is greater and the pressure rises. Alternatively the same pressure can be exerted in a given volume with a smaller mass of gas. Thus in the atmosphere where pressure increases can be released by free movement, the result of an increase in air temperature is that the density goes down. Density is also affected by humidity. Water molecules are heavier than those of atmospheric gases and increase the pressure due to molecular collisions. Thus in the presence of water vapour a given pressure can be sustained with a smaller mass of air and the density goes down.

8.3 International Standard Atmosphere

In order for meaningful comparisons to be made between various test results, it is important to eliminate variations due to atmospheric conditions. The International Standard Atmosphere (ISA) is a defined set of fixed conditions, somewhere within the spread of conditions found in practice. When a test is made, the actual atmospheric conditions are measured. Using the laws of physics, it is possible to calculate the effect of every difference between the actual conditions and ISA. If all results are corrected in this way, they can be presented with respect to ISA and as a result can immediately be compared with any other results obtained in the same way. Similarly, if the performance of a machine is defined in the flight manual with respect to ISA, it is possible to correct for the actual conditions and predict the real performance that can be expected.

In the ISA, pressure and temperature at mean sea level (MSL) are defined, along with standardized rates at which these change with height. Relative humidity (RH) is also defined to be zero. ISA MSL pressure is 1013.2 hPa. Temperature is 15°C and the density is 1.225 kg/m^3. The ISA lapse rate defines temperature as falling at 1.98°C per 1000 feet (which is a mongrel unit being part metric and part imperial). Although the height of the tropopause is a function of latitude because the earth is rotating, ISA defines the tropopause as 36 000 feet and above this the air temperature is −56.5°C. For many purposes the approximation of −2°C per 1000 feet can be used as the lapse rate.

Given the above, the pressure and density of the ISA can be calculated for any height above sea level. The density is important because it directly affects the power that can be produced by the engine(s) and the thrust produced by the rotors.

8.4 Pressure and density altitude

Altimeters do not measure height because they are basically pressure gauges with a creative scale. Only in ISA conditions does an altimeter measure height because the pressure will correctly match the scale reading. In all other conditions, the altimeter reads altitude, best defined as estimated height subject to an error. If sea level pressure at the location in question is known, the error can be minimized by adjusting the altimeter setting scale to that pressure. However, if instead the altimeter setting scale is set to the ISA MSL pressure of 1013.2 hPa, the altimeter reads pressure altitude. This is defined as the altitude in the ISA at which the same pressure exists. Pressure altitude is used primarily for flight in airways where it assures vertical separation (see section 7.7).

Whereas pressure is primarily a navigational tool, density is what governs performance. Engines produce power by releasing thermal energy due to combustion of hydrocarbon fuel in air. The mass of fuel that can be burned is directly proportional to the *mass* of the air available, whereas physical limitations in the construction of the engine set the greatest *volume* of air that can be drawn in. Consequently the density has a significant effect because it determines the mass of air per unit of volume the engine has drawn in. Lift is obtained when the rotor imparts downward momentum to the air. As the disc area is fixed, the volume of air the rotor can influence is also fixed for a given flight regime. Thus the degree to which the rotor can impart momentum is controlled by air density.

In the ISA the density at a given height will always be the same because ISA defines MSL pressure and temperature and specifies zero humidity. Thus using the standard figures, the pressure and temperature at any height can be deduced and the density must follow from that. In practice, the pressure and temperature will differ from ISA and the local density will not be ISA density. The concept of density altitude was introduced to allow for such changes. Density altitude is the altitude in the ISA at which the density is the same as the present density. In order to assess the performance of a machine under real conditions, the pilot will calculate the density altitude for those conditions and then consult the flight manual where the performance is specified as a function of density altitude.

Figure 8.1 shows a standard density/altitude chart that converts pressure altitude entered on the diagonal lines to density altitude measured along the vertical axis. The horizontal axis is air temperature and the single line falling to the right shows the ISA air temperature falling with height at the standard lapse rate with pressure altitude. Thus the chart shows that the line corresponding to zero pressure altitude crosses the line corresponding to zero density altitude at 15°C (ISA MSL conditions). In order to use the chart, the local pressure altitude is located where it crosses the ISA lapse rate line. Reading vertically down from this pressure altitude will give the ISA temperature for that height. If the actual temperature is the same, the pressure altitude and density altitude are the same. However, in practice a different temperature will often be found. If the diagonal line sloping from bottom left to top right and corresponding to the pressure altitude is followed until it intersects a vertical line drawn up from the actual temperature, the vertical position of the intersection determines the density altitude. The figure shows an example. In some flight manuals, the conversion to density altitude may be incorporated in certain performance charts. An example will be given in the next section.

The effect of humidity is not generally considered in density altitude charts, but a rule of thumb used by pilots is to add 1000 feet to the density altitude if the humidity is high. Typically helicopters have no means to measure humidity and the pilot's estimate has to be used.

8.5 Power management

Helicopter operation is largely concerned with power management. The power plant supplies power and the airframe demands power. Both supply and demand are

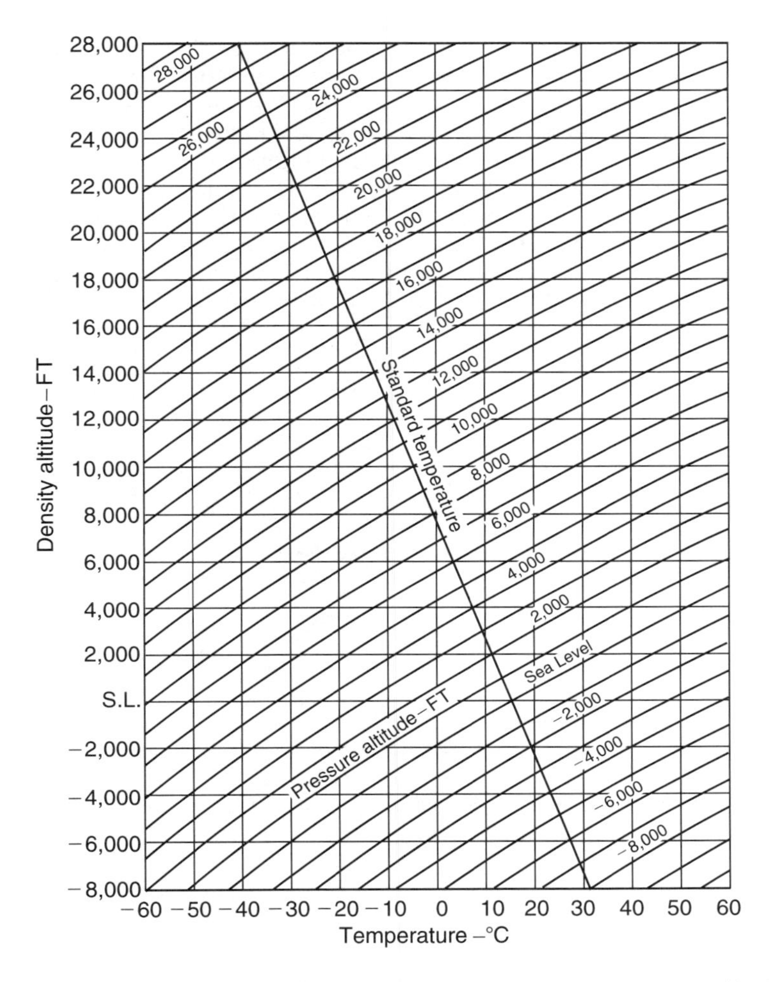

Fig. 8.1 A density/altitude chart. This allows density altitude to be obtained from pressure altitude. The chart corrects for ISA standard lapse rate and allows compensation for non-ISA temperature. See text.

subject to considerable variations. Sustained flight is not possible where demand exceeds available power. In the opposite case, flight in equilibrium is possible with a power surplus, usually called a *power margin*, which can be used for climbing or for manoeuvres.

In general the engine(s) provide power which has a physical limit to the transmission which has a physical torque limit. For a constant RPM machine such as a helicopter rotor, power is proportional to torque and engines may be described in terms of torque available. The engine power/torque available will be a function of density altitude. The rate at which fuel can be burned is limited primarily by the temperature that the power turbine blades can withstand. At low altitude, the compressor can produce plenty of mass flow and the combustion takes place with excess air that limits the power turbine temperature. As altitude increases the air density falls and the mass flow reduces. At some point the amount of fuel burned must be reduced to prevent excessive turbine temperature. From then on available power must fall. However, the fall in power is not as steep as might be expected because intake air temperature falls with altitude and this offsets the loss of density due to reduced pressure to some extent. As a result the engine power/torque available curve may show the characteristic knee of Figure 8.2. At low density altitudes the torque must be kept within the constant transmission limit whereas above a certain density altitude the available torque falls.

The rotors in turn require power that is a complex function of density altitude and airspeed. The power will be delivered to the main and tail rotors and will be absorbed by various forms of drag. Power required to hover is clearly important and the worst case will be at zero airspeed. Figure 8.3 shows a typical chart relating hover power

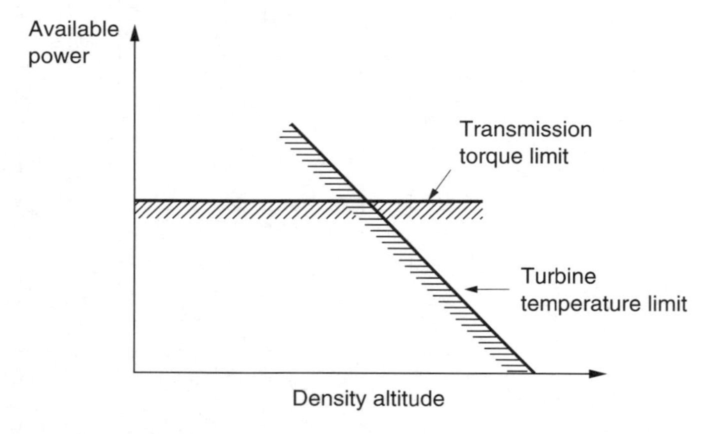

Fig. 8.2 Power available may be constant at low altitude owing to a transmission limit until turbine temperature limits dominate at higher density altitude.

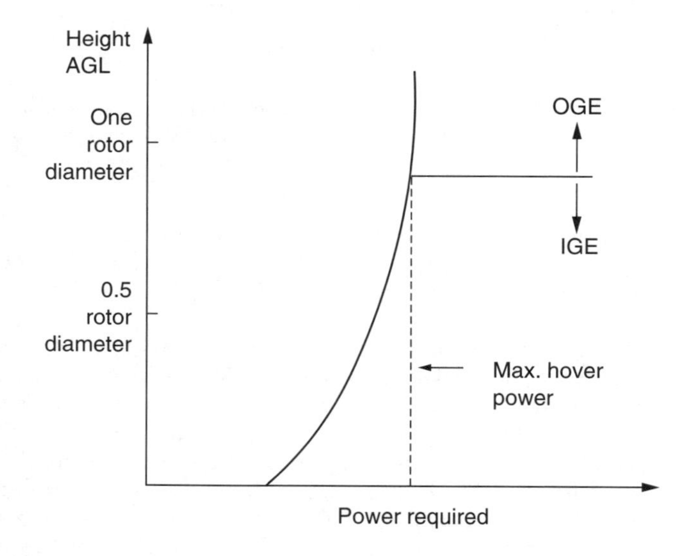

Fig. 8.3 Simple hover power chart.

to skid height for fixed density altitude. Above about one rotor diameter the power becomes constant and the machine is considered to be out of ground effect (OGE). In the flight manual, a chart similar to that shown in Figure 8.4 may be given. This allows hover performance to be predicted for any combination of all-up weight (AUW)

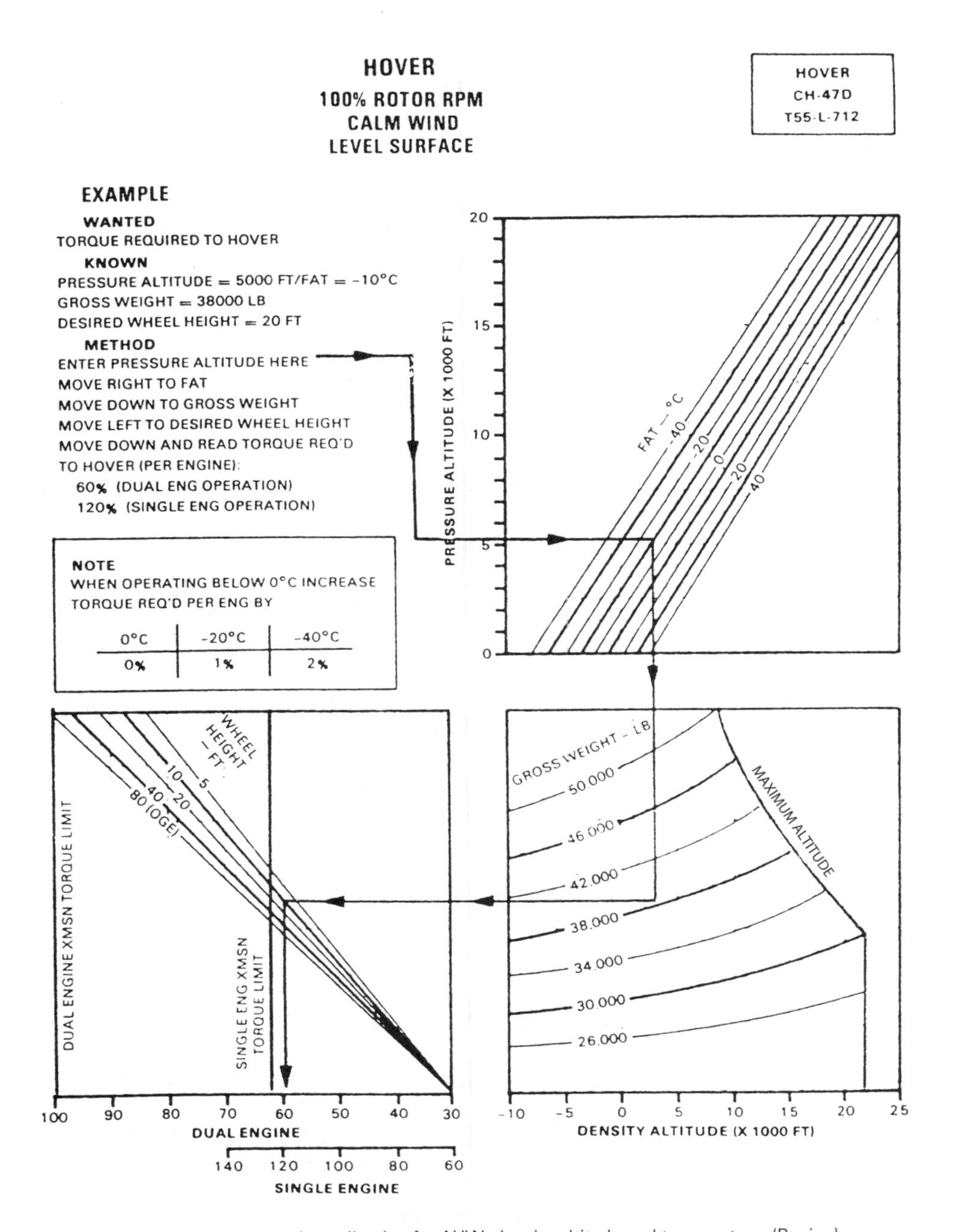

Fig. 8.4 Complex hover power chart allowing for AUW, density altitude and temperature. (Boeing)

and density altitude. The top right section of the chart allows density altitude to be computed by finding a point where a horizontal line from pressure altitude on the left scale intersects the free air temperature line. A vertical line from this point will give a reading on the density altitude scale. This line will also intersect the AUW curve on the bottom right chart. Moving left from this point will give a reading on the hover OGE torque scale.

The bottom left chart converts the hover OGE torque to the torque needed at various skid/wheel heights. The HOGE line is at an angle of 45° and does not change the torque required. However, the lines for various heights of IGE have different slopes and reduce the torque accordingly.

The chart can be used in different ways. In the above example the pressure altitude, air temperature and AUW were used to find the torque needed to hover. However, from AUW and known altitude of the destination it is possible to deduce the maximum air temperature at which a vertical landing could be made. In the case of an engine failure, the chart can also be used to show that above a certain AUW hovering even in ground effect is impossible and a rolling landing on a runway would be needed.

In practice, power needed to hover is affected by windspeed. In most cases the power needed will be reduced, except if the wind is from a particular direction and at a particular speed that causes the main and tail rotors to rob one another of inflow. This phenomenon was considered in Chapter 5. The tandem rotor helicopter is less sensitive to wind direction in the hover.

Turning next to forward flight, Figure 8.5(a) shows how the various power requirements typically change with airspeed. The rotors will require induced power and profile power and the hull will require parasitic power to overcome drag. The induced power will be high in the hover and falls as airspeed increases. This is because forward flight gives the rotor access to a larger mass of air per unit time and production of thrust becomes more efficient. The induced power will also increase with AUW. Profile power will also rise with AUW and increases slightly with airspeed as the profile drag on the advancing blade increases more than the drag on the retreating blade lessens. In contrast, the parasite power is substantially independent of AUW, but increases with airspeed. The drag force will increase as the square of airspeed, but the power is the product of the drag force and the airspeed and so will increase as the cube of the airspeed. At very high speeds the rotor disc must be inclined forwards substantially so that the horizontal component of thrust from the rotor can become large to overcome hull drag. This nose-down attitude may make the hull drag even greater. As main rotor thrust increases the tail rotor thrust will also need to increase to counteract torque and so the tail power will also increase at high speeds.

Figure 8.5(b) shows that there is an airspeed, typically around 50 knots, at which the total power is minimum. At this airspeed the induced power has fallen but the parasite power has not yet risen seriously. The figure also shows the available power, which is relatively constant, although in some machines it rises with airspeed owing to the ram effect of dynamic pressure at the intake. The power required reaches the available power on the right and this would determine the highest airspeed that could be reached, assuming no structural limits or retreating blade stall effects. At all lower airspeeds, a power margin exists and so the machine can take off vertically. At the minimum power airspeed the machine can stay airborne for the longest time. Clearly the greatest power margin exists at this airspeed and thus the maximum rate of climb would be obtained at full power. The minimum rate of descent in autorotation will also be obtained at this airspeed.

Figure 8.5(b) contrasts the conditions for a normally loaded machine at sea level, with the conditions for a high AUW, where the available power only exceeds the required

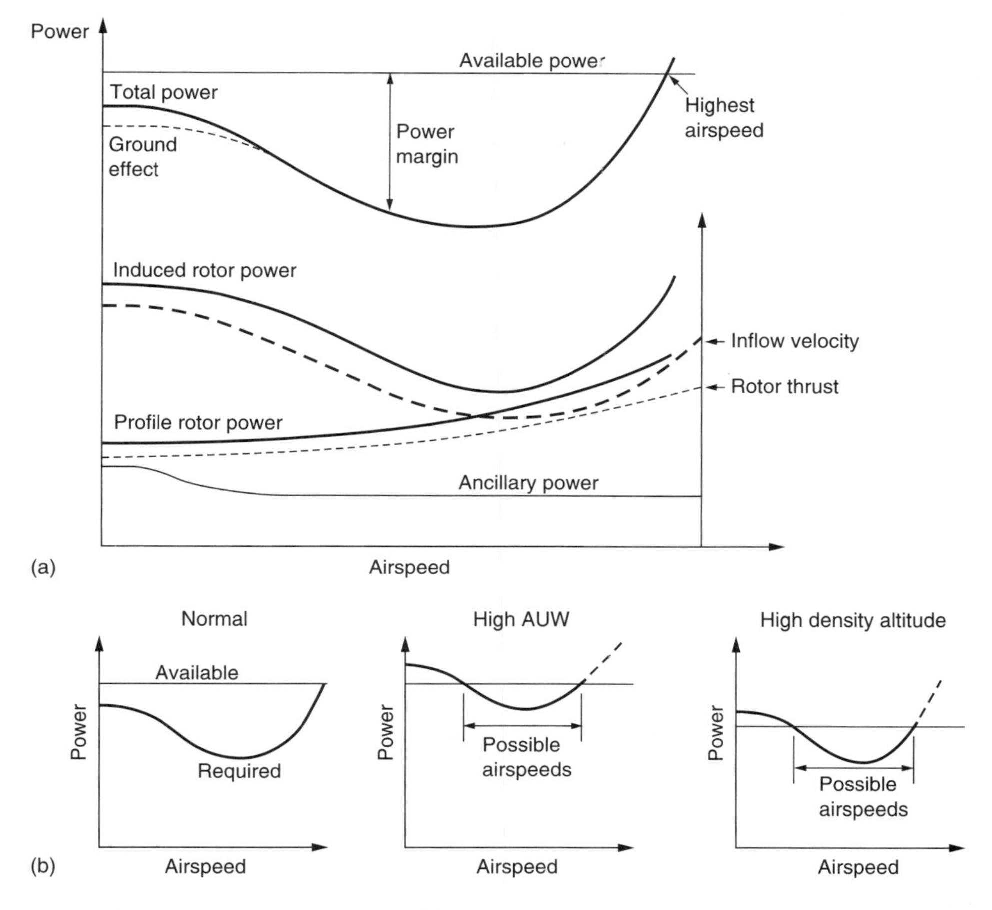

Fig. 8.5 Power as a function of airspeed. At (a) the various contributions from different sources have different functions of airspeed. At (b) flight is only possible where available power exceeds required power. With high AUW the power needed exceeds that available over part of the speed range. High density altitude reduces the power available and also limits the speed range.

power over a range of airspeeds and the conditions for a high density altitude where the available power has fallen. Once more level flight is only possible over a range of airspeeds.

8.6 Flying for maximum range

The airspeed that will yield the greatest range is not obvious. The problem can be approached by considering that greater distance can be flown at higher speed, provided it does not require disproportionately more power. The key here is the proportionality. If an increase in airspeed of, say, 5% needs a power increase of 4%, then obviously it is worth going faster, whereas if a speed increase of 5% requires a power increase of 6% the range will fall. Consequently an approximation to the best range in still air is

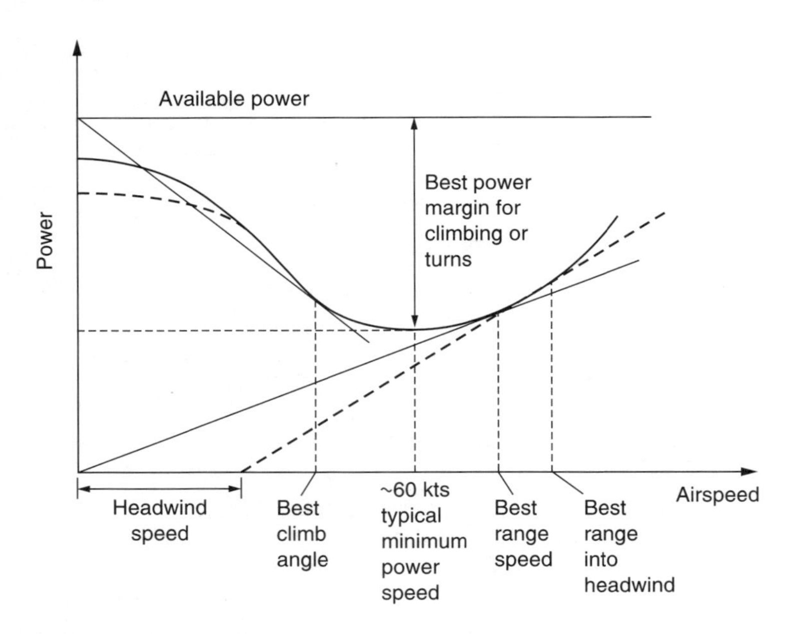

Fig. 8.6 Maximum range speed. In still air this is found by a line from the origin tangent to the power curve. In the case of wind the origin is offset by the windspeed.

obtained by finding the airspeed at which the rate of power increase is exactly equal to the rate of airspeed increase. Figure 8.6 shows that this speed can be found graphically where a line from the origin is tangential to the power curve. The presence of wind changes the situation. In this case it is necessary to find the airspeed at which the rate of power increase is exactly the same as the rate of groundspeed increase. Figure 8.6 shows that this can be found graphically by offsetting the origin by an amount given by the component of wind velocity in the direction of flight. For a headwind, the maximum range speed will be faster than that needed for still air.

There is a further complication in practice, because a real engine does not consume fuel proportionally to power produced. This is particularly true of turbine engines that sustain a loss of efficiency due to driving the compressor. This loss becomes a smaller proportion of total power as power produced increases and so it is a characteristic of a turbine that efficiency increases with power. Consequently the best range speed according to Figure 8.6 is not strictly correct for a turbine helicopter. To be precise, the machine should be flown at the airspeed where the rate of fuel consumption increase is exactly the same as the rate of groundspeed increase. Figure 8.7 shows fuel consumption against power normalized at minimum power speed. At (a) is the curve for a turbine helicopter. This has the same trend as the power curve that is also shown, but does not rise as steeply above minimum power speed because the engine is being used more efficiently. It will be seen that the airspeed for the best range is somewhat higher than the figure obtained in Figure 8.6. In practice the power curve, or the fuel consumption curve, remains close to the line from the origin over a range of airspeeds so it is not necessary to be rigid about the airspeed as the loss of range due to a moderate departure from the ideal is not serious. In practice the actual airspeed can usefully be faster than the theoretical maximum range speed by an amount that increases fuel flow by 1%. This gives a useful reduction in journey time with a negligible range penalty.

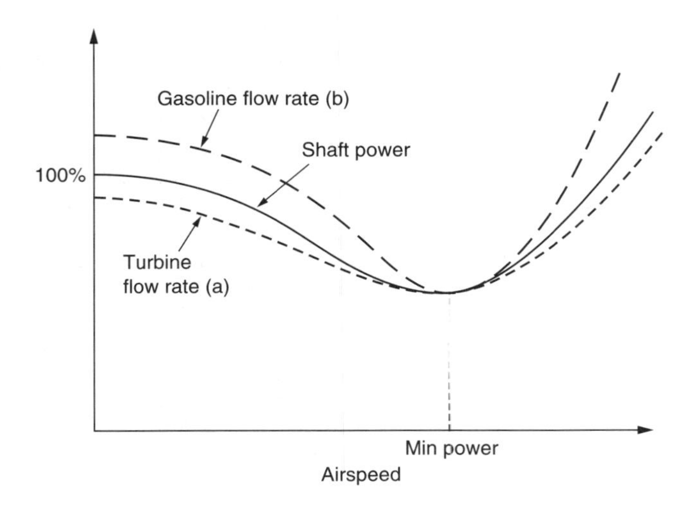

Fig. 8.7 Fuel flow for a turbine helicopter (a) and gasoline powered helicopter (b) superimposed on power required. Note these are not the same as the efficiencies of the two types of engine change differently with load. The turbine is inefficient at low power and most efficient at full power. The piston engine is most efficient at less than full power.

Gasoline-engine helicopters behave differently. At high power, the mixture may have to be rich so that some additional cooling is provided by fuel evaporation and this reduces efficiency. In cruise, the mixture can be leaned out to make the engine more efficient and improve range. Thus a gasoline engine may be most efficient at some fraction of full power rather than at full power as is the case for the turbine. As a result the fuel consumption curve rises more steeply than the power curve away from minimum power speed as shown in Figure 8.7(b). Thus the spread of speeds for maximum range with a piston engine will be smaller than for a turbine. The gasoline engine does not run on excess air and so available power will fall off rapidly with altitude. However, if a turbocharger is fitted, engine power can be maintained at altitudes sufficient to reduce parasite drag.

Where the utmost range is to be achieved, flying at an appropriate altitude may be useful. There are two related factors to consider. The first is that all real helicopters need an available load factor significantly above unity so they can manoeuvre safely. The greatest efficiency is obtained by running the rotor blades at a C_L of around 0.5, suggesting that the available ISA MSL load factor of a typical helicopter will be around 2. By definition cruise takes place at unity load factor where the lift to drag ratio changes little with C_L. The second is that helicopters need engines whose power is appropriate for the load factor and in cruise these will be running inefficiently at part power. For the best range anything that reduces the load factor and improves engine efficiency should be considered. Reducing the air density may do both. In the case of a rotor running at constant RRPM, reduced air density will mean that a greater angle of attack is needed. However, if this changes C_L from 0.5 to 0.6, the increases in induced drag and profile drag will be minor. However, there is a significant reduction in parasite drag.

A turbine engine has no actual throttle and so operates with excess air, like a Diesel engine. Unlike a Diesel engine, the compressor suffers profile drag. An efficient way of throttling a turbine engine back to cruise power is to reduce the density of the inlet air.

In a gasoline engine there is a throttle that forcibly limits the mass flow through the engine and reduces the inlet manifold pressure. The reduced manifold pressure opposes the motion of the piston on the induction stroke and causes *pumping loss*. Pumping loss can be reduced by supplying air with lower density and opening the throttle to compensate. Consequently flying at an appropriate altitude reduces the total power needed and makes both gasoline and turbine engines more efficient. In a gasoline engine there is another option available to increase efficiency, which is to use heated induction air. This is normally provided to prevent icing in humid conditions, but hot air is less dense than cold air and so throttles the engine back to cruise power with less pumping loss thereby increasing efficiency and range.

Figure 8.8(a) shows the fuel flow curves for a CH-47D at two AUWs at sea level and +10°C. Figure 8.8(b) shows the corresponding curves for the same AUWs but at 10 000 feet pressure altitude and the air temperature 20°C lower to allow for the lapse rate. It will be clear that the fuel flow is significantly reduced for the lower AUW by climbing to the higher altitude as the reduction in parasite power exceeds the increase in profile and induced power. For the higher AUW, C_L becomes too large and the increase in induced and profile power outweighs the reduction of parasite power and fuel flow is higher.

If altitude is taken to extremes, it would depend on the design of the machine and the AUW whether the engine ran out of power before the rotor stalls or vice versa. In practice a somewhat lower altitude can be found at which a significant improvement in range can be had. On a long ferry flight with auxiliary tanks, the weight of the machine reduces significantly as fuel burns off and the optimum altitude will increase. The best range will be obtained by adhering to a calculated altitude profile throughout the flight. It should be noted that at very high altitudes, reduced air density results in reduced cyclic authority.

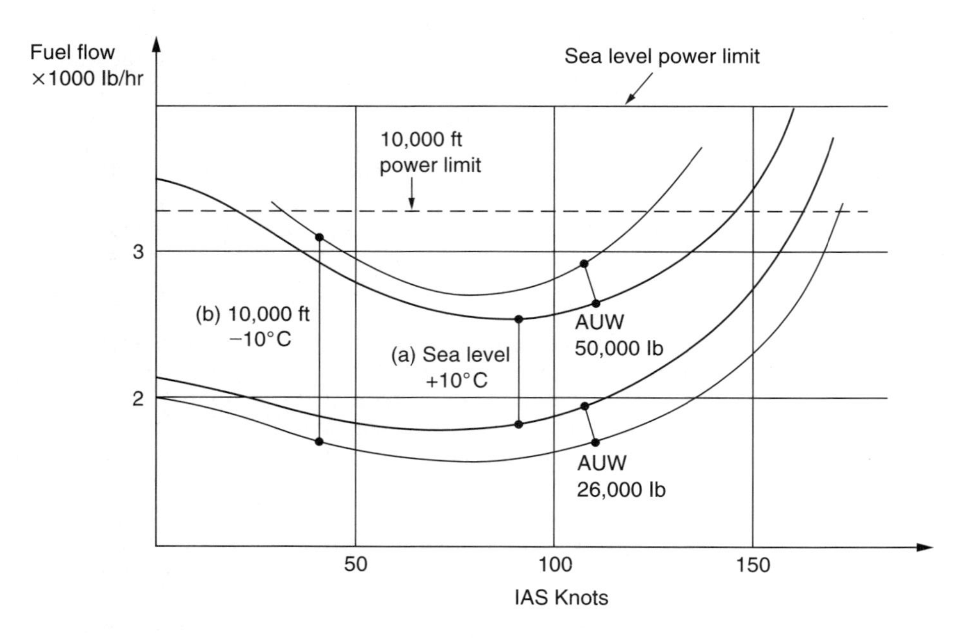

Fig. 8.8 Fuel flows curves for a CH-47D for two different AUWs. (a) At sea level. (b) At 10 000 feet (courtesy Boeing).

8.7 Climbing and descending

At airspeeds above 30 knots or so, the parasite power and the profile power and induced power of the rotors are affected only slightly by a change of disc inclination due to climb or descent. Hull drag will change slightly. Consequently it is reasonable to assume that any power margin in Figure 8.5 is available for climbing. Figure 8.9 shows a graph of power margin against airspeed. Effectively the upper curve is the difference between available power and power needed at constant altitude as a function of airspeed. By subtracting the available power from this curve, a lower, zero-power curve is obtained. The curves must terminate on the right at V_{ne} and as a result the region in which flight is possible is bounded by the two curves. The upper curve shows the rate of climb available if the entire power margin is converted into potential energy. The lower curve shows the rates of descent possible in autorotation.

Figure 8.9 shows the best rate of descent without power is at the same speed as the best rate of climb with power. However, in the case of power loss the goal is to reach a safe landing area rather than to descend at the minimum rate. There will be more possibilities if the machine is autorotated at the airspeed that gives maximum range. That airspeed is also derived in Figure 8.9. As before, the origin needs to be offset for head or tail winds. In autorotation the power comes from the release of potential energy as the machine descends. In theory, the heavier the machine, the more power is liberated, so it should be able to fly further. Under certain circumstances this will be observed. Where an emergency dictates maximum range in autorotation, the pilot

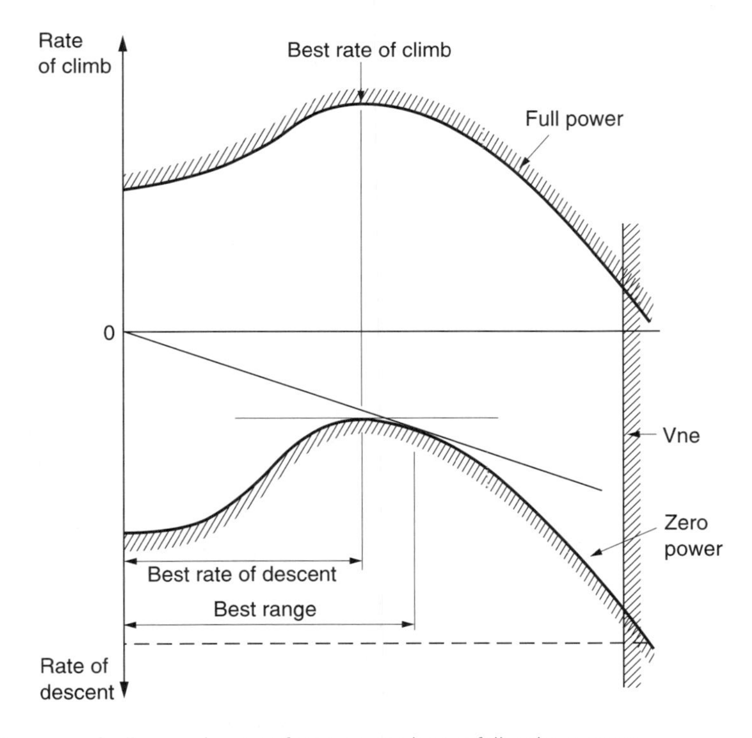

Fig. 8.9 Power margin diagram shows performance envelope at full and zero power.

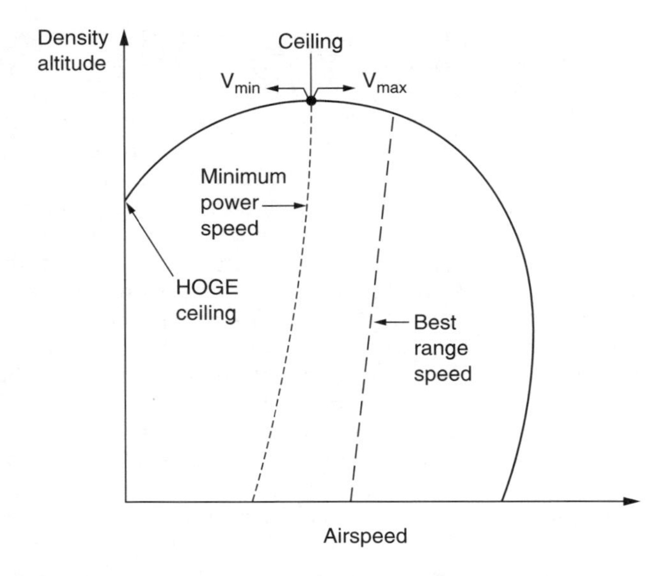

Fig. 8.10 Minimum and maximum airspeed as a function of density altitude. Note that at the operational ceiling there is only one airspeed available.

can increase rotor efficiency by reducing RRPM. This will reduce profile drag and require a larger angle of attack so that the lift to drag ratio will improve. Whilst very low RRPM is dangerous because it may not be possible to recover, a reduction of 10% is quite safe and has a dramatic effect on autorotation range. Prior to landing RRPM can be recovered by gradual application of back cyclic. In some machines the use of low RRPM may result in increased vibration if the rotor detuning has been optimized for a narrow RRPM range.

Figure 8.9 applies at one density altitude only and obviously climb or descent will change that. Figure 8.10 shows an alternative way of depicting the performance. This is a chart of minimum and maximum airspeed as a function of density altitude. Minimum power and maximum range speeds may also be shown as a function of density altitude. At sea level minimum airspeed is zero showing that vertical climb is possible. At some density altitude the HOGE power will exceed available power and then the minimum speed becomes non-zero. At sea level the maximum airspeed is power limited but may peak at some density altitude where reduced air density is reducing profile drag and improving rotor L/D. At higher density altitude maximum airspeed will fall again. As density altitude increases, the available power falls down into the power-required curve and the differences between minimum power speed, minimum airspeed and maximum airspeed become smaller. Where these speeds coincide, the absolute ceiling has been reached. As the absolute ceiling can only be reached asymptotically, the service ceiling is often taken to be a slightly lower density altitude at which a rate of climb of 100 fpm remains.

Given the existence of a minimum airspeed, potentially a helicopter could take off vertically at sea level with a full load and fly to some high altitude destination at best rate of climb speed where an attempted vertical landing would result in a crash. The fact that this does not happen in practice is due to a combination of the availability of power curves in the flight manual and piloting techniques that establish what the power margin is before attempting a landing.

In the case of a light piston-engine helicopter, the payload is small and the altitude performance modest. The flight manual may be quite sparse in the power curve department and safety is primarily assured by piloting technique. In the case of a heavy military transport helicopter, the payload may exceed the empty weight of the machine and flight in a wide combination of AUWs and density altitudes is possible. The flight manual may contain many pages of fuel flow curves for various AUWs, pressure altitudes and air temperatures. In the case of the CH-47D there are over 100 such curves.

When plenty of power is available, the steepest climb will be vertical. However, under certain combinations of payload and density altitude, vertical climb may not be possible. In this case obstacle clearance becomes an issue and finding the airspeed at which the steepest climb can be obtained is important. Even if there is a power margin, knowing the speed at which the steepest climb occurs is useful. The climb gradient is the ratio of climb speed to forward airspeed. Finding the optimum airspeed requires finding a point on the power curve where the surplus power is proportional to airspeed. Figure 8.6 shows that this can be done graphically by finding the point on the power curve where a line from the available power at zero airspeed is tangent.

In most cases the steepest climb performance will be required for obstacle clearance on take-off where there is insufficient power margin for HOGE. In this case the correct procedure is to make a running take-off by accelerating as hard as possible in ground effect and making no attempt to gain height until the best climb gradient airspeed is obtained. At this airspeed the attitude of the machine is adjusted so that no further increase in airspeed occurs. The machine will then climb at constant airspeed so that power margin goes entirely towards climbing. The flight manual may contain charts that allow obstacle clearance ability to be predicted for various AUWs and density altitudes.

8.8 Power management in multiple-engine machines

It is not generally the case that multiple engines are fitted to obtain more power. Turboshaft engines are available with phenomenal power output and a single engine could lift any but the most extreme helicopter. Instead the goal of multiple engines is to provide some degree of resilience to engine failure. In the case of passenger carrying civil helicopters, multiple engines and suitable operational procedures together allow safe flight to be maintained in the case of a single engine failure at any time. In such cases a landing would be made as soon as practicable after the failure. In military machines, engine failure may result from hostile acts and has to be considered more probable than natural failure. Frequently it will be a goal that the mission shall be completed despite the loss of one engine. A prompt landing is not viable if the machine is over enemy territory or water and flight may have to be sustained.

Turbine engines are light in weight for the power they produce, so there is no technical difficulty in providing further engines, although this will reflect in purchase and running costs. If the remaining engine(s) is to be able to provide enough power for flight should one fail, then clearly under normal conditions all engines will be operating at a fraction of their maximum power and will be relatively inefficient because of the profile drag in the compressors. Turbine engines are very reliable and failures are relatively uncommon. Consequently instead of providing excessive continuous power, it is more sensible to design engines that can be overrated for short periods of time. The continuous power rating of an engine is based largely on the temperature and rotational forces the blades at the hot end can withstand between overhauls. The oil-and air-cooling system will be designed to cope with these conditions. If these conditions are exceeded, the blade life will be reduced as higher than normal temperatures would increase the rate at which the blades creep or extend. The oil temperature may also rise. However, such conditions can be tolerated for a short time and an engine can be given a higher 'contingency' rating which it could only reliably tolerate for a few minutes. There may also be an even higher emergency rating that might only be sustainable for half a minute. If an engine failure is experienced during a critical phase of flight, the remaining engine would exceed its normal operating limits by entering contingency power to make up a good deal of the power lost.

If an engine enters one of these conditions, an indicator that can only be reset on the ground operates and a timer runs to measure the degree of overload. The use of these contingency power ratings in some cases may put the engine under such stress that it will need immediate and substantial parts replacement, particularly in the area of the blades at the hot end, but the frequency with which this happens is so low that the saving in fuel when the engines are working normally is of more consequence.

As machines have multiple engines for safety, it should be clear that adequate power should be available with one engine failed. With all engines running, more power is available than is needed. In this case the machine is flown normally by remaining within the transmission torque limits, whereas in the case of an engine failure the machine is flown within the ratings of the remaining engines. Figure 8.11 shows the example of a twin-engine machine in which the remaining engine will provide 140%

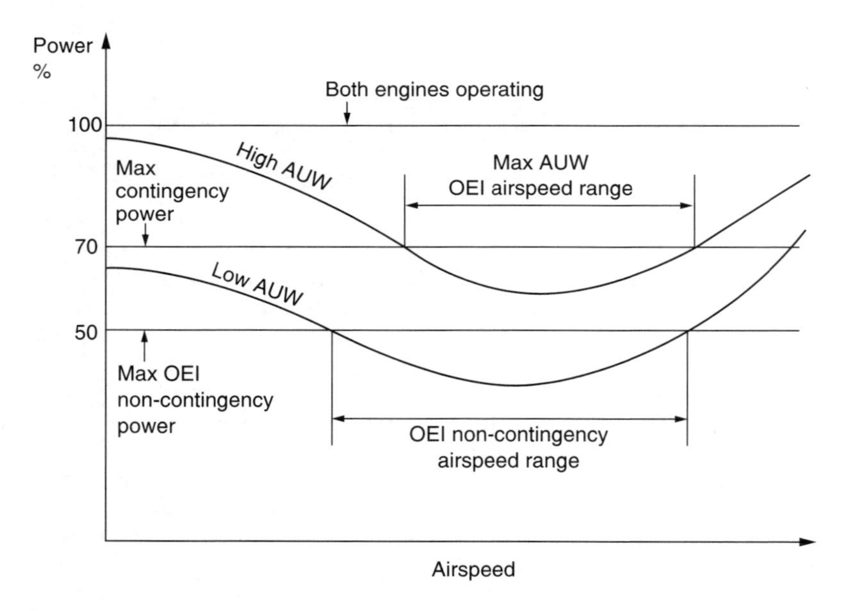

Fig. 8.11 One engine inoperative (OEI) conditions. If the remaining engine can provide 70% of transmission limit, with low AUW, nearly all of the flight envelope is available at contingency power, and a useful cruise envelope is available at non-contingency, or 50% of transmission limit. Such a machine could fly a long way after a failure. At high AUW, the machine can only stay airborne using contingency power, so it has to land before the time limit expires.

of its continuous torque rating during contingency. This means that the helicopter has 70% of its usual torque available to deal with the failure and for landing. In the case of a heavily loaded helicopter, 70% of torque may not be enough under some conditions and as a result the helicopter will have a form of avoid curve with one engine inoperative within which it cannot maintain height. This condition can be dealt with by appropriate piloting. Pilots of twin-engine machines prefer to stay out of the OEI avoid region so that an unexpected engine failure can be dealt with readily.

8.9 The flight manual

Elementary safety considerations suggest that the operation of aircraft must be subject to regulations and procedures many of which are ultimately part of the law of the land. The flight manual forms part of that framework as in most countries it is a legal requirement to carry the flight manual in the aircraft to which it relates. All qualified pilots are aware of the general procedures and regulations regarding aviation, but the flight manual lists those facts specific to a certain model of aircraft.

Neglecting prototypes and experimental machines, aircraft in general operation must be certified. This means that, independently of the designer and manufacturer, the relevant authorities have examined the design of the machine and carried out flight tests. The examination is intended to ensure that all relevant regulations regarding the construction have been met and that as far as reasonably possible any areas of weakness or mechanical unreliability have been eliminated. The flight tests are intended to show that the aircraft has no handling peculiarities that might be beyond the skill of the average pilot and to establish the limitations of flight conditions. The compilation of the flight manual and the accuracy of its contents will thus be seen to be part of the certification process.

Part of the flight manual is known as the approved section. This part contains the information that is mandatory for safe operation. It will contain a table of contents and a mechanism for logging the incorporation of any subsequent updated or amended sections so that it can readily be established that the manual is complete and up to date. The approved section contains four basic types of information: limitations, normal procedures, emergency procedures and performance data.

Limitations set out in the flight manual are intended to prevent the machine being flown in inappropriate circumstances, or, when it is flown, to prevent it being subject to excessive stresses. Many helicopters are not certified for flight in icing conditions. Many are not certified for IFR. Some can only safely be operated with a pilot and a co-pilot. Few are certified for aerobatics.

V_{ne} (airspeed not to be exceeded) primarily protects the blades and will often be specified as a function of altitude. The torque limit primarily protects the transmission. Turbine temperature limits protect the engine. The AUW of the machine must not exceed the figure in the flight manual. All-up weight can remain the same if the relative proportions of fuel and payload are changed. The distribution of weight is very important in a helicopter having a single main rotor. If the CM is some distance from the rotor shaft, cyclic control authority may be diminished in some directions and large alternating stresses are set up in the rotor. Figure 8.12(a) shows that the CM must be maintained within a conical region surrounding the shaft axis. In the case of a teetering rotor head, the apex of the cone coincides with the teetering bearing. In rotor heads with inbuilt flapping stiffness, the apex of the cone (b) will be above the rotor. Figure 8.12(c) shows that in a tandem rotor helicopter the fore-and-aft CM position must lie between two vertical limit lines. However, the position of these lines is a function of AUW.

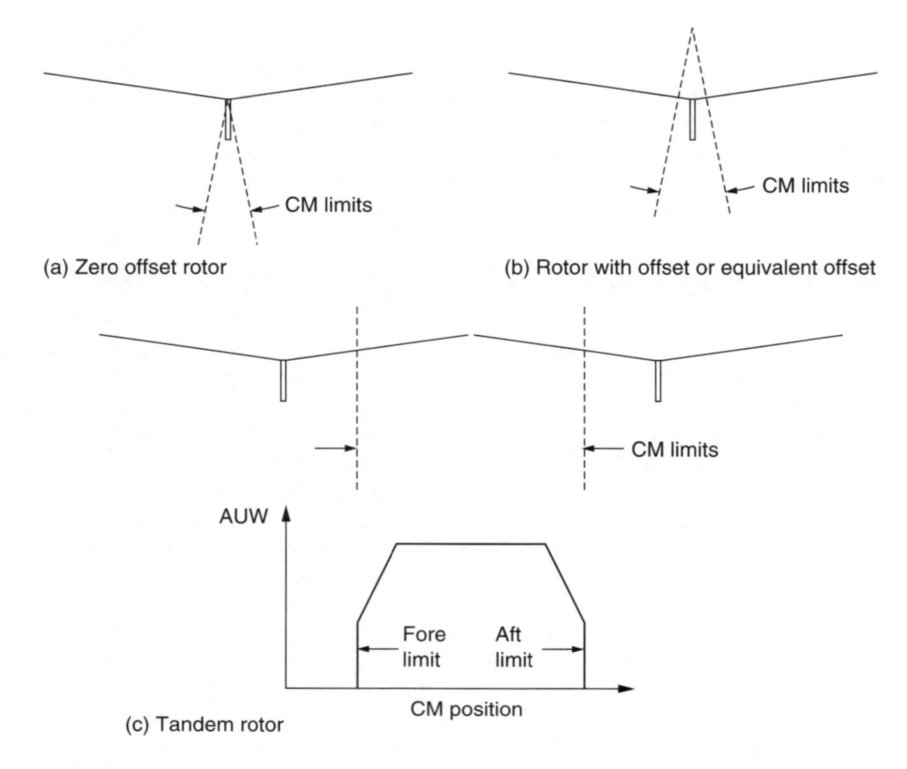

Fig. 8.12 (a) Allowed locations for CM with zero-offset head lie in a cone below the rotor head. Distance behind the rotor may be less than in front. (b) Cone apex is higher with offset or hingeless head. (c) Tandem rotor CM limits lie between two vertical lines which move closer as AUW increases.

When lightly loaded, the CM of a tandem is subject to remarkable freedom, but as AUW approaches the maximum value the CM position is more constrained to prevent an excessive load being carried by one rotor. Note that a tandem rotor helicopter will also have lateral CM position limits and heavy payloads must be positioned on the hull centreline. In general the certification authorities are not concerned with how the position of the CM is to be calculated, only where it is. The manufacturer's data will usually contain details on how the CM position is established.

In helicopters having zero-offset rotor heads, further limitations may be applied. The carrying of loads attached to the skids may be prohibited. There may also be a section describing the consequences of low-*g* or pushover manoeuvres designed to avoid mast bumping or boom strikes that are a characteristic hazard with this class of machine.

A windspeed limit for rotor starting may be found. Above that limit, blade sailing may be excessive, leading to droop stop pounding, blade damage or boom strikes.

In addition to the approved section, the flight manual will also contain a section provided by the manufacturer to assist in the operation of the helicopter. Civil flight manuals tend to be relatively sparse because aviation manufacturers have become targets for litigation in recent years. In contrast military flight manuals can be exceedingly comprehensive, particularly in areas such as fuel flow with respect to AUW and density altitude so that the best performance can be obtained from the machine.

8.10 Stability

An unconditionally stable aircraft is one that, if disturbed from trimmed flight, can be placed in any attitude and then, if the controls are released, it will return to straight and level flight on its own. The ability to return to straight and level flight is known as static stability, whereas dynamic stability is concerned with the manner, or promptness, with which the return occurs. Free flying model aeroplanes and some full-size aeroplanes are unconditionally stable. In forward flight on the upward part of the power curve, many helicopters are stable because the tail rotor acts like a fin and the tail surfaces have sufficient pitch authority. However, as the goal of the helicopter is to hover, a great deal of flying will take place on the back of the power curve where stability is impossible to achieve without pilot intervention or some artificial mechanism. Thus to be fair, it is not that helicopters are fundamentally less stable than aeroplanes, more that helicopters can explore flight regimes that are not available to aeroplanes in which control is simply harder.

Stability is a topic where there may be conflict between performance and safety and between complexity and economy. It is important to consider the entire system in order to form a balanced view. The airframe itself will have certain stability behaviour, but this will be modified by the actions of the pilot. Technically a helicopter airframe may be unstable, but piloting these unstable devices gives many of us a great deal of pleasure. This is because the actions of the pilot turn the unstable airframe into a stable system. In fact the unstable airframe with an intelligent pilot and/or control system forms the most manoeuvrable combination because it is axiomatic that the more stable the airframe is made, the less willing it is to deviate from its course.

Pilots very quickly become instinctive and fly at constant altitude with correct RRPM and no slip without conscious thought. It doesn't make much difference to the pilot if the machine deviates from his intended course because of a gust or because it is has a degree of instability. The result is the same; he operates the controls to bring it back on course. In good conditions, the pilot has so little difficulty in doing this that he can fly a machine with significant stability problems without necessarily realizing. However, if visibility becomes poor so that the horizon is obscured, the pilot's instinctive ability to stabilize the machine is impaired. Now the machine that is fun to fly VFR because it is so manoeuvrable becomes a menace in IFR because the workload involved in interpreting the artificial horizon and other instruments to make constant course corrections may be too great.

Today most helicopters are designed on the assumption that they will be used in IFR, but this has not always been the case. In practice machines certified for VFR only may have subsequently undergone a number of modifications in order to obtain IFR certification. These modifications frequently involve stability and pilot workload. In some cases modifications to stability can mean that a machine formerly requiring two pilots for IFR can then be certified for single pilot IFR flight.

The stability of the airframe can also be augmented artificially. Several of these techniques were shown in Chapter 7. In principle an unstable airframe with an artificial stability system gives the best of both worlds because the pilot can retain the manoeuvrability of the unstable airframe whilst enjoying the artificial stability. In practice some thought has to be given to what will happen if the artificial stability system fails in IFR. Either the artificial system has to be made redundant so that it continues to function in the presence of a failure or the airframe has to be made sufficiently stable that pilot workload remains reasonable if the augmentation system fails. Aerodynamic stability improvements are very reliable because they usually involve the addition of passive structures or changes to the shape of the existing structure.

Stability in helicopters is a complex issue in comparison to aeroplanes because helicopters are more asymmetrical, have more coupling between their degrees of freedom and greater structural elasticity owing to the use of hinges and/or flexible blades. The inherent flexibility between the rotor and the hull is responsible for many peculiarities of helicopter behaviour. In many cases the rotor tries to do one thing and the hull tries to do the opposite and it is unclear which one will dominate. It also follows from this that the stability characteristics of helicopters having zero-offset rotor heads will differ from those having real or effective offset.

In all aircraft, stability can be subdivided into specific areas. Speed stability is the ability of an aircraft to return to trimmed airspeed after a disturbance. Pitch stability is the ability to return to the correct pitch attitude after a disturbance. Directional stability is the ability to keep the nose at the front and lateral stability is the ability to return the wings/rotor to a level attitude after a disturbance. If all of these are acceptable the machine may still have a problem due to the interaction of the lateral and directional stability mechanisms. This is known as spiral stability. Helicopters may further be assessed by their stability in the hover.

Speed stability of a main rotor is extremely good owing to the phenomenon of flapback described in Chapter 3. If forward speed increases for any reason, the asymmetry of lift between the advancing and retreating blades results in a rolling couple which the main rotor will precess into rearward pitch. This tilts back the rotor thrust vector tending to reduce the airspeed. Thus the rotor itself has good speed stability. However, the hull is suspended below the rotor and the hull drag effectively acts at some distance below the rotor head, causing a downward pitching moment on the hull. Thus an increase in airspeed causes the hull to pitch down. Depending on the type of rotor head, this downward pitching moment will result in different hull attitude changes. As the swashplate is controlled from the hull, hull pitchdown will cause the rotor to pitch down.

Thus the helicopter itself may not display speed stability if the effect of the hull dominates the effect of the rotor. In practice the helicopter may need a tail plane to provide speed stability. If set at a suitable negative angle of incidence, the tail plane produces a downthrust that increases with airspeed and compensates for the hull drag.

In practice the rotor on its own has excess speed stability and large amounts of forward cyclic would be needed to increase speed. Thus some of the pitchdown due to hull drag can be used beneficially to reduce the cyclic travel needed. The tail plane area and incidence can be selected in such a way that the speed stability is just on the positive side of neutral. This means that airspeed increases will always result in the cyclic stick being trimmed forwards.

If the helicopter is disturbed on its pitch axis in forward flight, the RAF seen by the blades will change, but the advancing and retreating blades will see different changes. If a nose-up pitch disturbance is experienced, Figure 8.13 shows that the angle of attack of both blades will increase, but the increase seen by the advancing blade is greater. This produces a rolling couple to the retreating side that is precessed into a further nose-up pitching moment by the gyroscopic action of the rotor. Thus it will be seen that a rotor in forward flight is unstable in pitch. This is a further reason for the use of a tail plane in the helicopter. Dynamic stability here can be good because the tail plane has aerodynamic damping.

With a zero-offset rotor head, any tilt of the rotor disc in pitch is opposed because no couples are passed across the head and the hull and the swashplate attitude are little changed. If the rotor disc pitches but the swashplate doesn't there will be an application

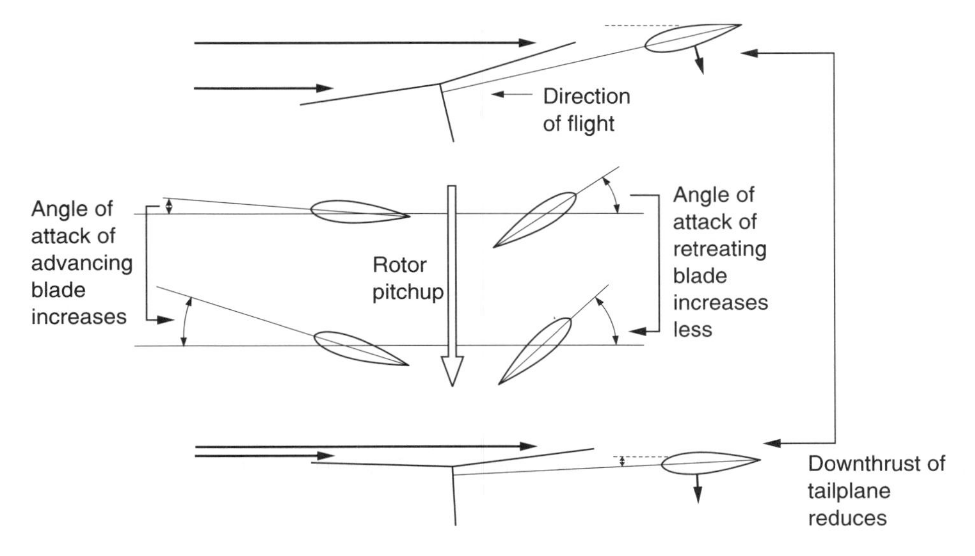

Fig. 8.13 The origin of pitchup instability. See text.

of cyclic pitch opposing the pitching. In the case of a hingeless rotor head, couples can be passed across the rotor head and these can cause the hull to follow the rotor. Thus a helicopter with a hingeless rotor head will tend to be less stable in pitch. We expect that because such rotor heads offer greater manoeuvrability.

Helicopters with a conventional tail rotor have good directional stability. As was seen in Chapter 5, the tail rotor acts like a fin and has good yaw damping.

In aeroplanes, lateral stability is obtained by the use of dihedral or sweepback. If the aeroplane is rolled by a disturbance, it will slip in the direction of the roll and the relative airflow will have a sideways component. In the case of dihedral, the wing on the low side will see an increased angle of attack whereas the other wing will see a reduced angle of attack resulting in a righting couple. In the case of sweepback, the aspect ratio of the wing on the low side will improve and that of the other wing will worsen, achieving the same result.

The mechanism in the helicopter is quite different. To find out what happens it is necessary to consider all moments acting about the roll axis which passes through the CM. In fact moments may be analysed about any axis, including one passing through the rotor head as Figure 8.14 shows. Essentially if a helicopter is disturbed in its roll axis it will slip and the main rotor will see a lateral component of airspeed. This results in sideflap that tends to oppose the roll and return the helicopter to a level attitude. Thus the rotor alone acts as if it had dihedral. However, the hull is suspended below the rotor and the displaced CM acts like a pendulum to create a righting moment. In addition, the lateral component of airspeed will cause a side thrust that tilts the hull in the same direction as the roll. Figure 8.14 shows that although the rotor is laterally stable, the hull is aerodynamically unstable. The outcome depends upon the vertical position of the CM, the aerodynamics of the hull and on the type of rotor head, but in most cases the stability of the rotor dominates and helicopters tend to be highly stable, possibly too stable, in roll. Dynamic stability is good because the rotor attitude has aerodynamic damping. The zero-offset rotor head is the least stable because there

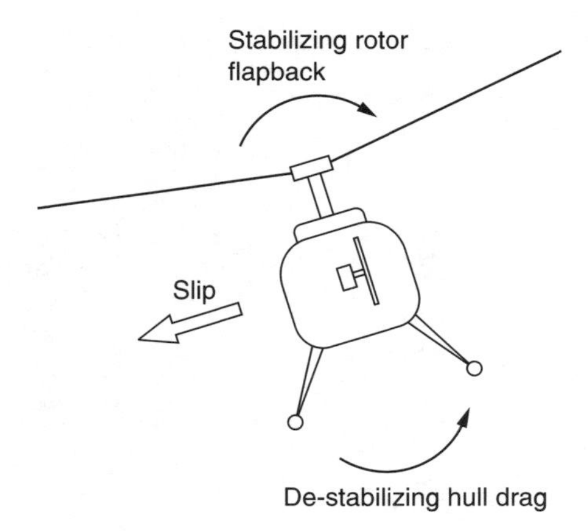

Fig. 8.14 Lateral stability. In the rotor alone, lateral stability is automatic due to flapback, but the hull is unstable laterally because side slip tilts it further in the direction of roll.

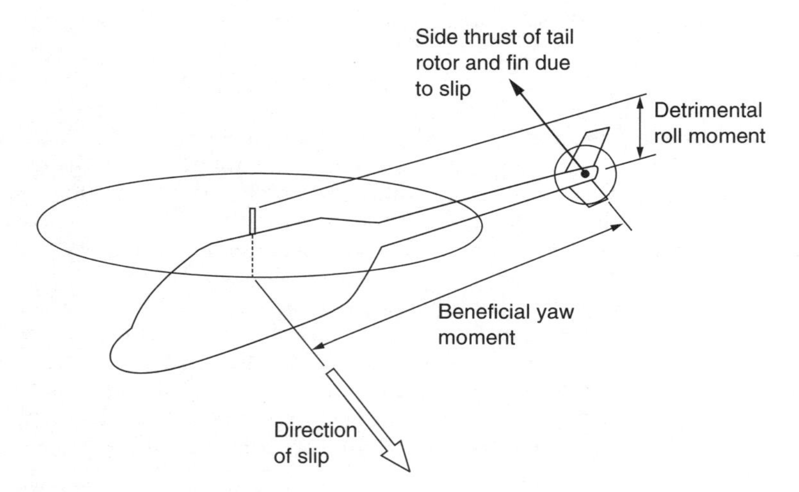

Fig. 8.15 Forces acting on a helicopter that has been disturbed in roll and which is slipping. The ratio of yaw to roll moments determines the spiral dive or Dutch roll characteristics.

is no couple from the rotor to stop the hull being rolled by lateral drag. This has the effect of rolling the swashplate. This applies cyclic in a destabilizing sense. The hingeless rotor is the most stable because lateral drag cannot roll the hull so readily against the stiffness of the head. Thus lateral stability is better with the hingeless rotor: precisely the opposite of the case for longitudinal stability.

Figure 8.15 shows that in a helicopter with slip, the sideways component of airspeed will act on the hull, the tail rotor and on any fins. The fin and tail rotor are intended to create a yawing moment that turns the machine to face the direction towards which it is slipping. However, if the fin and tail rotor have a large area and are mounted low on the machine, the rolling moment they produce may exceed the stabilizing rolling moment from the main rotor. The helicopter may enter a spiral dive if disturbed in roll. Spiral instability may be remedied by lowering the CM, reducing fin area or by moving it upwards or forwards or by adding side area to the hull high up ahead of the mast. Upward relocation of the tail rotor and/or increasing the main rotor flapping offset will also be beneficial.

On the other hand if the fin and tail rotor are high mounted, they may hardly oppose the dihedral effect of the main rotor at all and the helicopter may have too much lateral stability such that if disturbed it suffers a phenomenon called a dutch roll.

Dutch roll describes a flight path like a corkscrew. Following a slip, excessive lateral stability creates a restoring roll couple so powerful that there is insufficient momentum for the helicopter to overshoot the correct attitude. The helicopter then oscillates about its path near-sinusoidally in pitch, roll and yaw but with phase differences between the motions. The motion may be damped heavily, so that it dies out or it may stabilize with certain amplitude. In some cases the amplitude will grow indefinitely. In most cases pilot intervention prevents such a divergence. However, for IFR operation, Dutch roll is considered as detrimental as spiral instability. Dutch roll may be remedied by raising the CM, by lowering the tail rotor, by reducing the main rotor flapping offset, by increasing fin area, by moving the fin back and/or down or by adding side area to the hull low down. Fences may be seen on the bottom of the hull for this purpose.

Hover stability is marginal in most helicopters because the hull mass is located below the rotor. It might be thought that this would act like a pendulum and give stability but this is not what happens. The hull has mass and is more or less flexibly suspended from the rotor according to the type of rotor head. As a result the attitude of the hull will always lag the attitude of the rotor. Figure 8.16 shows that if a helicopter is trimmed to stationary hover and then the controls are locked, it will begin a divergent oscillation. The rotor on its own is stable in the hover because if it moves in any horizontal direction, flapback will tend to arrest the motion. However, it can be seen from the figure that the hull swings from side to side under the rotor, but because the swashplate is referenced to the hull, that swinging causes cyclic inputs. These are not phased to damp the motion, but in fact augment the motion.

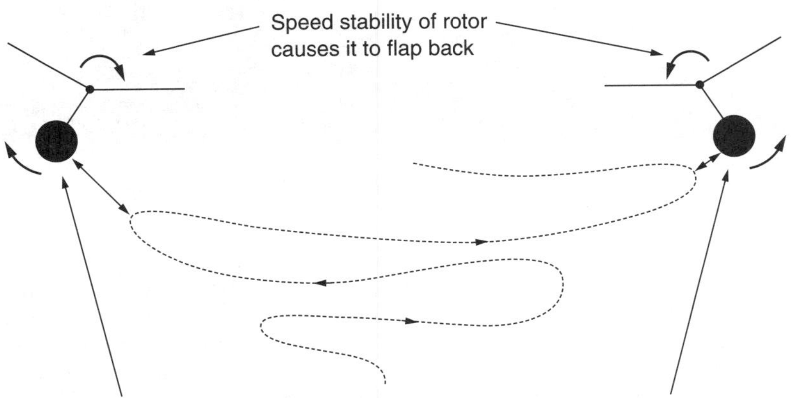

Fig. 8.16 Helicopters are unstable in hover and, without pilot intervention, will wander to and fro as shown here.

Fig. 8.17 The two Hiller VZ-IEs of 1957 were prototypes of the proposed Hiller Ring-Wing Coleopter, a craft capable of high-speed horizontal flight. (Hiller Museum)

Hover stability can be obtained by putting the hull above the rotor. This causes enormous practical difficulties and it is usually better to accept the hover instability. However, Stanley Hiller designed a number of one-man helicopters (Figure 8.17) in which the pilot was above the shrouded rotor assembly and these were very stable. A machine built by DeLackner had the same attributes.

9

Other types of rotorcraft

Although the single main rotor helicopter with a tail rotor predominates, there are many other possibilities, sometimes more complex, which give advantages under certain conditions that outweigh the complexity. This chapter considers those possibilities. Some of these are flawed and have not been developed, whereas some were ahead of their time and deserve to be revisited. Some have been a success from the outset, whereas others have required a great deal of development.

9.1 The gyroplane

The gyroplane was the first successful type of rotary wing craft and many of the constructional techniques used in early helicopters were developed there. The rotor is not mechanically driven in normal flight but instead the machine is driven forwards by a conventional aircraft engine and airscrew in either a tractor or pusher arrangement. Figure 1.10 showed that the tip path axis of the gyroplane rotor is tilted aft so that there is an upward component of the relative airflow along the rotor axis. Effectively the rotor is in translational autorotation. This was sometimes called automatic gyration hence the trade name 'autogyro' which Juan de la Cierva's gyroplanes were given. The name is now used almost interchangeably with gyroplane. The aft tilt of the rotor means that the rotor thrust is not vertical and has an aft-facing component comparable to the drag of a wing, and similarly comprising profile drag and induced drag. At constant airspeed in level flight the thrust of the propeller is in equilibrium with the weight, hull drag and the rotor thrust. The propeller is doing work against the induced drag of the rotor and thus delivers power to produce lift.

Driving the rotor in this way has the advantages that there is no mechanical transmission and no torque reaction. No tail rotor is necessary. However, the gyroplane is unable to hover because it needs constant upward inflow. If the airspeed is brought to zero, the machine will descend in a vertical autorotation. The gyroplane is an STOL aircraft that cannot stall and in the proper hands it is a very safe form of transport. However, in the absence of a tail rotor, yaw control at very low airspeeds is generally unsatisfactory and some accidents have occurred due to uncorrected yaw on touchdown. A generous amount of vertical tail area is required, for example the triple fins of the Air and Space Model U-18. As the rotor is driven in the same way as a wing, it is quite common for gyroplanes to include a fixed wing. This has the advantage of improving the load factor at high speeds.

The inability of the gyroplane to hover is its greatest shortcoming. Before the helicopter was developed, the gyroplane occupied a niche in which fixed-wing aircraft were unsuitable. Once the true helicopter was developed, the gyroplane could not compete and its market contracted dramatically. The basic problem is that the rotating wings introduce some of the high stress, the mechanical complexity and the high maintenance regime of the helicopter, for which there is no compensating ability to hover. In gyroplanes having pusher propellers, the visibility from the cockpit is truly excellent. It has had some success as a recreational machine and in applications such as aerial photography where slow stable flight is an advantage, but has effectively been eclipsed by the helicopter.

The gyroplane rotor experiences the same advancing/retreating and gyroscopic phenomena as the helicopter, in fact most of these were first discovered in the gyroplane. One does, however, tire of reading quite erroneous descriptions of how Juan de la Cierva's machines rolled over because of the lift difference between advancing and retreating blades. It is well established that the lift difference between advancing and retreating blades may result in a roll couple, but this is subject to the gyroscopic phase lag of 90° that results in flapback. Clearly the advancing/retreating asymmetry was not the origin of de la Cierva's difficulty.

Figure 9.1 shows that de la Cierva's early machines were basically aeroplanes having conventional aircraft controls but with a rotor added. The early rotors were hingeless and acted like gyroscopes. The problem was simply that when the conventional elevator was operated, a pitching moment was applied, but the rotor would precess this into a roll. Equally the operation of the ailerons would result in a rolling moment that would precess into a pitching motion. This was not a novel phenomenon; some World War I aircraft having rotary engines were notoriously gyroscopic and displayed precession whereby use of the elevator would cause an amount of yaw.

Fig. 9.1 An early de la Cierva autogyro having wings and aircraft-type control surfaces.

De la Cierva's free flying models were successful but they had fixed surfaces with no controls. As a result the problem was only revealed on the full-size prototypes, several of which were destroyed before flapping hinges were adopted. Blades mounted on flapping hinges cannot transfer rotor moments to the hull. With hinges fitted, the application of the conventional elevator caused the hull correctly to pitch with respect to the rotor. As was shown in Figure 4.22 this would cause the control axis to deviate from the tip path axis, resulting in an application of cyclic pitch to the rotor causing it to follow the attitude of the hull. The machine would no longer roll over when elevator controls were applied, and it would roll when the ailerons were used.

However, even if flapping hinges are fitted, this is still not enough. In translational flight, a flapping rotor will suffer from flapback due to asymmetry of lift, and will also suffer inflow and coning roll. The tip path axis will be aft of and to one side of the hub axis. The flapback is countered by propeller thrust, whereas the inflow and coning roll must be countered by steady application of opposing aileron in order to fly straight. Flapping hinges alone could not and did not control inflow and coning roll. De la Cierva soon discovered that the gyroplane is capable of very slow flight where conventional aircraft-type ailerons and elevators are ineffective. These were soon abandoned in favour of the tilting hub method of cyclic control described in section 4.15.

In a fixed pitch gyroplane application the tilting hub allows a fairly simple mechanical system. The rotor shaft is gimbal mounted and the blades have flapping hinges (and thus need dragging hinges and dragging dampers). Tilting the hub is possible because of the flapping hinges and this causes a cyclic pitch application. In stable forward flight the control axis (which is the hub axis) will be forward of the tip path axis because of flapback, and to one side because of inflow/coning roll. In other words the hub must be tilted with respect to the tip path axis.

The reader is cautioned that many of the popular explanations of blade flapping in gyroplanes are hopelessly flawed. The most common flaw is the assertion that the flapping hinges were essential to the gyroplane and by association to the helicopter. This is nonsense and the origin of it is easy to see. Many technical treatises on the gyroplane and the helicopter, for example the classic work of Gessow and Myers, analyse with respect to the control axis. To enable straight and level flight the control axis must be tilted and of course all rotors, hinged or hingeless, flap with respect to the control axis. The same erroneous account is seen in any number of general readership books. It seems that someone somewhere confused the tip path axis and the control axis and wrote an incorrect account that subsequent authors simply repeated without question.

The cyclic control requirements of the helicopter and the gyroplane are identical. As was seen in Chapter 4, the tilting hub with flapping hinges gives exactly the same cyclic control as a hingeless head with feathering hinges and a swashplate. Gyroplanes have been built with no flapping hinges and helicopters have been built with tilting hubs.

In early gyroplanes the blades were started by pulling a rope wound round a drum and brought up to speed by taxiing. Later machines drove the rotor from the engine to start it, using the ground contact to oppose torque reaction. The drive was disengaged before flight. The simple gyroplane takes off in the same way as an aircraft by accelerating along a runway, although for a much shorter distance. It can lift off as soon as the blades are turning fast enough. Using a flare, a gyroplane could lose most of its forward speed and need only a very short landing run. This allowed it to land in places from which it could not take off again. The solution was the so-called jump-take-off gyroplane. This would use the engine to drive the rotor on the ground in fine pitch until its RPM was somewhat in excess of that required for flight. By disconnecting the engine from the rotor and applying a higher pitch, the machine would jump into the air using only stored energy in the rotor, which would suffice until forward speed had been attained.

Cierva autogyros achieved jump-start using a delta-one (see section 4.7) hinge in what was known as an autodynamic head. The inclination of the dragging hinge pin caused an interaction between the drag angle and the pitch angle. This kept the blades in low pitch as they dragged back due to engine torque. As soon as the drive was disconnected the blades would swing forwards and increase their pitch and the machine would jump into the air.

Raoul Hafner built a jump-take-off gyroplane in which the rotor head had cyclic control via a swashplate and a collective pitch control. The pilot could control the jump and also set the RRPM in flight. The machine could land vertically because the pilot could lower the collective lever to speed up the rotor during the descent and then raise it again to use the stored energy to arrest the descent just before touchdown. Hafner's rotor head was essentially a helicopter head. Hafner went on to design the Bristol Sycamore, a practical and successful helicopter and the first British rotary wing aircraft to be given a civil Certificate of Airworthiness.

The upward inflow of the gyroplane suggests a rather different rotor design to that of the helicopter. The blade twist should be reversed compared to that of the helicopter, i.e. the root should have less pitch than the tip, although the amount of twist needed is smaller. In practice, many gyroplanes are built with no twist at all, for economy.

The rotor of the gyroplane is not supplying propulsive thrust and so the loads on the blades are reduced. Retreating blade stall is a much-reduced problem, especially if a fixed wing is provided, and as a result many early gyroplanes cruised at speeds considered high even in modern helicopters, especially those having wings.

Simple gyroplanes have two-bladed underslung teetering heads. If these have fixed pitch, the two blades can be rigidly fixed together and gimbal mounted. Cyclic pitch is applied by a single rod from the swashplate. This makes for a very light machine. However, the combination of low weight with a zero-offset rotor requires more caution on the part of the pilot, just as it does in the helicopter.

A gyroplane with a teetering head is vulnerable to a combination of low RPM and low *g*. Inexperienced pilots may try to dive like a fixed-wing aircraft using forward stick when they should descend by reducing power and applying aft stick. Forward cyclic will reduce RRPM. The correct recovery from low RPM/low *g* is a very gentle progressive application of rear cyclic. A rapid or panic application of rear cyclic at low *g* means that the disc tilts back without the machine following and there may be a blade strike. There have been some accidents because of this. In contrast a gyroplane with a hingeless rotor could be remarkably aerobatic.

9.2 The winged helicopter

The winged helicopter is conventional except for the addition of fixed wings (Figure 9.2). When this is done, the wing produces a proportion of the lift, and so the rotor produces less lift. However, the wing also produces drag. There will be the usual induced and profile drag as well as interference drag between the wing and the rotor.

This drag has to be balanced by thrust from the rotor. As a result the rotor tip path axis has to tilt forward further than it would without the wing. Although the vertical component of rotor lift is reduced, the horizontal component will be increased. The span of a practical wing will be less than the rotor diameter and so the wing will be less efficient than the rotor. Thus the winged helicopter is unlikely to be more efficient

Fig. 9.2 A winged helicopter obtains some of its lift in flight from the wing. However, in the hover, the wing may cause a download. This Sikorsky S-67 prototype flew like a fighter but was not produced.

than the pure helicopter, except at moderately high speed where the induced drag of the wing is less and the direction of the wing reaction has less rearward inclination.

The reason for fitting a wing is shown in Figure 9.3. This shows how the load factor of a rotor and a wing vary with airspeed. The wing obviously shows a square law trend with airspeed, but the rotor does not. Although the advancing blade shows this trend, the extra lift cannot be used because it would cause imbalance with the reduced lift from the retreating blade.

The simple helicopter is generally implemented as a low load factor machine because provision of a high load factor implies very poor L/D ratio in the hover. Thus if a high load factor is required in forward flight so that the helicopter can make tight banked turns, a good solution is to add a wing, as was done in the Sikorsky S-67 and the Boeing 347, neither of which entered production.

Clearly the wing will be a drawback in the hover and climb where it will cause a download. It is also a problem in autorotation where it will produce an upload. Following an engine failure at high speed, the wing will carry on lifting and it will be difficult to maintain rotor speed. A winged helicopter may also be less controllable, depending on the rotor technology. Figure 4.13 showed that a teetering head can only produce control by directing the rotor thrust so that it no longer coincides with the CM. If the thrust is reduced because the wing is providing some of the load, the control power is reduced in proportion. A hingeless rotor will suffer less from this effect because it can transfer moments to the mast.

Figure 9.4 shows the solutions Boeing tried out with the experimental 347. This was basically a CH-47 with a stretched hull, the stretch section being fitted with a wing.

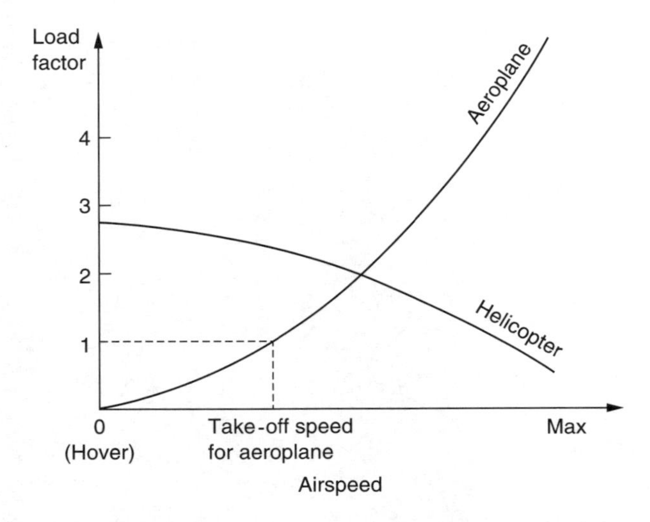

Fig. 9.3 The wing produces lift as the square of the airspeed whereas the rotor does not. Thus a winged helicopter can have a higher load factor at speed.

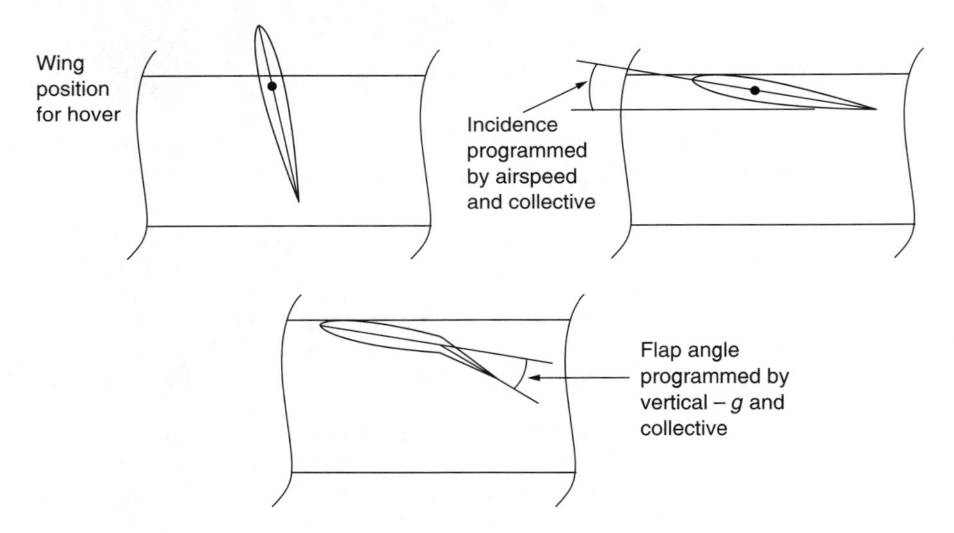

Fig. 9.4 The wing of the Boeing 347 could be set to 85° to reduce download in hover and had flaps to vary the camber according to flight conditions.

It was also tested with four-blade rotors. In the hover the wing could be tilted 85° to reduce download. In translational flight the goal of the wing was to offload the rotors and improve load factor. The wing had flaps that could be used to adjust camber. The outer parts of the flaps could be used differentially to act as ailerons. In Figure 9.4 it will be seen that the wing incidence was programmed from airspeed and collective pitch in such a way that the appropriate degree of rotor unloading was obtained. The flap setting was computed from the collective setting and from a *g*-sensor measuring acceleration normal to the cockpit floor. In a high-*g* manoeuvre, such as a steep turn, increased rotor thrust would initiate the acceleration, but the g-sensor would add wing camber by lowering the flaps so that the wing lift would increase in proportion to the rotor thrust. Whilst the 347 was very manoeuvrable indeed, all of the necessary components added to the weight and complexity.

If wings are successfully to be applied to helicopters, they need to be designed in at the outset. If the wing structure can also do duty carrying fuel and/or supporting the undercarriage, the extra weight is reduced. Most helicopters have relatively poor hull shapes and finish and the hull drag is high. If an improved hull design is used, in which the usual external accessories are flush mounted, and a retracting undercarriage is employed, the drag of the hull can be reduced by an amount that makes a significant contribution to reducing the extra thrust needed by the wing.

In some machines the undercarriage retracts into sponsons outside the hull in order to avoid encroachment into the hull space. Such structures may also be used for fuel. Although not intended to be wings, it is quite common to give these an airfoil section to obtain at least some lift in return for the drag.

9.3 The compound helicopter

The compound helicopter, or gyrodyne, is one in which the rotor does not produce any forward thrust in cruise. Instead the thrust is provided by other means. Experiments were performed using turbojets, but as might be expected from momentum theory, these were a very inefficient way of producing thrust at helicopter speeds. Conventional propellers give much better results.

In gyrodynes the pitch of the thrust propeller is increased as airspeed increases so that it continues to provide forward thrust. In cruise, the cyclic control of the main rotor is used to trim the control axis so that there is sensibly no flapping and the tip path axis remains at right angles to the direction of flight. This has a number of advantages. The thrust required from the rotor is reduced as it is only carrying the weight of the machine and not overcoming drag as well. This may be reduced further by the use of stub wings. Rotor power is reduced and this reduces the weight of the transmission. However, the main advantage is aerodynamic. As the rotor remains edge-on to the airflow, there is no need to increase the blade pitch to account for inflow as speed rises. This minimizes the effect of retreating blade stall and reduces vibration, allowing a higher airspeed to be reached.

The objections to the use of a wing in an otherwise conventional helicopter mostly disappear in the compound helicopter and are replaced by advantages. The thrust to overcome the drag of the wing does not come from the rotor, but instead is supplied by the propeller. The benefit of a higher load factor is obtained along with potential for higher speed. The winged helicopter has difficulty if the engine fails at high forward speed because the wing lift prevents the rotor obtaining upflow. However, in the gyrodyne, following an engine failure at speed the propeller can be set to the windmill brake state and used to create shaft power by slowing the machine down. This power will maintain the rotor speed until the forward speed drops sufficiently to lose wing lift.

In the Fairey Gyrodyne shown in Figure 1.11 the anti-torque tail rotor is replaced by a forward-facing variable-pitch propeller mounted on a short wing at one side of the main rotor. In the hover the anti-torque propeller pulls forwards and the main rotor has to be tilted backwards using the cyclic control to prevent forward drift. The Gyrodyne captured the world speed record in 1948, but development was halted following a fatal fatigue failure having nothing to do with the compound concept.

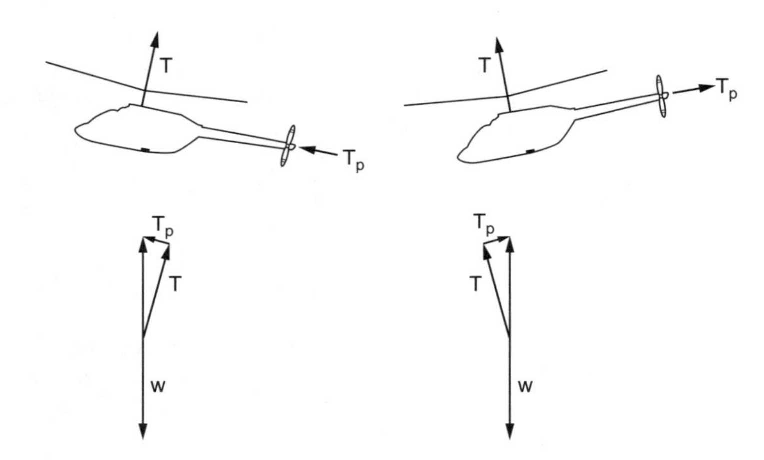

Fig. 9.5 With a propeller a gyrodyne can change its hover attitude using cyclic to balance prop thrust.

The Gazda Helicospeeder had a tail rotor mounted on a swivel. In the hover this would be oriented conventionally, but in forward flight the pilot could turn the tail rotor to point backwards and act as a pusher propellor. This machine still exists and has been restored by Stanley Hiller. Sikorsky also tried a similar device, which they called a Rotoprop, on a modified S-61.

In the Lockheed Cheyenne (Figure 1.12) a tail rotor and a rear-mounted pusher propeller are both provided. The tail rotor is on the end of the tail plane to clear the pusher. The propeller pitch can be controlled by a twist grip on the collective lever. In the hover the pusher propeller is typically operated in flat pitch and the tail rotor counteracts the torque as normal. However, the pusher prop can be used as a hover attitude control. Figure 9.5 shows that with prop thrust balanced by fore or aft cyclic the machine can hover nose up or nose down.

Accelerating into forward flight is not done using the propeller as it is far more effective to tilt the rotor disc as per pure helicopter practice. However, as speed builds up, the disc attitude is levelled and more power is fed to the prop. As the Cheyenne has a hingeless rotor it can provide sufficient control power even when it is off-loaded by the wing and no other controls are necessary. The Cheyenne was probably the highest performance helicopter ever built. It could do 215 knots in level flight, turn continuously at 60° bank angle, could pull 2.6 *g* and remain stable in negative *g*. The fact that the Cheyenne didn't enter production again had nothing to do with the compound concept.

In the Piasecki Pathfinder the rear prop is mounted in a ring duct equipped with deflectable vanes to provide anti-torque thrust for hovering. Whilst adequate for civil operations it is unlikely that such an arrangement would meet military crosswind hover requirements. The Pathfinder's ring duct also has adjustable horizontal vanes. In cruise these can be trimmed so that the hull is absolutely level for minimum drag even if the CM is not aligned with the rotor shaft. The rotor is then providing axial thrust only and so alternating loads on the mast are minimized.

The compound helicopter offers a tangible way of improving the performance of the helicopter, particularly in the area of forward speed. Compound helicopters have not yet reached the market, but this is not because of any fundamental performance problem. If there is a rotary wing technology deserving to be revisited with modern materials and control systems then this is it. A tail-mounted pusher propeller for forward flight would be naturally complemented by a NOTAR system for hovering, leading to a very clean external appearance with low drag.

9.4 The convertiplane

A convertiplane changes its operating principle according to the flight regime, most obviously between hover and cruise. The Fairey Rotodyne (Figure 1.11) is a convertiplane. This machine has two turboprop engines that in addition to driving variable pitch propellers can also produce significant amounts of compressed air. In the hover the compressed air is fed along the rotor blades and fuel is burned at the tips to drive the rotor. There is no torque reaction and yaw control is obtained using differential propeller pitch.

The tip jet drive is inefficient, but is only used for take-off and landing. Once forward speed has been obtained, the rotor drive is switched off and the rotor is tilted back so the machine becomes a gyroplane. The rotor is partially unloaded by the wing. Landing is achieved by reigniting the tip jets and reverting to helicopter mode, although landing on a runway in gyroplane mode is also possible.

The winged McDonnell XV-1 has a single piston engine driving either a propeller or an air compressor. In order to hover the compressor output is fed along the blades to tip jets as in the Rotodyne. No yaw control could be obtained with a single prop, and instead a pair of small fixed pitch tail rotors is installed, powered by small reversible hydraulic motors. The XV-1 can then take off as a helicopter. In forward flight the engine power can be diverted to the prop so that the machine becomes a compound gyroplane.

The XV-1 has a substantial wing area and full conventional aircraft controls and in cruise these will become functional. The fore-and-aft cyclic control is then automatically adjusted to control rotor speed. This is reduced to 50% of the hover RPM, so that the wing is carrying about 80% of the machine weight. Thus the XV-1 can fly as a helicopter, a gyroplane or an airplane.

Following the progression of unloading the rotor in forward flight, the next step in the view of some designers is to dispense with the rotor altogether. There have been numerous proposals. A two-blade rotor is stopped transversely in flight to become a wing, a four-blade rotor becomes an X-wing, a rotor folds and retracts into the hull. All of these proposals suffer from the same practical difficulty – as the rotor slows down it loses the centrifugal stiffening effect. As has been seen earlier in this book, as a rotor slows down it ceases to be gyroscopic and the phase lag starts to reduce from 90°. This causes a control problem. A more serious difficulty shown in Figure 9.6 is that at some stage a near-stationary rotor blade will have significant sweep forward and this is an unstable condition. In conventional wings and blades flutter is avoided by bringing the mass centroid towards the leading edge. However, at low rotor speed the reverse flow region would envelop the retreating blade such that leading and trailing edges interchange so this technique cannot be used. The blade and hub would need to be immensely rigid in torsion and bending to survive the loads. As a result no practical machine has emerged and it is unlikely that one will.

The only in-flight rotor-folding technique that appears viable is to tilt the rotor backwards and to feather the blades. As the rotor slows down it will cone up 90° and ultimately stop with the blades trailing in the slipstream. However, if the rotor is going to be tilted, why not use it as a propeller?

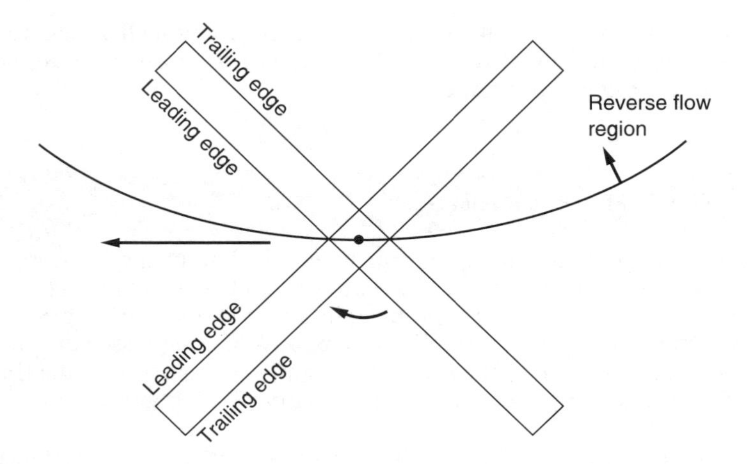

Fig. 9.6 The flaw in stopped rotor proposals. As rotor speed falls, the reverse flow region becomes enormous and leading and trailing edges interchange.

Tip jet convertiplanes are interesting and educational, but are unacceptably noisy as production items. More significantly, developments in fixed-wing STOL technology have delivered simpler solutions in many applications. Stopping the rotor in flight is simply too difficult. Consequently most of the recent progress made in convertiplanes has been in machines that tilt the rotors in some way to become propellers creating forward thrust. This is an elegant approach because there are no redundant propellers in the hover and no redundant rotors in cruise.

A prop rotor or tilt rotor machine can take off like a helicopter, but by turning the rotors about a horizontal axis forward flight will be achieved. When the rotor axes are horizontal there will be no rotor lift at all and the wing must produce 100% of the lift. In this mode the tilt rotor is an aeroplane with large propellers. Compared to a helicopter, drag is reduced considerably and the top speed and range are significantly improved. The load factor also becomes that of an airplane and so the tilt rotor should be highly manoeuvrable.

A rotor that provides enough hover thrust will be somewhat overspecified in forward flight and will be inefficient because it will work at a low L/D ratio with a high profile drag penalty. The solution currently adopted is to make the cruise more efficient by reducing the rotor size. This increases the disc loading in hover, and drives up the downwash velocity and the power required, but this power is not used for very long. It may be an advantage to use different RPM in the hover and in forward flight. With the rotor always in axial flow, the alternating forces experienced by the helicopter blade do not exist and the rotor does not need to be detuned against vibration. It is thus quite easy to change the rotor RPM for cruise.

The ideal would be a variable diameter rotor and although the forces involved are considerable this is not considered insoluble. With high disc loading and a wing, the tilt rotor is not expected to have good autorotation performance and it would appear inadvisable to rely on a single engine.

The Bell-Boeing Osprey (Figure 9.7) tilt rotor is a true helicopter in the hover. It has contra-rotating rotors that have cyclic and collective pitch. There are two engines, one in each wing tip nacelle. The entire engine and transmission tilts with the rotor. To give protection against engine failure the two rotors are connected by a cross shaft so that

Fig. 9.7 The Bell-Boeing Osprey tilt rotor is a true side-by-side helicopter in the hover.

either engine can drive both rotors. In the wing centre section the cross shaft drives accessories.

In the hover, the pitch axis is controlled with fore-and-aft cyclic applied equally to both rotors. Yaw is controlled by applying fore-and-aft cyclic in opposite directions. The roll control requires differential collective pitch. In addition to the cyclic and collective controls, the pilot has a control that can alter the angle of the two engine nacelles. These are synchronized so they will always have the same angle. From the hover, thrust is increased and the nacelles are brought down. Initially the download on the tail plane will cause the nose to rise, but as speed builds up and the dynamic pressure on the wing and tail increases, the machine becomes an aeroplane and responds to its control surfaces. The cyclic control of the rotor pitch is unnecessary in forward flight.

The wing forms an obstruction in the hover and in the Osprey the leading and trailing edges will both fold to present the least area to the downwash. As the angle of relative airflow seen by the wing changes, first the leading edge and then the trailing edge will fold back to the forward flight position. A disadvantage of the tilt rotor is that in the hover the lift is applied at the wing tips and the wing is highly stressed.

One of the enabling technologies that turned a research tool into the Osprey is an extremely complex electronic control system that recognizes the flight conditions and automatically directs the pilot's control movements to swashplates or control surfaces or both as necessary. This could probably not have been implemented mechanically. The extensive use of composites has led to an airframe light enough to allow a useful payload.

An alternative to the tilt rotor is the tilt wing in which the wing and rotors form a single assembly that tilts. As the wing chord is vertical in the hover the obstruction of the downwash will be reduced. In this attitude the wing forms a deep beam and will naturally be very stiff. The wing needs no folding flap mechanism and therefore will be less expensive. It may be possible to use the wing ailerons as a yaw control in the hover, leading to an approach in which the rotors have no cyclic control at all, but become variable pitch propellers. Hover roll control is then by differential collective, but control of the pitch axis in the hover requires an additional rotor at the tail which is only powered for the hover. The requirement for a horizontal tail rotor somewhat offsets the saving in complexity due to eliminating the use of cyclic pitch on the main rotors.

Whilst such an approach may be suitable for a civil machine, it is unlikely that the more stringent requirements of the military could be met, specifically the hover performance in a wind where the vertical wing is a drawback. The tilt wing also requires more caution in the transition from aircraft to helicopter mode where the prop rotor thrust and wing angle have to be carefully adjusted as a function of the airspeed.

9.5 Multi-rotor helicopters

Multi-rotor helicopters have advantages in some niches of the market, but are significantly less common than the conventional type partly because of the added complexity and partly because additional skills are needed to develop them. It is often the case that one individual has perceived a particular advantage of a configuration and perfected it for a particular niche. That niche having been filled, no other manufacturer then seeks to compete. As a result each type of machine will be associated with a particular manufacturer. Thus the tandem is associated with Vertol/Boeing, the coaxial with Kamov and the synchropter with Kaman.

Figure 9.8(a) shows that there are a number of possibilities with two rotors. The rotors may or may not overlap, and they may be placed in tandem or transversely to the direction of flight. If they do not overlap, the rotors may turn in the same direction or may contra-rotate. If the rotors turn the same way, they must be inclined away from the vertical to counteract torque as shown in Figure 9.8(b). Nicolas Florine successfully flew a machine of this kind in Belgium in 1933 as Figure 9.9 shows. In those days helicopter dynamics were poorly understood and Florine chose same-way rotation because the gyroscopic stability of the rotors would not cancel. Florine used hingeless rotors and the same-way rotation would also have minimized the stress on the hull. Test flights proved Florine right as in its day it was probably the most stable helicopter in existence. The Cierva Air Horse shown in Figure 9.10 had three rotors which all turned the same way. Torque cancellation was obtained using the Florine principle.

9.6 The side-by-side configuration

The side-by-side arrangement of contra-rotating rotors has some advantages in forward flight because it improves the aspect ratio of the 'disc' and reduces the induced power needed. Heinrich Focke's Fw 61 was a machine of this type and was probably the first helicopter in the world to progress from the flying test bed into a truly flyable machine. This machine achieved a degree of notoriety in 1938 when it was extensively demonstrated as a Nazi success symbol flown by Hanna Reitsch. This should not

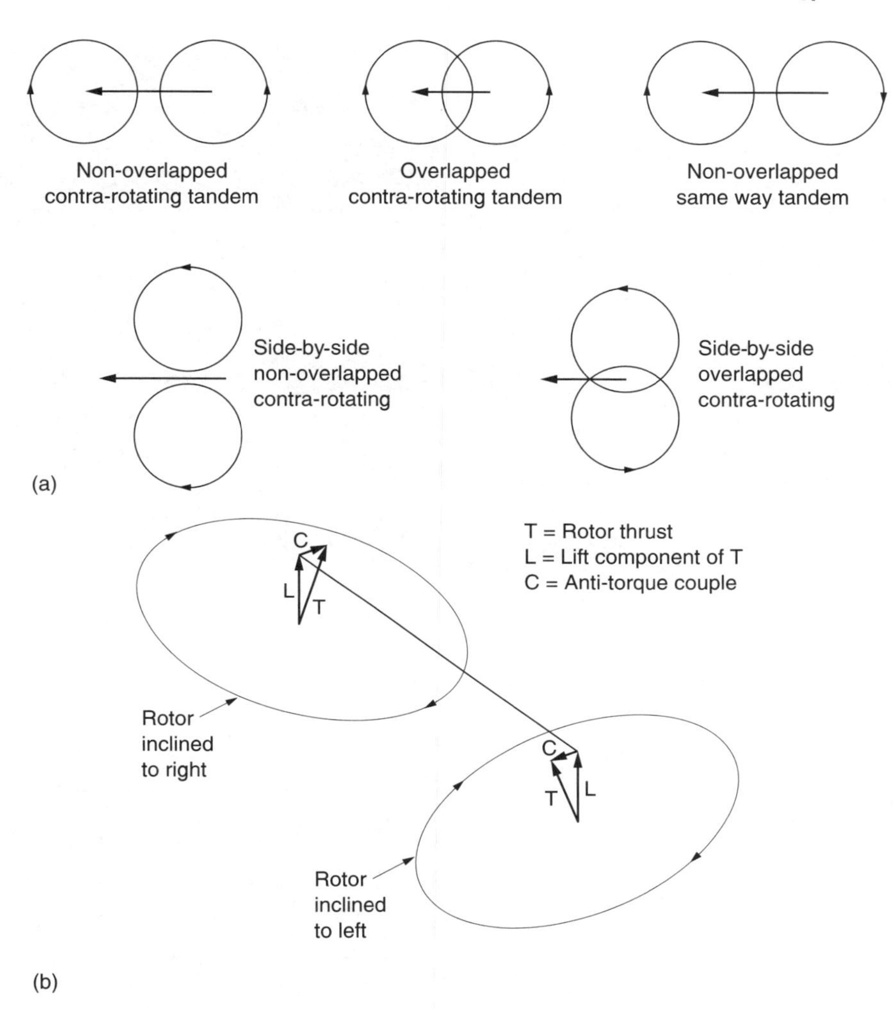

Fig. 9.8 (a) Possible configurations of multi-rotor helicopters. See text. (b) Florine helicopter did not contra-rotate but used inclined rotors for torque cancellation.

detract from Focke's technical achievement. The Fw-61 used the hull of a Focke-Wulf fixed-wing plane, with a forward-mounted radial engine carrying a small airscrew for cooling. Behind the engine was a gearbox that routed power out to each rotor along slim drive shafts.

Focke's next machine was the Fa 223 which first flew in 1940 and reached production in 1942. By this time Focke had left Focke-Wulf and formed a company with Gerd Achgelis, hence the designation Fa. This was a large side-by-side machine powered by a centrally mounted 1000 hp radial engine. With 12 metre diameter rotors it had a payload of 2000 kg, meaning that it could almost have lifted its Sikorsky contemporary. The Fa 223 reached 7500 metres, flew at 200 kph and was a great technical success, although few orders were fulfilled because of bombing raids on the factories. After the

Fig. 9.9 The Florine helicopter flew very well but development was hampered by lack of funding.

Fig. 9.10 The Cierva Air Horse had three rotors turning the same way using the Florine principle for torque cancellation. (AugustaWestland)

Fig. 9.11 Focke-Achgelis Fa-223 was powerful and capable. The weights below the rotor head are a vibration absorber for the cyclic control. (Steve Coates)

war the Fa 223 became the first helicopter to fly the English channel when the RAF recovered a captured machine (Figure 9.11).

The largest helicopter in the world was also a side-by-side. The Mil V-12 (Figure 9.12) used the drive trains and rotors of the Mil 6, itself a huge machine, with the rotation of one system reversed. Side-by-side helicopters have the disadvantage that the pylon structure needed to carry the rotors invariably causes drag. Considerable masses are cantilevered at the end of these pylons and it is also difficult to obtain enough structural rigidity.

9.7 Coaxial helicopters

The coaxial helicopter configuration was frequently seen in the experimental period as the contra-rotation offered torque cancellation. However, the coaxial machine also has some idiosyncrasies in autorotation and once the single rotor machine was perfected the coaxial was simply too complicated except for special applications. For use on naval vessels, where hangar space is limited, the absence of the tail rotor allows a more compact hull for a given load capacity. The Kamov machines demonstrate that amply. The coaxial configuration has also been used with success on unmanned helicopters.

Provided that the two rotors are not too far apart, the coaxial rotor system in the hover behaves very much like a single rotor with the same total solidity. This is not true if the vertical separation is significant, because if the wake of the upper rotor has started to contract the lower rotor will encounter increased inflow velocity and will

Fig. 9.12 Mil V-12 was the biggest helicopter ever built but did not enter production.

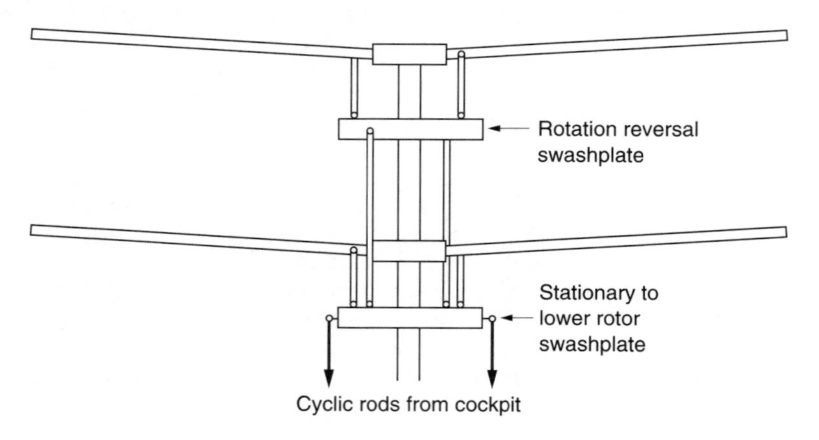

Fig. 9.13 Control system of coaxial helicopter is inevitable complicated, needing two swashplates. Collective controls usually run inside the mast.

thus need more power. However, this power loss may be cancelled by the fact that the lower rotor recovers the swirl energy of the upper rotor.

The coaxial helicopter needs two swashplates as can be seen in Figure 9.13. The first connects the stationary cyclic controls to the lower rotor, and the second allows the cyclic control to pass from the lower rotor to the upper rotor. The cyclic control thus tilts both rotors in the same sense. The collective control will apply the same pitch change to both rotors. Yaw control is commonly obtained by differentially changing the collective pitch. This requires rotating mixers to add the cyclic inputs from the swashplates to the collective inputs that generally pass up the centre of the mast. The resulting mechanism is incredibly complicated as a study of a Kamov rotor head will show.

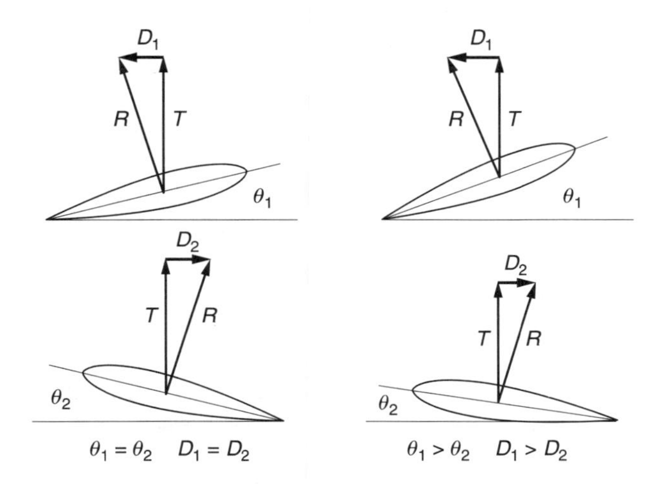

Fig. 9.14 In powered flight, with equal pitch on both rotors there is no net torque. Increasing the pitch of the upper rotor causes its blade reaction to rotate back whilst that of the other rotor rotates forward and a yaw results.

Figure 9.14 shows how the yaw control works. The machine is in a power-on condition so the blade reactions will be tilted back with respect to the shaft axis. The vector sum of these two reactions is vertical so there is no net torque and no yaw effect. Figure 9.14 shows that increasing the pitch of the upper blade will increase the thrust and tilt it back against the direction of rotation. At the same time the pitch of the lower blade is reduced. This will reduce the thrust and tilt it forwards, with the direction of rotation. The vector sum will now be non-vertical so that there is a net torque turning the hull the same way as the rotor whose pitch has been reduced.

In autorotation, the yaw control of a coaxial machine using differential collective is problematic as it may have no effect, or a reverse effect. This was initially not appreciated, and caused some hair-raising moments. Figure 3.14 introduced the autorotation diagram and Figure 9.15 shows an autorotation diagram appropriate for coaxial helicopters. Here the horizontal axis is the average angle of attack of the two rotors. In autorotation, the blade pitch will be chosen for the best rotor L/D at the bottom of the drag bucket. In most helicopters, the lower collective stop is set at the intersection of the α–θ line and the drag bucket. However, when differential collective is used for yaw, the pitch of one of the rotors can be even less and so operation can be to the left of the intersection. The problem is that when operating at $L/D_{\text{max.}}$, an increase in pitch or a decrease in pitch will have the same effect: the rotor experiences more drag and slows down. At some collective settings there is no yaw control at all. To make matters worse, helicopter designers often select blade sections having a relatively flat bottomed drag bucket. If the collective stop is slightly lower than the setting for $L/D_{\text{max.}}$, increasing the pitch will reduce drag and reducing pitch will increase it. As a result the yaw control reverses.

To overcome the problem, coaxial helicopters need a significant amount of vertical tail area to give yaw stability in autorotation. The fins are often slatted to prevent stalling at high angles of attack. In some machines a mechanism is fitted so that the action of the pedals is automatically reversed when the collective is fully lowered.

An alternative yaw control mechanism for coaxial helicopters was seen on the Gyrodyne helicopter. This consisted of drag brakes in the blade tips that could be

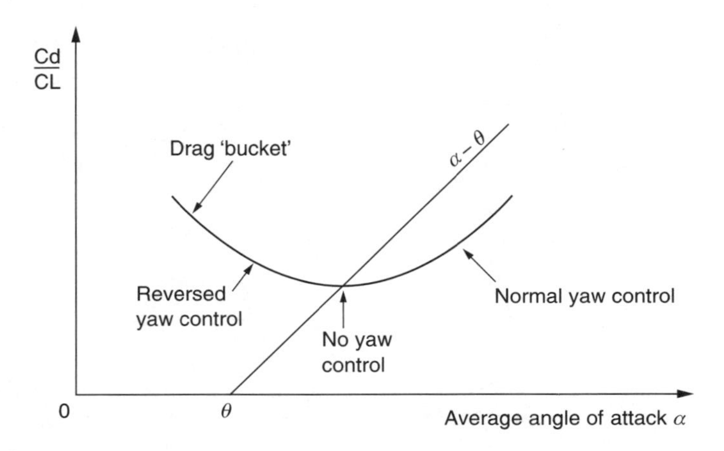

Fig. 9.15 Yaw control in coaxial helicopters is problematic in autorotation. Changing the blade pitch may not have the desired effect. According to where in the drag bucket the rotor is operating, the yaw control may have no effect or may even reverse.

extended in one or other rotor. This approach has the advantage that there is no reversal in autorotation.

Another suggestion for yaw control is the use of a friction brake between one or other of the rotor shafts and the hull. A moment's thought will reveal that with a conventional transmission this would not work. If the coaxial helicopter simply has two shafts driven in opposite directions at the same speed by the same input shaft, applying a brake to one of these has no yaw effect at all, but simply slows down both rotors. In order to use brakes, the two rotors have to be driven with a differential gearbox that allows one rotor to speed up as the other slows down. Subject to this complexity, the friction brake system has the advantage of significant reduction of rotating linkage as there is then no need for differential collective or tip drag brakes. There is also no control reversal.

In forward flight the coaxial helicopter has the advantage that the retreating blade of one rotor is on the opposite side to the retreating blade of the other rotor. This means that if a suitably rigid rotor head assembly can be built, the lift trough due to a retreating blade can be overcome by the lift from the advancing blade of the other rotor and higher forward speed becomes possible.

9.8 The synchropter

The synchropter is a helicopter having two contra-rotating synchronized rotors mounted side by side that are deeply meshed. Figure 1.17 showed that the rotor shafts are set on pylons and are tilted outwards so that the blades of one rotor can pass over the head of the other. In machines having two-bladed rotors the blades are phased 90° apart so that one blade is safely oriented fore and aft when the other passes overhead. The synchropter principle was invented by Anton Flettner and used on his successful Kolibri (hummingbird) helicopter.

Although there is no tail rotor, there is no consequent safety advantage. The outward tilt causes the rotors to pass close to the ground at the sides. Synchropters are usually placarded with warnings to ground personnel to approach from the front. Aerodynamically the synchropter is similar to the coaxial helicopter in that the efficiency loss of a lower blade working in the downwash of an upper blade is offset by swirl recovery. The lateral separation of the rotors means that the disc area is greater than the area of a single rotor and this reduces the disc loading and the induced power needed, although some of the advantage is lost because the rotor thrusts do not align. The horizontal components of the rotor thrusts are in opposition and some power is wasted. In forward flight the rotor separation gives a better disc aspect ratio.

From a performance standpoint, the twin rotor heads cause a drag penalty and the synchropter is not appropriate for high speed. However, as the synchropter is practically limited to two-blade rotors, it naturally suggests low solidity and disc loading which is the ideal for low speed, high altitude work or heavy lifting.

The synchropter is controlled in a similar way to the coaxial helicopter in that differential collective pitch is used as a yaw control. This will be subject to reversal in autorotation as for the coaxial helicopter. Some yaw control is also possible by the use of differential fore-and-aft cyclic, although the short distance between the rotor heads makes this ineffective. The usual fore-and-aft cyclic control affects both rotors equally, but the lateral cyclic may be adapted so that the discs tilt outwards more than they tilt inwards in order to preserve clearance between the heads and the blades. Generally synchropters have poor yaw control and require a large amount of fin area.

The outward tilt of the rotors means that the torque cancellation is not perfect. It can be seen from Figure 9.16 that although the torque cancels in the vertical axis, there is a component of rotor torque in the horizontal axis which is the same for both rotors and therefore adds. If the synchropter is arranged to have the advancing blades on the inside, the torque reaction will tend to pitch the hull nose up until the CM is far enough ahead of the masts to counter the torque. In the hover this will require the application of forward cyclic so that the rotor thrust remains vertical. In forward flight the forward CM is a useful stability aid. As forward speed builds up, the drag on the hull will act below the rotor heads and produce a couple tending to tilt the hull nose down. The rotor torque will counter this. Consequently there is a correct way for synchropter rotors to turn, i.e. with the advancing blades on the inside.

The Flettner 282 Kolibri (Figure 1.4) was not only the world's first synchropter, it was also the world's first production helicopter, beating Sikorsky's R-4 by several months as well as being technically superior. The Kolibri had very closely meshed two-bladed rotors. Figure 9.17 shows that these were fully articulated with friction disc dragging dampers. There was substantial flapping hinge offset so that the rotor blades would not flap into contact with the opposite head. This allowed space for the feathering hinge to be mounted inboard. The swashplates and pitch links were a compact and elegant arrangement. A 160 hp BMW radial aircraft engine mounted in the conventional aircraft attitude drove the transmission through an inclined shaft. The foot pedals

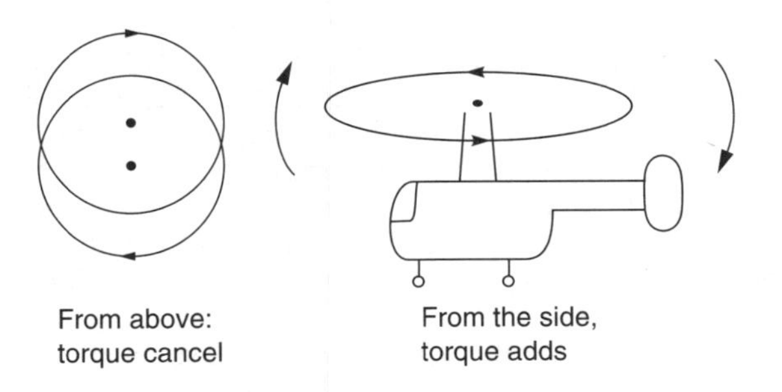

Fig. 9.16 In the synchropter there is a component of torque along a horizontal axis.

Fig. 9.17 The transmission and rotor heads of the Flettner Kolibri. (Steve Coates)

operated the rudder as well as the rotor yaw mechanism. The pilot could also change the angle of incidence of the tail plane.

The Kolibri was an advanced machine; it had adequate power and was highly manoeuvrable. In a mock dogfight with a Focke-Wulf 190 the fighter was unable to get the Kolibri in it sights. Following the end of the war, a significant amount of advanced German technology found its way to various allied countries. In the USA, Charles Kaman recognized the potential of the synchropter principle and combined it with the servo-tab control system to produce a highly successful series of machines. The Kellet Co. made some synchropters with three-bladed rotors but these had blade contact problems and did not reach production.

Kaman's early synchropters were piston engine powered, culminating in the HOK-1 into which was shoehorned a 600 hp radial aircraft engine, which still wasn't powerful enough for some purposes. The piston engine simply isn't suitable for helicopters above a certain size. The world's first turbine engine helicopter was a Kaman synchropter, as was the world's first twin turbine helicopter. Kaman's most well-known synchropter is the H-43B, aka Husky. It was also known as the Pedro, after a call sign originally used at a Texas air base. The Husky was basically a redesign of the piston engined H-43A using a turbine and it became a classic piece of industrial design. The compact turbine engine was placed on the roof behind the transmission, resulting in a large, open cabin accessed by rear clamshell doors as well as sliding doors at the side. The servo-tab control system meant that the forces needed from the pilot were quite low even though there was no power assistance.

The yaw control was aided by relatively wide rotor head spacing and an automatic yaw reverser to make the pedals work in the correct sense in autorotation. The Husky also had an impressive array of vertical fin area to maintain directional stability when the yaw control was fragile. The provision of a large fin area is a problem in coaxial helicopters and synchropters as Figure 9.18 shows. There is a wedge of space in which the fins can exist. The top of the wedge is defined by the rearmost flapping of the rotors and the bottom of the wedge is defined by where the ground will be in a flared landing.

The difficulty is that the further back the fin area is, the more effective it is. However, the further back the fin, the smaller it has to be. The only solution is to use a number of parallel fins. The early Huskys had a habit of knocking the tops off their fins. This was solved with a redesign, although expedience dictated making the top and bottom ends of the fins from glass fibre that would disintegrate in any extreme manoeuvre without causing any further damage.

Another difficulty with fins in helicopters is the wide range of directions the airflow can come from. This makes it hard to maintain the optimum aspect ratio of the fin. The Kaman Husky solved this problem very nicely by mounting the fins on a floating elevator that would align itself with the airflow. Figure 9.19 shows that the fins were then always correctly oriented.

The Husky had very low disc loading giving it excellent lifting ability and high altitude performance. In 1961 a Husky reached almost 33 000 feet! The torque cancelling of the synchropter meant that pilot workload was quite moderate and the lack of a tail rotor meant it could land in difficult terrain. The Husky distinguished itself as a rescue helicopter, particularly where crashed aircraft had caught fire. The Husky would hover overhead, using the downwash to blow flames clear of the fire fighters and victims.

The Husky was superseded by conventional turbine helicopters that were faster, but a number of surplus machines found their way into logging and aerial work where speed was not of the essence. Their success in this niche led Kaman to produce the

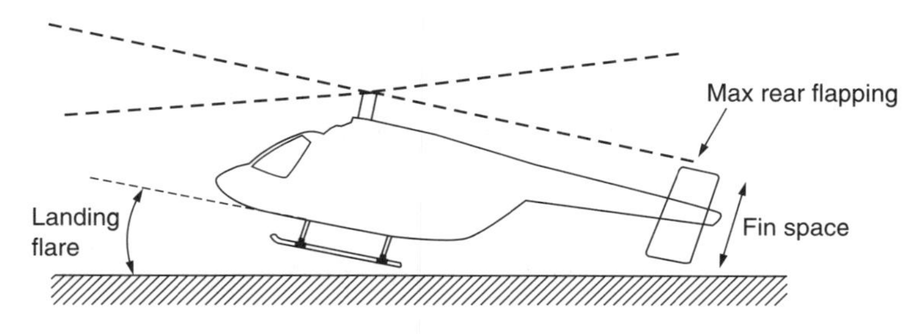

Fig. 9.18 Finding enough fin area is difficult in a synchropter.

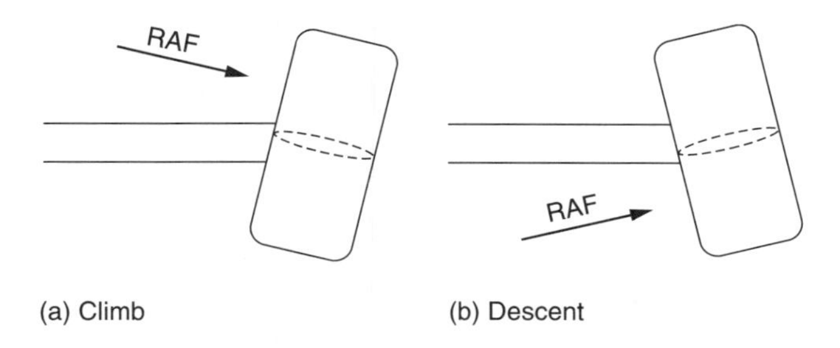

Fig. 9.19 The fins of the Husky are mounted on a floating tail plane so that they always align with airflow to offer their best aspect ratio.

K-Max: basically the mechanics of a Husky with a minimalist single-seat hull optimized for underslung load working. The K-Max is the only synchropter currently in production. As the K-Max is a single-seater, a number of Huskys have been refurbished to act as trainers.

9.9 The tandem rotor

Once the stress-reducing benefits of blade articulation became known, contra-rotation could be adopted in the tandem. This largely cancels torque reaction. Claims will often be seen that because there is no tail rotor to waste power, the tandem must be more efficient than the conventional helicopter. This is not necessarily true because the two rotor discs may overlap reducing the total disc area. In fact the real advantages of the tandem are that the position of the centre of mass is much less critical than in single rotor machines and that hovering is much less sensitive to wind direction. In military operations the ability to hover on any heading, load in a hurry and simply fly off saves valuable time. Later Chinooks have three load hooks in line beneath the hull so that supplies can be taken to three locations in one flight. This would be impossible with a single rotor machine because it could not handle the CM travel as the loads were dropped off one by one. Figure 9.20 shows that if the CM position is not in the centre the trim will be operated so that the rotor at the 'heavy' end produces more thrust and the other rotor produces less. A secondary effect of this condition will be that there is not a perfect torque balance and some yaw control will be required.

It is quite possible to construct a tandem rotor machine in which the two rotor discs are separated. The Florine machines and the Piasecki 'Dogship' (Figure 9.21) had this configuration. This approach results in a long and possibly heavy hull and a physically large machine requiring a lot of hangar space: a drawback in naval operations. As a result in modern tandems the two discs will be significantly overlapped such that the distance between the rotor heads is less than the rotor diameter. The rotors must be synchronized by a shaft joining the front and rear gearboxes so that, like the synchropter, the two rotors can share the same airspace without clashing.

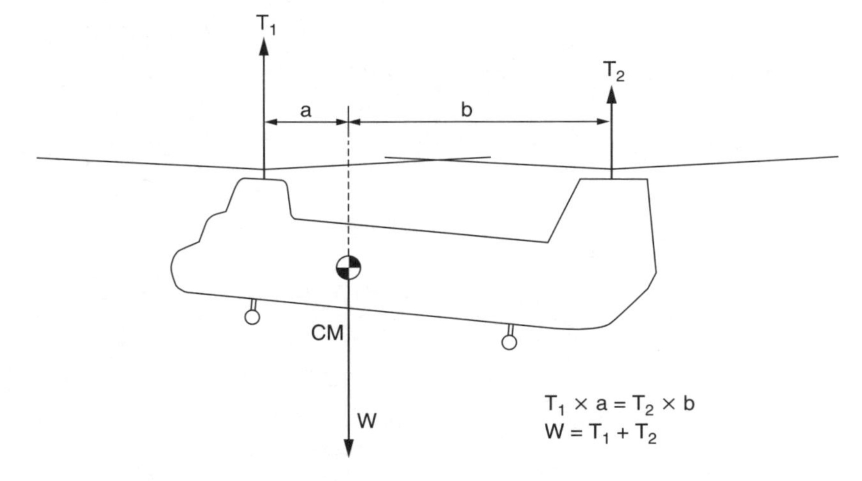

Fig. 9.20 The tandem configuration can deal with longitudinal CM offsets using a different amount of thrust on each rotor.

Fig. 9.21 The Dogship was Piasecki's first tandem rotor helicopter and was widely demonstrated to attract interest in the principle. It was the forerunner of the 'flying bananas' and the Chinook. (Boeing)

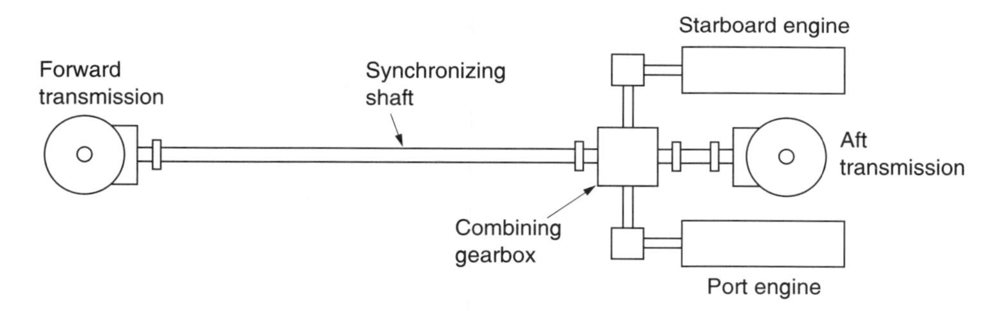

Fig. 9.22 The transmission system of the Chinook showing the synchronizing shaft between the rotors and the twin engines.

Chapter 5 described the interaction between the main and tail rotor of a conventional helicopter. In the hover it was seen to be highly variable with wind direction, whereas in translational flight there was a beneficial weathercocking effect. In the tandem the interference between the two rotors in the hover is beneficially almost independent of wind direction, whereas in forward flight it is detrimental to stability. Consequently it is simplistic to claim that one configuration is better than another. In some areas the conventional approach is superior, but in other areas the tandem wins.

Figure 9.22 shows the transmission arrangement of the CH-47 (Chinook) in which the two engines drive the synchronizing shaft via one-way clutches so that either engine can drive both rotors. It is interesting to note that except for the use of piston engines, the Florine III of 1935 had the same transmission layout. As the dominant masses and the rotor forces are both concentrated at the ends of the hull, the structure needs to be stiff against bending and torsion. The loss of disc area due to the overlap of the two discs causes some loss of aerodynamic efficiency, generally of the same order as the loss due to driving the tail rotor in a conventional machine. However, the shorter hull will be lighter and have less area exposed to downwash so the overall lifting capacity remains about the same.

In forward flight the tip path axes must be tilted forward just as in a single rotor machine to obtain a forward component of thrust. Figure 9.23(a) shows that in practice

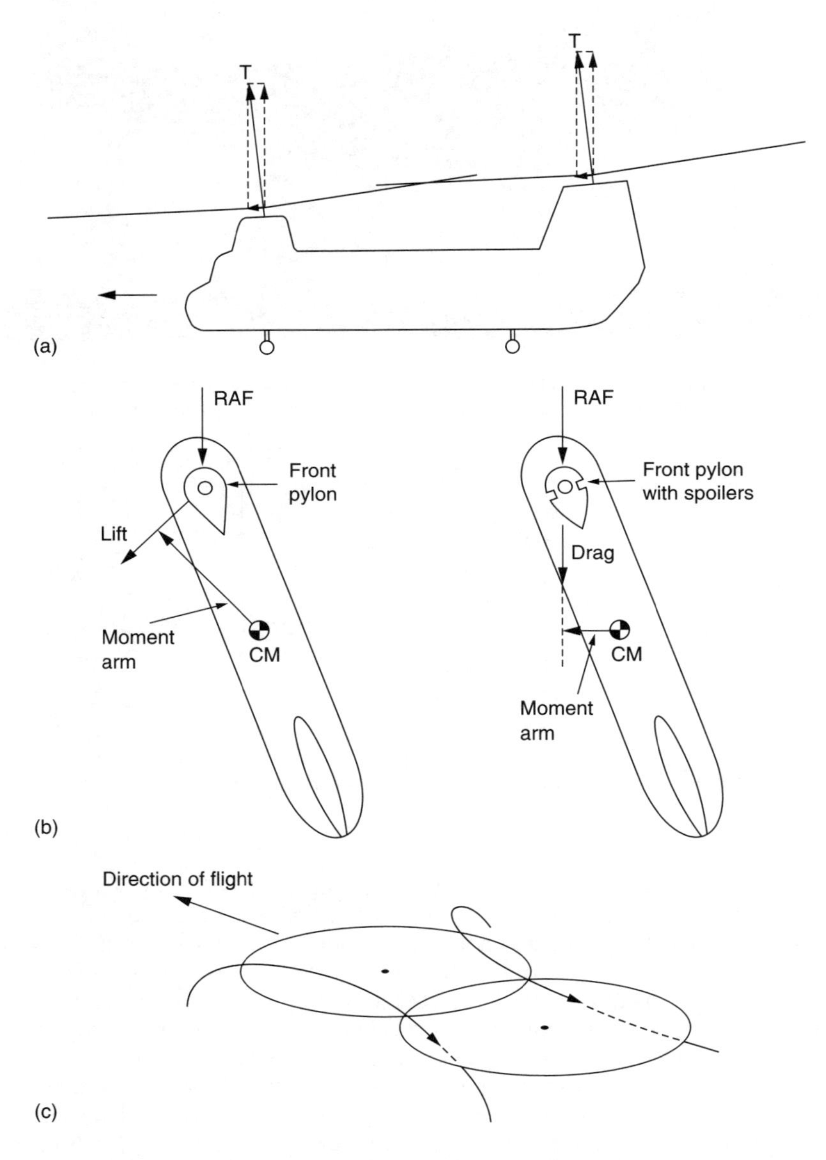

Fig. 9.23 Characteristics of tandem rotor helicopters. (a) The rear rotor is mounted on a pylon acting as a fin. (b) The front pylon is aerodynamically treated so that it creates drag rather than lift in a sideslip. (c) The rear rotor works in the vortex structure of the front rotor and has higher inflow.

the rotors are mounted so that they remain substantially in the same plane in translation. This allows the rear rotor to operate in relatively clean air. In order to keep the cabin level and to provide blade clearance, the rear rotor is mounted on a pylon. It is a characteristic of this approach that ground clearance of the front rotor disc is relatively small. However, the poor clearance exists directly in the pilot's field of view. With no tail rotor to weathercock and a short moment arm, a large amount of fin area is usually needed and even then tandem rotor machines are aerodynamically marginal in yaw.

This is not all bad because many operations require the machine to hover at an arbitrary orientation to the prevailing wind and the tandem does that very well. The rear pylon is always shaped to act as a substantial fin and the front pylon may have aerodynamic treatment to stop it acting as a fin. This may consist of irregularities designed to cause early flow separation when the relative airflow is off the fore-and-aft axis. Figure 9.23(b) shows that the resulting drag causes a smaller yaw component than if the pylon caused lift.

Figure 9.23(c) shows that the rear rotor is affected by the front rotor vortices in translational flight. Effectively the rear rotor is operating with a greater inflow velocity than the front rotor. In a machine with a central CM, both rotors will need to produce the same amount of lift. To develop the same thrust the rear rotor will need to use a higher blade pitch and will absorb more torque because of the higher inflow. The swirl of the forward rotor also acts asymmetrically on the hull. For these reasons the torque is not precisely balanced in translation. One prototype tandem helicopter could not initially fly in a straight line in translational flight because these effects were more powerful than the yaw control. Once the nature of the problem was understood a solution was not far behind.

There are ways to mitigate the problem. An obvious suggestion would be to move the CM forward. Whilst this would work, it somewhat negates the advantage of CM position freedom. In practice, the two rotors can be set at a different angle. Figure 9.24 shows that the front rotor shaft may be inclined forward with respect to the rear. This is done on the CH-47 in which there is a 5° difference between the angles of the shafts. This increases the inflow seen by the front rotor and decreases the lift component of its thrust, whereas the inflow of the rear rotor is reduced and the lift component of its thrust is greater. Another possibility is that the fin can be built with an integral camber so that in translational flight it produces a side thrust. The cambered fin of the CH-46 is obvious in a rear view.

The interference between the rotors can be minimized by flying the machine sideways as this increases the diameter of the stream tube and improves the aspect ratio,

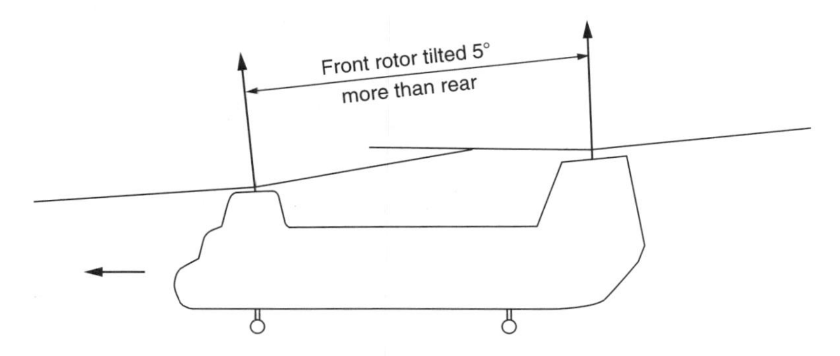

Fig. 9.24 The front and rear rotor shafts may be inclined at different angles.

giving greater translational lift. This technique works well in a steep climb. The drag of the broadside hull will be a problem as speed builds up. However, some advantage can still be gained by flying with sideslip; 15° is a typical figure. Some piston-engine tandems used as plane guards during carrier operations were operated in this way as standard procedure. In a twin-engine machine, sideslip operation may be used to handle one-engine-inoperative conditions. The direction of sideslip is important as the downwash from the front rotor is not symmetrical.

The tandem rotor market is one many manufacturers have stayed away from. In the UK, the most successful tandem was the Bristol Belvedere (Figure 9.25) designed by Raoul Hafner which started life mechanically as a pair of Sycamore mechanics with a synchronizing shaft and which subsequently evolved into a capable twin turbine. Yakovlev built a large tandem machine in the USSR but this was not successful. In the USA Bell built the HSL but it did not enter production. The name most closely associated with the tandem is that of Frank Piasecki whose early fragile canvas covered creations, dubbed 'flying bananas' from the bent hull needed for rotor clearance, matured into the enduring CH-46 and CH-47.

The CH-46 is about as attractive as a tandem can get and when it was first built its good lifting capacity and amphibious capability made it a favourite with the US Navy who use it for vertical replenishment (VERTREP) which is military-speak for moving material from one ship to another suspended below a helicopter. The army liked the concept but wanted something with a bigger cabin and the CH-47 was born.

The CH-47 Chinook is a superb piece of industrial design by any reference. The basic design dates from 1961 and because it is fundamentally well founded it is simply upgraded from time to time. It is amazingly versatile and genuinely amphibious. In other words it can operate normally from water rather than just floating in an emergency. The absence of a tail rotor means that it can land in scrub.

Fig. 9.25 The Bristol Belvedere was a twin turbine military machine that was successful in service despite frequent fires caused by its absurd engine starting system. (AugustaWestland)

For easy loading, the Chinook is designed around a constant cross-section cabin with a rear ramp. The fuel and undercarriage are housed in two long side sponsons that provide stability on water and cause no intrusion into the load space. For similar reasons the engines are placed above the cabin, and the synchronizing shaft and control pushrods run in a hump or spine above the cabin. The immense rear pylon/fin is used to house the engine combining transmission, the oil cooler, the rear gearbox and swashplate actuators, the APU and almost anything else that needs a home with the result that the hull is kept free for payload.

The control system of the tandem rotor helicopter is more complex than in a tail rotor-type machine. The rotor heads need to be articulated for reasons that will be seen. Figure 9.26(a) shows that the swashplates tilt simultaneously for roll control and tilt differentially for yaw control and to oppose any residual torque imbalance. They move up and down together for thrust control and differentially for fore-and-aft (pitch) control and to handle CM shifts. All of this is achieved in a Chinook in a diabolical mechanical mixer located in the control closet behind the port pilot's seat which takes the inputs from the pedals, cyclic stick and thrust lever (tandemspeak for collective) in the cockpit floor and adds and subtracts in various ways to obtain four pushrod outputs which go into the roof, one to control each side of each swashplate. The conceptual mechanism is shown in Figure 9.26(b). The autopilot is incorporated in an unusual fashion. Electrical autopilot signals and mechanical pilot inputs are supplied to small duplicated hydraulic actuators in the control closet located before the control mixer. These are called integrated lower control actuators (ILCA) and they operate the control mixer. The mixer pushrod outputs are then led to duplicated high pressure hydraulic actuators at the swashplates. These actuators have mechanical inputs. This approach allows a conventional autopilot to output yaw, pitch, roll and thrust commands to the airframe. If the autopilot had to control the swashplate actuators directly, an autopilot mixer would have been needed.

When the cyclic stick is pushed forward, differential collective pitch increases the thrust of the rear rotor relative to the front rotor (Figure 9.27(a)). The hull is tilted forward about the pitch axis (b) and this makes both rotor shafts tilt relative to the tip path axes. This hub tilt causes a cyclic feathering effect such that both rotors pitch forward to follow the hull (c). This is the control mechanism used in hovering.

In order to move into forward flight, forward differential collective is maintained so the machine gathers speed. The lift asymmetry due to translational flight in both rotors will cause the tip path axes to flap back with respect to the two control axes as shown in Figure 9.28(a). At a given airspeed, this results in a larger change in cabin floor angle, reduced blade clearance at the rear of the front rotor, more hull parasite drag and a large amplitude of flapping and dragging at the rotor heads because of the large differences between the tip path and shaft axes.

Thus in practice a tandem helicopter requires means to apply forward cyclic control to both swashplates in translational flight as a function of airspeed. Figure 9.28(b) shows that this results in closer alignment between the tip path and shaft axes, reducing the amount of flapping and dragging and improving the clearance of the front rotor. In the Chinook the fore-and-aft cyclic controls are not connected to the cyclic stick. Instead they are driven by an automatic system which is airspeed sensitive and starts to apply forward cyclic at speeds above about 70 knots. The application of forward cyclic to reduce flapping will also tend to increase the airspeed. To compensate for this, a series actuator, known as the DASH actuator, in the fore-and-aft cyclic pushrod is also operated by the automatic system in such a way that the pilot sees a fairly constant positive gradient between cyclic stick position and airspeed which is not impaired by the automatic forward cyclic control of the swashplates. The DASH actuator can be seen in Figure 9.26(b).

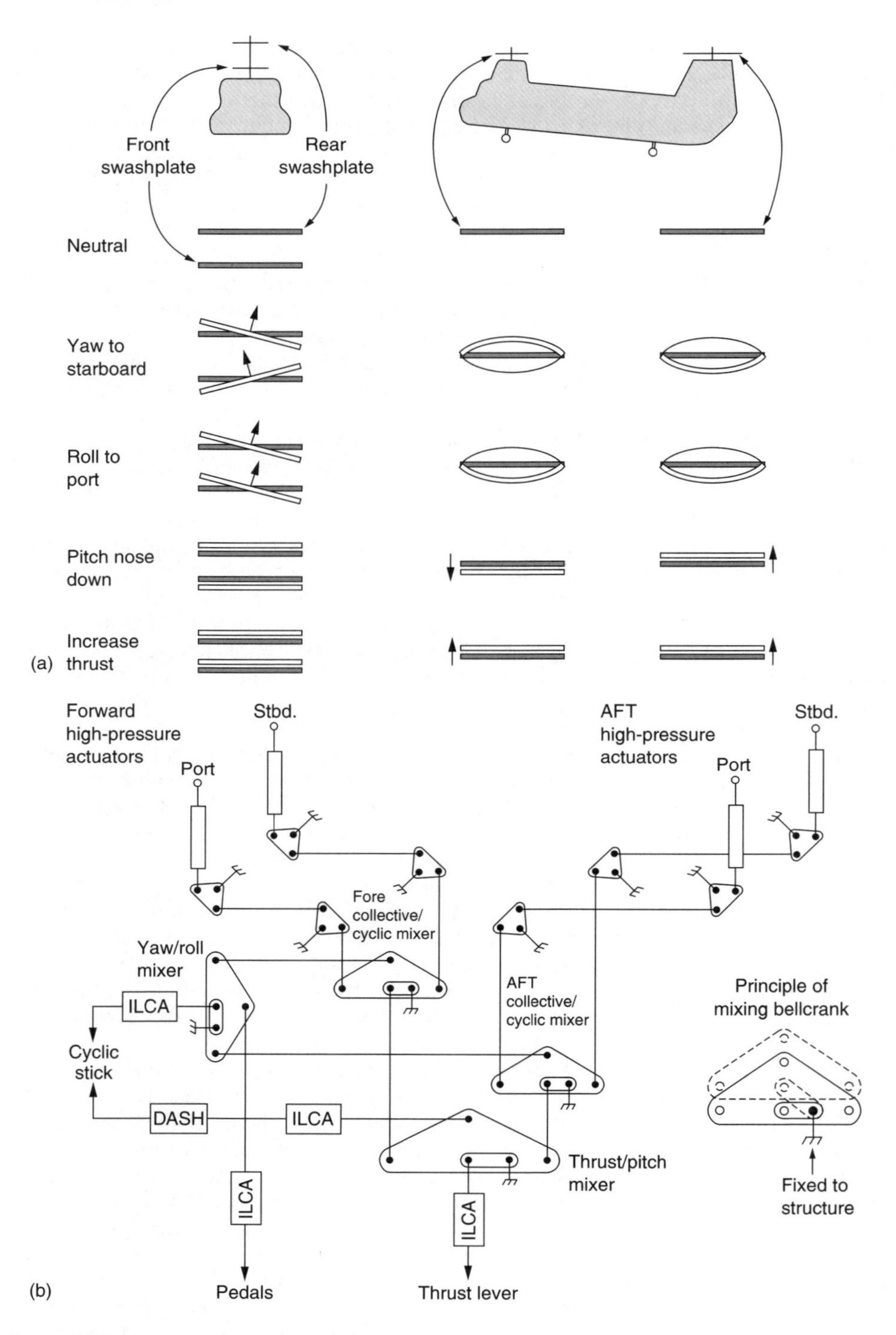

Fig. 9.26 Tandem control principles. (a) All control is by the two swashplates which tilt together for roll, differentially for yaw and which rise together for collective and rise differentially for pitch. (b) The mixer mechanism needed in a tandem to combine all of the pilot's controls into the two swashplates.

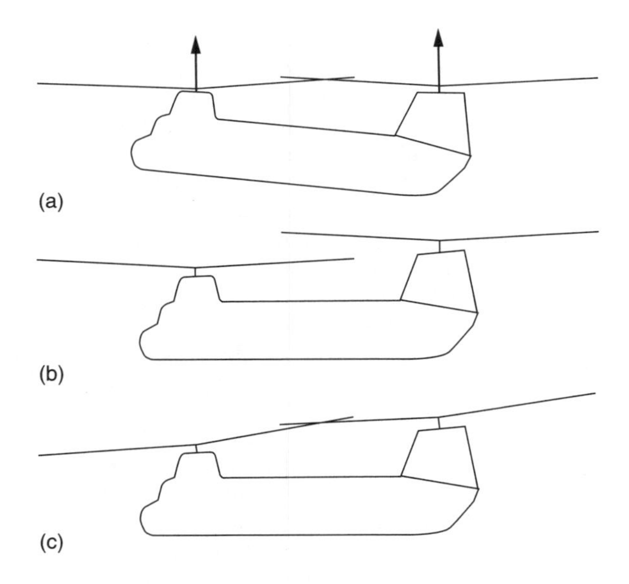

Fig. 9.27 (a) Pitch control increases lift on end of the hull and reduces it on the other. (b) The hull pitches. (c) Flapping hinges cause the rotors to follow the hull attitude.

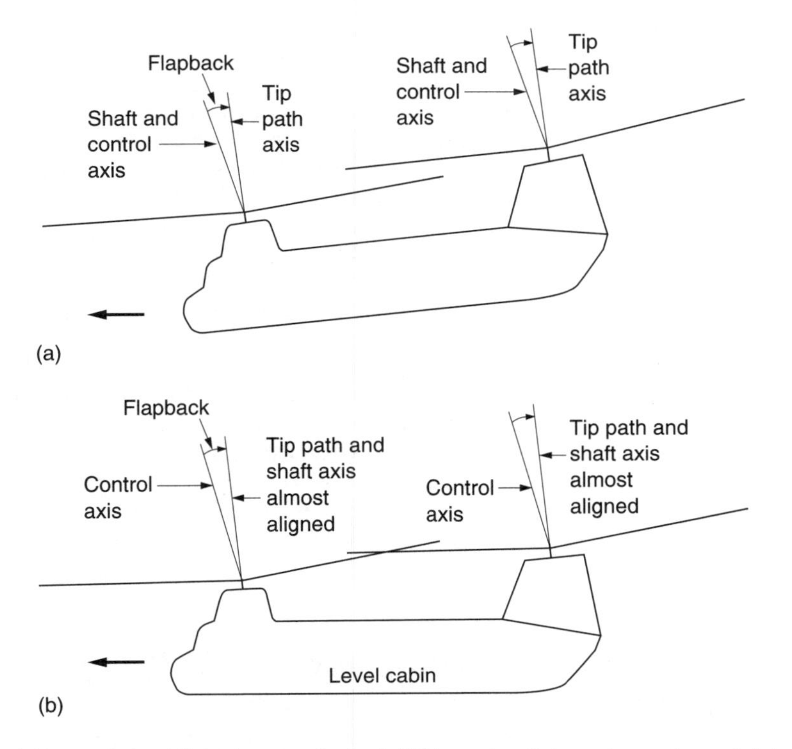

Fig. 9.28 (a) In translational flight the rotors flap back. (b) Forward cyclic is used on each swashplate to reduce flapback and level the hull.

The yaw control is by differential tilt of the rotor thrust and clearly if there is no rotor thrust there will be no yaw couple. If rotor thrust reverses the yaw control will reverse. Whilst this is not a problem in most flight conditions, a zero-*g* pushover could result in loss of yaw control. A crop-dusting turn or return-to-target manoeuvre is also difficult for a tandem because when the hull is pointing straight up after the first quarter loop the rotors need to be set to flat pitch and this removes the yaw authority. Consequently the tandem will perform a variation on the manoeuvre in which the yaw is commenced during the quarter loop. Momentum carries the yaw on during the flat pitch phase.

The tandem is generally designed to have a level cabin in translational flight and when at rest on the ground. This means that in the hover the cabin will slope down to the rear and the machine must touch down with the rear wheels first. This allows an interesting taxi mode where the rear wheels are carrying a significant weight but the front wheels are clear of the ground. Forward speed can be controlled by rear wheel brake and steering is possible by applying the brakes differentially.

The waterborne characteristics of a tandem are worth a few comments. The contra-rotation means that the rotors can be started and stopped whilst afloat without the machine rotating as a single main rotor machine would. The built-in tip path plane tilt causes a tendency to forward motion when the hull is floating level. When afloat the acrylic cockpit chin windows are partially submerged, and whilst this is not a problem whilst stationary, water taxiing at speed will put them under tremendous stress. To avoid this, pilots will lift the nose in a water taxi. The floor of the cabin is sealed to act as a raft for use on water and the ramp can be lowered whilst afloat to deploy or recover small boats or swimmers.

In the tandem rotor helicopter the rotor heads are designed to meet different requirements. The yaw control is obtained by tilting the two rotor discs laterally in opposite directions. Effective yaw requires that the tip path axes actually make a significant tilt and the flapping hinges are needed to allow this to happen. If conventional heads with significant flapping hinge offset are used, the result of an application of the yaw control is a serious twisting couple applied to the fuselage. Thus for a tandem rotor helicopter, the flapping hinge offset needs to be significantly reduced. This need not reduce the roll response, because a roll is obtained by applying lateral cyclic to both rotors in the same sense. As there are two rotors, each rotor only needs to apply half the roll moment. In order to prevent heavy underslung loads reducing the lateral cyclic authority the main load hook is on rollers and can traverse a curved lateral beam whose centre of curvature is near a line joining the rotorheads.

In the Chinook, the flapping hinges are provided with a relatively small offset for the above reasons. If the dragging hinges were coincident with the flapping hinges there would be a large drag angle change between the power-on and autorotation states and there would be little space for drag dampers. Instead the dragging hinges are placed outboard of the feathering hinges. This arrangement works very well and the only drawback is that when the blades are stationary gravity causes them to swing about the dragging bearing. These are not necessarily vertical, being outboard of the feathering bearings.

9.10 Remotely piloted and radio-controlled helicopters

In some cases putting a pilot in a helicopter is not the right thing to do. Without a pilot the helicopter can be much smaller, or it can lift more. The pilotless helicopter can also

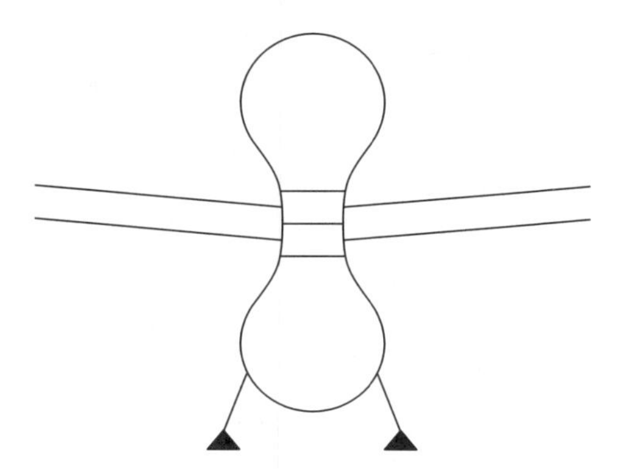

Fig. 9.29 UAV with coaxial rotors and a rotationally symmetrical body can attain high speeds by near-axial flight.

be used for tasks in which there is a significant risk. Pilotless aircraft are often called *drones* and the term is also applied to pilotless helicopters.

Small remotely controlled helicopters can be used for reconnaissance and may escape detection or prove very difficult targets if they are detected. Small helicopters have also been developed for crop dusting, notably in Japan. However, the limitations of the helicopter apply to small ones equally. Small fixed-wing drones can be launched from a simple ramp and recovered by parachute and will have greater range than the inefficient helicopter. Consequently drone helicopters are very rare indeed and fill extremely specialized niches.

The drone helicopter may depend entirely on a remote pilot, or it may be entirely autonomous so that it can complete an entire mission automatically. Early drones were totally pilot controlled and had to stay within sight of the pilot. Subsequently the drone could relay a television picture of the view from the cockpit so that the drone could be flown out of sight. As electronics developed, it became possible to integrate the helicopter controls with autopilot functions and control the whole with a navigation system. This led to the UAV (unmanned air vehicle) concept that can be applied to the helicopter as well as it can to the aeroplane.

In the absence of a pilot the UAV designer has much more freedom. The contra-rotating coaxial rotor is popular as it avoids the use of very small and delicate tail rotors. Figure 9.29 shows a typical UAV having coaxial rotors. The hull is axially symmetrical and high speed forward flight can be achieved by tilting the whole machine over. Tilt-rotor UAVs have also been developed.

9.11 Radio control principles

Radio control is simply the use of radio signals to relay the position of control sticks without any physical connection. In practical applications there will be a need to have several control channels simultaneously available and each one will need to be proportional, which means that small or large movements of the control stick will be remotely reproduced by the movement of a servo.

The foundation of this technology was a patent by A.H. Reeves in the 1930s which showed that any type of information could be carried by varying the width or distribution of pulses. In a helicopter, the actual flying controls require four simultaneous control channels. These will be lateral cyclic, fore-and-aft cyclic, collective and tail rotor. Subsidiary controls will be needed to start the engine and rotors and to adjust RRPM.

In an analog or pulse width modulation (PWM) system, each control is connected to a transducer such as a potentiometer or LVDT (see Chapter 7) that controls the length of a pulse. There will be one pulse for each control channel, followed by a reset or synchronizing pulse. The pulse signal is used to amplitude or frequency modulate a carrier wave in a radio transmitter. A receiver tuned to the transmitter will be able to demodulate the carrier signal and recover the pulse train. The reset pulse sets a pulse counter to zero. Each control pulse is then counted so that each pulse is routed to the correct channel.

A PWM system is analog because the infinitely variable position of a control stick results in an infinitely variable pulse length. In practical systems the resolution of the system may be reduced because of noise in the radio link. The digital or PCM (pulse code modulation) system converts the position of each control stick to a binary number of fixed word length. In an eight-bit system, there can be 256 different numbers. The binary numbers from the various control channels are multiplexed into a single transmission in which the number corresponding to each channel has a fixed place in the sequence. The start of the sequence is denoted by a fixed synchronizing pattern.

Index